Nitrogen Management in Crop Production

Nitrogen Management in Crop Production

Nand Kumar Fageria, PhD

National Rice and Bean Research Center of EMBRAPA
Santo Antônio de Goiás, Brazil

CRC Press
Taylor & Francis Group
Boca Raton London New York

CRC Press is an imprint of the
Taylor & Francis Group, an **informa** business

CRC Press
Taylor & Francis Group
6000 Broken Sound Parkway NW, Suite 300
Boca Raton, FL 33487-2742

First issued in paperback 2016

© 2014 by Taylor & Francis Group, LLC
CRC Press is an imprint of Taylor & Francis Group, an Informa business

No claim to original U.S. Government works

Version Date: 20140311

ISBN 13: 978-1-138-03416-7 (pbk)
ISBN 13: 978-1-4822-2283-8 (hbk)

Library of Congress Cataloging-in-Publication Data

Fageria, N. K., 1942-
    Nitrogen management in crop production / author: Nand Kumar Fageria.
        pages cm
    Includes bibliographical references and index.
    ISBN 978-1-4822-2283-8 (hardcover : alk. paper)
    1. Nitrogen in agriculture. 2. Crops and nitrogen. I. Title.

S587.5.N5F34 2014
631.8′4--dc23   2014007346

Visit the Taylor & Francis Web site at
http://www.taylorandfrancis.com

and the CRC Press Web site at
http://www.crcpress.com

This book is dedicated with much love and affection to my grandchildren Sofia, Maia, and Anjit.

Contents

Preface

The world's population is projected to be over 9 billion by the year 2050 and about 10 billion by the end of the twenty-first century. To feed the increasing world population, global food security will remain a worldwide concern in the twenty-first century and beyond, and food production needs to be continuously increased. Furthermore, sustainable development is undoubtedly one of the central approaches in safeguarding food security, which has led to the urgent demand for knowledge on fertilizer management in crop production. Agricultural development should fully exploit the capacity of soils and other resources but also help to maintain its fundamental function in the environment, hydrology, and the Earth's ecosystem. The use of chemical fertilizers is fundamental to achieving the maximum yield in modern agricultural systems. Among essential plant nutrients, nitrogen is one of the most important yield-limiting nutrients. It participates in many physiological and biochemical processes in plants. It is mainly responsible for determining yield and yield components in cereals and legumes. In addition, it is also responsible for the activation of many enzyme activities. The role of nitrogen in chlorophyll formation and photosynthesis in plants is well known. However, the recovery efficiency of this element when applied as a chemical fertilizer is less than 50% in most cropping systems because a large part of the applied nitrogen is lost by leaching, volatilization, denitrification, and erosion in the soil–plant system. Hence, a large part of the nitrogen applied to crop plants is not used by plants. This may create environmental problems. Hence, judicious use of nitrogen in crop production is highly desirable from economic and environmental points of view. The judicious use of nitrogen requires knowledge of the appropriate methods of nitrogen application. In addition, effective source and timing of nitrogen application during crop growth cycle are another important nitrogen management strategy. Furthermore, the use of an adequate rate is also important to avoiding loss and reducing the cost of production. Besides these practices, the use of nitrogen-efficient crop genotypes and the use of legumes that will fix atmospheric nitrogen in sufficient amounts are important strategies from social, economic, and environmental points of view.

The eight chapters of this book cover all these aspects or concerns. In addition, this book also covers the latest information from the literature at the international level to make it usable under most agroecological regions of the world. Most of the discussions are provided with experimental results to make them as practical as possible. Colored photos of nitrogen deficiency symptoms of important crop species such as rice, dry bean, wheat, soybean, and corn have been included in the third chapter. These photos will serve as a guide to farmers/extension workers, students of agronomy, and research scientists to identify nitrogen deficiency problems in important field crops. Furthermore, organic matter in the soil determines the physical, chemical, and biological properties that are directly related to nutrient uptake and use efficiency in crop plants. A chapter on organic matter is included to discuss its role in the sustainability of cropping systems. A large number of tables and figures have been included to make the book easy to read and comprehend. This book can be used as reference material by soil scientists, agronomists, breeders, plant physiologists, plant pathologists, environmental scientists, extension workers, agroindustrial administrators, farmers, and students of agricultural sciences.

I could not have written such a comprehensive book without the help of many people. I sincerely thank all of them. My sincere thanks to the National Rice and Bean Research Center of EMBRAPA for providing the necessary facilities, cooperation, and a friendly atmosphere that created a favorable academic environment in which to work and write this book. I also thank the Brazilian Scientific and Technological Research Council (CNPq) for providing financial support to my many research projects; several results of which are included in the chapters of this book. I thank the staff of Taylor & Francis Group/CRC Press, especially Randy Brehm and Kate Gallo, for their excellent handling of numerous issues and for their dedication to producing a high-quality book.

Finally, I thank my wife Shanti, daughter Savita, daughter-in-law Neera, sons Rajesh and Satya Pal, and grandchildren Anjit, Maia, and Sofia for their patience, encouragement, and understanding, without which I could not have possibly found the time and energy required for the writing of this book.

Nand Kumar Fageria
National Rice and Bean Research Center of EMBRAPA
Santo Antônio de Goiás
Brazil

Author

Nand Kumar Fageria, **PhD**, has been a senior research soil scientist at the National Rice and Bean Research Center, Empresa Brasileira de Pesquisa Agropecuária (EMBRAPA) since 1975. Dr. Fageria has been a nationally and internationally recognized expert in the area of mineral nutrition of crop plants and a research fellow and ad hoc consultant of the Brazilian Scientific and Technological Research Council (CNPq) since 1989. Dr. Fageria was the first to identify the zinc deficiency in upland rice grown on Brazilian Oxisols in 1975. He has developed crop genotype screening techniques for aluminum and salinity tolerance and nitrogen, phosphorus, potassium, and zinc use efficiency. Dr. Fageria also established adequate soil acidity indices such as pH, base saturation, Al saturation, and Ca, Mg, and K saturation for dry beans grown on Brazilian Oxisols in conservation or no-tillage system. He also determined the adequate and toxic levels of micronutrients in soil and plant tissues of upland rice, corn, soybean, dry bean, and wheat grown on Brazilian Oxisols. Dr. Fageria determined the adequate rate of N, P, and K for lowland and upland rice grown on Brazilian lowland soils, locally known as "Varzea" and Oxisols of "Cerrado" region, respectively. He also screened a large number of tropical legume cover crops for acidity tolerance, N, P, and micronutrients use efficiency. Dr. Fageria characterized the chemical and physical properties of Varzea soils of several states of Brazil, which can be helpful in better fertility management of these soils for sustainable crop production. Dr. Fageria also determined the adequate rate and sources of P and acidity indices for soybean grown on Brazilian Oxisols. Results of all these studies have been published in scientific papers, technical bulletins, book chapters, and congress or symposium proceedings.

Dr. Fageria is the author/coauthor of 12 books and more than 320 scientific journal articles, book chapters, review articles, and technical bulletins. His four books titled *The Use of Nutrients in Crop Plants*, published in 2009; *Growth and Mineral Nutrition of Field Crops*, third edition, published in 2011; *The Role of Plant Roots in Crop Production*, published in 2013; and *Mineral Nutrition of Rice*, published in 2014 are bestsellers in the list of CRC Press. Dr. Fageria has been invited several times by the editor of *Advances in Agronomy*, a well-established and highly regarded serial publication, to write review articles on nutrient management, enhancing nutrient use efficiency in crop plants, ameliorating soil acidity by liming on tropical acid soils for sustainable crop production, and the role of mineral nutrition on the root growth of crop plants. He has been an invited speaker at several national and international congresses, symposiums, and workshops. He has been a member of the editorial board of the *Journal of Plant Nutrition* and *Brazilian Journal of Plant Physiology* and also a member of the international steering committee of the symposium on plant–soil interactions at low pH since 1990. He is an active member of the American Society of Agronomy.

1 Functions of Nitrogen in Crop Plants

1.1 INTRODUCTION

Nitrogen (N) is one of the most abundant elements in nature. A large amount of N is present in the atmosphere, in the lithosphere, and in the hydrosphere. Its role as an essential plant nutrient is indisputable. N is one of the most yield-limiting nutrients in crop production worldwide. The increased use of fertilizer has been a major factor explaining perhaps one-third to one-half of the yield growth in developing countries since the *green revolution* (Bruinsma, 2003; Fischer et al., 2009). Developing countries now account for 68% of the total fertilizer use (Fischer et al., 2009). Its use has continued to increase by 3.6% per year over the past decade, which would still account for a significant share of yield growth. Using a measure of agricultural area standardized for land quality, the amount of fertilizer used per irrigated equivalent hectare is also now higher in developing countries than in industrial countries (Fischer et al., 2009).

Globally, fertilizer use has plateaued due to a decline in its use in industrial countries and a dramatic fall in the countries of the former Soviet Union after those countries moved toward a market economy. In developing countries, the increase in fertilizer use has been surprisingly consistent across most regions. Asia still has the highest and the fastest increase, but fertilizer use intensity is comparable in Latin America and the Middle East/North Africa too. However, fertilizer use per ha in sub-Saharan Africa is abysmally low due to reasons such as high prices and poor markets, which have been well documented (Morris et al., 2007). Low fertilizer use explains in large part the lagging productivity growth in that region (Fischer et al., 2009).

Sainju (2013) reported that a major nutrient required in ample amount to sustain crop yield and quality is N. N along with water is the key to the realization of the yield potential of modern crop cultivars or genotypes. In the absence of N inputs, modern crop cultivars yield little more than traditional counterparts. N was the key component of increasing the yield of most annual crops at an unprecedented rate, mainly that of rice, corn, barley, and wheat during the 1960s, when a term *green revolution* (1960–1980) was used. A dwarf gene was incorporated into the modern cultivars of these crops to reduce the height. This made sure that these cultivars did not fall down with the addition of high rate of fertilizers, especially N. Frey (1971) reported that modern high-yielding crop cultivars have been bred to respond to the high fertilizer rates (particularly N) necessary to support the increased demands of cropping systems. As a result, agricultural production is more dependent than ever before on heavy fertilization (Radin and Lynch, 1994). These newly developed cultivars were also having high yield potential and were responsive to a higher rate of N. In addition, the importance of N has increased in the recent years due to the high energy requirement for the production of chemical fertilizers containing N. As long as energy costs were low, N was cheap and there were no concerted efforts to increase the efficiency of its use by crop plants or to seek alternate sources of supply. The high cost of energy today changed this situation and increased attention has been given to the efficient use of N in crop production.

The deficiency of N is reported in most crops worldwide (Fageria, 2009, 2013, 2014; Fageria et al., 2011a). Figure 1.1 shows the response of lowland rice genotypes to N fertilization grown on a Brazilian Inceptsol. Similarly, Figure 1.2 shows the linear increase in dry matter yield of three lowland rice genotypes. Figure 1.3 shows the improvement in grain yield, shoot dry weight, and the number of pods with the application of N to dry bean grown on a Brazilian Oxisol. The

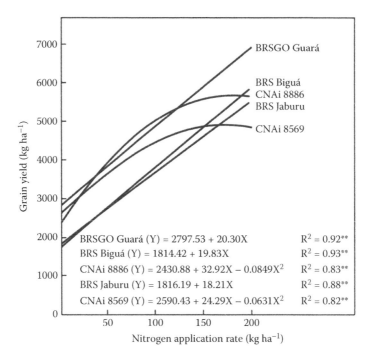

FIGURE 1.1   Response of five lowland rice genotypes to N fertilization.

main reasons for N deficiency in crop plants are its uptake in large amount compared to other essential plant nutrients. The uptake of N by most of the annual crops is the highest possible. It is sometimes equal to or slightly less than potassium by some crops such as rice (Fageria et al., 2011a). The decrease in organic matter and the loss of the top soil layer by wind and water erosion are other reasons for N deficiency in crop plants. Furthermore, the dynamic nature of N and its propensity for loss from soil–plant systems creates a unique and challenging environment for its efficient management (Fageria and Baligar, 2005). Agricultural productivity gains since the 1950s resulted in the development of farming systems that rely heavily on external inputs of energy and chemicals to replace the management and on-farm resources (Oberle, 1994; Porter et al., 2003).

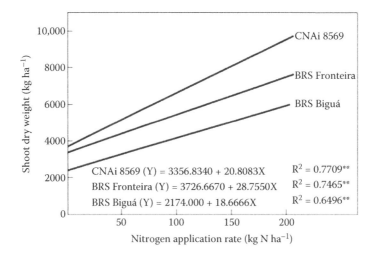

FIGURE 1.2   Relationship between N and shoot dry weight of three lowland rice genotypes.

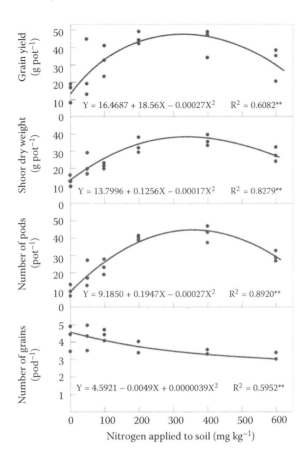

FIGURE 1.3 Relationship between N and yield and yield components of dry bean.

High quantities of inorganic fertilizer, particularly N, have been used to increase the world food production (Follett, 2001).

The increased crop productivity has been associated with a 20-fold increase in the global use of N fertilizer application during the past five decades (Glass, 2003), and this is expected to increase at least threefold by 2050 (Good et al., 2004). Although the N fertilizer use increased globally, this increase was not uniform in all countries or geographical regions (Vitousek et al., 2009) (Table 1.1). N use in North America and China increased significantly during the past few decades. In contrast, Denmark recorded a decrease in fertilizer use, and Africa is underusing N fertilizers (Liu et al., 2008; Malik and Rengel, 2013). Conventional breeding efforts in the past few decades have significantly increased crop yield and, as a corollary to this, also improved N use efficiency (NUE) (Kant et al., 2010). For example, a comparison between corn hybrids from the 1970s and the 1990s has shown that hybrids from the 1990s exhibited a higher yield response to increasing N supply (O'Neill et al., 2004).

N has long been recognized as a critical nutrient for the productivity of annual crops (Miller, 1939). In tropical America, N deficiency is a major soil constraint over 93% of the region occupied by acidic soils (Sanchez and Salinas, 1981). N fertilizer, along with irrigation, dramatically increased food production in developing countries during the green revolution (1960–1980) (Follett, 2001). Stewart et al. (2005) reported that the average U.S. corn yield was predicted to decline by 41% without N fertilizer, or in their words, N fertilizer was responsible for 41% of corn yield. Furthermore, the importance of N fertilization has been proved by the response of upland and lowland rice (*Oryza sativa* L.) (Fageria, 2001, 2013, 2014; Fageria and Baligar, 2001, 2005; Fageria

TABLE 1.1
Use of N, P, and K (10⁶ Tons) in Various Regions of the World

Region	Nitrogen		Phosphorus		Potassium	
	2000	2010	2000	2010	2000	2010
Africa	2.46	3.01	0.41	0.50	0.39	0.32
Eastern	0.33	0.58	0.11	0.14	0.07	0.07
Middle	0.03	0.02	0.00	0.00	0.02	0.01
Northern	1.45	1.72	0.16	0.23	0.10	0.06
Southern	0.43	0.39	0.08	0.07	0.12	0.09
Western	0.22	0.29	0.05	0.06	0.09	0.08
America	17.19	20.86	3.55	4.18	6.83	7.99
Northern	12.03	13.47	1.94	1.97	3.97	3.80
Central	1.75	1.56	0.19	0.13	0.28	0.27
Southern	3.25	5.65	1.40	2.06	2.51	3.88
Caribbean	0.17	0.18	0.03	0.02	0.07	0.03
Asia	46.72	67.03	7.73	13.02	6.52	10.48
Central	0.76	0.83	0.06	0.06	0.02	0.02
Eastern	23.21	35.75	4.11	7.65	3.40	4.79
Southern	15.26	21.34	2.45	4.20	1.53	3.27
Western	2.28	1.99	0.46	0.32	0.14	0.12
South Eastern	5.22	7.12	0.66	0.79	1.43	2.28
Europe	13.21	13.47	1.79	1.57	3.97	3.51
Northern	2.35	2.22	0.25	0.21	0.69	0.53
Southern	2.69	2.03	0.59	0.38	0.91	0.66
Eastern	3.54	5.00	0.39	0.69	0.94	1.51
Western	4.63	4.21	0.55	0.28	1.43	0.81
Oceania	1.19	1.51	0.69	0.58	0.29	0.20
World	80.79	105.89	14.17	19.86	18.00	22.50

Source: Adapted from FAO (Food and Agricultural Organization of the United Nations). 2013. Statistical database. FAO, Rome, Italy. Available at http://faostat.fao.org/Rome.

et al., 2011a), potato (*Solanum tuberosum* L.) (Hutchinson et al., 2003), dry bean (*Phaseolus vulgaris* L.) (Fageria, 2002), wheat (*Triticum aestivum* L.) (Fowler, 2003; López-Bellido et al., 2003), maize (*Zea mays* L.) (Cerrato and Blackmer, 1990; Kaizzi et al., 2012a), cassava (*Manihot esculenta* Crantz.) (Nguyen et al., 2002), faba bean (*Vicia faba* L.) (López-Bellido et al., 2003), cotton (*Gossypium hirsutum* L.) (Chua et al. (2003), sorghum (*Sorghum bicolor* L. Moench), (Kaizzi et al., 2012b), canola (*Brassica napus* L.) (Jackson, 2000; Vasilakoglou et al., 2012), and pearl millet (*Pennisetum glaucum* L. R. Br.) (Maman et al., 1999; Pandey et al., 2001) to N fertilization reported in various regions of the world.

Fertilizers account for almost half of the energy used in world agriculture, and the manufacture of N fertilizer is about 10 times more energy intensive than that of P and K fertilizers (Evans, 1993; Hasegawa, 2003). A survey of nutrient use in the United States indicated that N fertilizer from commercial sources was applied to 97% of the area planted with maize (Bausch and Diker, 2001). Biermacher et al. (2012) reported that according to the Oklahoma Cooperative Extension Service, N fertilizer typically accounts for up to 30% of the total production cost in the region. These authors also reported that over the past several years, prices for synthetic sources of N have trended increasingly higher. In fact, the price of N fertilizer increased by 120% between 2000 and 2007 in the United States (USDA, 2009). However, Economic Research Service (2010) reported that, in the United States alone, N fertilizer prices have more than doubled since 1990, reaching historic highs in mid-2008 due to high fertilizer demand and the inability of manufacturers to increase production levels (Huang et al., 2008). Furthermore, inorganic N fertilizers make up >20% of the operating expenses in maize production in Canada (Tollenaar, 1989), and it is imperative that the NUE be improved to increase the net income for growers. Similarly, N is one of the most limiting nutrients for cereal production in many West African countries (Pandey et al., 2001).

N deficiency is also responsible for the low yield of most food crops in the African continent. For example, corn is an important crop for the smallholder farmers in sub-Saharan Africa, but the yield has not increased significantly and per capita food production has declined since the 1980s (Greenland et al., 1994; Muchena et al., 2005). The main contributing factors are poor inherent soil fertility, particularly N and P deficiencies (Bekunda et al., 1997) exacerbated by soil fertility depletion (Vlek, 1993; Lynam et al., 1998), and other biophysical factors (Kaizzi et al., 2012a).

By 2050, the world's population is likely to be 9.1 billion, the CO_2 concentration 550 ppm, the ozone concentration 60 ppb, and the climate warmer by an average of 1.8°C (Jaggard et al., 2010). The global demand for crops is projected to increase 100–110% from 2005 to 2050, resulting in the expansion and intensification of agricultural land, with greater N inputs to increase yields (Tilman et al., 2011; Wilson et al., 2013). At the same time, there is added pressure on agricultural lands to produce energy, for example, U.S. energy legislation requires that 136 billion L of biofuel be used per year by 2022, 60 billion of which must be produced from nonfood feedstocks (Wilson et al., 2013). In addition, in the future, the demand of N will certainly increase because of an increasing demand by the world's population for food and fiber and its superb role in increasing crop yields. Today, commercial fertilizer N supplies approximately 45% of the total N input for global food production, and the world use is approximately 100 million metric tons (FAO, 2010). It is projected that annual total global N use will grow to approximately 112 million metric tons in 2020, and approximately 171 million metric tons in 2050 (Ladha et al., 2012). Half of the world's food production comes from fertilizer. Stewart et al. (2005) reported that, at the global scale, crop yields have increased by at least 30–50% as a result of fertilization. Furthermore, the use of adequate amount of N fertilizers in cropping systems is part of the solution to world food security. Philips and Norton (2012) reported that global wheat production has increased over two and a half times since 1960 to 2010, as a result of better farming practices, improved cultivars, and balanced nutrition. At the same time, fertilizer use in all agriculture has risen 4.3 times to keep up with the growing food demand. It is estimated that growers use around 15% of the fertilizer consumed to produce the current 650 million metric ton of wheat (Philips and Norton, 2012). These authors further reported that increase in fertilizer use mirrors the gains in productivity, although to maintain production it will require continual review of nutrient inputs. The challenge will be to ensure that future growth in food production is met by careful and targeted use of fertilizers. Looking into the importance of fertilizers and N in particular in crop production, the objective of this introductory chapter is to provide an overview of the functions of N in crop plants.

1.2 FUNCTIONS

Nitrogen plays an important role in the growth and development of plants. Its functions in the plants are extensive. N is mainly responsible for the growth and development of morphological traits in the cereals and legumes. In addition, N is also responsible for many physiological and biochemical functions in plants. The important biochemical functions of N include an essential constituent of enzymes, chlorophyll, nucleic acids, storage proteins, cell walls, and a vast array of other cellular components (Harper, 1994). Most of these reactions are responsible for improving the growth and development of crop plants. N is essential in the structure and function of amino acids, amides, nucleotides, nucleic acids, pigments, and some hormones (Hull and Liu, 2005; Bauer et al., 2012). The NO_3^- ion undergoes transformation after it is absorbed and is reduced to the amine form (NH_2^-). It is then utilized to form amino acids. Twenty amino acids are precursors of polypeptide chains comprising all protein (Bennett, 1993). Two other amino acids, glycine and glutamate, are precursors of N bases. Amino acids are essential to protein formation and are considered its building blocks. They are a part of the nucleic acids (DNA and RNA) that respectively hold the genetic information and direct protein synthesis (Bould et al., 1984). Because N is contained in the chlorophyll molecule, a deficiency of N will result in a chlorotic condition of the plant (Bennett, 1993). N is also a structural constituent of cell walls (Schrader, 1984). N is important in sucrose synthesis and in many reactions involving the utilization of sucrose as an energy source for plant growth and cell maintenance (Zinati et al., 2001). Hence, N deficiency in plants will have a profound influence on growth and development and consequently yield.

1.2.1 IMPROVEMENT IN GROWTH, YIELD, AND YIELD COMPONENTS AND RELATED TRAITS IN CEREALS AND LEGUMES

Before discussing the role of N in determining yield and yield components of crops, it is necessary to discuss what is yield and the relationship between yield and yield components. Yield is one of the most important measurements of a crop plant's economic value (Fageria, 2009). Yield is defined as the amount of a specified substance produced (e.g., grain, straw, and total dry matter) per unit area (Soil Science Society of America, 2008). Yield is generally expressed in kg ha^{-1} or Mg ha^{-1} (or metric ton ha^{-1}). In modern cropping systems, sustainability of yield is very important. Yield sustainability is defined as a continual, annual, or periodic yield of plants or plant material from an area, implied management practices that will maintain the productive capacity of the land, be economically feasible, and maintain environmental integrity of the ecosystem (Soil Science Society of America, 2008).

The yield in crop plants is genetically determined and also influenced by environmental factors (Fageria, 2013, 2014). Among environmental factors, the supply of essential plant nutrients in adequate amount and proportion is very important. Figure 1.4 shows grain yield and shoot dry weight or straw yield among 12 lowland rice genotypes. They were grown under similar field conditions for two consecutive years. However, genotypes showed significant variability in the grain yield and shoot dry weight (Figure 1.4). The shoot dry weight varied from 6602 kg ha^{-1} for genotype CNAi 8569 to 4041 kg ha^{-1} for genotype BRS Biguá, with an average value of 4980 kg ha^{-1}. Similarly, a variation in grain yield was from 4828 kg ha^{-1} for genotype BRSGO Guará to 3638 kg ha^{-1} for genotype BRS Jaburu, with an average value of 4287 kg ha^{-1}. A variation in grain yield of

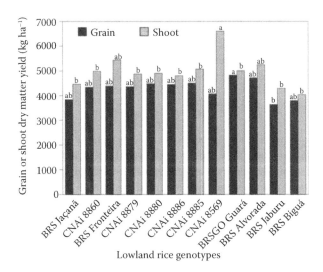

FIGURE 1.4 Grain and shoot dry weight of 12 lowland rice genotypes. Values are averages of 2 years field experimentation. (Adapted from Fageria, N. K., A. B. Santos, and V. A. Cutrim. 2008. *J. Plant Nutr.* 31:788–795. With permission.)

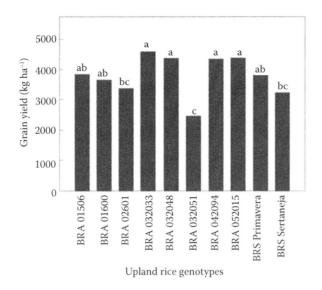

FIGURE 1.5 Grain yield of 10 upland rice genotypes.

upland rice was also present in Figure 1.5. Genotype BRA 032051 produced the lowest grain yield of about 2500 kg ha^{-1}, whereas genotypes BRA 032033 produced the highest grain yield of nearly 5000 kg ha^{-1}. These results clearly show genetic variation in grain yield among lowland and upland rice genotypes. Fageria and Barbosa Filho (2001) and Fageria et al. (2003) have reported differences in shoot and grain yield of lowland rice genotypes. Variation in shoot dry weight and grain yield among genotypes may be associated with differences in the amount of intercepted photosynthetically active radiation by the canopy, the radiation use efficiency (RUE), and grain harvest index (GHI) (Kiniry et al., 2001; Fageria and Baligar, 2005).

Grain yield in cereals and legumes is formulated or computed by yield components. In cereals, yield forming components are the number of panicle or head per unit area, grain per panicle, weight of thousand grain, and spikelet or grain sterility. For example, in rice, the grain yield can be calculated by using these yield components with the help of the following equation (Fageria, 2009):

$$\text{Grain yield (Mg ha}^{-1} \text{ or metric ton ha}^{-1}) = \text{number of panicles m}^{-2} \times \text{number of grains per panicle} \\ \times \% \text{ of filled grains} \times \text{weight of 1000 grain (g)} \times 10^{-5}$$

According to this equation, grain yield in cereal or especially in rice is a combination of various yield components. To obtain a good yield, an appropriate balance among these yield components is required. For example, to produce a yield of 6 Mg ha^{-1} or metric ton, the following combination of these yield components is necessary: 400 panicle m^{-2}, 80 grains per panicle, 85% of filled grains, and 22 g weight of 1000 grains. Putting these values of the yield components in the above equation, the calculated yield will be as follows:

$$\text{Grain yield (Mg ha}^{-1}) = 400 \times 80 \times 0.85 \times 22 \times 10^{-5} = 6$$

It is clear from this equation that grain yield in rice can be improved with the improvement in the number of panicles per unit area, improvement in grain per panicle, improvement in the filled spikelets, and also improvement in the 1000 grain weight. These components cannot be increased indefinitely because after a certain increase, there is negative interaction between some of these yield components. For example, when the number of panicles increased more than a certain level, the size and weight of the grain will decrease, and grain or spikelet sterility will increase (Fageria,

1989). In this case, the source (photosynthesis by leaves) may be the limiting factors. In other words, the photosynthetic capacity of a plant will not be able to properly fill the large number of grains produced per unit area or plants.

In legumes, the yield is mainly expressed as the product of pods per unit area or per plant, seeds per pod, and seed weight. Grain yield in legumes can be computed by using the following equation (Fageria, 2009):

$$\text{Grain yield (Mg ha}^{-1}) = \text{number of pods} \times \text{number of seeds per pod}$$
$$\times \text{weight of 1000 seeds (g)} \times 10^{-5}$$

Based on the above equation, Fageria (1989) calculated or discussed the yield of cowpea using the following yield components: number of pods per unit area or $m^{-2} = 155$; number of seeds per pod = 7; weight of 1000 seeds = 140 g. Putting these values of yield components in the above equation, the yield will be

$$\text{Yield (Mg ha}^{-1} \text{ or metric ton ha}^{-1}) = 155 \times 7 \times 140 \times 10^{-5} = 1.5$$

Studies have reported mechanisms of high grain yield in super-high-yielding rice (Xiong et al., 2013). It was reported that in several field studies, some super-high-yield rice cultivars produced 6.4–20% higher grain yield than ordinary rice check cultivars (Wang et al., 2002; Wu et al., 2007). The high grain yield of super-high-yield rice was attributed to large sink size (Wu et al., 2007; Zhang et al., 2009). The sink size can be increased by increasing the panicle number per unit area or spikelet number per panicle or both (Ying et al., 1998). It was also reported that super-high-yield rice had increased biomass production, high leaf area index (LAI), during the grain filling period, longer leaf area duration, a higher photosynthetic rate at the single leaf level, slower leaf senescence, and the tolerance of photoinhibition as compared to ordinary rice (Chen et al., 2002; Katsura et al., 2007; Lin et al., 2002; Wang et al., 2005, 2006, Zhang et al., 2009).

1.2.1.1 Plant Height

Plant height is the distance from the ground level to the tip of the tallest leaf for seedlings or juvenile plants. For mature plants, it is the distance from the ground level to the tip of the tallest panicle, ear, or head in cereals and branch legumes. Short and sturdy culms, more than any other character, favor lodging resistance. Lodging is the permanent displacement of the stems from their upright position. There are three types of lodging: breaking of the stem, bending of the stem, and rolling, in which the whole plant is uprooted from the ground and falls over (Fageria et al., 2006). A tall crop variety has greater bending moment than a short one because of culm height. Early lodging of long, thin culms disturbs leaf arrangement, increases mutual shading, interrupts transport of nutrients and photosynthates, causes grain sterility, and reduces yield (Jennings et al., 1997). Further, strong winds and rains during reproductive and grain filling stages of growth can cause lodging in annual crops. Lodging during the grain filling growth stage can reduce grain quality.

A short and sturdy culm also promotes favorable grain-to-straw ratios, adequate N responses, and high yield capacities. Increased N application is essential for higher yields, but causes elongation of the lower internodes, making the crop more susceptible to lodging. In addition, planting date, row spacing, and seeding rate affect lodging in soybean (*Glycine max* L. Merr.) (Willmot et al., 1989) as well as plant height, branch production, and basal pod height. A marked increase in harvest index and grain production per day has been associated with reduced plant height and earlier maturity (Evans et al., 1984). The heritability of dwarfism in cereals is high and is easy to identify, select for, and recombine with other traits. While yield gains due to the introduction of dwarfing genes into cereals have been remarkable, little evidence exists for concomitant improvements in photosynthetic rate, crop growth rate (CGR), or kernel weight.

TABLE 1.2

Plant Height, Grain Yield, and Lodging Rating of Traditional and Modern Upland Rice Cultivars under Brazilian Conditions

Cultivar	Plant Height (cm)	Grain Yield (kg ha^{-1})	Lodging Rating[a]
Traditional			
Rio Paranaíba	108	2780	3
Caiapó	105	2590	2
Guarani	98	2640	4
CNA 8054	108	2470	3
Average	105	2620	3
Modern			
Progresso	86	2620	1
CNA 8172	86	2860	1
CNA 8305	91	2990	1
BRS Canastra	88	2870	1
Average	88	2840	1

Source: Adapted from Morais, O P. 1998. Annual report of the project, "Breeding upland rice culti-vars". National Rice and Bean Research Center of Embrapa, Goiania, Brazil.

[a] Higher values mean relatively more susceptibility to lodging and lower values mean more resistant to lodging.

The creation of semidwarf cultivars spectacularly increased the yielding ability of many crops, such as rice and wheat. For example, Table 1.2 compares the yields of older and taller traditional cultivars of upland rice to modern, semidwarf cultivars of Brazil. Yields of the modern cultivars are higher, in part, because they are less susceptible to lodging than the old ones. Similarly, lowland rice cultivars used in the Philippines have, over the past 70 years, become shorter, in addition to having smaller, more upright leaves and reduced sensitivity to photoperiod (Fageria et al., 2006). Panicle weights initially became heavier, but later were made lighter with greatly increased panicle numbers.

In addition to lodging resistance, short-stature and sturdy culm cultivars give higher yields at close plant spacing as compared to the taller cultivars. However, a taller plant has an advantage in competing with weeds as compared to short-stature plant. In addition, grain yields decrease with increasing water depth. Under such conditions, intermediate stature (100–130 cm) is considered desirable over short stature (90–110 cm) for rice (Yoshida, 1981). Extremely dwarf height is also not good, because grain yield increases quadratically with the increasing plant height (Fageria, 2007, 2009, 2014). A marked increase in harvest index and grain yield per day has been associated with reduced plant height and earlier maturity (Evans et al., 1984). Evans et al. (1984) also reported that despite these changes, there has been no change in the photosynthetic rate, CGR, or spikelet weight. The plant height of 10 lowland rice genotypes grown under Brazilian conditions is presented in Figure 1.6. The plant height of these genotypes varied from 94 to 111 cm, with an average value of 102 cm. Similarly, the plant height of 19 upland rice genotypes is presented in Table 1.3. Plant height was significantly ($P < 0.01$) influenced by N and genotype treatments and across two N levels varied from 93 to 118 cm among genotypes with an average value of 103 cm (Table 1.3). There was a significant quadratic relationship between plant height (X) and grain yield ($Y = -196.3034 + 3.2542 X - 0.0097X^2$, $R^2 = 0.65**$) (Fageria et al., 2010). Hence, grain yield increases with increasing plant height but there is a limit of this increase. When plant height is too low, it produces less dry matter, and when it is too high, it may lodge and less responsive to N fertilization (Yoshida, 1981). This means that intermediate plant height is a better compromise and this is confirmed by the fact that

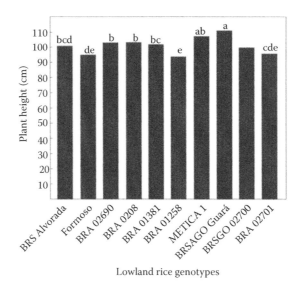

FIGURE 1.6   Plant height of 10 lowland rice genotypes. (Adapted from Fageria, N. K. 2007. *J. Plant Nutr.* 30:843–879. With permission.)

the maximum grain-yield-producing genotype CNAs 8993 (at high N level) did not have the maximum plant height, but this genotype had an intermediate plant height (Fageria et al., 2010).

Although plant height, is influenced by environmental factors, it is a genetically controlled plant trait and the heritability of dwarfism is high and easy to identify, select, and recombine with other traits (Jennings et al., 1997). These authors also reported that dwarf segregates have a fairly narrow range in height, presumably from minor gene action. Although a few are so short that they are undesirable, the great majority fall within the useful range from 80 to 100 cm with some reaching 120 cm under certain conditions (Jennings et al., 1997). During the 1960s, rice breeders made excellent progress in the development of dwarf cultivars that responded to heavy applications of N (Jennings et al., 1997).

1.2.1.2   Tillering

Tillers are the branches that develop from the leaf axils at each unelongated node of the main shoot or from other tillers during vegetative growth. The development of tiller buds after differentiation is greatly affected by environmental conditions, as well as by genotypic characteristics. The environment must be favorable for tiller development. It is essential that sufficient supplies of water, photosynthate, nutrients, and plant hormones are present, and that stress is minimal (Fageria et al., 2006). The addition of plant nutrients is particularly important when soils have low fertility. Nutrients required for the growth of tiller buds of rice must come from the main stem, since tiller buds have neither roots to absorb inorganic nutrients nor leaves to carry out photosynthesis. Once tillers have emerged from the subtending leaf sheaths, they can perform photosynthesis and produce carbohydrates. Tillers can also absorb soil nutrients through their own roots after the third leaf has completely emerged, since roots appear at the prophyll nodes of tillers at this stage of growth (Handa, 1995).

Tillering is an important trait in cereals in determining yield. It is one of the first development or growth stage in cereals and mainly depends on sowing density and the availability of water and N (Simane et al., 1993; Moragues et al., 2006). Tillering has special importance under biotic and abiotic stresses due to the compensation process. High tillering compensates for missing plants at low densities, but cultivars with limited tillering capacity lack this plasticity. However, under favorable environmental conditions, heavy tillering cultivars have no advantage over low tillering ones in relation to yield. Heavy tillering is not too advantageous in the direct-seeded rice, which is a common practice

TABLE 1.3

Plant Height of 19 Upland Rice Genotypes as Influenced by N and Genotype Treatments

Genotype	Plant Height (cm)
CRO 97505	106cde
CNAs 8993	104cde
CNAs 8812	93 g
CNAs 8938	99efg
CNAs 8960	112abc
CNAs 8989	99efg
CNAs 8824	94 g
CNAs 8957	105cde
CRO 97422	108bcd
CNAs 8817	109bcd
CNAs 8934	105cde
CNAs 9852	104cde
CNAs 8950	106cde
CNA 8540	93 g
CNA 8711	118a
CNA 8170	93 g
BRS Primavera	115ab
BRS Canastra	94fg
BRS Carisma	102def
Average	103
F-test	
N level (N)	**
Genotype (G)	**
N × G	NS
CV (%)	4

Source: Adapted from Fageria, N. K., O. P. Morais, and A. B. Santos. 2010. Nitrogen use efficiency in upland rice genotypes. *J. Plant Nutr.* 33:1696–1711.

Note: Means followed by the same letter in the same column are not significantly different at the 5% probability level by Turkey's test. Plant height values are across two N rates (0 mg N kg^{-1} and 400 mg N kg^{-1}).

**, NS: Significant at the 1% probability level and nonsignificant, respectively.

in mechanized agriculture in South America as well as in the United States. Under direct seeding, tillering capacity rarely affects grain yield within conventional seeding rates because the total panicle number per square meter depends more on the main culm than on tillers (Yoshida, 1981). Low seed rate is required more, however, by heavy-tillering cultivars as compared to low-tillering cultivars. Jennings et al. (1997) reported, however, that a combination of high tillering ability and compact or nonspreading culm arrangement is desirable. Compact culms that are moderately erect allow increased solar radiation to tillers and less mutual shading per unit of land area.

N fertilization significantly increases tillering in cereals, especially in rice (Fageria et al., 2003). Figure 1.7 shows the increase in tiller number with increasing levels of N in lowland rice grown on an Inceptisol of central Brazil. Murata and Matsushima (1975) reported that an N concentration of more than 35 g kg^{-1} (3.5%) is necessary for active tillering; at 25 g N kg^{-1} (2.5%), tillering stops, and below 15 g N kg^{-1} (1.5%), death of tillers takes place. Data in Table 1.4 show that N significantly increased tillering in lowland rice. About 66–96% of the variation in tillering was apparently due

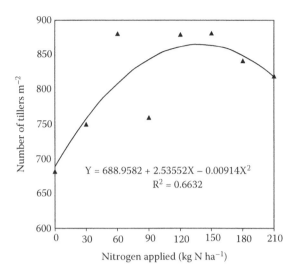

$$Y = 688.9582 + 2.53552X - 0.00914X^2$$
$$R^2 = 0.6632$$

FIGURE 1.7   Relationship between N application and number of tillers in flooded rice grown on an Inceptisol of central Brazil. (Adapted from Fageria, N. K. 1998. Annual report of the project "The study of liming and fertilization for rice and common bean in cerrado region". National Rice and Bean Research Center of Embrapa, Goiania, Brazil. With permission.)

to N fertilization depending on crop growth stage (Table 1.4). Tillering increased with the advancement of the crop growth and maximal values were achieved between 35 and 71 days after sowing, depending on the N rate, and then decreased thereafter. Grain yield in cereals is highly dependent upon the number of spikelet-bearing tillers produced by each plant (Power and Alessi, 1978; Nerson, 1980). The number of productive tillers depends on environmental conditions during tiller bud initiation and subsequent developmental stages. Numerous studies have shown that tiller appearance, abortion, or both are affected by environmental conditions, especially nutrient deficiencies (Black and Siddoway, 1977; Power and Alessi, 1978; Masle, 1985).

TABLE 1.4
Numbers of Tillers in Lowland Rice at Different N Rates during Crop Growth Cycle

N Rate	22 (IT)	35 (AT)	71 (IP)	97(B)	112 (F)	140 (PM)
kg ha^{-1}			m^{-2}			
0	506	681	652	541	499	468
30	516	749	715	547	516	495
60	574	880	772	601	571	531
90	599	759	751	597	561	522
120	632	876	812	623	573	569
150	619	862	883	660	580	592
180	557	880	903	662	588	572
210	565	819	934	666	590	581
R^2	0.82*	0.66*	0.96*	0.95**	0.91*	0.92*

Source: Adapted from Fageria, N. K. and V. C. Baligar. 2001. *Commun. Soil Sci. Plant Anal.* 32:1405–1429. With permission.

Note: Values are averages of 3 years field trial.

*,**Significant at the 5% and 1% probability levels, respectively.

TABLE 1.5

Correlation Coefficients (r) between Lowland Rice Grain Yield and Tiller Number at Different Growth Stages

Parameter	1st Year	2nd Year	3rd Year
Tiller number m^{-2} at IT	0.59**	0.41*	0.23^{NS}
Tiller number m^{-2} at AT	0.69**	0.43*	0.34*
Tiller number m^{-2} at IP	0.79**	0.59**	0.68**
Tiller number m^{-2} at B	0.67**	0.52**	0.46**
Tiller number m^{-2} at F	0.70**	0.37*	0.52**
Tiller number at PM	0.77**	0.48**	0.44*

Source: Adapted from Fageria, N. K. and V. C. Baligar. 2001. *Commun. Soil Sci. Plant Anal.* 32:1405–1429. With permission.

Note: IT, initiation of tillering; AT, active tillering; IP, initiation of panicle; B, booting; F, flowering; PM, physiological maturity.

*,**Significant at the 5% and 1% probability levels, respectively. NS, not significant.

The decrease in tiller number was attributed to the death of some of the last tillers as a result of their failure in competition for light and nutrients (Fageria et al., 2011a). Another explanation is that during the period of growth beginning with panicle development, competition for assimilates exists between developing panicles and young tillers. Eventually, the growth of many young tillers is suppressed, and they may senesce without producing seed (Dofing and Karlsson, 1993). A correlation between grain yield and the number of tillers per meter square at different growth stages is presented in Table 1.5. Tillering was related significantly with grain yield at all the growth stages; however, the highest correlation in all the 3 years of experimentation was obtained at the initiation of panicle growth stage. This means that the number of tillers determined at this growth stage had more significance than that at any other growth stage in lowland rice.

A tiller number is quantitatively inherited. Its heritability is low to intermediate depending on the cultural practices used and the uniformity of the soil. Although often associated with early vigor in short-statured materials, the tiller number is inherited independently of all other major characters. In many crosses, tiller erectness or compactness is recessive to a spreading culm arrangement (Jennings et al., 1997). Developing good plant types with high tillering capacity is rather simple. Many sources of heavy tillering are available in traditional tropical rice cultivars. When their culms are shortened, their tillering ability generally does not decrease and may increase (Jennings et al., 1997).

1.2.1.3 Shoot Dry Weight

Shoot dry weight is an important growth trait in crop plants. N has significant influence in increasing shoot dry weight of crop plants (Fageria, 2007; Fageria and Santos, 2008). Shoot dry weight of lowland rice increased significantly in the vegetative as well as reproductive growth stages (Table 1.6). The increase in shoot weight is mainly associated with an increase in leaf and culm weights during these growth stages. Rice plant weight mainly consists of organic matter such as protein and carbohydrate. Carbohydrates are composed of cell wall substances such as cellulose and reserved substances such as starch. The protein metabolism dominates in the vegetative growth stage and carbohydrate metabolism does in the reproductive growth stage (Murayama, 1995). The portion of inorganic matters in the weight of the rice plant is generally small. However, rice straw generally has high silicon content. In the mature shoot of rice, the silicon content can be as high as 10% (Murayama, 1995).

TABLE 1.6
Shoot Dry Matter Yield (kg ha⁻¹) of Lowland Rice at Different N Rates

	Days after Sowing					
N Rate	22 (IT)	35 (AT)	71 (IP)	97 (B)	112 (F)	140 (PM)
0	313	815	3065	5650	7694	5278
30	320	860	3709	6913	8953	6764
60	342	1230	3721	8242	11,056	7294
90	374	1044	4164	8695	10,758	7303
120	380	1229	4313	9570	13,378	8215
150	452	1207	4893	10,031	12,745	8624
180	351	1294	5077	11,290	13,682	9060
210	351	1130	5841	10,384	13,490	9423
R^2	0.56*	0.76*	0.97**	0.97**	0.94**	0.96**

Source: Adapted from Fageria, N. K. and V. C. Baligar. 2001. *Commun. Soil Sci. Plant Anal.* 32:1405–1429. With permission.

Note: IT, initiation of tillering; AT, active tillering; IP, initiation of panicle; B, booting; F, flowering; PM, physiological maturity. Values are averages of 3 years field trial.

*,**Significant at the 0.05 and 0.01 probability levels, respectively. Not significant.

Upland rice shoot weight decreased from flowering to physiological maturity (Fageria, 2007). Dry matter loss from the vegetative tissues during the interval from flowering to maturity was 35%, suggesting active transport of assimilates to the panicles, which resulted in a grain yield of 3811 kg ha⁻¹ (Fageria, 2007). Fageria et al. (2011a, 2006) reported more or less similar reduction in the shoot dry weight of upland rice from flowering to physiological maturity.

Increase in shoot weight is important because it is significantly associated with grain yield (Table 1.7; Figure 1.8). Shoot weight is a characteristic of genotypes and is also influenced by environmental factors. Differences have been observed in grain yield among plants or genotypes having the same amount of dry matter, because there exists differences in the utilization of photosynthates among them (Hayashi, 1995). Table 1.8 shows the relationship between N rate and shoot dry weight

TABLE 1.7
Correlation Coefficients (r) between Lowland Rice Grain Yield and Shoot Dry Matter Production during Different Growth Stages

Parameter	1st Year	2nd Year	3rd Year
Dry matter yield at IT	0.36*	0.37*	0.29[NS]
Dry matter yield at AT	0.71**	0.55**	0.42*
Dry matter yield at IP	0.63**	0.51**	0.63**
Dry matter yield at B	0.72**	0.81**	0.61**
Dry matter yield at F	0.81**	0.80**	0.57**
Dry matter yield at PM	0.78**	0.80**	0.53**

Source: Adapted from Fageria, N. K. and V. C. Baligar. 2001. *Commun. Soil Sci. Plant Anal.* 32:1405–1429. With permission.

Note: IT, initiation of tillering; AT, active tillering; IP, initiation of panicle; B, booting; F, flowering; PM, physiological maturity.

*,**Significant at the 0.05 and 0.01 probability levels, respectively. NS, not significant.

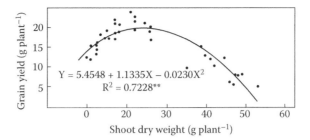

FIGURE 1.8 Relationship between shoot dry weight and grain yield of irrigated or lowland rice. (Adapted from Fageria, N. K., A. B. Santos, and A. M. Coelho. 2011b. *J. Plant Nutr.* 34:371–386. With permission.)

of 12 lowland rice genotypes. The increase in shoot dry weight was linear or quadratic with the application of N in the range of 0–200 kg ha^{-1}. The variation in shoot dry weight with the application of N was varied from 58% to 78%, depending on the genotypes (Table 1.8). N fertilization influences the shoot dry weight of lowland as well as upland rice genotypes (Fageria et al., 2003; Fageria and Baligar, 2005).

Grain yield in cereals is related to biological yield and GHI (Donald and Hamblin, 1976). The biological yield of a cereal crop is the total yield of plant tops and is an indication of the yield of the photosynthetic capability of a crop (Yoshida, 1981). The biological yield is a function of crop growth duration and CGR at successive growth stages (Tanaka and Osaki, 1983). GHI is the ratio of grain to the aboveground biological yield. The GHI is controlled by the partition of photosynthates between harvesting and nonharvesting organs during the crop growth cycle. Hence, the economic yield is closely related with the crop growth process. Grain yield or economic yield can be increased either by increasing total dry matter production or by increasing GHI. Some authors have speculated that a

TABLE 1.8

Relationship between N Rate (X) and Shoot Dry Weight (Y) of 12 Lowland Rice Genotypes

Genotype	Regression Equation	R^2
BRS Jaçanã	$Y = 2442.44 + 29.21X - 0.0595X^2$	0.5812**
CNAi 8860	$Y = 2978.50 + 31.49X - 0.0759X^2$	0.5401**
BRS Fronteira	$Y = 3467.83 + 20.25X$	0.7832**
CNAi 8879	$Y = 2860.37 + 31.59X - 0.0759X^2$	0.7365**
CNAi 8880	$Y = 3047.00 + 34.45X - 0.1054X^2$	0.6641**
CNAi 8886	$Y = 2876.00 + 34.49X - 0.1013X^2$	0.6548**
CNAi 8885	$Y = 2513.42 + 41.25X - 0.1044X^2$	0.7380**
CNAi 8569	$Y = 3726.67 + 28.76X$	0.7464**
BRSGO Guará	$Y = 2550.09 + 43.56X - 0.1267X^2$	0.7213**
BRS Alvorada	$Y = 2886.66 + 23.62X$	0.7191**
BRS Jaburu	$Y = 2577.11 + 25.69X - 0.0566X^2$	0.7837**
BRS Biguá	$Y = 2174.00 + 18.67X$	0.6496**

Source: Adapted from Fageria, N. K., A. B. Santos, and V. A. Cutrim. 2008. *J. Plant Nutr.* 31:788–795. With permission.

Note: Values are averages of 2 years.

**Significant at the 1% probability level.

further increase in grain yield in cereals such as rice through breeding can only be accomplished with an increase in total biological yield (Rahman, 1984) and thus the total straw yield. The highest GHI exhibited by California lowland rice cultivars under direct seeding was 0.59 (Roberts et al., 1993).

Hasegawa (2003) reported that higher yields of rice cultivars were associated with higher dry matter production and both increased dry matter and GHI equally contributed to yield increases. Peng et al. (2000) reported that the yield improvement of lowland rice cultivars released by International Rice Research Institute (IRRI) in the Philippines after 1980 was due to increases in biomass production. Similarly, Akita (1989) and Amano et al. (1993) reported that, when comparisons were made among the improved semidwarf cultivars, higher yield was achieved by increasing biomass production. Song et al. (1990) and Yamauchi (1994) reported that in hybrid rice having about 15% higher yield than inbred mainly because of an increase in biomass production rather than GHI. Dry matter production in rice has been reported to be significantly related to intercept photosynthetically active radiation (IPAR) (Kiniry et al., 2001). CGR depends on the amount of radiation intercepted by the crop and on the efficiency of conversion of intercepted radiation into dry matter (Sinclair and Horie, 1989). Hence, it can be concluded that the production of sufficient dry matter of shoot is important for improving the grain yield of rice.

López-Bellido et al. (2003) reported that high biomass is a prerequisite for achieving high faba bean seed yields. Loss and Siddique (1997), Thomson et al. (1997), and Mwanamwenge et al. (1998) also reported that seed yields of faba bean were positively correlated with the total dry matter production at harvest. Linear relationships between biomass and seed yields were reported for soybeans grown in Puerto Rico (Ramirez-Oliveras et al., 1997) and Australia (Mayers et al., 1991). Similarly, Board et al. (1996) and Rao et al. (2002) also reported strong positive correlations between yields and dry matter in soybeans grown in the United States. Dry matter production had highly significant associations with grain yields of plants grown under relatively high heat environments (Reynolds et al., 1994).

The shoot dry weight of dry bean was also significantly influenced by N fertilization. during growth cycle, except at 23 days after sowing (Table 1.9). The shoot dry weight was maximum at

TABLE 1.9
Dry Matter Yield of Shoot (kg ha⁻¹) of Dry Bean as Influenced by N Fertilization at Different Growth Stages

N Rate (kg ha⁻¹)	Days after Sowing				
	23	44	60	78	93
0	80.0	118.8	220.0	658.8	1191.3
40	93.1	135.6	493.8	971.3	1442.5
80	95.0	178.1	813.8	1796.3	3340.0
120	77.5	171.3	827.5	2576.3	4503.8
160	71.3	239.4	1260.0	2261.3	5345.0
200	71.3	337.5	1658.8	3240.0	6593.8
F-test	NS	*	**	**	**

Regression Analysis

N rate (X) versus dry matter yield 23 days (Y) = $84.5759 + 0.1284X - 0.0011X^2$, $R^2 = 0.1598^{NS}$

N rate (X) versus dry matter yield 44 days (Y) = 117.3500 exp. $(0.0023X) + 0.000011X^2$, $R^2 = 0.4808^*$

N rate (X) versus dry matter yield 60 days (Y) = 218.6368 exp. $(0.01734X) - 0.000038X^2$, $R^2 = 0.8328^*$

N rate (X) versus dry matter yield 78 days (Y) = 615.2077 exp. $(0.0147X) - 0.000034X^2$, $R^2 = 0.8399^*$

N rate (X) versus dry matter yield 23 days (Y) = 1012.8120 exp. $(0.0154X) - 0.000032X^2$, $R^2 = 0.7135^*$

*,**, NS: Significant at the 5% and 1% probability levels and nonsignificant, respectively.

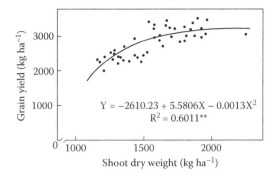

FIGURE 1.9 Relationship between shoot dry weight and grain yield of dry bean.

200 kg N ha^{-1} at all growth stages compared with lower N rates, except at 23 days growth stage. The increase in shoot dry weight at 44 days growth was 184% with the application of 200 kg N ha^{-1} compared with without N application or control treatment. Similarly, the shoot dry weight increase was 654% at 60 days after sowing, 392% at 78 days after sowing, and 453% at 92 days after sowing with the application of 200 kg N ha^{-1} compared with control treatment. It is reported by Fageria (2002) that among all essential plant nutrients, N is quantitatively most important for dry bean growth in most of the bean-producing regions around the world. N nutrition influences leaf growth and leaf area duration, and hence the carbohydrate source size, the photosynthetic rate per unit leaf area, and hence source activity, and the number and size of vegetative and reproductive storage organs, and hence sink capacity (Marschner, 1995). Pelegrin et al. (2009) studied the response of dry bean to N application on an Oxisol of central Brazil and reported that the bean crop response to N was highly significant. The bean plant's efficiency in using N fertilizer is rather low, being less than 70% of the applied fertilizer. The remaining fertilizer is lost through volatilization and leaching, losses being greater with broadcast application (Thung and Rao, 1999).

The variability in the shoot dry matter yield with N fertilization increased with the advancement of plant age up to 78 days of plant growth (Table 1.9). Maximum variability in shoot dry weight with N application was at 60 (83%) and 78 (84%) days plant growth (Table 1.9). This means that N requirement for dry bean is maximum between 60 and 78 days growth period. The data in Table 1.9 also show that dry matter increase was slow between the 23- and 44-day growth period and it was almost linear between 44 and 93 days plant growth. Shoot dry weight was having significant quadratic association with the grain yield of dry bean (Figure 1.9). Hence, improving the shoot dry weight is an important aspect in increasing the grain yield of this legume crop.

1.2.1.4 Panicle Number, Panicle Length, and Pod Number

Panicle density, panicle length, and pod number per unit area are important yield components in cereals and legumes, respectively. These traits have a significant positive association with grain yield in cereals and legumes (Fageria, 2007; Fageria and Santos, 2008). Figure 1.10 shows that the grain yield of upland rice increased significantly and quadratically with increasing panicle density of upland rice. Similarly, the grain yield of dry bean was having significant quadratic association with pod number (Figure 1.11). The N and genotypes treatments and their interactions significantly affected panicle number per pot and panicle length (Table 1.10). Panicle number varied from 3.7 to 10.3 per pot at low N level with an average value across the genotypes of 7.9 per pot. Similarly, panicle number at higher N level varied from 14.3 to 29.0 per pot with an average value of 21.4 per pot. Overall, the increase in panicle number per pot was 171% with the application of N as compared to treatment without N fertilization. The number of panicles (X) were having significant (P < 0.01) quadratic relationship with the grain yield (Y = $-32.9349 + 7.2697X - 0.1467X^2$, $R^2 = 0.81^{**}$). This

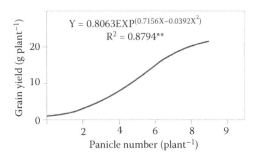

FIGURE 1.10 Relationship between panicle number and grain yield of upland rice.

means that 81% variability in grain yield was accounted as due to panicle number. Panicle number per unit area is considered as one of the important yield components in increasing upland rice yield (Fageria, 2007, 2009). Panicle length varied from 13.9 to 20.8 cm with an average value of 17.7 cm at low N level. Similarly, at high N level, panicle length varied from 18.4 to 26.6 cm with an average value of 22 cm. Overall, the increase in panicle length at higher N level was 24% compared with low N level. Panicle length (X) was having significant (P < 0.01) quadratic relationship with grain yield ($Y = -110.5796 + 9.9081X - 0.1299X^2$, $R^2 = 0.43^{**}$). Yoshida (1981) reported that the number of spikelets or grain per unit area of rice crop is positively correlated with the amount of N absorbed by the end of spikelet initiation stage or by flowering.

In small grains, N generally increases the number of tillers, resulting in a greater number of ears per unit land area (Halse et al., 1969; Pushman and Bingham, 1976). In cases where the ear number is little affected, N may then cause an increase in the ear weight of up to 70 per cent over the control (Gasser and Iordanou, 1967; Spratt and Gasser, 1970). That can result from an increase in the number of spikelets per ear if the N is applied early (Holmes, 1973; Langer and Liew, 1973; McNeal et al., 1971; Pearman et al., 1977). Grain weight generally shows much less variation than grain number with the application of N (Novoa and Loomis, 1981).

The pod number of dry bean was significantly influenced by N and genotype treatments and their interactions (Table 1.11). Pod number varied from 4.3 to 13.7 per pot at zero N rates with an average value of 7.2 across 20 genotypes. The increase in pod number was 3.6-fold at 400 mg N treatment compared to zero N treatment across 20 genotypes. Although F-test showed a significant influence of genotype and N × G interaction, however, genotypes were not differentiated in relation to the pod number by Tukey's test at lower and higher N rates.

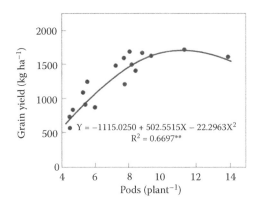

FIGURE 1.11 Relationship between pods and grain yield of dry bean. (Adapted from Fageria, N. K. 2007. *J. Plant Nutr.* 30:843–879. With permission.)

TABLE 1.10

Panicle Number and Panicle Length of 19 Upland Rice Genotypes as Influenced by Two N Rates

Genotype	Panicle Number per Pot		Panicle Length (cm)	
	0 mg N kg^{-1}	400 mg N kg^{-1}	0 mg N kg^{-1}	400 mg N kg^{-1}
CRO 97505	8.0ab	17.7fg	17.9bcd	23.8abcd
CNAs 8993	8.3ab	22.0a–f	16.7de	19.5efgh
CNAs 8812	10.3a	28.0ab	16.2def	19.1fgh
CNAs 8938	7.7abc	20.3c–g	17.5cde	20.6defgh
CNAs 8960	6.0bc	15.0fg	20.8a	26.6a
CNAs 8989	8.0ab	21.7a–g	16.3def	18.9fgh
CNAs 8824	9.3ab	26.3abcd	17.6cde	19.2efgh
CNAs 8957	6.7abc	17.0fg	19.6abc	25.4abc
CRO 97422	7.3abc	18.7efg	16.8de	23.9abcd
CNAs 8817	8.0ab	19.3defg	20.5ab	25.3abc
CNAs 8934	8.7ab	21.3b–g	17.6cde	21.6c–h
CNAs 9852	3.7c	14.3 g	18.6abcd	22.6b–g
CNAs 8950	6.7abc	19.7c–g	17.3cde	23.1a–e
CNA 8540	8.7ab	25.3a–e	13.9f	18.4 h
CNA 8711	8.7ab	21.0b–g	17.4cde	22.8a–f
CNA 8170	10.0ab	29.0a	18.8abcd	21.2d–h
BRS Primavera	7.7abc	17.0fg	19.7abc	25.7ab
BRS Canastra	8.0ab	25.3a–e	17.6cde	20.9d–h
BRS Carisma	8.3ab	27.0abc	15.3ef	18.7gh
Average	7.9	21.4	17.7	22.0
F-test				
N level (N)	**		**	
Genotype (G)	**		**	
N × G	**		**	
CV (%)	11		5	

Source: Adapted from Fageria, N. K., O. P. Morais, and A. B. Santos. 2010. *J. Plant Nutr.* 33:1696–1711. With permission.

Note: Means followed by the same letter in the same column are not significantly different at the 5% probability level by Tukey's test.

**Significant at the 1% probability level.

1.2.1.5 Spikelet Sterility

The spikelet sterility of lowland rice genotypes was significantly influenced by N and genotype treatments (Table 1.12). The N × genotype interaction was also significant for this trait. This means that spikelet sterility varied among genotypes with the variation in N levels. The variation in spikelet sterility was from 3.95% to 30.37% at the lower N level and from 4.93% to 24.53% at the higher N level. Overall, the spikelet sterility was 22.33% at the low N level and 19.08% at the higher N level. Similar results were obtained by Fageria and Baligar (2001) in lowland rice.

Spikelet sterility is an important yield component in rice (Figure 1.12), and reducing spikelet is one way to improve the yield. Overall, the filled spikelet percentage is about 85% in rice, even under favorable conditions (Yoshida, 1981). Hence, there exists a possibility of increasing rice yield by 15% if breeding eliminates spikelet sterility. The increase in photoassimilates during the spikelet

TABLE 1.11

Pod Number per Pot of 20 Dry Bean Genotypes as Influenced by N Fertilization

Genotype	0 mg N kg^{-1}	400 mg N kg^{-1}
Pérola	9.0a	16.3a
BRS Valente	5.3a	14.7a
CNFM 6911	13.7a	26.0a
CNFR 7552	5.3a	27.7a
BRS Radiante	6.3a	20.3a
Jalo Precoce	7.3a	20.7a
Diamante Negro	6.7a	25.7a
CNFP 7624	5.7a	30.3a
CNFR 7847	13.0a	34.7a
CNFR 7866	4.3a	15.3a
CNFR 7865	9.7a	25.7a
CNFM 7875	4.7a	29.3a
CNFM 7886	5.7a	31.7a
CNFC 7813	5.3a	29.0a
CNFC 7827	6.0a	34.3a
CNFC 7806	6.3a	24.0a
CNFP 7677	8.0a	31.7a
CNFP 7775	6.7a	34.3a
CNFP 7777	10.0a	28.0a
CNFP 7792	5.3a	15.7a
Average	7.2b	25.8a
F-test		
N rate (N)	**	
Genotype (G)	**	
N × G	**	

Source: Adapted from Fageria, N. K., L. C. Melo, and J. P. Oliveira. 2013. *J. Plant Nutr.* 36:2179–2190. With permission.

Note: Means followed by the same letter in the same column are not significantly different at the 5% probability level by Tukey's test. Average mean values were compared in the same line.

**Significant at the 5% probability level.

filling growth stage is one way to improve the spikelet filling rate. Of the 15% unfilled spikelets, however, about 5–10% are unfertilized and difficult to eliminate (Yoshida, 1981). When the filled spikelet number is more than 85%, the yield capacity or sink is limiting yield, and when the ripened spikelet number is less than 80%, assimilate supply or source is yield limiting (Murata and Matsushima, 1975).

Tanaka and Matsushima (1963) reported that the amount of carbohydrates stored in the shoot at the flowering stage improved spikelet filling by acting as a buffer substance in a case where a plant faced unfavorable conditions. Hayashi (1995) also reported improved spikelet filling by the large amount of carbohydrate accumulated during flowering, and cultivar differences exist in the amount of the carbohydrate accumulation at the flowering growth stage. Furthermore, Hayashi (1995) reported that an accumulation of a large amount of carbohydrates in the shoot before flowering also reduces spikelet degeneration.

TABLE 1.12
Spikelet Sterility of Eight Lowland Rice Genotypes as Influenced by Nitrogen Fertilization

Genotype	0 mg N kg^{-1}	304 mg N kg^{-1}
Javae	3.95c	4.93b
Rio Formoso	30.30ab	22.20a
CAN 6343	16.90bc	24.53a
CCNA 7550	25.22ab	22.87a
CAN 7556	31.90a	21.27a
CAN 7857	30.03ab	23.83a
CAN 8319	30.37ab	19.43ab
CAN 8619	10.00c	13.60ab
Average	22.33	19.08
F-test		
N level (N)	*	
Genotype (G)	**	
N × G	*	

Source: Adapted from Fageria, N. K. and M. P. Barbosa Filho. 2001. *Commun. Soil Sci. Plant Anal.* 32:2079–2090. With permission.

Note: Means followed by the same letter in the same column are not significantly different at the 5% probability level by Tukey's test.

*,**Significant at the 5% and 1% probability levels, respectively.

1.2.1.6 Grain Yield

Grain yield in cereals and legumes significantly increased with the addition of N fertilization (Watson et al., 1958; Halse et al., 1969; Spratt and Gasser, 1970; Langer and Liew, 1973; Novoa and Loomis, 1981; Fageria and Santos, 2008; Fageria et al., 2011a,b). Roth et al. (2013) reported a significant increase in the grain yield of corn with the addition of N. Benzian and Lane (1979) analyzed the relationship between the N supply, the grain protein content, and the yield in numerous experiments at Rothamsted and elsewhere. They found that greater N supply increased the grain protein content more or less linearly, while yield showed a diminishing return pattern. When N is strongly limiting,

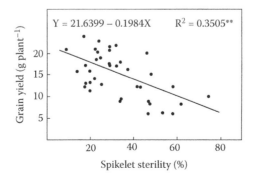

FIGURE 1.12 Relationship between spikelet sterility and grain yield of lowland rice. (Adapted from Fageria, N. K., A. B. Santos, and A. M. Coelho. 2011b. *J. Plant Nutr.* 34:371–386. With permission.)

TABLE 1.13

Relationship between N Rate (X) and Grain Yield of 12 Lowland Rice Genotypes

Genotype	Regression Equation	R^2
BRS Jaçanã	$Y = 1899.03 + 19.40X$	0.77**
CNAi 8860	$Y = 2642.53 + 17.01X$	0.74**
BRS Fronteira	$Y = 2630.31 + 26.79X - 0.0611X^2$	0.70**
CNAi 8879	$Y = 2511.89 + 29.09X - 0.0703X^2$	0.77**
CNAi 8880	$Y = 2213.08 + 39.69X - 0.1131X^2$	0.87**
CNAi 8886	$Y = 2430.88 + 32.92X - 0.0848X^2$	0.83**
CNAi 8885	$Y = 2516.87 + 28.93X - 0.0669X^2$	0.84**
CNAi 8569	$Y = 2590.43 + 24.28X - 0.631X^2$	0.82**
BRSGO Guará	$Y = 2797.53 + 20.30X$	0.92**
BRS Alvorada	$Y = 2912.43 + 18.17X$	0.66**
BRS Jaburu	$Y = 1816.19 + 18.21X$	0.88**
BRS Biguá	$Y = 1814.42 + 19.83X$	0.93**

Source: Adapted from Fageria, N. K., A. B. Santos, and V. A. Cutrim. 2008. *J. Plant Nutr.* 31:788–795. With permission.

Note: Values are averages of 2 years field experimentation.

**Significant at the 1% probability level.

small additions can result in greater yields but with decreased protein concentrations (Novoa and Loomis, 1981).

D'Andrea et al. (2006) reported that the grain yield of corn was significantly reduced by N deficiencies, and the response was mainly related to variations in kernel number per plant. Data in Table 1.13 show the influence of N on the grain yield of 12 lowland rice genotypes. Grain yield was significantly increased with the addition of N. Six genotypes were having significant linear increase in grain yield while the remaining six were having quadratic increase in grain yield. The variability in grain yield was 66–93% with the addition of N. The average grain yield of 12 lowland rice genotypes was plotted against the N rate (Figure 1.13). The increase in grain yield was significant and quadratic with the addition of N in the range of 0–200 kg ha^{-1}.

Maximum grain yield was obtained with the addition of 229 kg N ha^{-1} (Figure 1.13). In fertilizer experiments, 90% of the maximum yield is often considered as an economical rate (Fageria

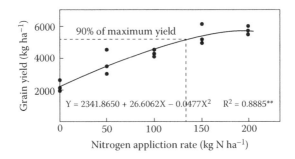

FIGURE 1.13 Grain yield of lowland rice as influenced by N application rate. Values are averages of 12 genotypes and 2 years field trial. (Adapted from Fageria, N. K., A. B. Santos, and V. A. Cutrim. 2008. *J. Plant Nutr.* 31:788–795. With permission.)

et al., 2003). A total of 90% of the maximum grain yield was obtained with the application of 136 kg N ha^{-1} (Figure 1.13). Singh et al. (1998) reported that maximum average grain yield of 20 lowland rice genotypes was obtained at 150–200 kg N ha^{-1}. Similarly, Dobermann et al. (2000) reported that 120–150 kg N ha^{-1} for field experiments in the dry season at the IRRI in the Philippines.

Data in Table 1.14 show the influence of the N rate and genotype on the grain yield of 20 dry bean genotypes. It was significantly influenced by the N rate and genotype treatments. Grain yield varied from 1.3 to 9.7 g pot^{-1} at low N level and at higher N rate variation was 10.4–31.8 g pot^{-1}. This means that dry bean genotypes respond differently to N fertilization at lower as well as at higher N rate application. Based on the grain yield data in Table 1.14, bean genotypes can be grouped into three broad groups in relation to the N response. In the first group are genotypes that produced lower yield at lower N rates but the yield increase was very high with the N fertilization. In this group are genotypes of CNFC 7827, CNFP 7677, CNFP 7775, CNFP 7624, CNFM 7886, CNFC 7813, CNFM 7886, BRS Radiante, CNFC 7806, CNFR 7552, Diamante Negro, and Jalo Precoce.

TABLE 1.14
Grain Yield of 20 Dry Bean Genotypes as Influenced by N Levels

Genotype	0 mg N kg^{-1}	400 mg N kg^{-1}
Pérola	8.3ab	22.5ab
BRS Valente	1.6ab	10.4b
CNFM 6911	9.7a	20.9ab
CNFR 7552	1.9ab	22.2ab
BRS Radiante	3.5ab	25.2ab
Jalo Precoce	4.4ab	19.6ab
Diamante Negro	4.1ab	21.1ab
CNFP 7624	1.4ab	28.9ab
CNFR 7847	8.5ab	24.3ab
CNFR 7866	2.0ab	16.6ab
CNFR 7865	6.2ab	21.1ab
CNFM 7875	1.3b	28.1ab
CNFM 7886	1.6ab	26.9ab
CNFC 7813	1.7ab	27.3ab
CNFC 7827	3.4ab	31.8a
CNFC 7806	2.9ab	22.3ab
CNFP 7677	4.4ab	30.1ab
CNFP 7775	3.8ab	29.5ab
CNFP 7777	7.8ab	27.2ab
CNFP 7792	2.5ab	12.0ab
Average	4.1	23.4
F-test		
N rate (N)	**	
Genotype (G)	**	
N × G	**	

Source: Adapted from Fageria, N. K., L. C. Melo, and J. P. Oliveira. 2013. *J. Plant Nutr.* 36:2179–2190. With permission.

Note: Within the same column, means followed by the same letter do not differ significantly at 5% probability level by Tukey's test.

**Significant at the 1% probability level.

In the second group are genotypes that produced reasonably well at lower as well as at higher rates of N. Genotypes fall in this group are Perola, CNFM 6911, CNFR 7847, CNFR 7865, and CNFP 7777. This group is most desirable because they produce high yield at low as well as at high levels of N. The third group includes genotypes that produced low yield at low as well as at higher N rates. The genotypes fall in this group are CNFP 7792, CNFR 7866, and BRS Valente. Crop species respond differently to soil and fertilizer N (Sinclair and Horie, 1989; Fageria and Baligar, 1993), including bean genotypes (Park and Buttery, 1989). Morphological, physiological, or biochemical mechanisms might be responsible for different response of crop genotypes to applied N (Fageria and Baligar, 1993; Marschner, 1995; Baligar et al., 2001).

1.2.1.7 Leaf Area Index

LAI is defined as the leaf area per unit soil area (cm^2 m^{-2}). This growth index can be calculated as follows (Fageria et al., 2006):

$$LAI = (A \times N)/10,000$$

where A is the leaf area (cm^2) and N is the number of tillers (cereals), branches (legumes), or plants per square meter. CGR is related to LAI and NAR (net assimilation rate) as follows:

$$CGR = LAI \times NAR$$

NAR is defined as the dry matter accumulation per unit of leaf area and expressed as grams per square meter of leaf per day (g m^{-2} leaf area per day), and can be calculated using the following equation (Brown, 1984):

$$NAR = (A/1)(dt/dW)$$

where A is the leaf area and dW/dt is the change in plant dry matter per unit time. The objective of measuring NAR is to determine the efficiency of plant leaves for dry matter production. The NAR values decrease with the advancement of plant growth due to the mutual shading of leaves and the reduced photosynthetic efficiency of older leaves (Fageria, 1992).

The source capacity of plants is primarily determined by the LAI, the leaf area duration, and the rates of photosynthesis, respiration, and amino acid synthesis (Novoa and Loomis, 1981). LAI is the main factor in biomass formation, and it varies in amount with plant population and nutrient supply. LAI is significantly increased with the addition of N in cereals and legumes (Watson et al., 1958; Langer and Liew, 1973; Pearman et al., 1977; Spiertz and Ellen, 1978; Fageria, 2007; Fageria and Santos, 2008). The greater LAI can be due to an effect on the leaf number or leaf size. Langer and Liew (1973) did not find any effect of N on the leaf number of the wheat main shoot. But the leaf number per plant will increase with N supply due to an increase in tiller number in cereals (Halse et al., 1969; Pearman et al., 1977). The area of each lamina is also increased by N due to the effect on both the cell number and the size (Novoa and Loomis, 1981). Leaf area duration is also extended by N fertilization (Langer and Liew, 1973; Pearman et al., 1977; Thomas et al., 1978).

LAI is one of the principal crop parameters affecting photosynthesis. LAI varies with environmental conditions, cultural practices, and stage of crop growth. In corn, N levels significantly increased LAI compared to control treatment (Uhart and Andrade 1995). For determinate crops, the best time to measure LAI is when it reaches its maximum at the beginning of reproductive growth; for indeterminate crops, the maximum LAI may occur well after flowering begins. Optimal leaf area varied with the crop species and genotypes within species. However, optimum values of LAI reported for different crops, such as for soybean about 3.2, for corn about 5, and for wheat 6–8.8 (Yoshida, 1972). Critical LAI values for the large leaved tropical legumes generally fall in the range of 3–4 (Muchow, 1985), but they can exceed 5 for small leaved pigeonpea (Rowden et al., 1981).

A very strong association, following an exponential function, between the LAI and the proportion of radiation intercepted by corn plants is documented in the scientific literature (Hipps et al., 1983; Trapani et al., 1992). Following these fundamental principles, a critical LAI threshold (close to 4.0 m^2 m^{-2}) was defined after which further increase in LAI is not reflected in the increase in the production of the radiation intercepted by corn (Maddonni and Otegui, 1996). Costa et al. (2002a) reported that genetically altering corn to increase the leaf number per plant could increase grain yield under short season conditions. These authors further reported that alterations resulting in leaf arrangements that maximize light interception and optimize RUE could further improve yield.

1.2.1.8   Root Growth

Nitrogen improves the root growth of cereals and legumes (Fageria and Moreira, 2011; Fageria, 2013). Root morphology is influenced by the amount of N fertilizer applied (Eghball et al., 1993). Eghball et al. (1993) showed that N stress in corn reduced root branching. Similarly, Maizlish et al. (1980) showed greater root branching in corn with increasing levels of applied fertilizer N. Costa et al. (2002b) reported that greater root length and root surface area were obtained at an N fertilizer rate of 128 kg N ha^{-1} compared with either the absence of fertilizer N or the higher rate of 255 kg N ha^{-1}. N fertilizer improves root growth in soils having low organic matter content (Gregory, 1994; Robinson et al., 1994). N fertilization may increase the crop root growth by increasing soil N availability (Weston and Zandstra, 1989; Garton and Widders, 1990). Sainju et al. (2001) observed that tomato (*Lycopersicon esculentum* Mill.) root growth was greater with hairy vetch and crimson clover cover crops and 90 kg N ha^{-1} than with no cover crops or N fertilization. N also improves the production of lateral roots and root hairs, as well as increasing rooting depth and root length density deep in the profile (Hansson and Andren, 1987). Hoad et al. (2001) reported that surface application of N fertilizer increases root densities in the surface layers of the soil.

N fertilization can increase the root length and root surface area and decrease root mass per unit area of corn (Anderson, 1987; Costa et al., 2002b). It is well known that roots tend to proliferate in nutrient-enriched soil zones (Drew et al., 1973; Qin et al., 2005). Russell (1977) refers to this as a compensatory response. The results of pot experiments showed that corn roots were longer and thinner in zones that were rich in N (Zhang and Barber, 1992, 1993; Durieux et al., 1994). Figures 1.14 and 1.15 show the root growth of two upland rice genotypes at two N levels. The root growth was much better with the addition of 300 mg N kg^{-1} soil as compared to control treatment.

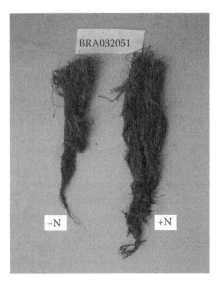

FIGURE 1.14   Root growth of upland rice genotype BRA 032051 without and with N (300 mg N kg^{-1}). Half of the N was applied at sowing and the remaining half at 45 days after sowing.

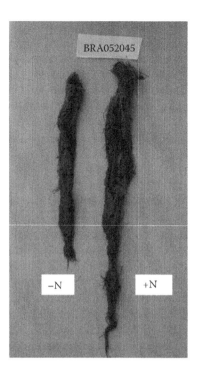

FIGURE 1.15 Root growth of upland rice genotype BRA 052045 without and with N (300 mg N kg⁻¹). Half of the N was applied at sowing and the remaining half at 45 days after sowing.

The author studied the influence of different N rates and three sources on root growth of lowland rice. The maximum root length (MRL), root dry weight, and the contribution of root to the total plant were significantly affected by N treatments (Table 1.15). The MRL varied from 16.25 to 27.25 cm, with an average value of 21.27 cm. In case of ammonium sulfate and common urea, the MRL decreased with increasing N rate but in case of polymer-coated urea it increased. The decrease in root length at higher N rate has been reported by Fageria and Moreira (2011). However, root length is genetically controlled and may vary from genotype to genotype. The root dry weight increased with the addition of N fertilizer and varied from source to source. The increase in root dry weight was quadratic in case of ammonium sulfate (Table 1.16) and maximum root dry weight was achieved with the application of 279 mg N kg⁻¹ soil. In case of common urea and polymer-coated urea, the increase in root dry weight was linear (Table 1.16) with an increasing N rate in the range of 0–400 mg N kg⁻¹ soil. Overall, the maximum root dry weight was higher at the highest N rate (400 mg N kg⁻¹) treatment. The improvement in the root dry weight of rice with the addition of rice has been reported by Fageria (2013).

Figures 1.16 through 1.18 show the root growth of lowland rice under different N rate and N sources. It can be seen from these figures that root growth was higher with the addition of N compared to control treatment with three N sources. In addition, root growth also varied with N source. The contribution of root in the total plant weight varied from 7.54% to 16.79%, with an average value of 12.50%. Fageria and Moreira (2011) reported that the contribution of the root in total plant weight varied with genotype and was also affected by fertilization, including N. However, the average contribution of the root in the total plant weight is less than 20%. Results of the present study fall within this limit.

The author studied the influence of N rates on the root growth of corn (Table 1.17). MRL was obtained at the lowest N rate and the minimum at the highest N rate. The decrease in root length was quadratic with increasing N rate in the range of 0–600 mg N kg⁻¹. The variation in the root length

TABLE 1.15

Maximum Root Length (MRL), Root Dry Weight (RDW), and Contribution of Root Dry Weight (CRDW) in the Total Plant Weight of Low Land Rice as Influenced by N Rate and Sources

Treatments[a]	MRL (cm)	RDW (g per plant[a])	CRDW (%)
0 (control)	22.25ab	1.45c	16.79a
AS 100	27.25a	2.34bc	12.92abc
CU 100	24.25ab	2.67bc	15.41ab
PU 100	19.00b	3.11b	16.02ab
AS 200	20.25ab	3.22b	12.80abc
CU 200	20.00ab	1.90bc	7.91c
PU 200	18.75b	2.23bc	8.75c
AS 400	16.25b	2.86bc	7.54c
CU 400	22.00ab	3.30b	10.92bc
PU 400	22.75ab	5.17a	15.98ab
Average	21.27	2.82	12.50
F-test	**	**	**
CV (%)	15.71	23.93	18.52

Note: Means followed by the same letter in the same column are not significant at the 5% probability level by Tukey's test.

[a] Treatments were AS = ammonium sulfate, CU = common urea, PU = polymer-coated urea and N rates were 0, 100, 200, and 400 mg kg^{-1} of soil.

**Significant at the 1% probability level.

was 88% with the variation in N rate. The root dry weight increased with increasing the N rate in a quadratic fashion. The maximum root dry weight was attained with the addition of 176 mg N kg^{-1}. Figure 1.19 shows corn root growth at different N rates. A low N supply generally leads to decreased root weight, suppression of lateral root initiation, increase in the C/N ratio within the plant, reduction in photosynthesis, and early leaf senescence (Malamy and Ryan, 2001; Martin et al., 2002; Malamy, 2005; Wingler et al., 2006).

1.2.2 COMPONENT OF ENZYMES

Nitrogen plays a significant role in the metabolism activities of plants. N is an indispensable elementary constituent of numerous organic compounds of general importance, amino acids, proteins, nucleic acids, and compounds of secondary plant metabolism such as the alkaloids (Mengel et al.,

TABLE 1.16

Relationship between N Rate and Root Dry Weight (RDW) of Lowland Rice

Variable	Regression Equation	R^2
N versus RDW (AS)	$Y = 1.39 + 0.0134X - 0.000024X^2$	0.46*
N versus RDW (UC)	$Y = 1.66 + 0.0038X$	0.51**
N versus RDW (UP)	$Y = 1.52 + 0.0083X$	0.70**

*Significant at the 5% probability level.
**Significant at the 1% probability level.

FIGURE 1.16 Root growth of lowland rice with ammonium sulfate source at different N rates.

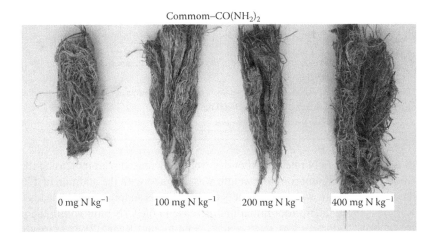

FIGURE 1.17 Root growth of lowland rice with common urea source at different N rates.

FIGURE 1.18 Root growth of lowland rice with polymer-coated urea source at different N rates.

TABLE 1.17

Maximum Root Length (MRL) and Root Dry Weight (RDW) of 61-Day-Old Corn Plants as Influenced by N Levels

N Rate (mg kg^{-1})	MRL (cm)	RDW (g per plant)
0	42.75	1.19
100	40.75	1.60
200	35.50	1.44
400	33.25	1.09
600	27.75	0.52
F-test	**	**
CV (%)	8.78	16.73

Regression Analysis

N rate versus MRL = $42.77 - 0.03X + 0.0000095X^2$, $R^2 = 0.88$**

N rate versus RDW = $1.29 + 0.0019X - 0.0000054X^2$, $R^2 = 0.77$**

**Significant at the 1% probability level.

2001). Rice (2007) reported that the essential nature of N as a plant nutrient is based primarily on its role as a component of enzymes, which are either entirely protein or largely composed of protein. Enzymes catalyze virtually every anabolic reaction (formation) and catabolic reaction (breakdown) within the plant and orchestrate the rate, timing, direction, and extent of most metabolic reaction pathways. Approximately, a large part of the total N (80–85%) in green plant material is sequestered in protein (Rice, 2007). The remaining is distributed in nucleic acids (about 10%) and across the soluble metabolic pool of amino acids (about 5%) and their amines and amides (Mengel et al., 2001). Unlike humans, plants synthesize all transaminases needed to produce the 20 specific amino acids required for the synthesis of all required plant proteins (Rice, 2007).

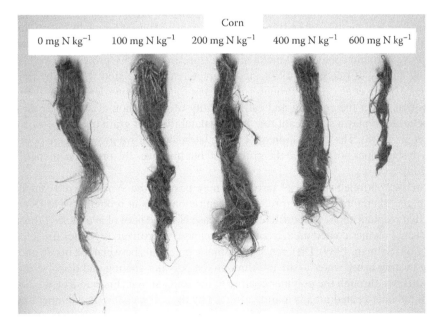

Corn

0 mg N kg^{-1} 100 mg N kg^{-1} 200 mg N kg^{-1} 400 mg N kg^{-1} 600 mg N kg^{-1}

FIGURE 1.19   Root growth of corn at different N levels.

1.2.3 Improvement in Grain Protein Content

Nitrogen is an important nutrient for improving N contents in the grain of cereals and legumes. Protein content in grains of cereals and legumes is an important index in determining the quality of grain for human consumption. For example, grain protein is a key quality measure for bread wheat, affecting gluten strength and bread loaf volume (Wall, 1979; Mallory and Darby, 2013). Mallory and Darby (2013) reported that ensuring adequate available N for grain protein development is a primary challenge for organic production of winter bred wheat (*Triticum aestivum* L.). Generally, wheat grain must have a protein concentration of 120 g kg^{-1} (12%) or greater to be considered suitable for bread flour (Mallory and Darby, 2013). Wheat grain that does not meet the acceptable level either receives a discounted price or must be sold into alternative, often lower-value markets. McKenzie et al. (2010) also reported that the most important wheat quality parameter is the grain protein concentration because it affects the milling and backing quality of the grain and because wheat growers generally receive a premium for high protein concentration.

Farmaha and Sims (2013) reported that because of higher protein concentration, hard red spring wheat is mainly used for blending with lower protein wheat for milling and baking in the state of Minnesota, USA. Farmers sometimes receive a premium payment if the protein concentration exceeds 140 g kg^{-1} (14%) but more frequently receive a discount payment if the protein content is <140 g kg^{-1}. The most critical factors controlling grain protein are genotypes and N availability (Peltonen and Virtanen, 1994; Wooding et al., 2000). Protein concentration depends on the plant's ability to translocate already accumulated N from vegetative plant parts to the developing grain or the ability to accumulate N, which may be mobilized directly to the developing grain near or soon after anthesis (Banziger et al., 1994). Some cultivars translocate a relatively high percentage of N to the developing grain during the reproductive period (Bhatia and Rabson, 1976). Therefore, the timing and amounts of N availability during the growing season of hard red spring wheat could have significant impacts on the N uptake amount and timing at which N is used to generate grain yield and protein (Farmaha and Sims, 2013).

In conventional systems, splitting applications of N fertilizer to include in season topdress applications has been shown to increase grain protein content and baking quality (Randall et al., 1990; Knowles et al., 1994; Peltonen and Virtanen, 1994; Mallory and Darby, 2013) as well as reduce N losses (Sowers et al., 1994). Randall et al. (1990) reported that topdressing N at heading increased grain protein, dough properties, and baking quality for irrigated wheat grown in Australia. Similarly, Nass et al. (2002, 2003) reported that topdressing N at stem elongation in wheat grown in eastern Canada improved protein content and met a milling standard of 13.5% protein.

The eating quality of rice is a multifaceted characteristic associated with many physiochemical properties (Chung et al., 2012). Protein, starch, and lipids are considered to be the main grain components that affect the cooking and eating quality of rice (Zhou et al., 2002). In particular, protein is believed to play a significant role in the palatability and grain quality of rice (Lin et al., 2005; Butt et al., 2008). The application of N in the late reproductive growth stage or at the grain-filling growth stage does not improve the grain yield but improves the protein content of rice grain (Fageria, 2014).

Conceptually, controlled-release N fertilizers may provide the synchronized supply of N for plant uptake, and therefore, increase protein concentration without reducing yield (Shaviv, 2001). The idea is that preplant applied N fertilizer will release N to the pool of available N throughout the growing season. Polymer-coated urea, a controlled-release N fertilizer, regulates the release of N to the soil solution (Salman, 1989). Differences in osmotic potentials between the inside and outside of the polymer coating cause water to diffuse through the polymer coating and dissolve urea, and the liquid urea diffuses through the polymer coating to the soil solution (Fujinuma et al., 2009). The N release from polymer-coated urea is mainly affected by the soil moisture and temperature with the release rates increasing as soil moisture and temperature increase (Gandeza et al., 1991; Fujinuma et al., 2009). McKenzie et al. (2010) and Farmaha and Sims (2013) found that the delayed N release from polymer-coated urea decreased the wheat grain yield as compared with the uncoated urea but increased protein concentration. Previous studies have reported that the maximum grain protein concentration is generally attained at N levels much higher than those required to reach maximum yield (Campbell et al., 1997; Fowler, 2003).

1.2.4 ESSENTIALS FOR PHOTOSYNTHESIS

Photosynthesis is the basic process underlying plant growth and production of the food, fuel, and fiber required to sustain human life (Tolbert, 1997). An understanding of photosynthesis is therefore necessary to appreciate processes that determine yield in agriculture, forestry, ecology, and many other fields. This extremely important biochemical process in green plants, which literally means building by light, probably evolved 3500 million years ago (Shopf, 1993). In general, photosynthesis is the process by which plants synthesize organic compounds from inorganic substances using light. During photosynthesis, carbon (C) from atmospheric carbon dioxide (CO_2) is fixed to become part of many organic molecules that constitute plant tissues (Fageria et al., 2006). Because of this, the total dry matter production of crop plants is correlated with photosynthetic rates integrated over plant growth cycles.

Photosynthesis is defined as a photochemical process involving the absorption of light energy by plant pigment and its conversion into stable chemical potential (e.g., adenosine triphosphate (ATP) and nicotinamide adenine dinucleotide phosphate, reduced (NADPH)) (Crop Science Society of America, 1992). Conversion of the light energy into chemical energy involves an electron flow between photosystem I (PS I) and photosystem II (PS II). This electron transport system is associated with the regeneration of NADPH and the establishment of an electrochemical gradient across the thylakoid membranes to produce ATP. The NADPH and ATP are used as electron and energy sources at various steps of CO_2 reduction and carbohydrate synthesis (Cakmak and Engels, 1999). The photosynthetic capacity of plants is determined by several factors, such as water availability, light conditions, and mineral nutrients. Mineral nutrients play several roles in the formation, partitioning, and the utilization of photosynthates. Therefore, mineral nutrient deficiencies, including N, substantially impair the production of dry matter and its partitioning between the plant organs (Marschner et al., 1996).

In green plants, in the photosynthetic process involving the photochemical reaction, carbohydrates are produced and O_2 and water are released according to the following equation:

$$6CO_2 + 12H_2O \,(\text{light energy}) \leftrightarrow C_6H_{12}O_6 + 6O_2 + 6H_2O$$

Based on the pathways of carbon metabolism and their behavior in CO_2 uptake, three types of higher plants are distinguished. These are known as C_3 plants, C_4 plants, and CAM (crassulacean acid metabolism) plants. In C_3 plants, the three-carbon phosphoglyceric acid is the first product (following the Benson–Calvin pathway). In C_4 plants, the four-carbon malic or aspartic acid (Kortschak–Hartt, Hatch–Slack pathway) is present. The CAM plants, fix CO_2 into four-carbon acids in the dark and reduce them during a subsequent light period (Crop Science Society of America, 1992). The C_4 plants have higher photosynthetic efficiency and produce more leaf area at lower leaf N contents as compared to C_3 plants (Sinclair and Horie, 1989). As the irradiance of the leaf increases, the photosynthetic rate of C_3 species reaches a maximum at a lower irradiance and at a lower value of photosynthesis than the C_4 species, and therefore has a lower RUE. Mitchell et al. (1998) found that the average RUE values during vegetative growth under optimum conditions were wheat 2.7 g MJ^{-1}, rice 2.2 g MJ^{-1}, corn 3.3 g MJ^{-1}, and soybean 1.9 g MJ^{-1}. Lindquist et al. (2005) reported in modern corn hybrids an RUE value of 3.8 MJ^{-1}, indicating an improvement in RUE with selection. The use of an adequate rate of N can improve leaf canopy and structure and can improve RUE.

TABLE 1.18
Examples of C_3 and C_4 Field Crops

C_3 Crops	C_4 Crops
Barley	Maize
Dry bean	Millet (pearl)
Cotton	Sorghum
Cowpea	Sugarcane
Oat	
Peanut	
Rice	
Soybean	
Sugarbeet	
Wheat	

Source: Adapted from Fageria, N. K. 1992. *Maximizing Crop Yields.* New York: Marcel Dekker; Fageria, N. K., V. C. Baligar, and C. A. Jones. 2011a. *Growth and Mineral Nutrition of Field Crops*, 3rd edition. Boca Raton: CRC Press. With permission.

Table 1.18 provides examples of C_3 and C_4 crops. The two types of photosynthetic pathways differ in chloroplast arrangement, primary photosynthetic enzymes, temperature response, water-use efficiency, light saturation, and response to CO_2 concentrations (Table 1.19). Mahon (1983) and Tanaka and Osaki (1983) determined the photosynthetic rates per unit leaf area for some important field crops (Table 1.20). No clear differences in photosynthetic rates exist among crop species, except for maize, which has about two times higher photosynthetic rates than many other crops. Variation in photosynthetic rates are often correlated with concentrations of nitrogenous compounds in leaves (Evans, 1983; Hirose and Werger, 1987). This correlation is explained by the fact that some 75% of all N in mesophyll cells of C_3 plants is associated with photosynthesis (Lambers, 1987). Further,

TABLE 1.19
Characteristics of C_3 and C_4 Plants

Characteristics	C_3	C_4
Photosynthetic efficiency	Low	High
Photorespiration	High	Low
Water utilization efficiency	Low	High
Optimum temperature for photosynthesis	10–25°C	30–45°C
Response to light intensity	Low	High
Response to CO_2 concentration	Low	High
Response to O_2 concentration	Low	High
Major pathway of photosynthetic CO_2 fixation	Reductive pentose phosphate cycle	C_4-dicarboxylic acid and reductive pentose phosphate cycle
Transpiration ratios	High	Low
Leaf chlorophyll a to b ratio	Low	High

Source: Adapted from Fageria, N. K. 1992. *Maximizing Crop Yields.* New York: Marcel Dekker; Fageria, N. K., V. C. Baligar, and C. A. Jones. 2011a. *Growth and Mineral Nutrition of Field Crops*, 3rd edition. Boca Raton: CRC Press. With permission.

TABLE 1.20
Photosynthetic Rate per Unit Leaf Area in Various Crop Plants

Crop Plant	Photosynthetic Rate ($mg\ CO_2\ dm^{-2}\ h^{-1}$)
Alfalfa	50–55
Bean (common)	30–40
Maize	60–80
Pea (field)	25–30
Potato	25–35
Rice (lowland)	40–50
Soybean	30–35
Sugarbeet	30–35
Sunflower	40–42
Wheat	28–39

Source: Adapted from Mahon, J. 1983. *Can. J. Plant Sci.* 63:11–21; Tanaka, A. and M. Osaki. 1983. *Soil Sci. Plant Nutr.* 29:147–158. With permission.

larger fractions of N (about 25%) become components of the enzyme Rubisco (ribulose-bisphosphate carboxylase). The rates of photosynthesis are therefore closely correlated with Rubisco activity in leaves (Evans, 1983). The net photosynthesis of individual rice leaves reaches maximum values of about 40–60 kilolux (800–1200 mol m^{-2} s^{-1}) near half full sunlight (Yoshida, 1981). However, the photosynthesis of well-developed canopies increase with increasing light intensity of up to full sunlight, and no indication of light saturation appears to be reached (Murata, 1961). In early stages of crop growth, the main determinant of photosynthesis is the extent of leaf area development. As the LAIs increase, so do the extents of light interception, and often exceed 95% for many cereal crops with LAI values of about 4. Once canopies close from leaf density, further increases in LAI have little effect on plant photosynthesis, which may be influenced by incident radiation and structures of canopies (Evans and Wardlaw, 1976).

Net photosynthesis rates of active, healthy single rice (C$_3$ plant) leaves are about 40–50 mg CO$_2$ dm^{-2} h^{-1} under light saturation conditions. Ishii (1988) noted no significant differences in photosynthetic rates of single rice leaves at heading stage of growth for 32 Japanese cultivars bred from 1880 to 1976. However, 25% increases in canopy photosynthesis were observed in new rice cultivars (bred between 1949 and 1976) as compared to old cultivars (bred from 1880 to 1913). Improvement in canopy photosynthesis was attributed to the higher efficiency of light interception by canopies because of changes in leaf angles. Old cultivars had droopy leaves with high mutual shading, while new cultivars had erect leaves with less mutual shading (Ishii, 1988). Similarly, Conocono et al. (1998) concluded that much of the large increases in rice yield over the past three decades could be attributed to improvements in canopy structure that enhanced canopy light interception and photosynthesis.

Photosynthesis provides 90–95% of plant dry weight (Kueneman et al., 1979). N is arguably essential to the entire photosynthetic apparatus, since it is contained in individual pyrrole subunits that form the tetrapyrrole ring structure common to all light-absorbing chlorophyll molecules (Rice, 2007). The tetrapyrrole ring gives chlorophyll its green color, and the term chloritic suggests an absence of green, a typical manifestation of N deficiency that reflects compromised chlorophyll function within chloroplasts (Rice, 2007).

Many studies have related the increase in CO$_2$ assimilation rate of many crops to increase in leaf N (Osman and Milthorpe, 1971; Rawson and Hackett, 1974; Yoshida and Coronel, 1976; Evans, 1983). Whenever a sufficiently broad range of leaf N contents has been examined, it has been consistently found that the relationship is nonlinear, the slope declining as the N content increases (Nevins and Loomis, 1970; Takano and Tsunoda, 1971; Wong, 1979; Lugg and Sinclair, 1981; Evans, 1983).

A strong positive correlation has been observed between the light saturated rate of photosynthesis of a leaf and its N content (Field and Mooney, 1986; Evans, 1989; Reich et al., 1994). That is, generally, higher N contents are associated with higher rates of maximum photosynthesis (Poorter and Evans, 1998). The reason for this strong relationship is the large amount of leaf organic N (up to 75%) present in the chloroplasts, most of it in the photosynthetic machinery (Evans and Seemann, 1989).

The rate of photosynthesis is also influenced by N nutrition (Novoa and Loomis, 1981). Burstrom (1943) observed an enhancement in apparent CO_2 assimilation with a rise in the nitrate content of wheat leaf laminae. Osman and Milthorpe (1971) found that the increase in the gross photosynthesis in that species was linear with N content. Murata (1969) reported a positive correlation between total protein N content and the photosynthetic activity of rice, and Watanabe and Yoshida (1970) observed a similar relationship in that species between chlorophyll content and photosynthesis rate.

N deficiency has been reported to reduce photosynthesis in higher plants (Barker, 1979). N and chlorophyll contents of leaves are closely correlated, and N deficiency brings about a sharp drop in the chlorophyll content of leaves. Increased resistance to CO_2 transfer into or within leaves has been demonstrated in N-deficient leaves (Ryle and Hesketh, 1969; Nevins and Loomis, 1970). These increased resistance may be due to metabolic changes in the mesophyll or to changes in stomatal aperture (Ryle and Hesketh, 1969). The inhibitory effect of O_2 on net photosynthesis is greater in N-deficient leaves than in those adequately fertilized with N (Natr, 1972). Possibly, N availability regulates the amount of substrate (glycolate) available for photorespiration through the formation of glycine and serine (Barker, 1979).

Generally, fast-growing plant species or genotypes within species have a lower N content and equal rate of photosynthesis, both expressed per unit leaf area, than slow-growing species and hence have a higher efficiency of photosynthesis per unit leaf N (Poorter et al., 1990; Boot et al., 1992). The higher photosynthesis efficiency of fast-growing species at optimum N supply is explained by two characteristics: (i) they invest less N in nonphotosynthetic components; and (ii) they exhibit a higher activation state of Rubisco and relatively higher activity of thylakoid reactions as compared to their capacity (Pons et al., 1994; Werf, 1996). Werf (1996) reported that there is a strong body of evidence that the total photosynthetic production of a crop during a season is strongly correlated with the integrated amount of intercepted light and not with the amount of dry matter production per unit of intercepted radiation energy, at least when determined over a whole growing season.

Roth et al. (2013) reported that N application in corn improved photosynthesis, especially at the later reproductive growth stage. Xia (2012) also reported similar effects of N on corn photosynthesis in Indiana, USA. In two studies comparing varied N rates on older and newer maize hybrids, photosynthesis increased with higher N rates, with a more significant difference later in the season (Ding et al., 2005; Echarte et al., 2008). In dryland agroecosystems, photosynthetic efficiency is mainly influenced by internal genetic factors and the external water and fertilizer environmental status (Shangguan et al., 2000; Jiang et al., 2004; Hou et al., 2013).

1.2.4.1 Leaf Nitrogen Content versus Photosynthesis

Several studies reported in the literature relating leaf N content and photosynthesis in crop plants. These studies involve wheat (Evans, 1983), rice (Uchida et al., 1982; Cook and Evans, 1983; Sinclair and Horie, 1989), soybean (Boote et al., 1978; Hesketh et al., 1981; Lugg and Sinclair, 1981; Sinclair and Horie, 1989), and corn (Wong et al., 1985; Sinclair and Horie, 1989). A strong positive correlation between leaf N content and carbon dioxide exchange rate or photosynthesis was reported (Harper, 1994). Sinclair and Horie (1989) reported that there was a significant quadratic association between leaf N content and carbon dioxide exchange (CER) for corn, rice, and soybean. The value of CER for corn was much higher (2.4 mg CO_2 m^{-2} s^{-1}) at about 1.2 g m^{-2} leaf N compared to soybean and rice (about 1.2 mg CO_2 m^{-2} s^{-1}). The substantial advantage in biomass accumulation under N-limited conditions for corn, compared with rice and soybean, is noteworthy (Harper, 1994). The higher photosynthesis rate in C_4 plants such as corn as compared to C_3 plants is well known (Fageria, 1992).

1.2.4.2 Improving Photosynthesis

Photosynthesis in crop plants is influenced genetically as well as determined by environmental conditions. Crop plants provided with adequate environmental conditions such as mineral nutrition, water availability, and temperature produce higher photosynthates as compared to those grown under adverse environmental conditions. In addition, photosynthesis can be improved with the use of genotypes having ideal architecture. In addition, photosynthesis can also be increased by decreasing photorespiration. Hence, here, two options (ideal architecture and biotic and abiotic stresses and decreasing photorespiration) will be discussed.

1.2.4.2.1 *Ideal Architecture and Biotic and Abiotic Stresses in Important Food Crops*

Ideal plant architecture for a higher yield can be achieved through plant breeding. It has been recognized that the spectacular yield increase of crops during the second part of the twentieth century has been attributed in almost equal measure to breeding and to the use of inputs or better management practices (Fageria et al., 2006). The most important morphological characters that have been bred into high-yielding cereal cultivars such as rice, wheat, and corn are short, stiff culm for lodging resistance, and erect leaves for increased interception of solar radiation. In addition to this, most of the crop cultivars released had resistance to biotic and abiotic resistance. The use of an adequate rate of nutrients and plant spacing or stands was also part of the improved management practices adopted.

Genetic improvement in grain yield has been intensively studied in wheat, barley, oat, corn, and rice (Austin et al., 1980; Wych and Rasmusson, 1983; Wych and Stuthman, 1983; Tollenaar, 1989; Feil, 1992; Peng et al., 2000). Most of these studies reported a positive historical cultivar trend in grain yield. Studies of historical cultivars often show that genetic improvement in yield potential has resulted from increases in GHI (Lawes, 1977; Austin et al., 1980; Riggs et al., 1981), which is associated with ideotype characters, for example, short stature in wheat and uniculm habit in corn and sunflower (*Helianthus annuus* L.) (Sedgley, 1991). Others reported that the improvement in yield potential has been associated with increases in the biomass yield in wheat (Waddington et al., 1986), corn (Tollenaar, 1989), oat (Payne et al., 1986), and soybean (Cregan and Yaklich, 1986). McEwan and Cross (1979), Wych and Rasmusson (1983), and Wych and Stuthman (1983) stated that the improvement in grain yield was related to both dry matter accumulation and GHI in wheat, barley, and oat. In the author's opinion, both the improvement (dry matter and GHI) contributed to yield increase in cereals and legumes and breeding and improved management practices have contributed to this increase. Several examples of crop improvement through breeding for ideal plant architecture and biotic and abiotic stresses are provided in this section.

1.2.4.2.1.1 *Rice*   Rice is a staple food for more than 50% of the world population (Li et al., 2013). China and India are the world's largest rice producer as well as consumer. In South America, rice is eaten everyday with dry bean by all sections of the population. Rice is mainly grown under two ecosystems, known as upland and lowland. Upland rice is defined as the rice grown on undulated well-drained soils, without water accumulation in the plots or fields, and totally depends on rainfall for water requirements (Fageria, 2011a, 2014). It is also known as aerobic rice because it is grown on well-drained soils. Upland rice is mainly grown in South America, Africa, and Asia. Brazil is the largest upland rice-producing country in the world. Lowland rice is mainly grown on

leveled lands; water accumulate in the fields or plots during most of the crop-growing cycle and most of the cases have controlled irrigation (Fageria et al., 2011a). Soils of the lowland rice are saturated during most of the growth cycle.

Increasing rice yield is the perpetual goal of rice breeding programs in most rice-growing regions or countries because of the growing food demand caused by population growth and reduction of areas devoted to rice production. Rice breeding has significantly increased in the past few decades. Breeding for both types of rice (upland and lowland) has been successful in improving the yield and grain quality (Peng et al., 2008). A typical example of breeding semidwarf high-yielding cultivars comes from IRRI. The tall tropical cultivar "peta" from Indonesia and the subtropical semidwarf cultivar "De-geo-woo-gen" from Taiwan were crossed to produce the semidwarf IR 8, which produced a record yield of 11 t ha^{-1} and responded well to N rates up to 150 kg ha^{-1} at several locations in tropical Asia (Chang, 1976). Dissemination of this improved plant type throughout Latin America was initiated in 1968 by the Colombian-based program of the International Center of Tropical Agriculture together with National Research Institutes in the region (Cuevas-Perez et al., 1995). Scientists at IRRI and several national breeding programs combined most of the desired features in the improved plant type, including reduced height (about 100 cm), leaf erectness, short, dark green leaves, stiff culms, early maturity, photoperiod insensitivity, N responsiveness, and high harvest index (Yoshida, 1981). The wide adoption of IR 8 and other high-yielding cultivars, such as IR 20 and IR 22, made it possible for the semidwarf cultivars to become important cultivars in Brazil, Colombia, Peru, Ecuador, Cuba, Mexico, Indonesia, Malaysia, the Philippines, India, Pakistan, Bangladesh, and South Vietnam. By 1972–1973, semidwarf cultivars occupied a large part of the area planted to high-yielding rice cultivars, including about 10% of the world's total area and 15% of the area in tropical Asia (Chang, 1976). Today, high-yielding semidwarf cultivars predominate in most lowland rice-producing areas.

Work is in progress at IRRI and many other international and national research centers to further improve plant type, grain quality, and pest resistance (Khush, 1995). Figure 1.20a–c shows the development of modern high-yielding rice cultivars from formerly prevalent traditional cultivars, and ideotype rice plants of the future. As shown in Figure 1.20b, the new plant type for irrigated rice was designed to attain yields of 12–13 t ha^{-1}. Yield improvement beyond 12 t ha^{-1} will require new plant architectures because of two major problems: the leaves responsible for grain filling will be shaded beneath a dense cover of panicles, and the immense weight of panicles will result

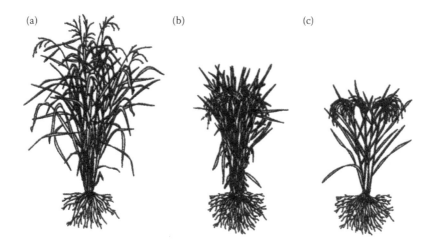

(a)   (b)   (c)

FIGURE 1.20 Traditional old (a), modern high-yielding (b), future ideotype (c), and plants of lowland rice. (Adapted from Fageria, N. K., V. C. Baligar, and R. B. Clark. 2006. *Physiology of Crop Production.* New York: The Haworth Press. With permission.)

in serious lodging (Figure 1.20b). A solution to both problems is to lower the panicle height in the canopy (Figure 1.20c). Such adaptations may be relevant to other rice ecosystems as well as to other cereals (Khush, 1995).

Peng et al. (2000) reported that during the past three decades, rice production in Asia more than doubled as a result of the adoption of modern cultivars, increased investments in irrigation, greater use of fertilizer, and some expansion in cultivated area. Peng et al. (2000) also studied the trend in the yield of rice cultivars/lines developed since 1966 by the IRRI in the Philippines. Regression analysis of yield versus year of release indicated an annual gain in rice yield of 75–81 kg ha^{-1}, equivalent to 1% per year. The highest yield obtained with the most recently released cultivars was 9–10 Mg ha^{-1}, which is equivalent to reported yields of IR8 and other IRRI cultivars obtained in the late 1960s and early 1970s at these sites (Peng et al., 2000). Peng et al. (2000) further reported that the increasing trend in yield of cultivars released before 1980 was mainly due to the improvement in GHI, while an increase in total biomass was associated with yield trends for cultivars/lines developed after 1980. These authors further reported that further increases in rice yield potential will likely occur through increasing biomass production rather than increasing GHI.

Rice panicle architecture is one of the most important agronomic traits affecting grain yield (Zhu et al., 2009; Li et al., 2013). The diversity of rice panicle architecture has long been the interest of many breeders (Zhu et al., 2009). Since the 1980s, many high-yielding *Japonica* rice cultivars with greater spikelet density than traditional cultivars were bred in northeastern China (Yang, 1984). Because the vertical angle of the panicle stems of these cultivars is generally <40° at maturity, they are known as high-yielding potential cultivars or erect panicle cultivars. In comparison, traditional cultivars have a panicle stem angle that is typically >40° (Xu et al., 2005). These cultivars are known as curved panicle type. Erect panicle rice cultivars have about 45% more spikelets per panicle than curved panicle-type cultivars (Zhou et al., 2006).

Wu (2009) reported that there were two breakthroughs in rice breeding in the last century. The first one was the development of the dwarf rice in the 1960s, which raised over 20% of rice yield per unit urea. The second one was the development of hybrid rice using three-line or cytoplasmic male sterile (CMS) system in the 1970s, which added another 20% increment in the average yield (Yuan, 1997). Yuan (1998) reported that rice yield remained almost stagnant between 1970s and 1990s since the introduction of the three-line system. According to plant physiologists, it is feasible to increase the rice yield potential to as high as 21.6 Mg ha^{-1} if the solar energy is efficiently harvested and converted (Cao and Wu, 1984; Wu, 2009).

1.2.4.2.1.2 Wheat   The global wheat production in 2009 stood at 685 million metric tons harvested from 225 million ha, with an average yield of 3.04 Mg ha^{-1} (FAO, 2011). To meet global demand, wheat production needs to increase to about 900 million metric tons by the year 2020, with an average yield of 4 Mg ha^{-1} (Ortiz et al., 2007). Because there is little scope of increasing the land area, a major yield increase should come from increased productivity. Wheat production is constrained worldwide by many biotic and abiotic stresses (Singh et al., 2008; Sharma et al., 2012). Breeding had played a significant role in the wheat yield improvement in the past and will continue to be a powerful tool to improve wheat yield in the future. However, genetic yield potential of a cultivar can be achieved when improved management practices are adopted. Among the crop management practices, the use of essential plant nutrients, especially N, is fundamental (Fageria, 2009, 2014).

The modification of wheat plant architecture through breeding is another example of ideal plant type for improving the yield of an important cereal (Byerlee and Curtis, 1988). Wheat yield has increased in a most noteworthy way during the twentieth century, with an average global increase of 250% during the past 50 years (from 1 to 2.5 Mg ha^{-1}). This is remarkable when one considers that wheat yields remained practically unchanged during the first half of the century (Slafer et al., 1994, 1996). Better plant architecture through plant breeding and better management practices are responsible for this accomplishment (Calderini and Slafer, 1998).

A detailed analysis of the past decade or so (Slafer et al., 1994) indicated that worldwide wheat yields might be asymptotically approaching a ceiling, as average yields did not increase from 1990 to 1995. However, yield potential data of CIMMYT (Centro Internacional de Mejoramento de Maize Y Trigo; Mexico) cultivars developed from the 1960s (Fageria et al., 2006) do not indicate a plateau. Indeed, the average increase per year was 0.9% (Braun et al., 1997). This genetic progress for increasing the yield potential was closely associated with an increase in photosynthetic activity through the ideal plant canopy. Both photosynthetic activity and yield potential increased over the 30-year period by 25% (Fageria et al., 2006).

Similarly, an estimated 50% of the increase in U.S. wheat yields from 1954 to 1979 can be credited to genetic improvement (Schmidt, 1984). The introduction of semidwarf cultivars of wheat had a large impact on the productivity of wheat in the Corn Belt and the Great Plains of the United States, and is probably the major source of genetic gain in both regions (Feyerherm et al., 1988). Semidwarf cultivars were planted on more than 90% of the area planted with wheat in the Corn Belt by 1979, when the genetic gain in productivity was as high as 74% (Siegenthaler et al., 1986). In the Great Plains states (Kansas, Nebraska, and Oklahoma), semidwarf cultivars occupied only 9%, 1%, and 38%, respectively, of the area planted with wheat, and genetic improvement for yield was only 45%. However, the area planted with semidwarf cultivars by 1984 increased to 70%, 38%, and 76%, respectively, in the three states, and the genetic improvement increased to 61% (Feyerherm, 1988).

Since the beginning of the 1960s, grain yield of winter wheat has increased by about 120 kg ha^{-1} year^{-1} in France. This increase was mainly related to the genetic improvement of cultivars and the use of adequate levels of N (Gouis and Pluchard, 1997). Average yields of wheat in the United Kingdom rose from 3 to 8 Mg ha^{-1} while the world average has risen from 1.08 to 2.7 Mg ha^{-1} (Jaggard et al., 2010). Reilly and Fuglie (1998) reported that the average yields of 11 crops in the United States had increased by between 1% and 3% per year during the past half century and the trend was linear or exponential, showing no sign that the rate was slowing down. A large study by Hafner (2003) showed that the national average yields of wheat, rice, and corn in 188 countries were mostly increasing, that the increase had been predominantly linear, and that the biggest producers' yields had increased at more than 33.1 kg ha^{-1} year^{-1}. In the developed and developing countries, much of this increase has been due to the use of N fertilizers, the use of adequate crop protection measures, and also planting high-yielding crop cultivars (Jaggard et al., 2010).

Sharma et al. (2012) quantified the genetic yield gains in CIMMYT's (International Maize and Wheat Improvement Center) spring bread wheat (*Triticum aestivum* L.) in the Elite Spring Wheat Yield Trial (ESWYT) distributed over the past 15 years (1995–2009) as determined by the performance of entries across 919 environments in 69 countries. Across locations in all countries, mean yields of the five highest-yielding entries showed an annual gain of 27.8 kg ha^{-1} (0.65%) compared to Attila, a check cultivar. These authors concluded that there is no evidence for stagnation in the genetic yield gain in the CIMMYT-developed elite lines for irrigated wheat environments worldwide but the genetic yield gain maintains a linear yield potential increase.

1.2.4.2.1.3 Corn The global food supply–demand model predicts that the global demand for corn will increase from 526 million metric tons to 784 million metric tons from 1993 to 2020, with most of the increased demand coming from developing countries (Duvick and Cassman, 1999). Assuming no increase in maize production area, an annual growth rate in corn yield of about 1.5% will be needed to meet this demand. However, from 1982 to 1994, the yield growth rate for corn was about 1.2% worldwide and only 1.0% in developed countries as a group, which account for the majority of total corn production. The situation has not been changed in corn yield increase in the developing countries in the past decade. Hence, more efforts will be required in the future to improve corn yield genetically as well as improved management practices to meet the world demand for this cereal.

Corn is an important grain crop worldwide. The United States is the highest corn-producing country. Corn production in the United States during the 2013 cropping season was about 350 million metric tons. China is the second largest corn-producing country in the world, with an annual sowing area of about 25 million hectares, and production of 130 million metric tons (Ci et al., 2013). Six cycles of cultivar replacement have occurred in China since the 1950s. Among corn cultivars released from the 1950s to the 2000s in China, the newer hybrids possess increased grain yield and relatively greater tolerance to compound stress (Ci et al., 2011). Many agronomic traits have been improved and kernel number per year and kernel weight from the 1950s to the 2000s are important advances in the Chinese corn improvement program (Ci et al., 2013).

The continuous increase in corn grain yield in the world's primary growing areas during the past few decades was mainly driven by the development of crowding stress-tolerant hybrids that allowed for dramatic increase in plant population and therefore in production per unit area (Duvick, 2005a; Robles et al., 2013). Differences in grain yield between older and newer corn hybrids have been shown to be a function of plant population, rather than plant yield potential (Tollenaar and Lee, 2002; Duvick, 2005a). This characteristic usually renders modern hybrids as density dependent (Berzsenyi and Tokatlidis, 2012). High densities are required for maximum yield potential, but optimum densities differ among hybrids and across seasons (Fasoula and Fasoula, 2000; Tokatlidis and Koutrouba, 2004; Fasoula and Tollenaar, 2005).

Increased plant population leads to greater LAI and increased interception of PAR from the midvegetative to early grain fill stages (Tollenaar and Aguilera, 1992; Maddonni and Otegui, 2004; Novacek et al., 2013). Increased plant population has been shown to decrease the number of ears per plant (Tollenaar et al., 1992) and number of kernels per ear (Westgate et al., 1997; Maddonni and Otegui, 2006), but has less influence on kernel weight (Begna et al., 1997; Westgate et al., 1997; Maddonni and Otegui, 2006). Corn grain yield is also influenced by other crop management practices, including reduced row spacing to obtain a more equidistant plant spacing, thereby reducing interplant competition (Bullock et al., 1988), increasing the interception of solar radiation (Andrade et al., 2002), and decreasing weed competition (Teasdale, 1995, 1998; Begna et al., 2001). In addition, transgenic insect resistance has complemented plant population increases and improved grain yield (Novacek et al., 2013).

Assefa et al. (2012) reported that corn yield has increased from about 1.5 Mg ha^{-1} in the early 1900s to 8.5 Mg ha^{-1} in the beginning of the 2000s in the United States. These authors analyzed data from corn trials conducted in Kansas from 1939 to 2009. On average, corn yields increased at a rate of 90 kg ha^{-1} year^{-1} in dryland and 120 kg ha^{-1} year^{-1} in irrigated trials. Changes in hybrid technology and changes in crop management factors, such as a decrease in planting and harvesting date by about a quarter of a day per year increased planting density at the rate of 597 plants ha^{-1} year^{-1}, and increased N and P fertilizer rates by 2.6 and 0.40 kg ha^{-1} year^{-1}, respectively, were found for the same time period in dryland corn (Assefa et al., 2012). These authors further reported that, in addition, climate changes contributed to yield increases in the past through increased total rainfall, average monthly minimum and maximum temperature in March, and decreased maximum temperature from July through September. Increases in plant density generally have a large positive impact on the incident solar radiation intercepted and as a consequence, on CGR around silking as well as a final yield (Tollenaar and Aguilera, 1992). Corn yield in the United States has also increased due to earlier planting dates (Kucharik, 2008) and more extensive use of irrigation (Cassman, 1999).

Genetic improvement in the grain yield of maize hybrids in North America and Europe during the past three to five decades has been extensively documented (Tollenaar and Aguilera, 1992). Grain yield improvement of maize hybrids appeared to be the result of increased dry matter accumulation. Increased dry matter may be attributable to the increased absorption of incident PAR and/or improved efficiency of converting absorbed PAR into dry matter. Some evidence indicates that modern hybrids absorb more of the seasonal incidence PAR than the older hybrids. Maximum LAI for modern hybrids was larger than for older ones, and leaves of modern hybrids stay green longer during the final phase of the life growth cycle (Tollenaar and Aguilera, 1992). Full season maize landraces adapted to the lowland tropics are typically tall, leafy, and prone to lodging, and have low harvest indices (Goldsworthy et al., 1974). During the initial stages of maize improvement at CIMMYT, reduction in the plant height was a priority (Fischer and Palmer, 1984), and reduced plant height has continued only as a secondary trait in breeding activity. Johnson et al. (1986) reported that 15 cycles of recurrent selection for reduced height in the lowland tropical maize population reduced plant stature by 37%, crop duration by 7%, and increased the proportional allocation of total biomass to husks, ears, and silks at the 50% silking stage. At the same time, researchers observed that grain yield, harvest index, and optimum plant density for grain yield each increased by 50–70%. Lodging was also substantially reduced. Similarly, Edmeades and Lafitte (1993) reported that lodging in maize declined from 39% to 10% with 18 cycles of recurrent selection.

Duvick (2005a,b) reviewed the literature regarding the contribution of breeding to yield advances in corn. He concluded that corn yields have increased continually wherever hybrid corn has been adopted, starting in the U.S. corn belt in the early 1930s. Plant breeding and improved management practices have produced this gain jointly. On average, about 50% of the increase is due to management and 50% due to breeding (Duvick, 2005b). The two factors interact so closely that neither of them could have produced such progress alone. However, genetic gains may have to bear a larger share of the load in future years (Duvick, 2005b). Trait changes that increase resistance to a wide variety of biotic and abiotic stresses are the most numerous, but morphological and physiological changes that promote efficiency in growth, development, and partitioning are also recorded. Newer hybrids yield more than their predecessors in unfavorable as well as favorable growing conditions (Duvick, 2005a,b). Table 1.21 shows data related to yield increase from 1961 to 2002 in different countries or continents.

Duvick (2005b) reported that phenotypic changes may indicate improvements in the efficiency of grain production (i.e., smaller tassels may release more energy for grain production). Stress tolerance is greatly improved, newer hybrids outyield the older ones not only in high-yield environments but also when trials are subjected to abiotic stress (heat and drought), or to biotic stress (i.e., insect pests). In addition, newer hybrids yield more than older hybrids because of continuing improvement in the ability of the hybrids to withstand the stress of higher plant density, which in turn is owed to their greater tolerance to locally important biotic and abiotic stresses (Duvick, 2005b).

In corn, changes in cultural practices such as weed and pest control, timelines of planting, and increased efficiency of harvest equipment have helped to raise yield over the years. Application of N fertilizers increased following World War II. They increased significantly from 60 kg ha^{-1} in 1964 to

---

TABLE 1.21
Corn Yield Increase in Different Regions/Continents from 1961 to 2002

Country/Continent	Average of 1961 (Mg ha^{-1})	Average of 2002 (Mg ha^{-1})	Annual Gain (kg ha^{-1} year^{-1})
European Union (15)	2.5	9.1	169
USA	3.9	8.2	109
Canada	4.6	7.6	69
China	1.2	5.0	103
South America	1.4	3.4	48
South Asia	1.0	1.7	20
Eastern Europe	1.8	4.2	42
Southern Africa	0.7	1.3	8
World	1.9	4.3	61

Source: Adapted from FAO 2004. Statistical Databases available at http://apps.fao.org/default.htm; Duvick, D. N. 2005a. *Adv. Agron.* 86:83–145.

140 kg ha^{-1} in 1985 but leveled off in the 1980s (Duvick, 2005b). Plant density increase has contributed significantly in corn yield in the United States. Plant density averaged about 30,000 plants ha^{-1} in the 1960s, 60,000 plants ha^{-1} in the 1980s, and densities at present are typically 80,000 plants ha^{-1} or higher in the major corn-producing regions of the country (Duvick, 2005a,b). Duvick (2005b) concluded that compared with the older hybrids, today's hybrids produce approximately the same amount of grain per plant, but on considerably more plant per unit area.

1.2.4.2.1.4 Soybean Soybean is an important legume crop worldwide. There are several uses of soybean such as oil for human consumption, feed for animals, and food for human consumption. Brazil, the United States, and Argentina are the largest soybean-producing countries in the world. These three countries are also the largest soybean-exporting countries in the world. The soybean production area is second only to that of corn in U.S. agriculture (Villamil et al., 2012). Soybean yield has improved in the United States through breeding as well as adopting improved management practices. North American soybean breeding efforts began early in the twentieth century and have resulted in a dramatic improvement in soybean yield (Carter et al., 2004). In 1924, the national estimates of U.S. soybean yield was just 740 kg ha^{-1}, but by 2010, the national yield estimate had almost quadrupled to 2925 kg ha^{-1}, translating into a linear yield of 23.4 kg ha^{-1} year^{-1} (USDA-NASS, 2012; Fox et al., 2013). Though the U.S. soybean yield trends have shown a steady 1.5% annual increase since the 1920s, some evidence indicates that yields of soybean and other crops may have reached a plateau in highly productive regions worldwide (Egli 2008; Villamil et al., 2012). Results from experimental sites within the Midwest region show that planting date (De Bruin and Pedersen, 2008a), seeding rate and row spacing (De Bruin and Pedersen, 2008a,b; Cox and Cherney, 2011), and crop rotation and tillage practices (Pedersen and Lauer, 2004) are among the crucial agronomic practices that farmers can manipulate to maximize soybean yields.

Contributions to this rise in soybean yield over time include improved crop genetics, increases in atmospheric CO_2, extended growing seasons as evidenced by the northern shift of USDA growing zones (Kaplan, 2012), and the optimizing of the production environment through changes in cultural practices, advances in planting and harvesting equipment, and improved herbicides and pesticides (Specht et al., 1999; Ustun et al., 2001; Egli, 2008; Fox et al., 2013). Breeding also contributed to lodging and disease resistance, improvements in photosynthetic efficiency, and higher nutrient use efficiency (Wilcox, 2001; Cober et al., 2005; De Bruin and Pedersen, 2009). Kumudi (2002) reported that the contribution of genetic improvement to overall soybean yield improvement has been estimated to range from 0.5% to 0.7% per year in continental North America. Published estimates of the annual gain in yield attributable to genetic improvement averaged about 15 kg ha^{-1} year^{-1} prior to the 1980s but now averaging about 30 kg ha^{-1} year^{-1} in both the public and private sectors.

In Canada, Morrison et al. (2000) tested 41 cultivars released over seven decades of breeding and selection, and found that the yield improvement had an association with a decrease in protein concentration and some reduction in lodging. Recent cultivars, compared with older cultivars, had a lower maximum leaf area, and higher photosynthetic rate and stomatal conductance per unit area (Morrison et al., 1999, 2000). Further studies on population density revealed that modern cultivars were more tolerant of population stress than older ones (Cober et al., 2005). Morrison et al. (2008) also reported that seed isoflavone concentration significantly increased over 58 years of soybean breeding for yield in the short-season region, and modern cultivars were more environmentally influenced for isoflavone concentration than older cultivars. Isoflavone was positively associated with N fixation and disease resistance (Zhang and Smith, 1995; Dixon, 2001).

Soybean yield increased significantly in China through breeding and adopting improved crop management practices. Xue et al. (2006) reported that the average soybean yield per hectare in the 1990s increased by 71.4%, compared with that of the 1950s, and the annual yield increase averaged 13.4 kg ha^{-1}. Similarly, Jin et al. (2010) also investigated improvements in the soybean yield in Northeast China from 1950 to 2006. These authors concluded that a positive correlation between seed yield and year of cultivar release was indicated with a 0.58% average annual increase. The seed number per plant was the most important contributor to yield gain, with a 0.41% increase per year. A 33% increase in the photosynthetic rate, 10.6% increase in plant dry weight and 19.0% increase in harvest index were found, while LAI decreased by 17.3%. Jin et al. (2010) also reported that in modern cultivars, plant height was reduced, which gave resistance to lodging, with the lodging score dropping from 3.2 in 1951 to 1.0 in 2006. Plant resistance to disease and pest infestation were also improved. Furthermore, yield stability was increased over years, which could be attributed to the stable pod production across different environments.

1.2.4.2.1.5   Peanut   Peanuts also known as groundnuts (*Arachis hypogaea* L.) are grown in the temperate and tropical climates. It is one of the world's most important oilseeds crops, along with soybean, cottonseed, rapeseed, and sunflower. Although originating in South America, the vast majority of peanut is produced in Asia and Africa. Approximately 94% of the peanut is produced in the developing countries, mostly under rainfed conditions (Dwivedi et al., 2003). There are many biotic and abiotic stresses to peanut production; these include drought, low and high temperatures, low fertility, diseases, and insects (Wynne and Gregory, 1981; Knauft and Wynne, 1995; Dwivedi et al., 2003). Genetic improvement and adopting adequate management practices have contributed significantly in peanut yield increase. However, there is a big gap in yield among developed and developing countries (Dwivedi et al., 2003).

Over 276 peanut cultivars were released between 1920 and 2000 for cultivation in various countries in Asia, Africa, and the Americas. Each has a specific adaptation to its respective region of production and cropping system (Dwivedi et al., 2003). A yearly genetic gain of nearly 15 kg per hectare has been reported for large-seeded Virginia type cultivars released from the 1950s to the 1970s in the United States (Mozingo et al., 1987). The higher-yielding cultivars developed during the 1950s, 1960s, and 1970s had an average yield increase of 3.4%, 10.2%, and 18.55, respectively, over the standard NC 4. However, since the 1970s, there has been increased emphasis on improving the pest resistance and quality traits so that the yield potential of cultivars released since that time has not surpassed those of the highest-yielding cultivars released during the 1970s (Dwivedi et al., 2003).

1.2.4.2.1.6   Cotton   Alteration of plant architecture in narrow-row cotton using management and genetic strategies to improve light penetration into the canopy and increase crop yields (Reta-Sanchez and Fowler, 2002) and N use efficiency. The modification of plant architecture such as reduced plant height, short branches, and modified leaf shape increased light penetration into the canopy. However, certain characteristics were more efficient in modifying light distribution into the canopy. Kerby et al. (1980) proposed that cotton plants with a combination of normal-type leaves near developing bolls and erect mutant-type leaves at the top of the canopy could be more efficient in increasing production and use of assimilates in narrow row cotton. Reta-Sanchez and Fowler (2002) reported that plants with modified leaf shape at the upper part of the canopy and normal leaves at the medium and lower part of the canopy increased light availability only at node 16. This behavior suggests the necessity of an earlier and greater canopy modification using other plant characteristics such as short branches (Kerby and Buxton, 1981) and modified leaf shape (Kerby et al., 1980; Wells et al., 1986; Peng and Krieg, 1991). Reta-Sanchez and Fowler (2002) also reported that the combination of leaf shape, reduced plant height, and short branches gave a greater light penetration through the canopy than the control. These authors further reported that plant with reduced plant height, short branches, and modified leaves grown at 97,000 plants ha^{-1} reached high values of light interception (90–97%), with LAI ranging from 3.7 to 5.2.

Bridge et al. (1971) and Bridge and Meredith (1983) reported that yield gains due to the genetic improvement of cotton averaged 10.2 and 9.5 kg ha^{-1} year^{-1} since about 1910 in the United States. These yield advances have been accompanied by higher lint percentages, smaller seed bolls, and higher micromere values (Fageria et al., 2006). Wells and Meredith (1984) indicated that the major component contributing to increased yields was increases in the number of fruits. This agrees with Evans' (1980) description of how yield was increased with smaller but more numerous fruits in other major crops.

1.2.4.2.1.7 Dry Bean   Vandenberg and Nleya (1999) summarized traits that might optimize canopy structure in common bean at harvest, which could be modified through breeding. These were (1) long internodes in the lower stem; (2) consistent internode elongation under a wide range of environmental conditions; (3) reduction of stem stunting during early season growth; (4) increased stem length; (5) increased stem strength, particularly in the more basal internodes; (6) reduction of pod length without decreasing seed size; (7) increase in pod curvature so that pod tips do not extend below the combine cutterbar; (8) long upright peduncles; (9) commencement of flowering at the upper nodes; (10) high fertility at the upper nodes; and (11) a sufficient number of main stem nodes to maximize productivity in the available growing season.

1.2.4.2.2 Decreasing Photorespiration

A large part of the plant photosynthetic product is lost in photorespiration. If this loss is minimized, there is a possibility of increasing photosynthesis in crop species or genotypes within species. This may be achieved by converting C_3 into a C_4 or by improving the specificity of Rubisco for CO_2 (Long et al., 2006). Long et al. (2006) discussed this issue in detail and readers may refer this article.

1.2.5 Leaf Nitrogen Content versus Radiation Use Efficiency

The interception of the radiation of the sun by plant leaves is an important factor in determining dry matter yield. Radiation use efficiency (RUE = dry matter produced per unit of intercepted light) is widely used in the analysis of crop growth and the calculation of biomass production in crop simulation models (Sinclair and Muchow, 1999; Milory and Bange, 2003). For simulation purposes, the intercepted radiation over a time period is multiplied by RUE to estimate the dry matter production for that period (Milory and Bange, 2003). RUE is related to the N content of the plants, especially in the leaves (Sinclair and Muchow, 1999). Leaf N content and light intensity affect leaf photosynthesis (Milroy and Bange, 2003).

The sensitivity of leaf photosynthetic rates to changes in leaf N was explored by Sinclair and Horie (1989) as a potentially important source of variation in RUE and reviewed by Sinclair and Muchow (1999). The theoretical study of Hammer and Wright (1994) provided a detailed analysis of the linkage between RUE and leaf N content. In the model of Sinclair and Horie (1989), the maximum leaf CO_2 assimilation rate was argued to be a direct function of leaf N content per unit leaf area. They assumed that the leaf N content was uniform throughout the leaf canopy. From this model, Sinclair and Horie (1989) calculated substantial differences among crop species in the relationship between RUE and leaf N content. At high leaf N contents, the calculated RUE response curves were saturated and there were only small changes in RUE with changes in leaf N. The maximum values of RUE from this analysis for each species were about 1.8 g MJ^{-1} for corn, 1.5 g MJ^{-1} for rice, and 1.3 g MJ^{-1} for soybean.

Sinclar and Muchow (1999) reported that, for C_3 species, the maximum RUE was generally calculated to be in the range of 1.4–1.5 g MJ^{-1} intercepted solar radiation. When N content in the leaves decreased to a minimum value, the RUE was very sensitive and decreased linearly in corn, rice, and soybean (Sinclair and Horie, 1989). Similar decrease in RUE was also reported by Hammer and Wright (1994) and Sands (1996) at lower N content in the leaves of peanuts. In sunflower, Gimenez et al. (1994) reported that increased soil N fertility resulted in both an increased specific leaf N content and increased RUE. A curvilinear relationship between RUE and leaf N content of sunflower was reported by Fisher (1993), Hall et al. (1995), and Bange et al. (1997). Sinclair and Muchow (1999) concluded that, based on the review of the literature, RUE achieved a saturated value at high leaf N contents and decreased curvilinearly with decreasing leaf N content below the saturation leaf N content.

1.2.6 Improves Grain Harvest Index

GHI is the ratio of grain yield to total biological yield. This index is calculated with the help of the following equation:

$$\text{GHI} = \frac{\text{grain yield}}{\text{grain + straw yield}}$$

The values for GHI in cereals and legumes are normally <1. Although GHI is a ratio, it is sometimes also expressed in percentages. GHI is an important trait in improving grain yield in cereals and legumes. The term GHI was introduced by Donald (1962), and since then has been considered to be an important trait for yield improvement in field crops. Donald and Hamblin (1976) discussed relationships between GHI and yield, and concluded that this was an important index for improving crop yields. Fageria et al. (2011b) studied the relationship between GHI and grain yield in lowland rice under greenhouse conditions (Figure 1.21). The relationship was significant and quadratic. Sixty-eight percentage variation in grain yield was due to GHI. Similarly, the author also studied the relationship between GHI and grain yield of lowland rice under field conditions (Figure 1.22). The relationship was significant and quadratic between these two traits. Generally, GHI has positive associations with grain yield (Rao and Bhagsari, 1998; Rao et al., 2002), and N is important for improving GHI. Fageria and Baligar (2005) have reported that GHI is an important trait in improving the grain yield of annual crops. N deficiency and N excess affect assimilate partitioning between vegetative and reproductive organs in crops (Donald and Hamblin, 1976).

Thomson et al. (1997) reported greater seed yields of faba bean with higher GHIs. Morrison et al. (1999) examined physiological differences associated with seed yield increases of soybean in Canada within groups of cultivars released from 1934 to 1992. These authors concluded that the increase in the seed yield with the year of release was significantly correlated with increases in the harvest index (0.5% per year), photosynthesis, and stomatal conductance, as well as decreases in LAI. They further concluded that present-day cultivars are more efficient at producing and allocating carbon resources to seeds than were their predecessors.

Snyder and Carlson (1984) reviewed GHI for selected annual crops and noted variations from 0.40 to 0.47 for wheat, 0.23 to 0.50 for rice, 0.20 to 0.47 for bunch-type peanut (*Arachis hypogaea* L.), and 0.39 to 0.58 for dry bean. The GHI values of modern crop cultivars are commonly higher than old traditional cultivars for major field crops (Ludlow and Muchlow, 1990). Cox and Cherney (2001) reported average GHI values of 0.50 for 23 forage corn hybrids. Miller et al. (2003) reported

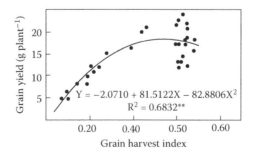

FIGURE 1.21 Relationship between grain harvest index and grain yield of lowland rice. (Adapted from Fageria, N. K., A. B. Santos, and A. M. Coelho. 2011b. *J. Plant Nutr.* 34:371–386. With permission.)

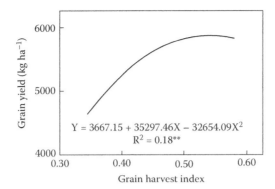

FIGURE 1.22   Relationship between grain harvest index and grain yield of lowland rice.

GHI values of 0.39 for pea (*Pisum sativum* L.), 0.37 for lentil (*Lens culinaris* Medik.), 0.41 for chickpea (*Cicer arietinum* L.), 0.28 for mustard (*Brassica juncea* L.), and 0.38 for wheat grown on loamy soil. Winter and Unger (2001) reported that sorghum GHI values varied from 0.39 to 0.45 depending on the type of tillage system adopted. Rice GHI values varied greatly among cultivars, locations, seasons, and ecosystems, and ranged from 0.35 to 0.62, indicating the importance of this variable for yield simulation (Kiniry et al., 2001). Rao et al. (2002) reported GHI values of soybean ranged from 0.37 to 0.45 with a genotypic mean of 0.43. Rao and Bhagsari (1998) reported similar ranges for GHI values for soybean grown in Georgia. Lopez-Bellido et al. (2000) reported that GHI values for wheat varied from 0.41 to 0.45 (mean value of 0.44) depending on tillage methods, crop rotation, and N rate.

The limit to which harvest index can be increased is considered to be about 0.60 (Austin et al., 1980). Hence, cultivar with low harvest indexes would indicate that further improvement in partitioning of biomass would be possible. On the other hand, cultivars with harvest indexes between 0.50 and 0.60 would probably not benefit by increasing harvest index (Sharma and Smith, 1986).

Genetic improvement in annual crops such as wheat, barley, corn, oat, rice, and soybean has been reported due to improvement in dry weight as well as GHI (Fageria and Baligar, 2005). Unkovich et al. (2010) reported that the GHI of crop plants has increased over time due to breeding for higher yield. These authors also reported that the rate of GHI increase due to cereal breeding in Australia is about 0.015 per decade, compared to about the 0.02 achieved in the United Kingdom. Short-stature, modern crop cultivars have a higher GHI than their taller forebears, although total dry matter production is most often very similar (Evans, 1993). A recurring theme in GHI has also been a shortening of the vegetative growth stage of crop plants, providing proportionately longer grain filling period (Siddique et al., 1989; Lopez Pereira et al., 2000; Unkovich et al., 2010).

Peng et al. (2000) reported that genetic gain in rice cultivars released before 1980 was mainly due to improvement in the GHI, while increases in total biomass were associated with yield trends for cultivars developed after 1980. The cultivars developed after 1980 had relatively high GHI values and further improvement in the GHI was not achieved. These authors also reported that further increases in rice yield potential would likely occur through increasing biomass production rather than increasing the GHI. Tollenaar et al. (1997) reported that the GHI values of corn were 0.41 at low N rates (no added N) and 0.45 at higher N rates (150 kg N ha^{-1}) across two hybrids. The GHI values of 10 upland rice genotypes were influenced by N fertilization (Table 1.22). Overall, increases in the GHI values were 19% at high N rates compared with low N rates. Similarly, Fageria and Santos (2008) also determined GHI of 20 dry bean genotypes under two N levels (Table 1.23). The GHI of these genotypes increased with the addition of N compared to control treatment. Overall, increase

TABLE 1.22

Grain Harvest Index of 10 Upland Rice Genotypes as Influenced by N Rate

Genotype	Low N Rate (0 mg kg⁻¹)	High N Rate (400 mg kg⁻¹)
CRO 97505	0.48	0.54
CNAs 8993	0.52	0.57
CNAs 8934	0.39	0.46
CNAs 8950	0.43	0.52
CNAs 8711	0.42	0.49
Primaveira	0.49	0.55
Canastra	0.36	0.44
Carisma	0.37	0.47
CNAs 8960	0.45	0.54
CNAs 8957	0.40	0.56
Average	0.43	0.51

Source: Adapted from Fageria, N. K. 2007. *J. Plant Nutr.* 30:843–879. With permission.

TABLE 1.23

Grain Harvest Index of 20 Dry Bean Genotypes Grown on Brazilian Oxisol

Genotype	0 mg N kg⁻¹	400 mg N kg⁻¹
Pérola	0.54	0.57
BRS Valente	0.27	0.42
CNFM 6911	0.50	0.53
CNFR 7552	0.29	0.43
BRS Radiante	0.36	0.49
Jalo Precoce	0.35	0.46
Diamante Negro	0.32	0.46
CNFP 7624	0.21	0.50
CNFR 7847	0.46	0.54
CNFR 7866	0.24	0.59
CNFR 7865	0.46	0.46
CNFM 7875	0.21	0.56
CNFM 7886	0.23	0.51
CNFC 7813	0.32	0.56
CNFC 7827	0.40	0.58
CNFC 7806	0.38	0.53
CNFP 7677	0.38	0.55
CNFP 7775	0.38	0.56
CNFP 7777	0.50	0.50
CNFP 7792	0.42	0.50
Average	0.36	0.52

Source: Adapted from Fageria, N. K. and A. B. Santos. 2008. *J. Plant Nutr.* 31:983–1004. With permission.

TABLE 1.24
Grain Harvest Index of Principal Field Crops

Crop Species	Grain Harvest Index
Corn	0.62
Upland rice	0.51
Lowland rice	0.47
Wheat	0.56
Barley	0.57
Sorghum	0.70
Oat	0.48
Triticale	0.46
Soybean	0.35
Chickpea	0.55
Faba bean	0.62
Field pea	0.58
Lentil	0.51
Mungbean	0.55
Peanut	0.57
Dry bean	0.52
Sunflower	0.48
Canola	0.41
Lupin	0.50
Vetch	0.47

Source: Adapted from Fageria, N. K. and A. B. Santos. 2008. *J. Plant Nutr.* 31:983–1004; Fageria, N. K., A. B. Santos, and V. A. Cutrim. 2007. *Pesq. Agropec. Bras.* 42:1029–1034; Unkovich, M., J. Baldock, and M. Forbes. 2010. *Adv. Agron.* 105:173–219. With permission.

in the GHI was 44% with the addition of 400 mg N kg^{-1} as compared to control treatment. These results indicate that higher GHI values can be obtained with proper N management in annual crops. The GHI of important field crops is presented in Table 1.24.

Improvement in GHI of corn genotypes with the addition of N has been reported by Costa et al. (2002a). These authors reported that GHI of corn hybrids varied from 0.47 to 0.53 at one location and from 0.52 to 0.62 at another location. Begna et al. (1997) reported GHI values of corn hybrids within this range. A higher variation in GHI among corn genotypes was observed at low N rates (0 and 85 kg N ha^{-1}) than at the intermediate or the higher N rates (170 and 255 kg N ha^{-1}). Roth et al. (2013) reported increase in the GHI of five corn hybrids with the addition of N in the range of 0–269 kg ha^{-1}.

1.2.7 ROLE OF NITROGEN IN INCREASING YIELD POTENTIAL OF IMPORTANT CEREALS IN THE FUTURE

Before discussing the role of N in increasing the yield potential of important cereals in the future, it is logical to define yield and yield potential. There are several definitions of yield potential in the literature. For example, potential yield is an estimate of the upper limit of yield increase that can be obtained from a crop plant (Fageria, 1992). Reynolds et al. (1999) defined genetic yield potential as the yield of adapted lines in a favorable environment in the absence of agronomic constraints. Evans and Fischer (1999) defined potential yield as the maximum yield that could be reached by a crop or

TABLE 1.25
Definitions of Yield Measures

Yield (Symbol)	Definition	Estimation
Average farm yield (FY)	Average yield achieved by farmers in a defined region over several seasons	Regional or national statistics, ground or satellite surveys of fields
Economically attainable yield given current markets and institutions (AYa)	Optimum (profit maximizing) yield given prices paid/received by farmers, taking account of risk and existing institution	On-farm experiments or sometimes crop models
Economically attainable yield assuming efficient markets/institutions (AYb)	Optimum yield given prices that would prevail in efficient markets with well-functioning risk insurance markets	On-farm experiments or sometimes crop models
Potential yield (PY)	Maximum yield with latest varieties, removing all constraints, including moisture, at generally prevailing solar radiation, temperature, and day length	Highly controlled on-station experiments or crop models calibrated with latest variables, well-monitored crop contests
Water-limited potential yield (PYw)	Maximum yield under normal rainfed conditions, removing all constraints for PY except for moisture	Highly controlled on station experiments or crop models or crop contests
Theoretical yield (TY)	Maximum theoretical yield for prevailing solar radiation based on prevailing knowledge of crop physiology and photosynthetic efficiency	An accepted estimate is given by the initial slope of the photosynthesis versus solar radiation response curve discounted for dark respiration

Source: Adapted from Fischer, R. A., D. Byerlee, and G. O. Edmeades. 2009. Can technology deliver on the yield challenge to 2050? Paper produced for the FAO expert meeting on how to feed the world in 2050, Rome, 24–26 June, 2009. Food and Agricultural Organization of the United Nations Economic and Social Development Department. Rome, Italy.

genotype in a given environment as determined, for example, by simulation models with plausible physiological and agronomic assumptions. Fischer et al. (2009) presented several definitions of yield measures (Table 1.25), including potential yield. The potential yield of several crops has grown substantially in the past several decades through breeding, along with improved management practices and this has driven farm yield growth (Fischer et al., 2009).

Improved crop nutrition, especially N has made a huge increase in yields in developed as well developing countries. For example, in wheat in the United Kingdom, the optimum dose of N fertilizer, now about 200 kg N ha^{-1}, increases yield about twofold (Jaggard et al., 2010). Between 1950 and 1980, average N dressings for winter wheat increased from 50 to 180 kg ha^{-1} but have risen only slowly since then (Jaggard et al., 2010). Today, it is rare for crops in countries with well-developed arable agriculture to receive suboptimal doses of N fertilizers, and applications are falling slightly as farmers fine-tune their agronomy (Jaggard et al., 2010).

1.2.7.1 Components of Potential Yield

Crop physiologists have developed useful equations for calculating or exploring potential grain yield and its components under radiation- or water-limited conditions (Monteith, 1977; Passioura, 1977). These equations can be written or described as below:

$$PY = \text{total aboveground dry weight (TDW)} \times HI \tag{1.1}$$

$$PY = \int PAR_i \times RUE \times HI \tag{1.2}$$

$$PYw = transpiration\,(T) \times TE \times HI \tag{1.3}$$

where $\int PAR$ is the integral of photosynthetically active radiation (PAR, MJ) intercepted by green tissue over the life of the crop, RUE, or RUE, is the efficiency with which PAR_i is converted into aboveground biomass (g MJ^{-1}). For PYw, T is the amount of water taken up and transpired by the plant (mm), and TE is the transpiration efficiency for creating dry weight (mg g^{-1} of kg ha^{-1} mm^{-1}). A parallel to Equation 1.3 for PY_N, N-limited potential yield, can be written as N absorbed and NUE. There are many variations of these identities (Mitchell et al., 1998), but they all point toward efficiency with which a limiting input (radiation, water, N) is captured, then used to create dry weight and how efficiently that biomass is converted into grain (HI).

Progress in the potential yield increase in the past few decades has largely come through better crop nutrition, especially N nutrition, producing greater leaf area of longer duration, hence increased PAR and modest increases in RUE (Muchow and Sinclair, 1994; Bange et al., 1997). Progress in breeding for increased PY over the past 50 years has been very significant, and is generally attributed to increase in HI, often via shorter stature in wheat, rice, and tropical corn (Johnson et al., 1986). An exception is temperate corn adapted to the United States, where HI has remained relatively stable under favorable conditions and PY has increased because TDW has increased (Duvick, 2005a).

1.2.7.2 Projections of Potential Yield of Major Cereals

A significant yield increase in cereals and legumes has been achieved with breeding and cultural practices in the past few decades. In the future, yield increase will be possible but it should be through a combination of breeding work and better management practices, including mineral nutrition; N. Loomis and Amthor (1999) stated that although achieving high yield is conceptually simple (i.e., maximize the extent and duration of radiation interception, use the captured energy in efficient photosynthesis, partition the new assimilates in ways that provide optimal proportions of leaf, stem, root, and reproductive structures and maintain those at minimum cost), the processes involved are complex. Fischer et al. (2009) gave a summarized discussion on projected yield potential of wheat, rice, and corn. The views of these authors are presented in this section and the importance of N is associated with this yield potential.

1.2.7.2.1 Wheat

A well-researched estimate of wheat yield potential for the United Kingdom (Sylvester-Bradley et al., 2005) based on reasonable assumptions, including an RUE of 2.8 g MJ^{-1} and HI of 0.6, while deploying stem dry matter as efficiently as possible to minimize lodging risk, resulted in 19 Mg grain yield ha^{-1} under most favorable environmental conditions (well watered); this could result in a 50% increase in average farm yields to around 13 Mg ha^{-1} by 2050 (Fischer et al., 2009). To achieve this yield level, significant increases in nutrient levels of N, P, and K are required. Among the three nutrients, especially N will be required in a higher amount due to its loss (>50%) in the soil plant system. According to Raun and Johnson (1999), about 33% of applied N by chemical fertilizers is recovered in cereals. Jaggard et al. (2010) also reported that in addition to the use of chemical fertilizers, especially N, much of the increasing yield of annual crops would be owing to increased light capture achieved by breeding for delayed senescence. This will be especially important, in the future, to counteract the effect of a warmer climate that would make grain crops mature earlier.

1.2.7.2.2 Rice

Mitchell et al. (1998) predicted that conventional selection could result in a tropical and subtropical rice potential yield of 11.3 Mg ha^{-1} for IR72 maturity. However, Fageria (2014) reported that potential yield of irrigated or lowland rice under favorable environmental conditions is about 16 Mg ha^{-1}. Still there is scope of increase potential yield by 15%, which means a yield of about 18.4 Mg ha^{-1}

can be obtained under favorable environmental conditions (solar radiation, water, fertility, control of diseases, insects, and weeds) using modern cultivars. Fischer et al. (2009) reviewed the literature and reported that potential yield increase can be significant if C_4 pathways of photosynthesis could be incorporated into rice. To obtain a yield of about 18 Mg ha^{-1}, there will be a need of more than 300 kg N ha^{-1}. Such a high amount of N should be applied fractionally during the rice growth cycle and the timing should be determined in field experiments under each agroecological condition.

1.2.7.2.3   Corn

Corn has C_4 photosynthetic pathway and the potential yield may be higher as compared to wheat and rice, which have C_3 photosynthetic pathways. Fischer et al. (2009) reported that the maximum yield recorded for corn was from Chile (>20 Mg ha^{-1}). The Chile central valley climate is more favorable than the U.S. corn belt. The grain yield of 20.9 Mg ha^{-1} was recorded for corn grown near Manchester, Iowa, which was obtained from a harvest area of 0.5 ha that was within 2 ha field (Murrell and Childs, 2000). However, Cassman et al. (2003) reported that, in Nebraska, the corn yield of 21–23 Mg ha^{-1} has been reported in contests but the Nebraska number is an average for the period 1983 to 2002.

In addition, an estimate of the theoretical yield potential of about 25 Mg ha^{-1} has been reported for corn grown under conditions encountered in central North America (Tollenaar, 1983). This estimate was based on the mean daily incident solar irradiance during the growing season and a number of assumptions (Tollenaar and Lee, 2002): (i) leaf photosynthetic efficiency is 0.067 mol CO_2 per mole photon or a 4.4% efficiency of conversion of intercepted photosynthetic active radiation into crop biomass, (ii) full light interception between 1 July and 30 September and some simplifying assumptions about light interception before and after this period, (iii) 50% GHI, and (iv) roots constitute 10% of total crop dry matter at physiological maturity. This estimate is similar to several high yields obtained by corn producers (Tollenaar and Lee, 2002). Such high levels of corn yield will require a very high amount of N to attain and sustain production. Hence, the importance of N is projected to be high as the yield level is expected or projected higher in future corn production.

1.3   CONCLUSIONS

Nitrogen is the essential nutrient required in greater amounts by plants and is often the most limiting nutrient for crop production under most agroecosystems. In the future, increasing crop production will depend more on N than any other nutrient. N prices have been increased significantly in the past few decades. Under these situations, increasing crop yields per unit area through the use of appropriate N management practices has become an essential component of modern crop production. In addition, the efficient use of N is essential to reduce the cost of crop production and also environmental pollution. N plays an important role in many morphological, physiological, and biochemical processes in the plants. It is mainly responsible for improving growth yield and yield components in cereals and legumes. The growth parameters that improve with the addition of N are plant height, shoot dry weight, and root growth. All these growth traits have significant positive association with grain yield in cereals and legumes. Similarly, yield components such as panicle number and number of pods have significant positive association with grain yield. However, spikelet sterility has significant negative association with grain yield. The influence of N is significant in the assimilation of carbon in the source organs (especially leaves) and translocate and the utilization of photosynthates in the sink organs (grains), which determines the growth rate and productivity of crop plants. It is essential for photosynthesis in higher plants. Adequate N content in the leaves improves RUE and consequently crop yields. In general, RUE is an independent measure that can be used to benchmark crop performance and highlight yield limitations. Radiation use is a valuable approach for interpreting large variations in crop yield from season to season and across locations resulting from climatic variations.

Nitrogen deficiency limits plant size and diminished metabolic activity in the leaves as a result of which photosynthesis is reduced. It is also among the components of many enzymes that participates in many metabolic processes in the growth and development of plants. The GHI is improved with the addition of an adequate amount of N in the crop plants. It has a significant positive association with grain yield. It can be concluded that the adequate N status of the plants is a major factor in determining the yield of field crops.

A greater response to N application by modern crop cultivars is related to ideal plant architecture, including strong and stiff stem and vertical arrangement of leaves, which are responsible for lodging resistance and higher interception of radiation and use efficiency. In addition, modern crop cultivars also have resistance to biotic and abiotic stresses, which is also responsible for permitting the use of higher N application rate and consequently improving yields. Plant breeding and the use of adequate management practices have contributed significantly to improving crop yields in the past few decades. There is still a large gap between potential yield and actual yield obtained by the farmers in developed as well as developing countries. Hence, in the author's opinion, the observed yield gap indicates that average crop yields can still be significantly improved by management and breeding. In crop management practices, N fertilization will play a significant role in increasing crop yields.

REFERENCES

Akita, S. 1989. Improving yield potential in tropical rice. In: *Progress in Irrigated Rice Research,* ed., International Rice Research Institute, pp. 41–73. Los Baños, the Philippines: International Rice Research Institute.

Amano, T., Q. Zhu, Y. Wang, N. Inoue, and H. Tanaka. 1993. Case studies on high yields of paddy rice in Jiangsu Province, China. I. Characteristics of grain production. *Jpn J. Crop Sci.* 62:267–274.

Anderson, E. L. 1987. Corn root growth and distribution as influenced by tillage and nitrogen fertilization. *Agron. J.* 79:544–549.

Andrade, F. H., P. Calvino, A. Cirilo, and P. Barbieri. 2002. Yield responses to narrow rows depend on increased radiation interception. *Agron. J.* 94:975–980.

Assefa, Y., K. L. Roozeboom, S. A. Staggenborg, and J. Du. 2012. Dryland and irrigated corn yield with climate, management, and hybrid changes from 1939 through 2009. *Agron. J.* 104:473–482.

Austin, R. B., J. Bingham, R. D. Blackwell, L. T. Evans, M. A. Ford, C. L. Morgan, and M. Taylor. 1980. Genetic improvements in winter wheat yields since 1900 and associated physiological changes. *J. Agric. Sci.* 94:675–689.

Baligar, V. C., N. K. Fageria, and Z. L. He. 2001. Nutrient use efficiency in plants. *Commun. Soil Sci. Plant Anal.* 32:921–950.

Bange, M. P., G. L. Hammer, and K. G. Rickert. 1997. Effect of specific leaf nitrogen on radiation use efficiency and growth of sunflower. *Crop Sci.* 37:1201–1207.

Banziger, M., B. Feil, J. E. Schmid, and P. Stamp. 1994. Utilizing of late applied fertilizer nitrogen by spring wheat genotypes. *Eur. J. Agron.* 3:63–69.

Barker, A. V. 1979. Nutritional factors in photosynthesis of higher plants. *J. Plant Nutr.* 1:309–342.

Begna, S. H., R. I. Hamilton, L. M. Dwyer, D. W. Stewart, and D. L. Smith. 1997. Effects of population density and planting pattern on the yield and yield components of leafy reduced-stature maize in short season area. *J. Agron. Crop Sci.* 179:9–7.

Begna, S. H., R. I. Hamilton, L. M. Dwyer, D. W. Stewart, D. Cloutier, L. Assemat, K. Foroutan-Pour, and D. L. Smith. 2001. Weed biomass production response to plant spacing and corn hybrids differing in canopy architecture. *Weed Technol.* 15:647–653.

Bekunda, M. A., A. Bationo, and H. Ssali. 1997. Soil fertility management in Africa; A review of selected research trials. In: *Replenishing Soil Fertility in Africa*, ed., R. J. Buresh, pp. 63–79. Madison, Wisconsin: SSSA.

Bennett, W. F. 1993. Plant nutrient utilization and diagnostic plant symptoms. In: *Nutrient Deficiencies and Toxicities in Crop Plants*, ed., W. F. Bennett, pp. 1–7. St. Paul, Minnesota: The American Phytopathological Society.

Benzian, B. and P. Lane. 1979. Some relationships between grain yield and grain protein of wheat experiments in South-East England and comparisons with such relationships elsewhere. *J. Sci. Food. Agric.* 30:59–70.

Berzenyi, Z. and I. S. Tokatlidis. 2012. Density dependence rather than maturity determines hybrid selection in dryland maize production. *Agron. J.* 104:331–336.

Bhatia, C. R. and R. Rabson. 1976. Bioenergetic consideration in cereal breeding for protein improvement. *Science* 194:1418–1421.

Braun, H. J., S. Rajaram, and M. V. Ginkel. 1997. CIMMYT's approach to breeding for wide adaptation. In: *Adaptation in Plant Breeding*, ed., P. M. A. Tigerstedt, pp. 197–205. Dordrecht: Kluwer Academic Publishers.

Bridge, R. R. and W. R. Meredith. 1983. Comparative performance of obsolete and current cotton cultivars. *Crop Sci.* 23:949–952.

Bridge, R. R., W. R. Meredith, and J. F. Chism. 1971. Comparative performance of obsolete varieties and current varieties of upland cotton. *Crop Sci.* 11:29–32.

Bauer, S., D. Lloyd, B. P. Horgan, and D. J. Soldat. 2012. Agronomic and physiological responses of cool-season turfgrass to fall-applied nitrogen. *Crop Sci.* 52:1–10.

Bausch, W. C. and K. Diker. 2001. Innovative remote sensing techniques to increase nitrogen use efficiency of corn. *Commun. Soil Sci. Plant Anal.* 32:1371–1390.

Biermacher, J. T., R. Reuter, M. K. Kering, J. K. Rogers, J. Blanton, Jr., J. A. Guretzky, and T. J. Butler. 2012. Expected economic potential of substituting legumes for nitrogen in bermudagrass pastures. *Crop Sci.* 52:1923–1930.

Black, A. L. and F. H. Siddoway. 1977. Hard red and durum spring wheat responses to seeding date and NP-fertilization on fallow. *Agron. J.* 69:885–888.

Board, J. E., W. Zhang, and B. G. Harville. 1996. Yield ranking for soybean cultivars grown in narrow and wide rows with late planting dates. *Agron. J.* 88:240–245.

Boot, R. G. A., P. M. Schildwacht, and H. Lambers. 1992. Partitioning of nitrogen and biomass at a range of N-addition rates and their consequences for growth and gas exchange in two perennial grasses from inland dunes. *Physiol. Plant.* 86:152–160.

Boote, K. J. R. N. Gallaher, W. K. Robertson, K. Hinson, and L. C. Hammond. 1978. Effect of foliar fertilization on photosynthesis, leaf nutrition, and yield of soybeans. *Agron. J.* 70:787–791.

Bould, C., E. J. Hewitt, and P. Needham. 1984. *Diagnosis of Mineral Disorders in Plants.* Vol. 1. New York: Chemical Publishing.

Brown, R. H. 1984. Growth of the green plants. In: *Physiological Basis of Crop Growth and Development,* ed., M. B. Tesar, pp. 153–174. Madison, Wisconsin: ASA.

Bruinsma, J. 2003. *World Agriculture towards 2015/2030: An FAO Perspective*, Rome, Italy.

Bullock, D. G., R. L. Nielsen, and W. E. Nyquist. 1988. A growth analysis comparison of corn grown in conventional and equidistant plant spacing. *Crop Sci.* 28:254–258.

Burstrom, H. 1943. Photosynthesis and assimilation of nitrate by wheat leaves. *Ann. Royal Agric. Coll. Sweden* 11:1–50.

Butt, M. S., F. M. Anjum, S. U. Rehman, M. T. Nadeem, M. K. Sharif, and M. Anwer. 2008. Selected quality attributes of fine basmati rice: Effect of storage history and varieties. *Int. J. Food Prop.* 11:698–711.

Byerlee, D. and B. C. Curtis. 1988. Wheat: A crop transformed. *Span* 30:110–113.

Cakmak, I. and C. Engels. 1999. Role of mineral nutrients in photosynthesis and yield formation. In: *Mineral Nutrition of Crops: Fundamental Mechanisms and Implications*, ed., Z. Rengel, pp. 141–168. New York: Howorth Press.

Calderini, D. F. and G. A. Slafer. 1998. Changes in yield and yield stability in wheat during the 20th century. *Field Crops Res.* 57:335–347.

Campbell, C. A., F. Selles, R. P. Zentner, B. J. McConkey, S. A. Brandts, and R. C. McKenzie. 1997. Regression model for predicting yield of hard red spring wheat grown on stubble in the semiarid prairie. *Can. J. Plant Sci.* 77:43–52.

Cao, Z. and X. Wu. 1984. *Plant Physiology.* Beijing: High Education Press.

Carter, T. E., R. L. Nelson, C. H. Sneller, and Z. Cui. 2004. Genetic diversity in soybean. In: *Soybeans: Improvement, Production, and Uses*, eds., H. R. Boerma and J. E. Specht, pp. 301–416. Madison: ASA, CSSA, and SSSA.

Cassman, K. G. 1999. Ecological intensification of cereal production systems: Yield potential, soil quality, and precision agriculture. *Proc. Natl. Acad. Sci. USA* 96:5952–5959.

Cassman, K. G., A. Dobermann, D. T. Walters, and H. Yang. 2003. Meeting cereal demand while protecting natural resources and improving environmental quality. *Annu. Rev. Environ. Resour.* 28:315–358.

Cerrato, M. E. and A. M. Blackmer. 1990. Comparison of models for describing corn yield response to nitrogen fertilizer. *Agron. J.* 82:138–143.

Chang, T. T. 1976. Rice. In: *Evolution of Crop Plants*, ed., N. W. Simmonds, pp. 98–104. Longman: London.

Chen, B. S., Y. H. Zhang, X. Li, and D. M. Jiao. 2002. Photosynthetic characteristics and assimilate distribution in super hybrid rice Liangyoupeijiu at late growth stage. *Acta Agron. Sin.* 28:777–782.

Chung, S. I., C. W. Rico, S. C. Lee, and M. Y. Kang. 2012. Separation of protein from rice grains with different eating qualities by two dimensional gel electrophoresis. *Agron. J.* 104:49–53.

Chua, T. T., K. F. Bronson, J. D. Booker, J. W. Keeling, A. R. Mosier, J. P. Bordovsky, R. J. Lascano, C. J. Green, and E. Segarra. 2003. In season nitrogen status sensing in irrigated cotton. I. Yield and nitrogen 15 recovery. *Soil Sci. Soc. Am. J.* 67:1428–1438.

Ci, X., M. Li, X. Liang, Z. Xie, D. Zhag, and X. Li. 2011. Genetic contribution to advanced yield for maize hybrids released from 1970 to 2000 in China. *Crop Sci.* 51:13–20.

Ci, X., D. Zhang, X. Li, J. Xu, X. Liang, Z. Lu, P. Bai et al. 2013. Trends in ear traits of Chinese maize cultivars from the 1950s to 2000s. *Agron. J.* 105:20–27.

Cober, E. R., M. J. Morrison, B. Ma, and G. Butler. 2005. Genetic improvement rates of short season soybean increase with plant population. *Crop Sci.* 45:1029–1034.

Conocono, E. A., J. A Egdane, and T. I. Setter. 1998. Estimation of canopy photosynthesis in rice by means of daily increases in leaf carbohydrate concentration. *Crop Sci.* 38:987–995.

Cook, M. G. and L. T. Evans. 1983. Nutrient responses of seedlings of wild and cultivated *Oryza* species. *Field Crops Res.* 6:205–218.

Costa, C., L. M. Dwyer, D. W. Stewart, and D. L. Smith. 2002a. Nitrogen effects on grain yield and yield components of leafy and nonleafy maize genotypes. *Crop Sci.* 42:1556–1563.

Costa, C., L. M. Dwyer, X. Zhou, P. Dutilleul, C. Hamel, L. M. Reid, and D. L. Smith. 2002b. Root morphology of contrasting maize genotypes. *Agron. J.* 94:96–101.

Cox, W. J. and D. J. R. Cherney. 2001. Influence of brown midrib, leafy, and transgenic hybrids on corn forage production. *Agron. J.* 93:790–796.

Cox, W. J. and J. H. Cherney. 2011. Growth and yield responses of soybean to row spacing and seeding rate. *Agron. J.* 103:123–128.

Cregan, P. B. and R. W. Yaklich. 1986. Dry matter and nitrogen accumulation and portioning in selected soybean genotypes of different derivation. *Theor. Appl. Genet.* 72:782–786.

Cuevas-Perez, F. E., L. E. Berrio, D. I. Gonzalez, F. Correa-Victoria, and E. Tulande. 1995. Genetic improvement in yield of semidwarf rice cultivars in Colombia. *Crop Sci.* 35:725–729.

Crop Science Society of America. 1992. *Glossary of Crop Science Terms*. Madison, Wisconsin: CSSA.

D'Andrea, K. E., M. E. Otegui, A. G. Cirilo, and G. Eyherabide. 2006. Genotypic variability in morphological and physiological traits among maize inbred lines—Nitrogen response. *Crop Sci.* 46:1266–1276.

De Bruin, J. L. and P. Pedersen. 2008a. Soybean seed yield response to planting date and seeding rate in the upper Midwest. *Agron. J.* 100:696–703.

De Bruin, J. L. and P. Pedersen. 2008b. Effect of row spacing and seeding rate on soybean yield. *Agron. J.* 100:704–710.

De Bruin, J. L. and P. Pedersen. 2009. Growth, yield and yield component changes among old and new soybean cultivars. *Agron. J.* 101:124–130.

Ding, L., K. J. Wang, G. M. Jiang, D. K. Biswas, H. Xu, L. F. Li, and Y. H. Li. 2005. Effects of nitrogen deficiency on photosynthetic traits of maize hybrids released in different years. *Ann. Bot.* 96:925–930.

Dixon, R. A. 2001. Natural products and plant disease resistance. *Nature* 411:843–847.

Dobermann, A., D. Dawe, R. P. Roetter, and K. G. Cassman. 2000. Reversal of rice yields decline in a long term continuous cropping experiment. *Agron. J.* 92:633–643.

Dofing, S. M. and M. G. Karlsson. 1993. Growth and development of uniculm and conventional tillering barley lines. *Agron. J.* 85:58–61.

Donald, C. M. 1962. In search of high yield. *J. Aust. Inst. Agric. Sci.* 28:171–178.

Donald, C. M. and J. Hamblin. 1976. The biological yields and harvest index of cereals as agronomic and plant breeding criteria. *Adv. Agron.* 28:361–405.

Drew, M. C., L. R. Saker, and T. W. Ashley. 1973. Nutrient supply and the growth of the seminal root system in barley. *J. Exp. Bot.* 24:1189–1202.

Durieux, R. P., E. J. Kamprath, W. A. Jackson, and R. H. Moll. 1994. Root distribution of corn: The effect of nitrogen fertilization. *Agron. J.* 86:958–962.

Duvick, D. N. 2005a. The contribution of breeding to yield advances in maize (*Zea mays* L.) *Adv. Agron.* 86:83–145.

Duvick, D. N. 2005b. Genetic progress in yield of united states maize (*Zea mays* L.). *Maydica* 50:193–202.

Duvick, D. N. and K. G. Cassman. 1999. Post-green revolution trends in yield potential of temperate corn in the North-central United States. *Crop Sci.* 39:1622–1630.

Dwivedi, S. L., J. H. Crouch, S. N. Nigham, M. E. Ferguson, and A. H. Paterson. 2003. Molecular breeding of groundnut for enhanced productivity and food security in the semi-arid tropics: Opportunities and challenges. *Adv. Agron.* 80:153–221.

Echarte, L., S. Rothstein, and M. Tollenaar. 2008. The response of leaf photosynthesis and dry matter accumulation to nitrogen supply in an older and a newer maize hybrid. *Crop Sci.* 48:656–665.

Economic Research Service. 2010. Fertilizer use and price. Available at http://www.ers.usda.gov/data/fertilizeruse (verified 3 September 2011). USDA, Economic Res. Serv., Washington, DC.

Edmeades, G. O. and H. R. Lafitte. 1993. Defoliation and plant density effects on maize selected for reduced plant height. *Agron. J.* 85:850–857.

Eghball, B., J. R. Settimi, J. W. Maranville, and A. M. Parkhurst. 1993. Fractal analysis for morphological description of corn roots under nitrogen stress. *Agron. J.* 85:287–289.

Egli, D. B. 2008. Soybean yield trends from 1972 to 2003 in mid-western USA. *Field Crops Res.* 106:53–59.

Evans, L. T. 1980. The natural history of crop yield. *Am. Sci.* 68:388–397.

Evans, J. R. 1983. Nitrogen and photosynthesis in the flag leaf of wheat (*Triticum aestivum* L.). *Plant Physiol.* 72:297–302.

Evans, J. R. 1989. Photosynthesis and nitrogen relationships in leaves of C_3 plants. *Oecologia* 78:9–19.

Evans, L. T. 1993. *Crop Evolution, Adaption and Yield*. Cambridge University Press: Cambridge.

Evans, L. T. and R. A. Fischer. 1999. Yield potential: Its definition, measurement, and significance. *Crop Sci.* 39:1544–1551.

Evans, J. R. and J. R. Seemann. 1989. The allocation of protein nitrogen in the photosynthetic apparatus: Costs, consequence, and control. In: *Photosynthesis*, ed., W. R. Briggs, pp. 183–205. New York: Liss.

Evans, L. T., R. M. Visperas, and B. S. Vergara. 1984. Morphological and physiological changes among rice varieties used in the Philippines over the last seventy years. *Field Crops Res.* 8:105–124.

Evans, L. T. and I. F. Wardlaw. 1976. Aspects of the comparative physiology of the grain yield in cereals. *Adv. Agron.* 28:301–359.

Fageria, N. K. 1992. *Maximizing Crop Yields*. New York: Marcel Dekker.

Fageria, N. K. 1989. *Tropical Soils and Physiological Aspects of Crops*. Brasilia, Brazil: EMBRAPA.

Fageria, N. K. 1998. Annual report of the project "The study of liming and fertilization for rice and common bean in cerrado region". National Rice and Bean Research Center of Embrapa, Goiania, Brazil.

Fageria, N. K. 2001. Nutrient management for improving upland rice productivity and sustainability. *Commun. Soil Sci. Plant Anal.* 32:2603–2629.

Fageria, N. K. 2002. Nutrient management for sustainable dry bean production in the tropics. *Commun. Soil Sci. Plant Anal.* 33:1537–1575.

Fageria, N. K. 2007. Yield physiology of rice. *J. Plant Nutr.* 30:843–879.

Fageria, N. K. 2009. *The Use of Nutrients in Crop Plants*. Boca Raton: CRC Press.

Fageria, N. K. 2013. *The Role of Plant Roots in Crop Production*. Boca Raton: CRC Press.

Fageria, N. K. 2014. *Mineral Nutrition of Rice*. Boca Raton: CRC Press.

Fageria, N. K. and V. C. Baligar. 1993. Screening crop genotypes for mineral stresses. In: *Workshop on Adaptation of Plants to Soil Stress*, ed., University of Nebraska, pp. 142–159. Lincoln, Nebraska: University of Nebraska.

Fageria, N. K. and V. C. Baligar. 2001. Lowland rice response to nitrogen fertilization. *Commun. Soil Sci. Plant Anal.* 32:1405–1429.

Fageria, N. K. and V. C. Baligar. 2005. Enhancing nitrogen use efficiency in crop plants. *Adv. Agron.* 88:97–185.

Fageria, N. K., V. C. Baligar, and R. B. Clark. 2006. *Physiology of Crop Production*. New York: The Haworth Press.

Fageria, N. K., V. C. Baligar, and C. A. Jones. 2011a. *Growth and Mineral Nutrition of Field Crops*, 3rd edition. Boca Raton: CRC Press.

Fageria, N. K. and M. P. Barbosa Filho. 2001. Nitrogen use efficiency in lowland rice genotypes. *Commun. Soil Sci. Plant Anal.* 32:2079–2090.

Fageria, N. K., L. C. Melo, and J. P. Oliveira. 2013. Nitrogen use efficiency in dry bean genotypes. *J. Plant Nutr.* 36:2179–2190.

Fageria, N. K., O. P. Morais, and A. B. Santos. 2010. Nitrogen use efficiency in upland rice genotypes. *J. Plant Nutr.* 33:1696–1711.

Fageria, N. K. and A. Moreira. 2011. The role of mineral nutrition on root growth of crop plants. *Adv. Agron.* 110:251–330.

Fageria, N. K. and A. B. Santos. 2008. Yield physiology of dry bean. *J. Plant Nutr.* 31:983–1004.

Fageria, N. K., A. B. Santos, and A. M. Coelho. 2011b. Growth, yield and yield components of lowland rice as influenced by ammonium sulfate and urea fertilization. *J. Plant Nutr*. 34:371–386.

Fageria, N. K., A. B. Santos, and V. A. Cutrim. 2007. Yield and nitrogen use efficiency of lowland rice genotypes as influenced by nitrogen fertilization. *Pesq. Agropec. Bras*. 42:1029–1034.

Fageria, N. K., A. B. Santos, and V. A. Cutrim. 2008. Dry matter and yield of lowland rice genotypes as influenced by nitrogen fertilization. *J. Plant Nutr*. 31:788–795.

Fageria, N. K., N. A. Slaton, and V. C. Baligar. 2003. Nutrient management for improving lowland rice productivity and sustainability. *Adv. Agron*. 80:63–152.

FAO (Food and Agricultural Organization of the United Nations). 2010. FAO statistical databases, http:/apps.fao.org.

FAO (Food and Agricultural Organization of the United Nations). 2011. Statistical database. FAO, Rome, Italy. http://www.fao.org (accessed 3 July 2011).

FAO (Food and Agricultural Organization of the United Nations). 2013. Statistical database. FAO, Rome, Italy. http://www.faostat.fao.org/rome.

Farmaha, B. S. and A. L. Sims. 2013. Yield and protein response of wheat cultivars to polymer-coated urea and urea. *Agron. J*. 105:229–236.

Fasoula, V. A. and D. A. Fasoula. 2000. Honeycomb breeding: Principles and applications. *Plant Breed. Rev*. 18:177–250.

Fasoula, V. A. and M. Tollenaar. 2005. The impact of plant population density on crop yield and response to selection in maize. *Mydica* 50:39–48.

Feil, B. 1992. Breeding progress in small grain cereals; A comparison of old and modern cultivars. *Plant Breed*. 108:1–11.

Feyerherm, A. M., K. E. Kemp, and G. M. Paulsen. 1988. Wheat yield analysis in relation to advancing technology in the Midwest United States. *Agron. J*. 80:998–1001.

Field, C. and H. A. Mooney. 1986. The photosynthesis–nitrogen relationship in wild plants. In: *On the Economy of Plant Form and Function*, ed., T. J. Givnish, pp. 25–55. Cambridge: Cambridge University Press.

Fischer, R. A. 1993. Irrigated spring wheat and timing and amount of nitrogen fertilizer. II. Physiology of grain yield response. *Field Crops Res*. 33:57–80.

Fischer, K. S. and A. F. E. Palmer. 1984. Tropical maize. In: *The Physiology of Tropical Field Crops,* eds., P. R. Goldsworthy and N. M. Fisher, pp. 231–248. New York: John Wiley & Sons.

Fischer, R. A., D. Byerlee, and G. O. Edmeades. 2009. Can technology deliver on the yield challenge to 2050? Paper produced for the FAO expert meeting on how to feed the world in 2050, Economic and Social Development Department, 24–26 June 2009, Rome, Italy.

Fowler, D. B. 2003. Crop nitrogen demand and grain protein concentration of spring and winter wheat. *Agron. J*. 95:260–265.

Follett, R. F. 2001. Innovative ^{15}N microplot research techniques to study nitrogen use efficiency under different ecosystems. *Commun. Soil Sci. Plant Anal*. 32:951–979.

Fox, C. M., T. R. Cary, A. L. Colgrove, E. D. Nafziger, J. S. Haudenshield, G. L. Hartman, J. E. Specht, and B. W. Diers. 2013. Estimating soybean grain for yield in the northern USA—Influence of cropping history. *Crop Sci*. 53:2473–2482.

Frey, K. J. 1971. Improving crop yield through plant breeding. In: *Moving off the Yield Plateau*, eds., J. D. Eastin and R. D. Munson, pp. 15–58. Madison, Wisconsin: ASA, CSSA, and SSSA.

Fujinuma, R., N. J. Balster, and J. M. Norman. 2009. An improved model of nitrogen release for surface applied controlled release fertilizer. *Soil Sci. Soc. Am. J*. 73:2043–2050.

Garton, R. W. and I. E. Widders. 1990. N and P preconditioning of small plug seedlings influences growth and yield of processing tomatoes. *HortScience* 25:655–657.

Gandeza, A. T., S. Shoji, and I. Yamada. 1991. Simulation of crop response to polyolefin-coated urea: I. Field dissolution. *Soil Sci. Soc. Am. J*. 55:1462–1467.

Gasser, J. K. R. and I. G. Iordanou. 1967. Effects of ammonium sulphate and calcium nitrate on the growth, yield and nitrogen uptake of barley, wheat, and oats. *J. Agric. Sci*. 68:307–316.

Gimenez, C., D. J. Connor, and F. Rueda. 1994. Canopy development, photosynthesis and radiation use efficiency in sunflower in response to nitrogen. *Field Crops Res*. 38:15–27.

Glass, A. D. M. 2003. Nitrogen use efficiency of crop plants: Physiological constraints upon nitrogen absorption. *Critical Rev. Plant Sci*. 22:453–470.

Good, A. G., A. K. Shrawat, and D. G. Muench. 2004. Can less yield more? Is reducing nutrient input into the environment compatible with maintaining crop production? *Trends Plant Sci*. 9:597–605.

Goldsworthy, P. R., A. F. E. Palmer, and D. W. Sperling. 1974. Growth and yield of lowland tropical maize in Mexico. *J. Agric. Sci. (Cambridge)*. 83:223–230.

Gouis, J. L. and P. Pluchard. 1997. Genetic variation for nitrogen use efficiency in winter wheat (*Triticum aestivum* L.). In: *Adaptation in Plant Breeding*, ed., P. M. A. Tigerstedt, pp. 2243–246. Dordrecht, The Netherlands: Kluwer Academic Publishers.

Greenland, D. J., G. D. Bowen, H. Eswaran, R. Rhoades, and C. Valentin. 1994. *Soil, Water and Nutrient Management Research: A New Agenda*. Bangkok, Thailand: International Board Soil Research and Management.

Gregory, P. J. 1994. Root growth and activity. In: *Physiology and Determination of Crop Yield*, ed., G. A. Peterson, pp. 65–93. Madison, Wisconsin: ASA, CSSA, and SSSA.

Hafner, S. 2003. Trends in maize, rice and wheat yields for 188 nations over the past 40 years: A prevalence of linear growth. *Agric. Ecosyst. Environ.* 97:275–283.

Hall, A. J., D. J. Connor, and V. O. Sadras. 1995. Radiation use efficiency of sunflower crops: Effects of specific leaf nitrogen and ontogeny. *Field Crops Res.* 41:65–77.

Halse, N. J., E. A. N. Greenwood, P. Lapins, and C. A. P. Boundy. 1969. An analysis of nitrogen deficiency on growth and yield of western Australia wheat crop. *Aust. J. Agric. Res.* 20:987–998.

Hammer, G. L. and G. C. Wright. 1994. A theoretical analysis of nitrogen and radiation effects on radiation use efficiency in peanut. *Aust. J. Agric. Res.* 45:575–589.

Handa, K. 1995. Differentiation and development of tiller buds. In: *Science of Rice Plant: Physiology*, Vol. 2, eds., T. Matsuo, K. Kumazawa, R. Ishi, K. Ishihara, and H. Hirata, pp. 61–65. Tokyo: Food and Agriculture Policy Center.

Hansson, A. C. and O. Andren. 1987. Root dynamics in barley, lucerne, and meadow fescue investigated with a minirhizotron technique. *Plant Soil* 103:33–38.

Harper, J. E. 1994. Nitrogen metabolism. In: *Physiology and Determination of Crop Yield*, eds., K. J. Boote, J. M. Bennett, T. R. Sinclir, and G. M. Paulsen, pp. 285–302. Madison, Wisconsin: ASA, CSSA, and SSSA.

Hasegawa, H. 2003. High yielding rice cultivars perform best even at reduced nitrogen fertilizer rate. *Crop Sci.* 43:921–926.

Hayashi, H. 1995. Translocation, storage and partitioning of photosynthetic products. In: *Science of the Rice Plant: Physiology*, Vol. 2, eds., T. Matsuo, K. Kumazawa, R. Ishii, K. Ishihara, and H. Hirata, pp. 546–5655. Tokyo: Food and Agriculture Policy Research Center.

Hesketh, J. D., W. L. Ogren, M. E. Hageman, and D. B. Peters. 1981. Correlations among leaf CO₂-exchange rates, areas and enzyme activities among soybean cultivars. *Photosynth. Res.* 2:21–30.

Hipps, L. E., G. Asrar, and E. T. Kanemasu. 1983. Assessing the interception of photosynthetically active radiation in winter wheat. *Agric. Meteorol.* 28:253–259.

Hirose, T. and M. J. A. Werger. 1987. Nitrogen use efficiency in instantaneous and daily photosynthesis of leaves in the canopy of a *Solidago altissma* stand. *Physiol. Plantarum* 70:215–222.

Hoad, S. P., G. Russell, M. E. Lucas, and I. J. Bingham. 2001. The management of wheat, barley, and oat root systems. *Adv. Agron.* 74:193–246.

Holmes, D. P. 1973. Inflorescence development of semidwarf and standard height wheat cultivars in different photoperiod and nitrogen treatments. *Can. J. Bot.* 51:941–956.

Hou, X., R. Li, Z. Jia, and Q. Han. 2013. Rotational tillage improves photosynthesis of winter wheat during reproductive growth stages in a semiarid region. *Agron. J.* 105:215–221.

Huang, W. Y., W. McBride, and U. Vasavada. 2008. Recent volatility in U.S. fertilizer prices. *Amber Waves* 7:28–31.

Hull, R. J. and H. Liu. 2005. Turfgrass nitrogen: Physiological and environmental impacts. *Int. Turfgrass Soc. Res. J.* 10:962–975.

Hutchinson, C., E. Simonne, P. Solano, J. Meldrum, and P. Livingston-Way. 2003. Testing of controlled release fertilizer programs for seep irrigated Irish potato production. *J. Plant Nutr.* 26:1709–1723.

Ishii, R. 1988. Varietal differences of photosynthesis and grain yield in rice. *Korean J. Crop Sci.* 33:315–321.

Jackson, G. D. 2000. Effects of nitrogen and sulfur on canola yield and nitrogen uptake. *Agron. J.* 92:644–649.

Jaggard, K. W., A. Qi, and E. S. Ober. 2010. Possible changes to arable crop yields by 2050. *Phil. Trans. R. Soc. B.* 365:2835–2851.

Jennings, P. R., W. R. Coffman, and H. E. Kauffman. 1997. *Rice Improvement*. Los Bãnos, the Philippines: International Rice Research Institute.

Jiang, D., Z. J. Xie, W. X. Cao, T. B. Dai, and Q. Jing. 2004. Effects of postanthesis drought and waterlogging on photosynthetic characteristics, assimilates transportation in winter wheat. *Acta Agron. Sin.* 30:175–182.

Jin, J., X. Liu, G. Wang, L. Mi, Z. Shen, X. Chen, and S. J. Herbert. 2010. Agronomic and physiological contribution to the yield improvement of soybean cultivars released from 1950 to 2006 in Northern China. *Field Crops Res.* 115:116–123.

Johnson, E. C., K. S. Fischer, G. O Edmeades, and A. F. E. Palmer. 1986. Recurrent selection for reduced plant height in lowland tropical maize. *Crop Sci.* 26:253–260.

Kaizzi, K. C., J. Byalebeka, O. Semalulu, I. Alou, W. Zimwanguyizza, A. Nansamba, P. Musinguzi, P. Ebanyat, T. Hyuha, and C. S. Wortmann. 2012a. Maize response to fertilizer and nitrogen use efficiency in Uganda. *Agron. J.* 104:73–82.

Kaizzi, K., J. Byalebeka, O. Semalulu, I. Alou, W. Zimwanguyizza, A. Nansamba, P. Musinguzi, P. Ebanyat, T. Hyuha, and C. S. Wortmann. 2012b. Sorghum response to fertilizer and nitrogen use efficiency in Uganda. *Agron. J.* 104:83–90.

Kant, S., Y. M. Bi, and S. J. Rothstein. 2010. Understanding plant response to nitrogen limitations for the improvement of crop nitrogen use efficiency. *J. Exp. Bot.* 6:1–11.

Kaplan, K. 2012. USDA unveils new plant hardiness zone map. USDA. 25 January 2012. http://www.ars.usda.gov/is/pr/2012/120125.htm (verified 18 September 2012).

Katsura, K., S. Maeda, T. Horie, and T. Shiraiwa. 2007. Analysis of yield attributes and crop physiological traits of Liangyoupeijiu a hybrid rice recently bred in China. *Field Crops Res.* 103:170–177.

Kerby, T. A. and D. R. Buxton. 1981. Competation between adjacent fruiting forms in cotton. *Agron. J.* 73:867-871.

Kerby, T. A., D. R. Buxton, and K. Matsuda. 1980. Carbon source-sink relationship within narrow-row cotton canopies. *Crop Sci.* 20:208–212.

Kiniry, J. R., G. McCauley, Y. Xie, and J. G. Arnold. 2001. Rice parameters describing crop performance of four U.S. cultivars. *Agron. J.* 93:1354–1361.

Knauft, D. A. and J. C. Wynne. 1995. Peanut breeding and genetics. *Adv. Agron.* 55:393–445.

Knowles, T. C., B. W. Hipp, P. S. Graff, and D. S. Marshall. 1994. Timing and rate of topdress nitrogen for rainfed winter wheat. *J. Prod. Agric.* 7:216–220.

Kueneman, E. A., D. H. Wallace, and P. M. Ludford. 1979. Photosynthetic measurements of field grown dry beans and their relation to selection for yield. *J. Amer. Soc. Hort. Sci.* 104:480–482.

Khush, G. S. 1995. Increased genetic potential of rice yield: Methods and perspectives. In: *Rice in Latin America: Perspectives to Increase Production and Yield Potential*, eds., B. S. Pinheiro and E. P. Guimarais, pp. 13–29. Document No. 60. Goiania, Brazil: EMBRAPA-CNPAF.

Kucharik, C. J. 2008. Contribution of planting date trends to increased maize yields in the central United States. *Agron. J.* 100:328–336.

Kumudi, S. 2002. Trials and tribulations: A review of the role of assimilate supply in soybean genetic yield improvement. *Field Crops Res.* 75:211–222.

Ladha, J. K., C. K. Reddy, A. T. Padre, and C. V. Kessel. 2012. Role of nitrogen fertilization in sustaining organic matter in cultivated soils. *Better Crops Plant Food* 96:24–25.

Lambers, H. 1987. Does variation in photosynthetic rate explain variation in growth rate and yield. *Neth. J. Agric. Sci.* 35:505–519.

Langer, R. H. M. and F. K. Y. Liew. 1973. Effect of varying nitrogen supply at different stages of the reproductive phase on spikelet and grain production and on grain nitrogen of wheat. *Aust. J. Agric. Res.* 24:647–656.

Lawes, D. A. 1977. Yield improvement in spring oats. *J. Agric. Sci.* 89:751–757.

Li, X. T., H. T. Cheng, N. Wang, C. M. Yu, L. Y. Qu, P. Cao, N. Hu, T. Liu, and W. Y. Lyu. 2013. Critical factors for grain filling of erect panicle type japonica rice cultivars. *Agron. J.* 105:1404–1410.

Lin, S. K., M. C. Chang, Y. G. Tsai, and H. S. Lur. 2005. Proteomic analysis of the expression of protein related to rice quality during caryopsis development and the effect of high temperature on expression. *Proteomics* 5:2140–2156.

Lin, X. Q., D. F. Zhu, and Y. P. Zhang. 2002. Achieving high yielding plant type in a super rice varieties optimizing crop management. *China Rice* 2:10–12.

Lindquist, J. L., T. J. Arkebauer, D. T. Walters, K. G. Cassman, and A. Dobermann. 2005. Maize radiation use efficiency under optimal growth condition. *Agron. J.* 97:72–78.

Liu, C., M. Watanabe, and Q. Wang. 2008. Changes in nitrogen budgets and nitrogen use efficiency in agroecosystems of the Changjang River basin between 1980 and 2000. *Nutr. Cycl. Agroecosys.* 80:19–37.

Loomis, R. S. and J. S. Amthor. 1999. Yield potential, plant assimilatory capacity, and metabolic efficiencies. *Crop Sci.* 39:1584–1596.

Long, S. P., X. G. Zhu, S. L. Naidu, and D. R. Ort. 2006. Can improvement in photosynthesis increase crop yields? *Plant Cell Environ.* 29:315–330.

Lopez-Bellido, L., R. J. Lopez-Bellido, J. E. Castillo, and F. J. Lopez-Bellido. 2000. Effects of tillage, crop rotation, and nitrogen fertilization on wheat under rainfed Mediterranean conditions. *Agron. J.* 92:1054–1063.

López-Bellido, R. J., L. Lopez-Bellido, F. J. Lopez-Bellido, and J. E. Castillo. 2003. Faba bean (*Vicia faba* L.) response to tillage and soil residual nitrogen in a continuous rotation with wheat (*Triticum aestivum* L.) under rainfed Mediterranean conditions. *Agron. J.* 95:1253–1261.

Lopez Pereira, M., N. Trapani, and V. O. Sadras. 2000. Genetic improvement of sunflower in Argentina between 1930 to 1995: Part III. Dry matter partitioning and grain composition. *Field Crops Res.* 67:215–221.

Loss, S. P. and K. H. M. Siddique. 1997. Adaptation of faba bean *(Vicia faba* L.) to dryland Mediterranean type environments. I. Seed yield and yield components. *Field Crops Res.* 52:17–28.

Ludlow, M. M. and R. C. Muchlow. 1990. A critical evaluation of traits for improving crop yields in water-limited environments. *Adv. Agron.* 43:107–153.

Lugg, D. G. and T. R. Sinclair. 1981. Seasonal changes in photosynthesis of field grown soybean leaflets. 2. Relation to nitrogen content. *Photosynthetica* 15:138–144.

Lynam, J. K. S., S. M. Nandwa, and E. M. A. Smaling. 1998. Editorial. *Agric. Ecosyst. Environ.* 71:1–4.

Maddonni, G. A. and M. E. Otegui. 1996. Leaf area, light interception, and crop development in maize. *Field Crops Res.* 48:81–87.

Maddonni, G. A. and M. E. Otegui. 2004. Intra-specific competition in maize: Contribution of extreme plant hierarchies amount plants affects final kernel set. *Field Crops Res.* 85:1–13.

Maddonni, G. A. and M. E. Otegui. 2006. Intra-specific competition in maize: Contribution of extreme plant hierarchies to grain yield, grain yield components and kernel composition. *Field Crops Res.* 97:155–166.

Mahon, J. 1983. Limitations to the use of physiological variability in plant breeding. *Can. J. Plant Sci.* 63:11–21.

Maizlish, N. A., D. D. Fritton, and W. A. Kendall. 1980. Root morphology and early development of maize at varying levels of nitrogen. *Agron. J.* 72:25–30.

Malamy, J. E. 2005. Intrinsic and environmental response pathways that regulate root system architecture. *Plant Cell Environ.* 28:67–77.

Malamy, J. E. and K. S. Ryan. 2001. Environmental regulation of lateral root initiation in Arabidopsis. *Plant Physiol.* 127:899–909.

Malik, A. I. and Z. Rengel. 2013. Physiology of nitrogen use efficiency. In: *Improving Water and Nutrient Use Efficiency in Food Production Systems*, ed., Z. Rengel, pp. 105–121, Ames, Iowa: John Wiley & Sons.

Mallory, E. B. and H. Darby. 2013. In season nitrogen effects on organic hard red winter wheat yield and quality. *Agron. J.* 105:1167–1175.

Maman, N., S. C. Mason, T. Galusha, and M. D. Clegg. 1999. Hybrid and nitrogen influence on pearl millet production in Nebraska: Yield, growth, and nitrogen uptake, and nitrogen use efficiency. *Agron. J.* 91:737–743.

Marschner, H. 1995. *Mineral Nutrition of Higher Plants*, 2nd edition. New York: Academic Press.

Marschner, H., E. A. Kirkby, and I. Cakmak. 1996. Effect of mineral nutritional status on shoot-root partitioning of photoassimilates and cycling of mineral nutrients. *J. Exp. Bot.* 47:1255–1263.

Masle, J. 1985. Competition among tillers in winter wheat: Consequences for growth and development of the crop. In: *Wheat Growth and Modeling*, eds., W. Day and R. K. Atkin, pp. 33–54. New York: Plenum Press.

Martin, T., O. Oswald, and I. A. Graham. 2002. Arabidopsis seedling growth, storage lipid mobilization, and photosynthetic gene expression are regulated by carbon: Nitrogen availability. *Plant Physiol.* 128:472–481.

Mayers, J. D., R. J. Lawn, and D. E. Byth. 1991. Adaptation of soybean (*Glycine max* L. Merrill) to dry season of the tropics. I. Genotype and environment effects on phenology. *Aust. J. Agric. Res.* 42:497–515.

McConkey, B. G., C. A. Campbell, R. P. Zentner, F. B. Dyck, and F. Selles. 1996. Long-term tillage effects on spring wheat production on three soil textures in the brown soil zone. *Can. J. Plant Sci.* 76:747–756.

McEwan, J. M. and R. J. Cross. 1979. Evolutionary changes in New Zealand wheat cultivars. In: *Proceedings of International Wheat Symposium, 5th,* 23–28 February 1978, ed., S. Ramanujam, pp. 198–203. New Delhi: Indian Society of Genetics and Plant Breeding.

McKenzie, R. H., A. B. Middleton, P. G. Pfiffner, and E. Bremer. 2010. Evaluation of polymer coated urea and urea inhibitor for winter wheat in southern Alberta. *Agron. J.* 102:1210–1216.

McNeal, F. N., M. A. Berger, P. l. Brown, and C. F. McGuire. 1971. Productivity and quality response of five spring wheat genotypes, *Triticum aestivum* L., to nitrogen fertilizers. *Agron. J.* 63:908–910.

Mengel, K., E. A. Kirkby, H. Kosegarten, and T. Appel. 2001. *Principles of Plant Nutrition*, 5th edition. Dordrecht: Kluwer Academic Publishers.

Milroy, S. P. and M. P. Bange. 2003. Nitrogen and light responses of cotton photosynthesis and implications for crop growth. *Crop Sci.* 43:904–913.

Miller, E. C. 1939. A physiological study of the winter wheat plant at different stages of its development. Kansas Agric. Exp. Station Tech. Bulletin 47.

Miller, P. R., Y. Gan, B. G. McConkey, and C. L. McDonald. 2003. Pulse crops for the northern Great Plains: I. Grain productivity and residual effects on soil water and nitrogen. *Agron. J.* 95:972–979.

Mitchell, P. I., J. E. Sheehy, and F. I. Woodward. 1998. Potential yield and the efficiency of radiation use in rice. IRRI discussion paper series No. 32. Manila, the Philippines, IRRI.

Monteith, J. L. 1977. Climate and the efficiency of crop production in Britain. *Phil. Trans. R. Soc.*, London Series B 281:277–294.

Moragues, M., L. F. Garcia del Moral, M. Moralejo, and C. Royo. 2006. Yield formation strategies of durum wheat landraces with distinct pattern of dispersal within the Mediterranean basin I: Yield components. *Field Crops Res.* 95:194–205.

Morais, O. P. 1998. Annual report of the project, "Breeding upland rice cultivars". National Rice and Bean Research Center of Embrapa, Goiania, Brazil.

Morris, M., V. A. Kelly, R. J. Kopicki, and D. Byerlee. 2007. *Fertilizer Use in African Agriculture: Lessons Learned and Good Practice Guide.* Washington DC: The World Bank.

Morrison, M. J., E. R. Cober, M. F. Saleem, N. B. Mclaughlin, J. Fregeau-Reid, B. L. Ma, W. Yan, and L. Woodrow. 2008. Changes in isoflavone concentration with 58 years of genetic improvement in short season soybean cultivars in Canada. *Crop Sci.* 48:2201–2208.

Morrison, M. J., H. D. Voldeng, and E. R. Cober. 1999. Physiological changes from 58 years of genetic improvement of short-season soybean cultivars in Canada. *Agron. J.* 91:685–689.

Morrison, M. J., H. D. Voldeng, and E. R. Cober. 2000. Agronomic changes from 58 years of genetic improvement of short-season soybean cultivars in Canada. *Agron. J.* 92:780–784.

Mozingo, R. W., T. A. Coffelt, and J. C. Wynne. 1987. Genetic improvement in large seeded Virginia type peanut cultivars since 1944. *Crop Sci.* 27:228–231.

Muchena, F., D. Onduru, G. Gachini, and A. D. Jager. 2005. Turning the tides of soil degradation in Africa: Capturing reality and exploring opportunities. *Land Use Policy* 22:23–31.

Muchow, R. C. 1985. Canopy development in grain legumes grown under different soil water regimes in a semi-arid environment. *Field Crops Res.* 11:99–109.

Muchow, R. C. and T. R. Sinclair. 1994. Nitrogen response of leaf photosynthesis and canopy radiation use efficiency in field-grown maize and sorghum. *Crop Sci.* 34:721–727.

Murata, Y. 1961. Studies on the photosynthesis of rice plant and culture significance. *Bull. Natl. Inst. Agric. Sci. (Tokyo)*, D,9:1–169.

Murata, 1969. Physiological responses to nitrogen in plants. In: *Physiological Aspects of Crop Yield*, eds., J. D. Eastin, F. A. Haskins, C. Y. Sullivan, and C. H. M. Van Bavel, pp. 235–263. Madison, Wisconsin: ASA.

Murata, Y. and S. Matsushima. 1975. Rice. In: *Crop Physiology: Some Case Histories*, ed., L. T. Evans, pp. 73–99. London: Cambridge University Press.

Murayama, N. 1995. Development and senescence of an individual plant. In: *Science of the Rice Plant: Physiology*, Vol 2, eds., T. Matsuo, K. Kumazawa, R. Ishii, K. Ishihara, and H. Hirata, pp. 119–178. Tokyo: Food and Agriculture Policy Research Center.

Murrell, T. S. and F. R. Childs. 2000. Redefining corn yield potential. *Better Crops* 84:33–37.

Mwanamwenge, J., S. P. Loss, K. H. M. Siddique, and P. S. Cocks. 1998. Growth, seed yield and water use of faba bean (*Vicia faba* L.) in a short season Mediterranean type environment. *Aust. J. Exp. Agric.* 38:171–180.

Nass, H. G., E. D. Caldwell, D. F. Walker, M. Price, and J. B. Sanderson. 2003. A protocol for spring milling wheat production in the Maritimes. *Can. J. Plant Sci.* 83:715–723.

Nass, H. G., Y. Papadopolous, J. A. MacLeod, C. D. Caldwell, and D. F. Walker. 2002. Nitrogen management of spring milling wheat underseeded with red clover. *Can. J. Plant Sci.* 82:653–659.

Natr, L. 1972. Influence of mineral nutrients on photosynthesis of higher plants. *Photosynthetica* 6:80–99.

Nerson, H. 1980. Effects of population density and number of ears on wheat yield and its components. *Field Crops Res.* 3:225–234.

Nevins, D. J. and R. S. Loomis. 1970. Nitrogen nutrition and photosynthesis in sugar beet (*Beta vulgaris* L.). *Crop Sci.* 10:21–25.

Nguyen, H., J. J. Schoenau, D. Nguyen, K. V. Rees, and M. Boehm. 2002. Effects of long term nitrogen, phosphorus, and potassium fertilization on cassava yield and plant nutrient composition in North Vietnam. *J. Plant Nutr.* 25:425–442.

Novoa, R. and R. S. Loomis. 1981. Nitrogen and plant production. *Plant Soil* 58:177–204.

Novacek, M. J., S. C. Mason, T. D. Galusha, and M. Yaseen. 2013. Twin rows minimally impact irrigated maize yield, morphology and lodging. *Agron. J.* 105:268–276.

Oberle, S. 1994. Farming systems options for U.S. agriculture: An agroecological perspective. *J. Prod. Agric.* 7:119–123.

O'Neill, P. M., J. F. Shanahan, J. S. Schepers, and B. Caldwell. 2004. Agronomic responses of corn hybrids from different eras to deficit and adequate levels of water and nitrogen. *Agron. J.* 96:1660–1667.

Ortiz, R., R. M. Trethowan, G. Ortiz Ferrara, M. Iwanaga, J. H. Dodds, J. H. Crouch, J. Crossa, and H. J. Braun. 2007. High yield potential, shuttle breeding and a new international wheat improvement strategy. *Euphytica* 157:365–384.

Osman, A. M. and F. L. Milthorpe. 1971. Photosynthesis of wheat leaves in relation to age, illumination and nutrient supply. II. Results. *Photosynthetica* 5:61–70.

Pandey, R. K., J. W. Maranville, and Y. Bako. 2001. Nitrogen fertilizer response and efficiency for three cereal crops in Niger. *Commun. Soil Sci. Plant Anal.* 32:1465–1482.

Park, S. J. and B. R. Buttery. 1989. Identification and characterization of common bean (*Phaseolus vulgaris* L.) lines well modulated in the presence of high nitrate. *Plant Soil* 119:237–244.

Passioura, J. B. 1977. Grain yield, harvest index, and water use of wheat. *J. Aust. Inst. Agric. Sci.* 43:117–120.

Payne, T. S., D. D. Stuthman, R. L. McGraw, and P. P. Bregitzer. 1986. Physiological changes associated with three cycles of recurrent selection for grain yield improvement in oats. *Crop Sci.* 26:734–736.

Pearman, I. S. M. Thomas, and G. N. Thorne. 1977. Effects of nitrogen fertilizer on growth and yield of spring wheat. *Ann. Bot.* 41:93–108.

Pedersen, P. and J. G. Lauer. 2004. Soybean growth and development response to rotation sequence and tillage system. *Agron. J.* 96:1005–1012.

Pelegrin, R., F. M. Mercante, I. M. N. Otsubo, and A. k. Otsubo. 2009. Response of common bean crop to nitrogen fertilization and inoculation in Mato Grosso do Sul state of Brazil. *R. Bras. Ci. Solo* 33:219–226.

Peltonen, J. and A. Virtanen. 1994. Effect of nitrogen fertilizers differing in release characteristics on the quality of storage proteins in wheat. *Cereal Chem.* 71:1–5.

Peng, S. and D. R. Krieg. 1991. Single leaf and canopy photosynthesis response to plant age in cotton. *Agron. J.* 83:704–708.

Peng, S., Laza, R. C., R. M. Visperas, A. L. Sanico, K. G. Cassman, and G. S. Khush. 2000. Grain yield of rice cultivars and lines developed in the Philippines since 1966. *Crop Sci.* 40:307–314.

Peng, S., G. S. Khush, P. Virk, Q. Tang, and Y. Zou. 2008. Progress in idiotype breeding to increase rice yield potenti. *Field Crops Res.* 108:32–38.

Phillips, S. and R. Norton. 2012. Global wheat production and fertilizer use. *Better Crops Plant Food* 96:4–6.

Pons, T. L., A. V. D. Werf, and H. Lambers. 1994. Photosynthetic nitrogen use efficiency of inherently slow and fast growing species: Possible explanations for observed differences. In: *A Whole-Plant Perspective of Carbon–Nitrogen Interactions*, eds., J. Roy and E. Garnier, pp. 61–78. The Hahue: SPB Academic Publishing.

Poorter, H. and J. R. Evans. 1998. Photosynthetic nitrogen use efficiency of species differ inherently in specific leaf area. *Oecologia* 116:26–37.

Poorter, H., C. Remkes, and H. Lambers. 1990. Carbon and nitrogen economy of 24 wild species differing in relative growth rate. *Plant Physiol.* 67:223–226.

Porter, P. M., D. R. Huggins, C. A. Perillo, S. R. Quiring, and R. K. Crookston. 2003. Organic and other management strategies with two and four year crop rotations in Minnesota. *Agron. J.* 95:233–244.

Power, J. F. and J. Alessi. 1978. Tiller development and yield of standard and semi-dwarf spring wheat varieties as affected by nitrogen fertilizer. *J. Agric. Sci.* 90:97–108.

Pushman, F. M. and J. Bingham. 1976. The effect of a granular nitrogen fertilizer and a foliar spray urea on the yield and breadmaking quality of winter wheats. *J. Agric. Sci.* 87:281–292.

Qin, R., P. Stamp, and W. Richner. 2005. Impact of tillage and banded starter fertilizer on maize root growth in the top 25 centimeters of the soil. *Agron. J.* 97:674–683.

Radin, J. W. and J. Lynch. 1994. Nutritional limitations to yield: Alternative to fertilization. In: *Physiology and Determination of Crop Yield*, eds., K. J. Boote, J. M. Bennett, T. R. Sinclir, and G. M. Paulsen, pp. 277–283. Madison, Wisconsin: ASA, CSSA, and SSSA.

Rahman, M. S. 1984. Breaking the yield barriers in cereals with special reference to rice. *J. Aust. Institute. Agric. Sci.* 504: 228–232.

Randall, P. J., J. R. Freney, C. J. Smith, H. J. Moss, C. W. Wrigley, and I. E. Galbally. 1990. Effect of additions of nitrogen and sulfur to irrigated wheat at heading on grain yield, composition and milling and baking quality. *Aust. J. Exp. Agric.* 30:95–101.

Rao, M. S. S. and A. S. Bhagsari. 1998. Variation between and within maturity groups of soybean genotypes for biomass, seed yield, and harvest index. *Soybean Genet. Newsl.* 25:103–106.

Rao, M. S. S., B. G. Mullinix, M. Rangappa, E. Cebert, A. S. Bhagsari, V. T. Sapra, J. M. Joshi, and R. B. Dadson. 2002. Genotype X environment interactions and yield stability of food-grade soybean genotypes. *Agron. J.* 94:72–80.

Ramirez-Oliveras, G., C. A. Stutte, and E. Orengo-Santiago. 1997. Hydrogen ion efflux differences in soybean roots associated with yields. *J. Agric. Univ. Puerto Rica* 81:159–180.

Raun, W. R. and G. V. Johnson. 1999. Improving nitrogen use efficiency for cereal production. *Agron. J.* 91:357–304.

Rawson, H. M. and C. Hackett. 1974. An exploration of the carbon economy of the tobacco plant. III. Gas exchange of leaves in relation to position on the stem, ontogeny and nitrogen content. *Aust. J. Plant Physiol.* 1:551–560.

Reich, P. B., M. B. Walters, D. S. Ellsworth, and C. Uhl. 1994. Photosynthesis-nitrogen relations in Amazonian tree species. I. Patterns among species and communities. *Oecologia* 97:62–72.

Reilly, J. M. and K. O. Fuglie. 1998. Future yield growth in field crops: What evidence exists? *Soil Till. Res.* 47:275–290.

Reta-Sanchez D. G. and J. L. Fowler. 2002. Canopy light environment and yield of narrow row cotton as affected by canopy architecture. *Agron. J.* 94:1317–1323.

Reynolds, M. P., M. Balota, M. I. B. Delgado, I. Amani, and R. A. Fischer. 1994. Physiological and morphological traits associated with spring wheat yield under hot, irrigated conditions. *Aust. J. Plant Physiol.* 21:717–730.

Reynolds, M. P., S. Rajaram, and K. D. Sayre. 1999. Physiological and genetic changes of irrigated wheat in the post green revolution period and approaches for meeting projected global demand. *Crop Sci.* 39:1611–1621.

Rice, R. W. 2007. The physiological role of minerals in the plant. In: *Mineral Nutrition and Plant Disease*, eds., L. E. Datnoff, W. H. Elmer, and D. M. Huber, pp. 9–29. St. Paul, Minnesota: The American Phytopathological Society.

Riggs, T. J., P. R. Hanson, N. D. Start, D. M. Niles, C. L. Morgan, and M. A. Ford. 1981. Comparison of spring barley varieties grown in England and Wales between 1880 and 1980. *J. Agric. Sci.* 97:599–610.

Roberts, S. R., J. E. Hills, D. M. Brandon, B. C. Miller, S. C. Scardaci, C. M. Wick, and J. F. Williams. 1993. Biological yield and harvest index in rice: Nitrogen response of tall and semidwarf cultivars. *J. Prod. Agric.* 6:585–588.

Robinson, D., D. J. Linehan, and D. C. Gordon. 1994. Capture of nitrate from soil by wheat in relation to root length, nitrogen inflow, and availability. *New Phytol.* 128:297–305.

Robles, M., I. A. Ciampitti, and T. J. Vyn. 2013. Response of maize hybrids to twin-row spatial arrangement at multiple plant densities. *Agron. J.* 104:1747–1756.

Roth, J. A., I. A. Ciampitti, and T. J. Vyn. 2013. Physiological evaluations of recent drought tolerant maize hybrids at varying stress levels. *Agron. J.* 105:1129–1141.

Rowden, R., D. Gardiner, P. C. Whiteman, and E. S. Wallis. 1981. Effects of planting density on growth, light interception, and yield of photoperiod insensitive *Cajanus cajan*. *Field Crops Res.* 4:201–213.

Russell, R. S. 1977. *Plant Root Systems*. New York: McGraw-Hill.

Ryle, C. J. A. and J. D. Hesketh. 1969. Carbon dioxide in nitrogen deficient plants. *Crop Sci.* 9:451–454.

Sainju, U. M. 2013. Tillage, cropping sequence, and nitrogen fertilization influence dryland soil nitrogen. *Agron. J.* 105:1253–1263.

Sainju, U. M., B. P. Singh, and W. F. Whitehead. 2001. Comparison of the effects of cover crops and nitrogen fertilization on tomato yield, root growth, and soil properties. *Sci. Hortic.* 91:201–214.

Salman, O. A. 1989. Polyethylene-coated urea. I. Improved storage and handling. *Ind. Eng. Chem. Res.* 28:630–632.

Sanchez, P. A. and J. G. Salinas. 1981. Low-input technology for managing Oxisols and Ultisols in tropical America. *Adv. Agron.* 34:279–406.

Sands, P. J. 1996. Modeling canopy production. III. Canopy light-utilization efficiency and its sensitivity to physiological and environmental variables. *Aust. J. Plant Physiol.* 23:103–114.

Schmidt, J. W. 1984. Genetic contributions to yield gains in wheat. In: *Genetic Contributions to Yield Gains of Five Major Crop Plants,* ed., W. R. Fehr, pp. 89–101. Madison, Wisconsin: CSSA.

Schrader, L. E. 1984. Functions and transformations of nitrogen in higher plants. In: *Nitrogen in Crop Production*, ed., R. D. Hauck, pp. 55–66. Madison, Wisconsin: ASA, CSSA, and SSSA.

Sedgley, R. H. 1991. An appraisal of the Donald ideotype after 21 years. *Field Crops Res.* 26:93–112.

Shangguan, Z. P., M. A. Shao, and J. Dyckmans. 2000. Nitrogen nutrition and water effects on leaf photosynthetic gas exchange and water use efficiency in winter wheat. *Environ. Exp. Bot.* 44:141–149.

Sharma, R. C. and E. L. Smith. 1986. Selection for high and low harvest index in three winter wheat populations. *Crop Sci.* 26:1147–1150.

Sharma, R. C., J. Crossa, G. Velu, J. Huerta-Espino, M. Vargas, T. S. Payne, and R. P. Singh. 2012. Genetic gains for grain yield in CIMMYT spring bred wheat across international environments. *Crop Sci.* 52:1522–1533.

Shaviv, A. 2001. Advances in controlled release fertilizers. *Adv. Agron.* 71:2–41.

Shopf, J. W. 1993. Microfossils of the early archean apex chert: New evidence of the antiquity of life: *Science* 260:640–646.

Siddique, K. H. M., R. K. Belford, M. W. Perry, and D. Tennant. 1989. Growth, development and light interception of old and modern wheat cultivars in a Mediterranean type environment. *Aust. J. Agric. Res.* 40:473–487.

Siegenthaler, V. L., J. E. Stepanich, and L. W. Briggle. 1986. Distribution of the varieties and classes of wheat in the United States, 1984. USDA Sta. Bull. 739. Washington DC: U.S. Gov. Print. Office.

Simane, B., P. C. Struik, M. Nachit, and J. M. Peacock. 1993. Ontogenic analysis of yield components and yield stability of durum wheat in water limited environments. *Euphytica* 71:211–219.

Sinclair, T. R. and T. Horie. 1989. Leaf nitrogen, photosynthesis, and crop radiation use efficiency. A review. *Crop Sci.* 29:90–98.

Sinclair, T. R. and R. C. Muchow. 1999. Radiation use efficiency. *Adv. Agron.* 65:215–265.

Singh, R. P., D. P. Hodson, J. Huerra-Espino, Y. Jin, P. Njau, R. Wanyera, S. A. Herrera-Foessel, and R. W. Ward. 2008. Will stem rust destroy the worlds wheat crop? *Adv. Agron.* 98:271–309.

Singh, U., J. K. Ladha, E. G. Castillo, G. Punzalan, A. Tirol-Padre, and M. Duqueza. 1998. Genotypic variation in nitrogen use efficiency in medium and long duration rice. *Field Crops Res.* 58:35–53.

Slafer, G. A., E. H. Satorre, and F. H. Andrade. 1994. Increase in grain yield in bread wheat from breeding and associated physiological changes. In: *Genetic Improvement in Field Crops*, ed., G. A. Slafer, pp. 1–68. New York: Marcel Dekker.

Slafer, G. A., D. F. Calderini, and D. J. Miralles. 1996. Generation of yield components and compensation in wheat: Opportunities for further increasing yield potential. In: *Increasing Yield Potential in Wheat: Breaking the Barriers*, ed., M. Reynolds, pp. 101–133. Mexico: CIMMYT.

Snyder, F. W. and G. E. Carlson. 1984. Selecting for partitioning of photosynthetic products in crops. *Adv. Agron.* 37:47–72.

Soil Science Society of America. 2008. *Glossary of Soil Science Terms*. Madison, Wisconsin: SSSA.

Song, X. F., W. Agata, and Y. Kawamitsu. 1990. Studies on dry matter and grain production of F_1 hybrid rice in China. I. Characteristics of dry matter production. *Jpn. J. Crop Sci.* 59:19–28.

Sowers, K. E., W. L. Pan, B. C. Miller, and J. L. Smith. 1994. Nitrogen use efficiency of split nitrogen applications in soft white winter wheat. *Agron. J.* 86:942–948.

Specht, J. E., D. J. Hume, and S. V. Kumudini. 1999. Soybean yield potential—A genetic and physiological perspective. *Crop Sci.* 39:1560–1570.

Spiertz, J. H. J. and J. Ellen. 1978. Effects of nitrogen on crop development and grain growth of winter wheat in relationship to assimilation and utilization of assimilates and nutrients. *Neth. J. Agric. Sci.* 25:182–197.

Spratt, E. D. and J. K. R. Gasser. 1970. Effects of fertilizer nitrogen and water supply on the distribution of dry matter and nitrogen between the different parts of wheat. *Can. J. Plant Sci.* 50:613–625.

Stewart, W. M., D. W. Bibb, A. E. Johnston, and J. T. Smyth. 2005. The contribution of commercial fertilizer nutrients to food production. *Agron. J.* 97:1–6.

Sylvester-Bradley, R., J. Foulkes, and M. Reynolds. 2005. Future wheat yield; Evidence, theory and conjecture. In: *Proceedings of the 61st Easter School in Agricultural Science*, ed., Nottingham University, pp. 233–260. Nottingham: Nottingham University Press.

Takano, Y. and S. Tsunoda. 1971. Curvilinear regression of the leaf photosynthetic rate on leaf nitrogen content among strains of *Oryza* species. *Jpn. J. Breed.* 21:69–76.

Tanaka, T. and S. Matsushima. 1963. Analysis of yield-determining process and the application to yield prediction and culture improvement of lowland rice. *Proc. of Crop Sci. Soc. Jap.* 32:35–38.

Tanaka, A. and M. Osaki. 1983. Growth and behavior of photosynthesized ^{14}C in various crops in relation to productivity. *Soil Sci. Plant Nutr.* 29:147–158.

Teasdale, J. R. 1995. Influence of narrow row/high population corn on weed control and light transmittance. *Weed Technol.* 9:113–118.

Teasdale, J. R. 1998. Influence of corn population and row spacing on corn and velvetleaf yield. *Weed Sci.* 46:447-453.

Thomas, S. M., G. N. Thorne, and I. Pearman. 1978. Effect of nitrogen on growth, yield and photorespiratory activity in spring wheat. *Ann. Bot.* 42:827–837.

Thomson, B. D., K. H. M. Siddique, M. D. Barr, and J. M. Wilson. 1997. Grain legume species in low rainfall Mediterranean-type environments: I. Phenology and seed yield. *Field Crops Res.* 54:173–187.

Thung, M. and I. M. Rao. 1999. Integrated management of abiotic stresses. In: *Common Bean Improvement in the Twenty First Century*, ed., S. P. Singh, pp. 331–370. Dordrecht: Kluwer Academic Publishers.

Tilman, D., C. Balzer, J. Hill, and B. L. Befort. 2011. Global food demand and the sustainable intensification of agriculture. *Proc. Natl. Acad. Sci.* 108:20260–20264.

Tokatlidis, I. S. and S. D. Koutrouba. 2004. A review study of maize hybrids dependence on high plant populations and its implications for crop yield stability. *Field Crops Res.* 88:103–114.

Tolbert, N. E. 1997. The C_2 oxidative photosynthetic carbon cycle. *Annu. Rev. Plant Physiol. Plant Mol. Biol.* 48:1–25.

Tollenaar, M. 1983. Potential vegetative productivity in Canada. *Can. J. Plant Sci.* 63:1–10.

Tollenaar, M. 1989. Genetic improvement in grain yield of commercial maize hybrids grown in Ontario from 1959 to 1988. *Crop Sci.* 29:1365–1371.

Tollenaar, M. and A Aguilera. 1992. Radiation use efficiency of an old and a new maize hybrid. *Agron. J.* 84:536–541.

Tollenaar, M., A. Aguilera, and S. P. Nissanka. 1997. Grain yield is reduced more by weed interference in an old than in a new maize hybrid. *Agron. J.* 89:239–246.

Tollenaar, M., L. M. Dwyer, and D. W. Stewart. 1992. Ear and kernel formation in maize hybrids representing three decades of grain yield improvement in Ontario. *Crop Sci.* 32:432–438.

Tollenaar, M. and E. A. Lee. 2002. Yield potential, yield stability and stress tolerance in maize. *Field Crops Res.* 75:161–169.

Trapani, N., A. J. Hall, V. O. Sadras, and F. Vilella. 1992. Ontogenic changes in radiation use efficiency of sunflower. *Field Crops Res.* 29:301–316.

Uchida, N., Y. Wada, and Y. Murata. 1982. Studies on the changes in the photosynthetic activity of a crop leaf during its development and senescence. II. Effect of nitrogen deficiency on the changes in the senescing leaf of rice. *Jpn. J. Crop Sci.* 51:577–583.

Uhart, S. A. and F. H. Andrade. 1995. Nitrogen deficiency in maize: I. Effects on crop growth, development, dry matter partitioning, and kernel set. *Crop Sci.* 35:1376–1383.

Unkovich, M., J. Baldock, and M. Forbes. 2010. Variability in harvest index of grain crops and potential significance for carbon accounting: Examples from Australian agriculture. *Adv. Agron.* 105:173–219.

USDA. 2009. Agricultural prices 2008 summary. Pr 1–3(09). National Agricultural Statistics service, Washington DC. http://usda.mannlib.cornell.edu/usda/nass/agriPriceSu//2000s/2009/AgricPricSu-08–05–2009.pdf (accessed 15 November 2009).

USDA-NASS. National agricultural statistics service. http://www.nass.usda.gov/Dataand statistics/index.asp. (accessed 18 September 2012).

Ustun, A., F. L. Allen, and B. C. English. 2001. Genetic progress in soybean of the US Midsouth. *Crop Sci.* 41:993–998.

Vandenberg, A. and T. Nleya. 1999. Breeding to improve plant type. In: *Common Bean Improvement in the Twenty-First Century*, ed., S. P. Singh, pp. 167–183. Kluwer Academic Publishers, Dordrecht, The Netherlands.

Vasilakoglou, I., K. Dhima, N. Karagiannidis, T. Gatsis, and K. Petrotos. 2012. Competitive ability and phytotoxic potential of four winter canola hybrids as affected by nitrogen supply. *Crop Sci.* 50:1011–1021.

Villamil, M. B., V. M. Davis, and E. D. Nafziger. 2012. Estimating factor contributions to soybean yield from farm field data. *Agron. J.* 104:881–887.

Vitousek, P. M., R. Naylor, and T. Crews. 2009. Nutrient imbalances in agricultural development. *Science* 324:1519–1520.

Vlek, P. L. G. 1993. Strategies for sustaining agriculture in sub-Saharan Africa. In: *Technologies for Sustaining Agriculture in the Tropics*, eds., J. Rogland and R. Lal, pp. 265–277. Madison, Wisconsin: ASA, CSSA, and SSSA.

Waddington, S.R., J.K. Ransom, M. Osmanzai, and D.A. Saunders. 1986. Improvement in the yield potential of bred wheat adapted to northwest Mexico. *Crop Sci.* 26:698-703.

Wall, J. S. 1979. The role of wheat proteins in determining baking quality. In: *Recent Advances in the Biochemistry of Cereals*, eds., D. L. Laidman and R. G. Wyn, pp. 275–311. Lindon: New York Academy.

Wang, Q. A., C. M. Lu, and Q. D. Zhang. 2005. Midday photoinhibition of two newly developed super rice hybrids. *Photosynthetica* 43:277–281.

Wang, Q., Q. D. Zhang, D. Y. Fan, and C. M. Lu. 2006. Photosynthetic light and CO_2 utilization and C_4 traits of two novel super rice hybrids. *J. Plant Physiol.* 163:529–537.

Wang, X., L. X. Tao, M. Y. Yu, and X. L. Huang. 2002. Physiological model of super hybrid rice variety Xieyou9308. *Chin. J. Rice Sci.* 16:38–44.

Watanabe, H. and S. Yoshida. 1970. Effects of nitrogen, phosphorus and potassium on photophosphorylation in rice in relation to the photosynthetic rate of single leave. *Soil Sci. Plant Nutr.* 16:163–166.

Watson, D. J., G. N. Thorne, and S. A. W. French. 1958. Physiological causes of differences in grain yield in varieties of barley. *Ann. Bot.* 22:321–352.

Wells, R. and W. R. Meredith, Jr. 1984. Comparative growth of obsolete and modern cotton cultivars. III. Relationship of yield to observed growth characteristics. *Crop Sci.* 24:868–872.

Wells, R., W. R. Meredith, Jr., and J. R. Williford. 1986. Canopy photosynthesis and its relationship to plant productivity in near-isogenic cotton lines differing in leaf morphology. *Plant Physiol.* 82:635–640.

Werf, A. V. D. 1996. Growth analysis and photoassimilate partitioning. In: *Photoassimilate Distribution in Plants and Crops*, eds., E. Zamski and A. A. Schaffer, pp. 1–20. New York: Marcel Dekker.

Westgate, M. E., F. Forcella, D. C. Reicosky, and J. Somsen. 1997. Rapid canopy closure for maize production in the northern US corn belt: Radiation use efficiency and grain yield. *Field Crops Res.* 49:249–258.

Weston, L. A. and B. H. Zandstra. 1989. Transplant age and N and P nutrition effects on growth and yield of tomatoes. *HortScience* 24:88–90.

Wilcox, J. R. 2001. Sixty years of improvement in publicly developed elite soybean lines. *Crop Sci.* 41:1711–1716.

Willmot, D. B., G. E. Pepper, and D. Nafzier. 1989. Random stand deficiency and replanting delay effects on soybean yield, yield components, canopy and morphological responses. *Agron. J.* 81:425–430.

Wilson, D. M., E. A. Heaton, M. Liebman, and K. J. Moore. 2013. Intraseasonal changes in switchgrass nitrogen distribution compared with corn. *Agron. J.* 105:285–294.

Wingler, A., S. Purdy, J. A. MacLean, and N. Pourtau. 2006. The role of sugars in integrating environmental signals during the regulation of leaf senescence. *J. Exp. Bot.* 57:391–399.

Winter, S. R. and P. W. Unger. 2001. Irrigated wheat grazing and tillage effects on subsequent dry land grain sorghum production. *Agron. J.* 93:504–510.

Wooding, A. R., S. Kavale, F. MacRitchie, F. L. Stoddard, and A. Wallace. 2000. Effects of nitrogen and sulfur fertilizer on protein composition, mixing requirements, and dough strength of four wheat cultivars. *Cereal Chem.* 77:798–807.

Wong, S. C. 1979. Elevated atmospheric partial pressure of CO_2 and plant growth. I. Interactions of nitrogen nutrition and photosynthetic capacity in C_3 and C_4 plants. *Oecologia* 44:68–74.

Wong, S. C, I. R. Cowan, and G. D. Farquhar. 1985. Leaf conductance in relation to rate of CO_2 assimilation: I. Influence of nitrogen nutrition, phosphorus nutrition, photon flux density, and ambient partial pressure of CO_2 during ontogeny. *Plant Physiol.* 78:821–825.

Wu, X. 2009. Prospects of developing hybrid rice with super high yield. *Agron. J.* 101:688–695.

Wu, W. G., H. C. Zhang, G. C. Wu, C. Q. Zhai, Y. F. Qian, and Y. Chen. 2007. Preliminary study on super rice population sink characters. *Sci. Agric. Sin.* 40:250–257.

Wych, R. D. and D. C. Rasmusson. 1983. Genetic improvement in malting barley cultivars since 1920. *Crop Sci.* 23:1037–1040.

Wych, R. D. and D. D. Stuthman. 1983. Genetic improvement in Minnesota adapted oat cultivars since 1923. *Crop Sci.* 23:879–881.

Wynne, J. C. and W. C. Gregory. 1981. Peanut breeding. *Adv. Agron.* 34:39–217.

Xia, Y. 2012. Photosynthesis-related physiological responses of field grown maize to plant density and nitrogen stress during vegetative and reproductive stages. Ph.D. dissertation. Purdue University, West Lafayette, Indiana, USA.

Xiong, J., C. Q. Ding, G. B. Wei, Y. F. Ding, and S. H. Wang. 2013. Characteristics of dry matter accumulation and nitrogen uptake of super high yielding early rice in China. *Agron. J.* 105:1142–1150.

Xu, Z., W. Chen, L. Zhang, and S. Yang. 2005. Design principles and parameters of rice ideal panicle type. *Chin. Sci. Bull.* 50:2253–2256.

Xue, N. Y., E. H. Li, and Y. Jiang. 2006. The evolution tendency of agronomic characters of soybean cultivars released in Heilongjian Province. *Soybean Sci.* 25:445–449.

Yamauchi, M. 1994. Physiological bases of higher yield potential in F_1 hybrids. In: *Hybrid Rice Technology: New Developments and Future Prospects*, ed., S. S. Virmani, pp. 71–80. Los Baños, the Philippines: International Rice Research Institute.

Yang, S. 1984. The primary discussion on the theories and methods of the ideotype breeding of rice. *Acta Agron. Sin.* 17:6–13.

Ying, J. F., S. B. Peng, Q. R. He, H. Yang, C. D. Yang, R. M. Visperas, and K. G. Cassman. 1998. Comparison of high yield rice in tropical and subtropical environments: I. Determinants of grain and dry matter yields. *Field Crops Res.* 57:71–84.

Yoshida, S. 1972. Physiological aspects of grain yield. *Annu. Rev. Plant Physiol.* 23:437–464.

Yoshida, S. 1981. *Fundamentals of Rice Crop Science.* Loa Bãnos, the Philippines: International Rice Research Institute.

Yoshida, S. and V. Coronel. 1976. Nitrogen nutrition, leaf resistance and leaf photosynthetic rate of the rice plant. *Soil Sci. Plant Nutr.* 22:207–211.

Yuan, L. 1997. Hybrid rice breeding for super high yield. *Hybrid Rice* 6:1–6.

Yuan, L. 1998. Hybrid rice development and use: Innovative approach and challenges. *Int. Rice Comm. Newsl.* 47:7–15.

Zhang, J. and S. A. Barber. 1992. Maize root distribution between phosphorus fertilized and unfertilized soil. *Soil Sci. Soc. Am. J.* 56:819–822.

Zhang, J. and S. A. Barber. 1993. Corn root distribution between ammonium fertilized and unfertilized soil. *Commun. Soil Sci. Plant Anal.* 24:411–419.

Zhang, F. and D. L. Smith. 1995. Preincubation of *Bradyrhizobium japinicum* with genistein accelerates nodule development of soybean at suboptimal root zone temperatures. *Plant Physiol.* 108:961–968.

Zhang, Y. B., Q. Y. Tang, Y. B. Zou, D. Q. Li, J. Q. Qin, and S. H. Yang. 2009. Yield potential and radiation use efficiency of super hybrid rice grown under subtropical conditions. *Field Crops Res.* 114:91–98.

Zinati, G. M., D. R. Christenson, and D. Harris. 2001. Spatial and temporal distribution of ^{15}N tracer and temporal pattern of N uptake from various depths by sugarbeet. *Commun. Soil Si. Plant Anal.* 32:1445–1456.

Zhou, S., J. Hou, Z. Xu, and H. Song. 2006. Analysis on seed setting rate and relationship between source and sink of erect panicle rice cultivars. *J. Shenyang Agric. Univ.* 37:141–143.

Zhou, Z., K. Robards, S. Helliwell, and C. Blanchard. 2002. Composition and functional properties of rice. *Int. J. Food Technol.* 37:849–868.

Zhu, K., D. Tang, C. Yan, Z. Chi, H. Yu, and J. Chen. 2009. Erect panicle gene encoded a novel protein that regulates panicle erectness in indica rice. *Genetics* 184:343–350.

2 Nitrogen Losses in Soil–Plant System

2.1 INTRODUCTION

Nitrogen (N) is the mineral nutrient that is most limiting for crop production around the world and is often applied as a fertilizer to maintain adequate soil levels for crop production. Geisseler et al. (2012) reported that adequate N supply is crucial to obtaining high yields in intensive crop production. While insufficient N application can have serious economic consequences for the farmer, excessive fertilization increases the risk of environmental pollution, especially groundwater pollution with NO_3^-, NH_3 volatilization, and emissions of N_2O. Even in a well-managed cereal production system, a substantial fraction (typically 40–60%) of N fertilizer inputs can be lost (Galloway et al., 2002). Dinnes et al. (2002) reported that N is essential for the growth and reproduction of all life-forms, and except for legume crops and virgin soils with relatively high soil organic matter, soil N must usually be supplemented to sustain food and fiber production.

Globally, about 100 Tg of N (100 million metric tons) is applied to farmland every year as a fertilizer (Gruber and Galloway, 2008; Kim et al., 2011). Among various N sources, urea has been the most preferred, mainly due to its effectiveness and cost, accounting for 50% of the total world consumption of N fertilizer (Vaio et al., 2008). A major part of the applied N is lost in the soil–plant system and its use efficiency in crop plants is low. N losses in a soil–plant system are among the most important mechanisms or processes responsible for the low recovery efficiency of applied chemical fertilizers. In important food crops, N recovery of applied chemical fertilizers never exceeds more than 50%. Many researchers have reported that 30–40% of the applied N is utilized by plants (Dobermann and Cassman, 2002; Cisse and Vlek, 2003; Shah et al., 2004; Fageria et al., 2011; Fageria, 2013, 2014). In fact, up to 50% of the surface-applied urea could be volatilized as NH_3 (Gioacchini et al., 2002) because urea broadcast onto the soil surface is quickly hydrolyzed into NH_3 and easily oxidized into NO_x. In spite of the obvious risk of high emissions, nearly half the amount of urea is required to be broadcast over the soil surface for a quick supply of N (Wang et al., 2008).

The most important N losses in the soil–plant system occur through the combination of ammonium volatilization, leaching of NO_3^- N from soil profile to lower depths where the roots cannot absorb it, losses through denitrification, losses through surface runoff and soil erosion, and also losses as NH_3 gas through the foliage of plants (Fageria and Baligar, 2005; Hull and Liu, 2005; Bauer et al., 2012; Sainju, 2013). These losses of applied N in cropping systems not only increase the cost of production but also increase the risk of environmental pollution by surface and groundwater contamination and the alteration of atmospheric composition. A marked loss of N applied to agricultural soils has raised concerns about the environmental impacts of N that escape from the rooting zone (Blackmer, 2000). Furthermore, a high concentration of NO_3^- N in water is associated with public health problems (Owens, 1994). High NO_3^- N concentration in water is associated with human disorders and diseases such as methemoglobinemia and cancer (Smith et al., 1990). *Methemoglobinemia,* also called blue-baby syndrome, occurs when nitrates are converted into nitrites in the guts of human infants and ruminant animals. The nitrites decrease the bloods ability to carry oxygen to the body cells. Since inadequately oxygenated blood lacks the red color, infants with the condition take on a bluish skin color (Brady and Weil, 2002). Brady and Weil (2002) reported that death from *methemoglobinemia* is quite rare; however, in many countries, there is a limit to the amount of nitrates in drinking water. In the United States, this limit is 45 mg L^{-1} nitrate or 10 mg L^{-1} N in the nitrate form (Follet and Walker, 1989; Brady and Weil, 2002). The European Economic Community (EEC) recommends 50 mg NO$_3^-$ per liter (11.5 mg N per liter) as an upper limit in potable water (Jurgens-Gschwind, 1989). In addition to NO$_3^-$ N, ammonia, which is in equilibrium with NO$_4^+$, depending on the pH, is also toxic to humans and animals (Hooda et al., 2000).

Any technology or nutrient management system that provides a strong control over nitrate formation or its direct release into the soil may reduce its adverse effects (Shaviv, 2001). It is the general feeling today that the nitrate directive in the EEC (Goodchild, 1998) and the efforts made in the United States (Livingston and Cory, 1998) are not effective enough in reducing the problems associated with nitrate. Additional efforts have been made to further improve or even increase the control over the release of nitrate into the environment (Goodchild, 1998; Joosten et al., 1998; Livingston and Cory, 1998; Wendland et al., 1998).

The loss of N from the soil–plant system to river and lake water may also create environmental problems. The N and other nutrients stimulate the growth of algae and other organisms, which sink to the bottom when they die. When these materials decompose by bacteria, other organisms use the oxygen from the water that could not sustain aquatic life. The fish and other aquatic species either die or move to other areas that are more adequately supplied with oxygen. This state of low oxygen in the water (<2–3 mg O$_2$ per liter) is known as *hypoxia*, and the process that brings it is called *eutrophication* (Brady and Weil, 2002).

Glass (2003) reported that, unfortunately, analyses of the costs of excessive fertilizer application of this sort generally fail to factor in environmental costs. In terms of dollars, these are exceedingly hard to evaluate and this difficulty is reflected in the paucity of papers dealing with this topic (Schlegel et al., 1996). In contrast, there is a large literature reporting the damaging effects on the environment associated with nutrient overloading. Many sources suggest that some forested lands in Europe and North America have been strongly impacted by N pollution and may be exhibiting symptoms of N excess (Fenn et al., 1998; Campbell et al., 2000). Perakis and Hedin (2002) reported that over 70% of the N in woodland rivers in Europe and North America is in the form of NO$_3^-$. In contrast, in these more remote South America streams in temperate forests in Chile and Argentina, NO$_3^-$ made up only 5% of the N, while 75% was in organic form. N budget calculations performed in the major subbasin of the Upper Mississippi River watershed show that fertilizer is the most important input of N to the basins (Glass, 2003).

N fertilizer consumption has increased more rapidly than that of P and K to support world food supplies (FAO, 2006; Jantalia et al., 2012). In addition, N fertilizer prices have increased significantly in the last few decades. In the United States alone, N fertilizer prices have more than doubled since the 1990s (Economic Research Service, 2010), reaching historic heights in mid-2008 due to high fertilizer demand and the inability of manufacturers to increase production levels (Huang et al., 2008). The findings of many studies (Glover et al., 2010; Dalal et al., 2011) support the contention that bringing the land under cultivation or crop production reduces soil total N content and increases N losses. In the future, the N demand for crop production will likely increase and the cost of production will also increase. Under these scenarios, improperly managed N fertilizers can form gases or soluble compounds with the potential to pollute air (NH$_3$, NO$_X$, N$_2$O, and N$_2$) or contaminate surface and groundwater with nitrate N (Aneja et al., 2003). Furthermore, it is essential to know how N losses occur in the soil–plant system to take necessary measures to reduce these losses and improve the efficiency of use of N in crop plants. Appropriate N management not only improves N use efficiency by crops but also reduces environmental pollution. The objective of this chapter is to discuss N losses mechanism or processes in the soil–plant system, factors that are responsible for these losses, and suggest management practices to reduce these losses.

2.2   CYCLE IN SOIL–PLANT SYSTEM

The knowledge of nutrient cycle in a soil–plant system is an important aspect to understanding its availability to plants and adopting management practices for maximizing its uptake and efficiency of use (Fageria and Baligar, 2005). N cycle in soil–plant system is very dynamic and complex due to the involvement of climatic, soil and plant factors, and their interactions. The N cycle is defined as the sequence of biochemical changes undergone by N, wherein it is used by a living organism, transformed upon the death and decomposition of the organism, and ultimately converted into its original oxidation state (Soil Science Society of America, 2008). A simple definition of the N cycle is interacting biological processes in the soil are termed the N cycle. Addition of N in the soils and uptake by plants is also a part of the N cycle. In agroecosystems, the N balance between the input and output determines the fate of N in these systems (Ghoshal, 2002; Singh et al., 2011). Therefore, N dynamics holds the key to designing suitable management strategies to achieve sustainable crop production.

Silgram and Shepherd (1999) reported that the N cycle in soils is largely microbially mediated, and a major component involves the transformation of organic N into the available mineral forms, primarily nitrate (NO_3-N) and ammonium (NH_4-N). The opposing process of immobilization essentially involves the conversion of mineral N into organic N by microorganisms, with the balance between mineralization and immobilization processes (net mineralization) determining the effect on the magnitude of the soil mineral N pool. The mineralization–immobilization balance is of pivotal importance as it controls the supply to and magnitude of the plant available mineral pool (Silgram and Shepherd, 1999).

One factor contributing to the low efficiency of N fertilization is the highly dynamic nature of the N cycle. A considerable part of the N available to crops may originate from the mineralization of organic materials such as soil organic matter, manure, or crop residues. Transformations from one N form into another, including mineralization, are mainly mediated by soil microorganisms, which are affected by a number of factors, including temperature, water content, oxygen availability, pH, supply of nutrients, soil texture, as well as organic matter content and quality (Robertson and Groffman, 2007; Geisseler et al., 2012). These dynamic interactions make it difficult to estimate the amount of N mineralized from organic sources and the temporal pattern of mineralization (Geisseler et al., 2012).

Figure 2.1 shows a simplified model of N cycle in a soil–plant system. The main components of N cycle are the N transformation process. In addition, it can be seen from this figure that the main N input sources of N to a soil–plant system are chemical fertilizers, organic manures or residues, biological N_2 fixation, and atmospheric N_2. Similarly, the main N depletion sources in a soil–plant system are leaching, denitrification, volatilization, surface runoff, and plant uptake. A detailed discussion of N cycle or processes in agricultural soils can be found in Stevenson (1986).

Soil organic matter plays a key role in the global C and N cycles (Curtin et al., 2012). Soils contain more than twice as much C as the atmosphere and three times the amount stored in living plants (Schlesinger, 1997). Organic N and its mineralization in soil–plant system is also part of N cycle. The mineralization of soil organic N is microbially mediated, with the rate of mineralization being strongly dependent on temperature and soil moisture. In addition, ammonified N may be fixed on the soil colloides depending on cation exchange capacity. The N immobilization also has temporary influence on N uptake to plants. The N immobilization is defined as the transformation of inorganic N compounds (NH_4^+, NO_3^-, NO_2^-, and NH_3) into the organic state. Soil organisms assimilate inorganic N compounds and transform them into organic N constituents of their cells and tissues, and the soil biomass (Jansson and Persson, 1982).

In the surface mineral soils, N content ranges from 0.2 to 5.0 g kg^{-1} with an average value of about 1.5 g kg^{-1} (Brady and Weil, 2002). More than 90% of the N in most soils is in the form of organic matter. The organic form of N protects the N from loss; however, it is also not available to crop plants. This organic form of N should be mineralized to NH_4^+ and NO_3^- before its uptake by plants. Mineral N seldom accounts for more than 1–2% of the total N in the soil (Brady and Weil, 2002). Mineralization is the conversion of an element from an organic form to an inorganic state as a result of microbial activity (Soil Science Society of America, 2008). Soil microorganisms play a key role in the mineralization of organic substances, which they decompose to obtain mineral

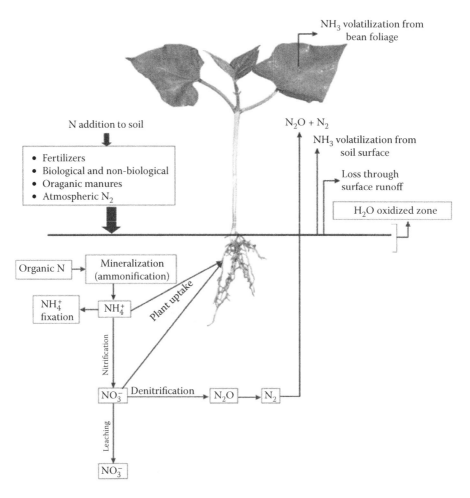

FIGURE 2.1 A simplified version of N cycle in soil–plant system. (From Fageria, N. K. 2009. *The Use of Nutrients in Crop Plants*. Boca Raton, Florida: CRC Press. With permission.)

nutrients, including N and energy. Apart from abiotic factors such as soil temperature, soil water content, and soil aeration, the properties of the organic amendments themselves affect the decomposition process or mineralization (Nett et al., 2012).

According to Stevenson (1982), mineralization is the conversion of organic forms of N into NH_4^+ and NO_3^-. The initial conversion into NH_4^+ is referred to as ammonification and the oxidation of this compound to NO_3^- is termed nitrification. Nitrification is one of the key processes determining the efficiency of fertilizer use by crops, as well as N losses from soil through leaching of NO_2^- and NO_3^- and emissions of N_2O and N_2 gas resulting from denitrification and anaerobic NH_4^+ oxidation, respectively (Nieder and Benbi, 2008; Wu et al., 2011). The nitrification process occurs in two phases in the soil–plant system and can be represented by the following equations:

$$2NH_4^+ + 3O_2 \Leftrightarrow 2NO_2^- + 2H_2O + 4H^+$$
$$2NO_2^- + O_2 \Leftrightarrow 2NO_3^-$$

In the process of nitrification, bacteria known as *Nitrosomonas* are involved in the process of conversion of ammonia into nitrites and the bacteria, which convert nitrites into nitrates, are known as *Nitrobacter*. Collectively, the nitrifying organisms are known as *Nitrobacteria*. Under optimal soil temperature, pH, and humidity, nitrification occurs at a very fast rate. Sahrawat (2008) and De Boer and Kowalchuk (2001) reported that the *Nitrosomonas* and *Nitrobacter* bacterial responsible for the nitrification process is pH sensitive, and thus, nitrification proceeds slowly in acid soils. Sahrawat (1982) studied nitrification in 10 soils (eight mineral and two Histosols) having a range in texture, pH (3.4–8.6), organic C (12.2–227.0 g kg^{-1}), and total N (0.9–1.2 g kg^{-1}). The amounts of NO_3-N produced at 30°C after 4 weeks of incubation of the soils varied from 0 to 123 mg kg^{-1} soil. Soils with pH < 5.0 did not nitrify at all; the organic soil with pH 5.6 produced only 5 mg kg^{-1} NO_3-N soil during the period. Soils having pH > 6.0 nitrified at a rapid rate and released NO_3-N ranging from 98 to 123 mg kg^{-1} soil.

Several studies have shown that oxidation of NH_4^+ to NO_3^- (nitrification) is one of the major cause of soil acidity in many ecosystems (Sahrawat, 2008; Rodriguez et al., 2008). The nitrification reaction releases protons (H^+), which result in acidification of the soil when ammonical and most organic N sources are converted into nitrate. According to the microbial-mediated two-step chemical reaction (Ross et al., 2011), conversion of one mole of NH_4^+ into NO_3^- results in the production of two moles of H^+ into the soil environment to cause soil acidity.

In addition, nitrification is an oxidation process and aeration of soil increases nitrification. Ploughing and cultivation are recognized means of promoting nitrification. Nitrification results in the release of H^+ ions, leading to soil acidification. Furthermore, the enzymatic oxidation of nitrification also releases energy. The utilization of NH_4^+ and NO_3^- by plants and microorganisms constitutes assimilation and immobilization, respectively.

The concentration of N_2O has been rising at an increasing rate, particularly since about 1960. Currently, N_2O concentrations in the atmosphere are increasing at about 0.3% per year (Seiler, 1986). The atmosphere contains about 1500 Tg (Terra gram = 10^{12} g) of N_2O. Perhaps, 90% of the emissions are derived from soils through biologically mediated reactions of nitrification and denitrification (Byrnes, 1990). Emission of N_2O may be perceived as a leakage of intermediate products in each of these processes. While soils are a major source of N_2O, they also serve as a sink (Freney et al., 1978; Fageria and Gheyi, 1999).

The fixation of NH_4^+ is an important component of the N cycle in soil–plant systems. Three specific processes are recognized as responsible for the fixation and retention of N applied to soil, namely, fixation by clay minerals, NH_3 fixation by soil organic matter, and biological immobilization of NH_4 by heterotrophic microorganisms (Nommik and Vahtras, 1982). Adsorption of NH_4^+ in a nonexchangeable form in the interlayer region of expanding 2:1—layer aluminosilicate clay minerals (fixation) may reduce the fertilizer—use efficiency of this nutrient when added to soils in which such minerals predominate (Stehouwerand and Johnson, 1991). This type of fixation was first reported early in this century and many studies have been conducted to investigate this phenomenon with respect to fertilizer use efficiency (Nommik and Vahtras, 1982). A greater fixation of NH_4^+ has been reported with anhydrous NH_3 (AA) than with other forms of NH_4^+-releasing fertilizers (Young and Cattani, 1962). This may be due to the enhanced reaction of NH_3 with more acidic water in 2:1 clay interlayer (Nommik and Vahtras, 1982). Many earlier studies reported that the increase in pH increased fixation of NH_4^+ (Nommik, 1957). This effect is generally attributed to decreased competition with H_3O^+ as pH increased in a lower pH range (2.5–5.5), and to decreasing the charge of interlayer hydroxy–Al polycations in a higher pH range (5.5–8.0). An increasing solution concentration of NH_4^+ has been shown to increase the amount of NH_4^+ fixed, while the percentage fixation (amount fixed/amount added) was decreased (Black and Waring, 1972). The removal of organic matter from mineral surfaces has been found to increase NH_4^+ fixation (Hinman, 1966).

Fixed NH_4^+ is involved in the N dynamics of soil, and may be an important component of the N fertility status of some agricultural soils. Between 18% and 23% of added $^{15}NH_4$ was fixed (specifically adsorbed) after 15 days of incubation in soils containing relatively high amounts of vermiculitic clay (Drury et al., 1989). Ammonium ^{15}N has been observed to be fixed and released in proportion to added $^{15}NH_4$ and vermiculite content (Keerthisinghe et al., 1984). Drury and Beauchamp (1991) reported that immobilization of 5.7% of the added $^{15}NH_4$ occurred in the high fixing soil, and 3.9% of the added $^{15}NH_4$ occurred in the low fixing soil. Extractable and fixed NH_4 fractions were interrelated pools. When fertilizer NH_4 was added to the soil, a proportional amount was fixed by the clay minerals. When nitrification and immobilization depleted extractable NH_4, fixed NH_4 was released. The fixed NH_4 pool appeared to be a slow-released reservoir, with fixed NH_4 released being slower than the rate of fixation. Juma and Paul (1983) reported that NH_4 fixation was enhanced when nitrification was inhibited with 4-amino-1-, 3-, 4-triacole (ATC), a nitrification inhibitor. Neeteson et al. (1986) observed that over one-half of added NH_4 disappeared and reappeared after 5 weeks. Since the clay minerals in these soils were not found to fix added NH_4, they postulated that the immobilization occurred as an osmoregulation mechanism and then NH_4 was remineralized when the microbes died and decomposed. One NH_4 is fixed by clay mica and it is protected against nitrification until it is released from fixation sites (Scherer and Mengel, 1986). It can be concluded that there are a number of ways by which NH_4^+ fixation can be regulated. He et al. (1990) reported that about 16% of the $^{15}NH_3$-N injected was fixed by the soil. Of the N fixed, 48% was accounted for by chemical fixation into the soil organic fraction, most of which was removed by sequential extraction and 52% was accounted for as clay-fixed NH_4. These authors also reported that NH_3-N fixed by organic matter occurs in forms that are less stabilized and more biologically available than the native soil N. Nommik and Vahtras (1982) presented an extensive review of the mechanisms and influential factors affecting NH_4 fixation.

A large quantity of N accumulates in seeds of cereals and legumes, which are not recycled and lost from the soil system. Fageria et al. (2011) reported that about 50% of the total N accumulated was in the grains of upland rice cultivars IAC 47 and IR 43. The remaining 50% was in the roots and shoots. In legume crops, grains remove still higher N. This means that in a N management strategy, its removal by a crop should be taken into account.

Animal agriculture is an important component of N cycle in a soil–plant system that is generally ignored when discussing this topic. Agricultural activity is estimated to be responsible for 90% of anthropogenic NH_3 emissions (Boyer et al., 2002), with the majority coming from livestock production and 12% coming from N fertilizer application (Ferm, 1998). As more animals are added to a fixed land base, and more feed is imported onto a farm, the excretion of manure nutrients can surpass the cycling capacity of local land, air, and water resources (Powell and Broderick, 2011). In many industrialized countries and regions of the world, higher incomes have greatly increased the demand for meat, milk, and other animal products. This trend of tremendous livestock expansion raises concern over excessive nutrient use, nutrient loss, and environmental contamination (Ma et al., 2010a,b). Dairy cows excrete urea N in urine, which is a source of NH_3 emission into the atmosphere. Of the total feed N consumed by lactating cows on commercial dairy farms, a general range of 20–35% is secreted in milk (Jonker et al., 2002). Ration formulation may influence not only feed N transformation into milk but also the proportion of N excreted in feces and urine (Powell and Broderick, 2011). As dietary crude protein increases and N intake exceeds the requirement, the efficiency of the feed N use (i.e., the relative amount of consumed feed N that is secreted as milk) declines and the excretion of urinary N increases (Broderick and Clayton, 1997; Nousiainen et al., 2004). Urea N comprises from 55% to 80% of the total urinary N, depending on the concentration of crude protein in the ration (Olmos Colmenero and Broderick, 2006). Urease enzymes that are present in feces and soil rapidly hydrolyze urea to ammonium, which can be quickly converted into NH_3 gas and lost to the atmosphere. Thus, the increase in urea N excretion due to excessive crude protein in dairy rations increases NH_3 emissions during the collection, storage, and land application of manure (Powell et al., 2008; Arriaga et al., 2010).

2.3 LOSSES THROUGH AMMONIA VOLATILIZATION

Ammonia volatilization is an important process of N loss from soil–plant systems when N fertilizers are surface applied, especially in alkaline soils (Harrison and Webb, 2001; Ma et al., 2010a,b; Singh et al., 2012a,b). This can result in low N use efficiencies by crops (Bouwman et al., 2002;

Rochette et al., 2008). Studies showed that up to 60% of applied fertilizer N could be lost through NH_3 volatilization in flooded rice soils (Xing and Zhu, 2000), depending on the type of N fertilizer, tillage practices, and soil properties (temperature and pH) (Duan and Xiao, 2000; Gioacchini et al., 2002; Hayashi et al., 2011; Zhang et al., 2011). The gaseous ammonia in the atmosphere can cause environmental problems such as soil acidification and changes in biodiversity (Eerden et al., 1998; Stevens et al., 2004; Emmett, 2007; Xu et al., 2013).

Ammonia volatilization represents an agronomic N loss with NO_x and SO_x creates particulate 2.5-μm aerosols that scatter light, resulting in haze (Asman et al., 1998; Sharma et al., 2007). Several studies have been conducted to estimate N loss through NH_3-N volatilization from inorganic N fertilizers applied to agricultural lands (Wang et al. 2004; Pacholski et al., 2006). The economic impact associated with NH_3-N emission from N fertilizers applied to agricultural land is estimated at U.S. \$11.6 billion annually worldwide (FAO/IEF, 2001). Ammonia volatilization also causes serious climatic and environmental problems (Gay and Knowlton, 2005). In addition, the emitted NH_3-N exacerbates global climate change (Singh et al., 2012a,b). Although the reaction of NH_3-N with OH radicals is relatively slow, the intermediate NH_2 radical can provide a substantial nitrous oxide (N_2O) source, thus contributing to the production of N_2O, one of the major greenhouse gases (GHGs) (Finlayson-Pits and Pitts, 2000). Dentener and Crutzen (1994) estimated that 4% (3×10^6 MT N year^{-1}) of the globally emitted NH_3-N can be oxidized by OH radicals to N_2O, mainly in the tropics. The redeposition of NH_3-N in nonagricultural soils can also lead to acidification (Singh et al., 2012a,b). Therefore, it is important to identify and select N sources with minimal NH_3-N losses to safeguard environmental quality and reduce crop production costs (Singh et al., 2012a,b).

Ammonium volatilization can be expressed by the following equation (Bolan and Hedley, 2003):

$$NH_4^+ + OH^- \Leftrightarrow NH_3 + H_2O$$

Ammonia volatilization from flooded rice soils is a major mechanism for N loss and a cause of low fertilizer use efficiency by rice. Reviews on NH_3 volatilization from flooded rice soils indicate that losses of ammonical-N fertilizer directly applied to floodwater may vary from 10% to 50% of the amount applied (Fillery and Vlex, 1986; Mikkelsen, 1987). Losses, however, are site and soil management specific; thus, disparities may exist in reported rates of volatilization, depending on rate-controlling factors and methods of measurement. Ammonia volatilization under flooded rice conditions is influenced by five primary factors: NH_4-N concentration, pH, temperature, depth of flooded water, and wind speed (Jayaweera and Mikkelsen, 1990; Fageria and Gheyi, 1999).

Volatilization losses from surface applications of urea-containing N sources can be related to soil and weather conditions following the application. Fox and Hoffman (1981) categorized NH_3 volatilization losses in Pennsylvania based on the rainfall amount and the length of time between surface application and rainfall. They concluded that 10 mm of rain accruing within 2 days after surface N application resulted in no NH_3 volatilization losses. The losses increased with increased time between application and rain and are substantial (>30%) if no rain falls within 6 days. Urban et al. (1987) reported that maximum NH_3 losses from surface-applied urea in a growth chamber occurred between 4 and 8 days after application.

Urea is the principal source of N for crop production around the world because of its high N content (46%), low relative cost, ease in handling, and compatibility with other fertilizer materials (Kissel et al., 2009). Ammonia volatilization occurs when urea is hydrolyzed in the presence of water and urease to form NH_4 (Torello et al., 1983; Kissel and Cabrera, 1988). The hydrolysis of urea can be expressed by the following equation (Fageria et al., 2010):

$$CO(NH_2)_2 + 3H_2O \Leftrightarrow 2NH_4^+ + 2OH^- + CO_2$$

$$NH_4^+ + 2O_2 \Leftrightarrow NO_3^- + 2H^+ + H_2O$$

Bolan and Hedley (2003) reported that the ammonification processes as shown in the first equation result in the consumption of H^+ ions (or release of OH^- ions). The consumption of H ion temporarily raises the soil pH around the reaction site (Kissel et al., 1988). The pH around the urea granule and the urine spot increases to alkaline conditions (around 7.5–8.0). The alkaline conditions induce the conversion of NH_4^+ ions into NH_3 gas, leading to the volatilization loss of ammonia (Fenn and Hossner, 1985). However, the subsequent conversion of NH_4^+ ions into NO_3^- ions results in the release of H^+ ions, leading to soil acidification as shown in the second equation above. Hence, the long-term effect of urea is soil acidification as NH_4^+ converts into NO_3 and releases H^+ ion (Olson and Kurtz, 1982).

Volatilization of NH_3 is also influenced by soil moisture, pH, texture, and temperature (Torello and Wehner, 1983; Bouwmeester et al., 1985; Ferguson and Kissel, 1986; Sommer and Olesen, 1991). In addition, conditions that favor the gaseous loss of NH_3 from surface-applied N include high crop residue, warm temperature (>13°C), a drying soil surface (water vapor loss from surface), and low cation-exchange capacity (as found in sandy soil) (Bouwmeester et al., 1985; Ferguson and Kissel, 1986; Clay et al., 1990; Sommer and Christensen, 1992). Most reported NH_3 volatilization losses from urea range from 15% to 40% of the applied N; however, some NH_3 losses from urea have been reported up to 83% of the applied N (Lightner et al., 1990; Grant et al., 1996; Sommer and Ersboll, 1996; Vaio et al., 2008; Kissel et al., 2009).

Huckaby (2012) studied ammonia (NH_3) volatilization from three N sources (urea—46% N, polymer-coated urea—41% N, and methylene urea—40% N) applied to turfgrass. The amount of N applied by three sources was 146 kg ha^{-1}. The ammonia gas losses were measured in the field plots for 10 days. Average data over 2 years showed that ammonia volatilization was 14.8% from urea, 5.8% with methylene urea, and 4% with polymer-coated urea. Periodic wetting and drying of the soil surface was shown to increase NH_3 volatilization from surface-applied urea because of the effect on urea hydrolysis (Titko et al., 1987). The major part of NH_3 volatilizes within 24 h (Bowman et al., 1987; Knight et al., 2007). Hence, the principal practice to reduce ammonia loss from surface-applied urea is irrigation immediately after its irrigation. Bowman et al. (1987) reported that irrigation reduced N loss from 36% of applied N (no irrigation) to 3% to 8% of applied N (1 cm water applied). Continued irrigation, to 4 cm of water, reduced NH_3 volatilization to about 1% of N applied (Bowman et al., 1987).

Ammonia volatilization from surface-applied urea has long been recognized as an important mechanism of N loss (Engel et al., 2011). Losses of up to 50% of the applied N rate have been reported in the field (Ryden and Lockyer, 1985), reducing the efficiency of the use of N fertilizer and contributing to the N enrichment of natural ecosystems through dry or wet deposition. Engel et al. (2011) reported that cumulative NH_3 losses from urea varied but averaged 20.5% of applied N across 12 field trials. The largest losses (30–44% of applied N) occurred after urea was applied to high water content soil surfaces followed by a period of slow drying with little or no precipitation.

These authors further reported that emissions occurred for a prolonged period, often lasting for >42 days. Periods (1–2 weeks) of high NH_3 flux (>30 g N ha^{-1} h^{-1}) frequently occurred when the mean daily soil temperature (1 cm depth) was −2°C to 5°C. These authors further reported that ammonia losses were moderated by applying urea to dry soil surfaces. The larger NH_3 loss occurred from both acidic and alkaline soils, suggesting that, in the field, the impact of soil water conditions at the time of fertilization and the magnitude of precipitation that followed were more important than the soil pH in defining the NH_3 loss potential (Engel et al., 2011).

Urea is an important source of N fertilizers in crop production around the world. Managing urea appropriately is critical to minimizing potential loss, especially through ammonia volatilization that has been shown to account for 20–80% of N loss in rice production (Beyrouty et al., 1988; Griggs et al., 2007; Norman et al., 2009). Ammonium volatilization occurs in the dry-seeded-delayed flood rice culture when urea is hydrolyzed to ammonium carbonate [$(NH_4)_2CO_3$] by the urease enzyme and ammonium carbonate decomposes to produce NH_3 and CO_2 (Dillon et al., 2012). The proportion of NH_3 to NH_4^+ is determined by the pH of the soil (Boswell et al., 1985). Soil and flooded water pH, soil and air temperature, cation-exchange capacity, H^+-buffering capacity, N source, wind speed, humidity, and NH_3 concentrations all affect the rate of NH_3 volatilization (Harper et al., 1983; Boswell et al., 1985; Bouwmeester et al., 1985).

2.4 LOSSES THROUGH DENITRIFICATION

Denitrification is a microbial process of converting NO_3^- into gases N_2O or N_2. This process occurs when soil is water saturated and microrganisms no longer have ready access to O_2. Most microorganisms depend on O_2 for energy conversion by utilizing O_2 as the last electron acceptor, thereby converting O_2 into CO_2. However, certain microorganisms can also utilize NO_3^- in the same way as O_2 in anaerobic conditions (due to water saturation), thereby utilizing O_2 as the last electron acceptor when converting NO_3^- into N_2O or N_2 (Gerik et al., 1998).

Denitrification is a major loss of N from soil–plant system and it is mainly an anaerobic bacterial respiration. Losses from irrigated soils in California ranged from 95 to 233 kg N ha^{-1} $year^{-1}$ (Ryden and Lund, 1980). The denitrification occurs in the following reductive ways:

$$NO_3^- \text{ (nitrate)} \Rightarrow NO_2^- \text{ (nitrite)} \Rightarrow NO \text{ (nitric oxide)} \Rightarrow N_2O \text{ (nitrous oxide)} \Rightarrow N_2 \text{ (dinitrogen)}$$

N_2O is produced from microbial transformation of N in soils and manures and is often enhanced where the available N exceeds plant requirements, especially under wet conditions (Mosier et al., 1991, 2006; Phillips, 2007; Buchkina et al., 2010; Collins et al., 2011). Manure applications may cause relatively high N_2O emissions when the soil contains NO_3 and the decomposition of organic C in manure enhances denitrification or nitrification (Moller and Stinner, 2009; Sistani et al., 2010; Collins et al., 2011).

Most denitrifying bacteria exist in the topsoil (0–30 cm) with the number exponentially decreasing down to 120–150 cm (Parkin and Meisinger, 1989). Denitrification is influenced by several factors such as soil pH, temperature, organic C supply, nitrate concentration, aeration, and water status (Aulakh et al., 1992). Owing to the influence of several physical and chemical factors, the exact quantity of N loss due to denitrification is difficult. However, Aulakh et al. (1992) reported that, overall, N losses due to denitrification might be about 30% in an agroecosystem. In addition, the estimation of global denitrification losses ranges from 83 Tg $year^{-1}$ (Stevenson, 1982) to 390 Tg $year^{-1}$ (Hauck and Tanji, 1982; Aulakh et al., 1992). A large part of N is ultimately returned to the atmosphere through biological denitrification, thereby completing the cycle (Stevenson, 1982).

Denitrification usually occurs in soil high in organic matter, under extended periods of waterlogged conditions and as temperature rise. While N_2 is an inert gas that poses no known environmental risk, N_2O is one of the GHGs that contribute to the destruction of the Earth's protective ozone layer (Fageria and Gheyi, 1999). Measured N_2O emissions from fertilized cropland fall within a wide range: 0.7–51.8 mg N_2O-N m^{-2} d^{-1} (Sehy et al., 2003; Rochette et al., 2004; Venterea et al., 2005; Drury et al., 2006; Molodovskaya et al., 2012). Among different vegetation covers, denitrification was significantly higher in grasslands than in either hardwood or pine forest areas (Lowrance et al., 1995).

About 6% of the total GHG emissions in the United States is contributed by agricultural activities (Greenhouse Gas Working Group, 2010; USEPA, 2011). The amount of CO_2 and N_2O emissions contributed by agriculture account for about 25% and 70%, respectively, of the total anthropogenic emissions (Cole et al., 1997). Fossil fuel consumption, land conversion into cropland, lime application, and N fertilization are major sources of agriculture CO_2 emissions while soil management practices contribute about 92% of the total N_2O emissions (USEPA, 2011).

A major part of the denitrification occurs in the surface soil. However, a small part also occurs in the subsoil. The predominant N_2O production process in the subsoil is the denitrification of NO_3^- and it has been suggested that, in some cases, rates up to 60–70 kg N ha^{-1} $year^{-1}$ may be possible

(Van Cleemput, 1998; Thomas et al., 2012). Water content and water-filled pore space is often a key determinant of subsoil denitrification (Van Groenigen et al., 2005). The denitrification and N_2O production have been demonstrated on both saturated subsoils (Wells et al., 2001) and in subsoils that are predominantly aerobic (Muller et al., 2004). In a few studies, the release and transport of N_2O from NO_3^--contaminated shallow groundwater into the overlying vadose zone has been estimated (Bottcher et al., 2011; Minamikawa et al., 2011). Typically, the low availability of C substrates is considered a major limiting factor (Yeomans et al., 1992; Brye et al., 2001; Murray et al., 2004). Nevertheless, it has been proposed that even low rates of denitrification may be important as a sink for leached NO_3^- where vadose zones are deep and, in some cases, hundreds of meters thick (Jarvis and Hatch, 1994).

2.5   LOSSES THROUGH LEACHING

N losses through leaching are one of the significant processes in soil–plant systems. Leaching is the physical movement of NO_3^- through the soil (Stevenson, 1986). As water moves through the soil, so does the NO_3^- in solution. In soils where water moves rapidly through the soil profile and where water (either from rainfall or irrigation) exceeds evapotranspiration, nitrate is commonly found below the rooting zone and is no longer available for plant uptake. Numerous studies have shown that fertilizer applications of N that exceed crop requirements can result in NO_3^- leaching and can contaminate groundwater (Baker and Johnson, 1981; Angle et al., 1993; McKenney et al., 1995). Soil texture (percentage of sand, silt, and clay) controls the magnitude and rates of many physical, chemical, and hydrological processes in soils. The process of NO_3-N leaching from the soil profile is more severe in light-textured soils as compared to heavy-textured soils. It is also related to the quantity of precipitation and irrigation frequency and the quantity of water applied through irrigation. Nitrate (NO_3^-) ion is negatively charged and easily leaches under heavy rainfall or irrigation. Owing to high rainfall, the N leaching is more in humid and subhumid area soils as compared to dry region soils. Singh et al. (2012a,b) reported that due to heavy texture and lower percolation rate (<1.0 mL h^{-1}) of the Greenville loam, it had significantly lower NO_3-N leaching losses as compared to the Lakeland sand.

The leaching of NO_3-N not only occurs from N applied to crops. However, it may also be a problem in intensively managed pastures. Intensively grazed pastures generate nitrate N (NO_3-N) and the leaching of this nitrate N can contaminate groundwater creating grave environmental concerns or problems globally (Jabro et al., 2012). The high levels of NO_3-N (>45 mg L^{-1}) are toxic to human infants (Lehrsch et al., 2001). The NO_3-N is also toxic to mammals, although in greater concentrations. Hence, increasing NO_3-N in the groundwater used for drinking is of concern around the globe (Strebel et al., 1989; Spalding and Exner, 1993; Lehrsch et al., 2001).

Stout et al. (1998) studied the effect of NO_3-N leached from urine-impacted areas. Their results support that NO_3-N levels beneath pastures exceed the 10 mg L^{-1} drinking water maximum acceptable contaminant level. About 25% of the N in urine and 2% of the n in feces leach beneath the root zone in the northeast United States (Stout, 2003). Numerous other studies support that large amounts of N in animal excreta may be leached from intensively managed grazed pasture (Stout et al., 2000; Di and Cameron, 2002; Stout, 2003; Decau et al., 2004; Sorensen and Rubek, 2012; Jabro et al., 2012).

The amount of N leached depends on the soil type, the source of N fertilizer, the crops, and the methods of fertilizer application. Also, nitrate is a major factor associated with the leaching of bases such as calcium, magnesium, and potassium from the soil. The nitrate and bases move out together. As these bases are removed and replaced by hydrogen, the soil becomes more acidic. N fertilizers containing strong acid-forming anions such as sulfate increase acidity more than other carriers without acidifying anions (Fageria and Gheyi, 1999). In addition, applied N that is not utilized by plants can leach as NO_3-N from the fields to ground and surface waters, causing environmental and health problems stretching beyond the original agricultural fields (NRCS, 1997; Gupta et al., 2000;

Chighladze et al., 2012). Nitrate N in the soil profile (0.1–1.2 m depth) accounted for 20–38% of the apparent fertilizer N recovery, thus, most likely resulting in the accentuation of N losses due to deep leaching of NO_3^- from the soil (Turpin et al., 1998).

2.6 LOSSES THROUGH SOIL EROSION

Soil erosion results in losses of agricultural productivity and continues to be a major issue, especially under conventional agricultural practices (Lal, 1999; Veum et al., 2012). As a result, conservation management practices have become increasingly popular due to a wide range of environmental benefits, including increased soil organic carbon (Lal et al., 1994), reduced erosion (Robinson et al., 1996), reduced runoff (Veum et al., 2009), and increased aggregate stability (Angers et al., 1993). From 1982 to 1997, agricultural soil erosion in the United States declined by approximately 1 billion metric tons per year, and a quarter of this decrease has been attributed to conservation management efforts (Wiebe and Gollehon, 2006).

Soil erosion is a natural phenomenon. But when the removal of the soil is faster than soil formation through bedrock weathering, soil erosion becomes a problem, often resulting in the reduced ability of soils to perform their functions (Mchunu et al., 2011). Serious soil erosion is now occurring in most of the world's major agricultural regions, and the problem is growing as more marginal land is brought into agricultural production (FAO, 2008). Since soil erosion is a major threat to the sustainability of soils and soil functions in the years to come, finding remediation to soil erosion is a key issue (FAO, 2008).

N loss through soil erosion is a big loss in the soil–plant system. The erosion may be related to wind erosion and water erosion or to both. Feng et al. (2011) reported that soil wind erosion and fugitive dust emission contribute to land degradation, loss of soil productivity, and poor air quality and visibility. Atmospheric dust also influences climate by altering the Earth's radiation balance (Tegen and Lacis, 1996). The potential for wind erosion is higher in arid and semiarid regions (Feng et al. 2011). During high wind events, soil entrainment is dominated by suspension processes (Kjelgaard et al., 2004; Sharratt, 2011). Suspension of fine particulates from relatively small emission source areas can impact communities downwind. Soil and mineral particles with a diameter of <60 μm are especially important to air quality because they contain significant amounts of soil nutrients (Zobeck and Fryrear, 1986) and contaminants (Pye, 1987). Of particular concern are these particles with mean aerodynamic diameters ≤10 and 2.5 μm that are stringently regulated by the U.S. Environmental Protection Agency (USEPA, 1990). Furthermore, wind and water erosion removes the topsoil layer that contains a high amount of organic matter. When the top layer organic matter provides a large amount of N to crop plants through mineralization. Hence, the loss of topsoil layer means the loss of a high amount of N in the organic form.

2.7 LOSSES THROUGH AMMONIA VOLATILIZATION FROM FOLIAGE

The absorption and loss of N through the plant canopy is also an important part of N cycling in soil–plant systems. Controlled as well as field studies showed that plants can absorb NH_3 from the air as well as lose NH_3 to the air by volatilization (Farquhar et al., 1980; Fageria and Baligar, 2005). The emission of NH_3 has considerably increased over recent decades. Factors influencing NH_3 losses include soil and plant N status and plant growth stage (Sharpe and Harper, 1997). Abundant supply favors NH_3 losses, especially if the supply is in excess of plant requirements (Fageria and Baligar, 2005). The loss of NH_3 through the plant canopy can occur during the whole growth cycle of a crop (Harper and Sharpe, 1995). However, some scientists have reported that the highest NH_3 volatilization rates for major agricultural crops occur during the reproductive growth stage (Francis et al., 1997). Absorption of atmospheric NH_3 has been associated with low plant N content and with high atmospheric NH_3 concentrations (Harper and Sharpe, 1995). Raun and Johnson (1999) reported that N losses under field conditions are generally attributed to the volatilization of NH_3 from leaves of N-rich plants.

2.8 FACTORS AFFECTING NITROGEN LOSSES

N losses from soil–plant system are influenced by climatic, soil, and plant factors and are also influenced by N sources and methods of application. Climatic factors that influence N losses are precipitation and temperature. Soil factors that influence N losses are soil pH and texture. N fertilizer sources and methods of application also influence N losses from soil–plant system. Plant factors associated with N losses are fallow versus cropped lands.

2.8.1 Precipitation

Quantity and intensity of precipitation or rainfall during crop growth significantly influences N losses from the soil–plant system. Heavy rainfall of longer duration in the early crop growth is responsible for the leaching of the NO_3-N from the plant root zone to the lower depths where it cannot be absorbed by the plant roots. In the early growth stage, root growth of the crop plants is not well developed and nitrified N is not assimilated. Hence, chances of loss through leaching are quite high. N deficiency symptoms in upland rice grown in Brazilian Oxisols are often observed by the author when rainfall is heavy after emergence of seedlings until the beginning of the tillering stage that depends on genotype and climatic conditions. However, in most genotypes, tillering starts at about 20 days after sowing in the central part of Brazil, locally known as Cerredo region (Fagreia and Knupp, 2013).

The development of different growth stages in upland rice is given in Table 2.1. Fageria and Knupp (2013) also reported that maximum root length followed a significant quadratic response with the advancement of the plant age from 19 to 120 days after sowing. There was a linear increase in the root length from tillering initiation to flowering. Thereafter, the root length was more or less constant or reached to the plateau (Fageria and Knupp, 2013). An excess of soil moisture is also responsible for the denitrification losses from the plant root zone or rhizosphere. Most of the N losses due to leaching by rainfall occur in the humid and subhumid regions of the world where the annual rainfall is high. For example, in the central part of Brazil, the average annual rainfall is about 1500 mm, mostly during the months of October to March. Hence, N leaching from the cropping system is very common in this region.

TABLE 2.1
Timing (DAF, Days after Sowing) of Upland Rice Growth Stages and Definitions

Growth Stage	DAF	Definition
Germination	5	Defined as the stage when the coleoptile tip first became visible.
Tillering initiation	19	Defined as the crop growth stage when the first tiller from the main shoot is visible.
Active tillering	45	Defined as the development stage at which maximum tillering rate per unit time during crop growth.
Panicle initiation	61	Defined as the initiation of the panicle.
Booting	85	Defined as the development stage at which the panicle is enclosed by the sheath of the uppermost leaf.
Flowering	95	Defined as the physiological stage at which flowers are visible on the panicles.
Physiological maturity	120	Defined as the growth stage at which grains are ripened and panicles are ready for harvest.

Source: From Fageria, N. K. and A. M. Knupp. 2013. *Plant Nutr.* 36:1–14. With permission.

2.8.2 TEMPERATURE

Temperature on soil surface influences N losses by volatilization, leaching, and denitrification. When N is top dressed in a crop during growth cycle and temperature is higher (>13°C), a high rate of N loss through NH_3 volatilization from soil-applied N fertilizer such as urea is expected (Jantalia et al., 2012). Similarly, nitrification processes are also high under high temperatures (>13°C) and when there is heavy rainfall for a prolonged time, more N is expected to be lost under these climatic conditions. Maximum nitrification in soils occurs at 30–35°C (Black, 1968; Resek et al., 1971). Mahendrappa et al. (1966) reported that attempts to agree on optimum conditions have largely been unsuccessful, since nitrifying bacteria found in a particular soil are the result of natural selection and adjustment to the climate. These authors found that soils from the northern states of the United States nitrified best at a temperature of 20°C and 25°C, while soils from the southern states nitrified best at 35°C.

Like nitrification, denitrification is also affected by temperature. Stanford et al. (1975) reported that denitrification rate minimal at 0–5°C increased 10-fold between 5°C and 10°C. Between 15°C and 35°C, the temperature coefficient of denitrification (Q_{10}) was approximately 2. In field studies with different tillage and cropping systems, Aulakh et al. (1983,1984) observed little influence of temperature on denitrification rate during the crop-growing season when temperatures ranged from 10°C to 30°C. However, in early spring and late fall when temperatures were 5°C or below, virtually, no denitrification was detectable even in wet soils with high NO_3 content. Temperature variations may change the proportion of N oxide gases produced during denitrification. In a temperature range of 15–30°C, the major product of denitrification in wet- or water-logged soils is N_2. Under similar conditions at a temperature of 4–8°C, N_2O and NO often predominate (Bailey, 1976; Aulkh et al., 1992).

The effect of temperature on the N losses process is influenced by influencing soil microbial community. N losses are reported to be higher under higher temperatures as compared to lower temperatures. Hence, soil microbial community related to both nitrification and denitrification is more sensitive to higher temperatures. Elevated temperatures can directly alter the soil microbial community functions by affecting the temperature-sensitive microbial enzyme activity (Von Lutzow and Kogel-Knabner, 2009; Gray et al., 2011). Elevated temperatures can also influence soil microbial abundance and composition by altering net primary production and therefore the pool of available substrates utilized for microbial growth (Gray et al., 2011). Elevated temperatures also increase the evaporative flux of water from the soil and therefore indirectly affect microorganisms through soil drying (Pendall et al., 2004; Filella et al., 2004). The effects of elevated temperatures on the total microbial biomass are variable (Pendall et al., 2004), and groups of microorganisms differ in their responses to elevated temperature (Gray et al., 2011). In a recent study, Engel et al. (2011) reported that the mean daily temperature of –2°C to 5°C did not provide protection against realizing large NH_3 emissions.

2.8.3 SOIL pH

Soil pH is an important factor in determining N losses from soil–plant system. Most of the NH_3 volatilization from applied N sources occurs at alkaline soil pH (>7.0). Many studies conducted in the United States and elsewhere have shown that heavy losses of N through ammonia gas occurred on heavily limed or calcareous soils (Larsen and Gunary, 1962; Terman and Hunt, 1964). Pesek et al. (1971) reviewed the literature on ammonia volatilization from various countries, including the Netherlands. These authors concluded that results from the Netherlands obtained with 176 soil samples from all over the world showed that the volatilization of ammonia is correlated better with $CaCO_3$ content than with pH of the soil.

Generally, denitrification decreases in acid soils and increases with increasing soil pH. Klemedtsson et al. (1977) and Muller et al. (1980) found a direct positive relationship between the rate of denitrification and the soil pH in acid soils from Sweden and 22 locations in Finland. The optimum soil pH for denitrification varies with species and the age of organisms and NO_3^- concentration (Delwiche and Bryan, 1976). However, it was reported by Aulakh et al. (1992) that most denitrifying microorganisms have an optimum pH level between 6 and 8.

2.8.4 TEXTURE

Soil texture is an important soil physical property, influencing N losses from a soil–plant system. Soil texture influences water infiltration rate in the soil profile and thereby influences NO_3 leaching. It is higher in light-textured soils as compared to heavy-textured soils. Soil texture influences denitrification processes significantly due to physical variations in soil structure, pore size, aggregation, and water infiltration rates that affect aeration, water-holding/absorption capacity, and microenvironment and may be due to other natural differences in the capacity of soils to supply NO_3 and organic C (Aulakh et al., 1991, 1992). Denitrification rate is reported to increase with increasing clay content in the soil (Aulakh et al., 1992).

2.8.5 SOIL WATER AND AERATION

Soil water content is very important for crop production. It is used for the determination of the soil water balance and the transport of chemical to plants and groundwater and for irrigation management. It is a basic hydrological condition that affects groundwater recharge, surface water flow, and transpiration (Xu et al., 2011). Soil water content and aeration may affect nitrification and denitrification processes and hence N losses from the soil–plant system. As soil water content increases above field capacity, nitrification drops abruptly and with complete waterlogging, nitrification ceases except for a small amount at the soil or water surface and denitrification may occur (Pesek et al., 1971). Similarly, when oxygen content of the soil air falls below 2%, nitrification rate drops abruptly (Amer and Bartholomew, 1951).

Aulakh et al. (1992) reported that aerobic microbial activities increase with soil water content until a point is reached where further water restricts the diffusion and availability of oxygen. The rate of O_2 diffusion through water is about 10^4 times less than through air. The maximum rates of predominantly aerobic processes such as respiration, nitrification, and mineralization occur at the highest soil water content at which O_2 remains nonlimiting. Aulakh et al. (1992) reported that aerobic microbial activities are greater at a soil water content equivalent to 60% of a soil's water-holding capacity (Linn and Doran, 1984). Several authors reported that denitrification is negligible at water content below 60% of maximum water-holding capacity, regardless of NO_3^- supply (Aulakh et al., 1991).

2.8.6 METHOD OF NITROGEN FERTILIZER APPLICATION

Methods of N fertilizer application significantly influence N losses from the rhizosphere. N fertilizers applied on the surface of the soil and not incorporated into the soil will lose more N than NH_3 volatilization as compared to its incorporation into the soil. Similarly, deep placement of N fertilizers into the soil in flooded rice will lose less than those applied on standing water in the field. Fenn and Miyamoto (1981) reported that urea needs to be incorporated to a depth of 5 mm or more in soil to minimize NH_3 volatilization, with mechanical or irrigation incorporation being suitable to minimize losses.

2.8.7 NITROGEN SOURCE AND PARTICLE SIZE

N source and particle size influence N losses from a soil–plant system. Sources that delay nitrification processes when applied to crops will be more efficient as compared to those sources that nitrify immediately after their application. Sources such as polymer-coated or sulfur-coated urea are better sources because of their slow N liberation quality compared to common urea. Comparisons of granular and finely divided fertilizer when broadcast and mixed with soil also show effects of particle size on nitrification (Hauck and Stephenson, 1965). The rate of nitrification was reported to be higher when particle size of a fertilizer was smaller (Pesek et al., 1971). The use of ammonium sulfate, diammonium phosphate, and monoammonium phosphate applied in an aqueous solution and in a range of particle sizes (1–4 mm) to a poorly buffered sandy soil, the rate of nitrification was higher than the smaller particle size (Pesek et al., 1971).

Jantalia et al. (2012) evaluated the NH_3-N loss from four urea-based N sources (urea, urea ammonium nitrate, superurea, and polymer-coated urea) for 2 years using corn as a test crop. Results of this study showed that super urea that contains a urea inhibitor had the lowest level of NH_3-N loss as compared to the other sources. Analyzed across years, the estimated NH_3-N losses for the N sources were in the order: polymer-coated urea = urea ammonium nitrate > urea > superurea. In addition, these authors also concluded that both years results showed that measurement time may need to be increased to evaluate NH_3-N volatilization from polymer-coated urea N sources such as polymer-coated urea. Wu et al. (2011) reported that ammonium rather than NO_3-based fertilizers is generally used in paddy fields or in flooded rice to prevent N loss from microbial denitrification.

2.8.8 PLANT FACTORS

Plant factors play a significant role in the loss of N from a soil–plant system. Pesek et al. (1971) reviewed the literature on the effects of plant factors in N losses of a soil–plant system. It was concluded by these authors that N loss in the cropped area was 20% and in the fallow area, it was 12% of the amount available in a lysimeters study. Similarly, Stefanson and Greenland (1970) in Australia found, from periodic sampling of gases over wheat, that amounts of N_2O and N_2 (cumulative) were consistently greater in the presence of than in the absence of growing plants, over a range of soil water contents from 50% to 80% of field capacity. The higher N loss in cropped land as compared to uncropped land may be related to denitrification that preferably occurs in the vicinity of roots. Apparently, root excretions are vitally concerned as hydrogen donors and in the development of anaerobic conditions in the rhizosphere (Pesek et al., 1971). McKenney et al. (1995) also reported that $NO + N_2O$ production under many crops (hairy vetch, red clover, annual ryegrass, reed canarygrass, and corn) was significantly higher as compared to control (without plants) treatment under both without aerobic and with aerobic incubation.

2.9 MANAGEMENT PRACTICES TO REDUCE NITROGEN LOSSES

Over the last few decades, N fertilization had played a significant role in increasing crop yields worldwide. In the future also, the importance of N in crop production will be indisputable due to the increasing demand of food, fiber, and energy. However, the energetic cost of N fertilizers is very high and N fertilization often represents the most expensive energy input in cereal-based cropping systems (Crews and Peoples, 2004). In addition, N is a highly mobile nutrient in soil–plant systems and lost easily and contributes to agricultural-related pollution through leaching, volatilization, and denitrification (Drinkwater et al., 1998; Limaux et al., 1999). Indeed, it has been estimated that, often, 50% or less of the N fertilizer applied to the soil is recovered by cereals and that this percentage decreases as the N fertilizer rate increases (Foulkes et al., 1998; Raun and Johnson, 1999; Blankenau et al., 2000; Giambalvo et al., 2010).

Fageria (2013) reported that N losses from the soil–plant system not only reduced N fertilizer efficiency but are also responsible for environmental pollution. For example, NO_3^- leaching in excess can contaminate body and ground water that may be a problem for human health. Similarly, the emission of N_2O gas into the atmosphere can cause ozone layer destruction. Franzluebbers (2007), Herrero et al. (2010), and Sainju (2013) reported that N fertilization can increase crop yields, but excessive application can degrade soil and environmental quality by increasing soil acidification, N leaching, and emissions of N_2O, a highly potent GHG with shared responsibility for global warming. When evaluating among management systems, it is important to consider options that have the potential to improve agronomic productivity, economic return, and/or environmental sustainability (Johnson et al., 2012). An increase in soil N storage reduces N losses through leaching, volatilization, denitrification, surface runoff, erosion, and N_2O emissions (Sainju et al., 2012b). Hence, adopting appropriate management practices to reduce N losses not only improves N use efficiency and reduces the cost of crop production but also reduces environmental pollution. These practices include use of an adequate rate of N, the use of an effective source of N, use of appropriate timing of N application, use of conservation tillage, adopting appropriate crop rotation, and planting N-efficient crop species and genotypes within species. A detailed discussion on these practices is given in Chapter 8. However, some part of the management practices associated with the reduction of emission of gases from the applied fertilizer is more pertinent to this chapter and is discussed here.

2.9.1 IRRIGATING AFTER TOPDRESSING NITROGEN FERTILIZERS

When N is top dressed during crop growth, the application of irrigation (wherever possible) water in adequate amounts (>15 mm) may significantly reduce N loss by volatilization. In dryland agriculture when the crop is totally dependent on rainfall, the application of N in top dressed can be done a little earlier than the rain is expected. Meteorological forecasting can be used to define the N topdressing timing under dryland farming system. In addition, applied N incorporated into the soil can also reduce N losses significantly. Research on effective irrigation depth has varied from 5 to 75 mm to minimize volatilization losses by NH_3 (Harper et al., 1983; Bouwmeester et al., 1985; Black et al., 1987; Mugasha and Pluth, 1995). Holcomb et al. (2011) reported that the application of NH_3 losses can be significantly reduced by ammonia losses by irrigation immediately after urea application to wheat crop. These authors developed a regression equation by plotting irrigation rate versus cumulative total N loss as NH_3 yielded the following equation ($R^2 = 0.91$):

$$NH_3\text{-}N \text{ loss (\% of N applied)} = 62.655\exp.(-0.1559 \times \text{irrigation rate in mm})$$

Using the model to solve for an 80% reduction in NH_3 loss resulted in a need for 10.3 mm of irrigation, a 90% reduction needs 14.8 mm of irrigation, and a 95% reduction needs 19.2 mm of irrigation. Holcomb et al. (2011) further reported that as the irrigation rate increases, NH_3 volatilization loss decreases but with a diminishing return, particularly at irrigation rates of >14.8 mm. Black et al. (1987) reported that a simulated rate of 16 mm of precipitation was enough to reduce NH_3 loss to 2%. Kissel et al. (2004) applied 24 mm of simulated rain immediately after a urea application and they found <1% loss, compared with 5% when simulated rain was applied at day 7.

Black et al. (1987) reported that precipitation was not effective after 48 h because hydrolysis had already occurred. Engel et al. (2011) reported that after application of urea on the soil surface, if precipitation events that follow are light (<8 mm), NH_3 loss may be reduced to 10–20% and if the events are heavy (>18 mm), the loss may be <10% of the applied N. Jantalia et al. (2012) reported that irrigation with 16–19 mm of water 1 day after N fertilization reduced $NH_3\text{-}N$ loss from surface-applied N fertilizers to a range of 0.1–4% of total N applied depending on N sources. Halvorson and Del Grosso (2012) reported that irrigated soils in semiarid climates have relatively low N_2O–N losses, provided irrigation is well managed to avoid water-logged conditions and potential for denitrification.

2.9.2 USE OF ORGANICALLY ENHANCED FERTILIZERS

There are several sources of N as inorganic fertilizers. However, there are three most common sources that are largely used as a source of N in crop production worldwide. These sources are urea, ammonium sulfate, and anhydrous ammonia. Generally, N sources that contain nutrients in complex organic compounds are more efficient environmentally and agronomically due to their slower N release (Singh et al., 2012a,b). Singh et al. (2012a,b) reported that organically enhanced fertilizers (fertilizer manufactured by using sterilized and chemically converted organic additives extracted from municipal wastewater biosolids) could be an attractive N source. These authors further reported that the environmental and food security benefits of organically enhanced fertilizer result from both recycling of sterilized and converted organic wastes (C, amino acids, and micronutrients) and minimizing N losses from land to atmosphere and from land to water, which are characterized by a disrupted N cycle and meet the standards on nutrient management set by Natural Resources Conservations Service of the USA. However, the use of such products at a commercial scale in crop production is still debatable.

2.9.3 Use of Slow Nitrogen Release Fertilizers

Many types of controlled-release or slow-release fertilizers (S-coated, thermosetting resin-coated, and thermoplastic resin-coated fertilizers) have been developed and evaluated for increasing crop yield, fertilizer N use efficiency, and decreasing N losses (Shaviv, 2001; Shoji, 2005; Yang et al., 2012). Enhanced efficiency of N fertilizers that control N release has been available in the U.S. fertilizer market for several years, but their use has been limited due to their higher cost (Snyder, 2008; Stewart, 2008). However, increasing N fertilizer prices, heightened environmental awareness, increasing area under conservation tillage, and improved manufacturing technology have led to an increased interest in the enhanced efficiency of N fertilizers (Rochette et al., 2009a,b; Halvorson et al., 2010b).

Yang et al. (2012) reported that controlled release of urea can improve activities of N metabolism enzymes in leaves, and the efficiency of use of N of rice. Similarly, Cao et al. (2008) reported that enhanced N recovery by rice is accomplished in part through increased activity of the N assimilatory enzymes. Sun et al. (2009) reported that GS (glutamine synthetase) content of leaves could be adopted as an index to assess N accumulation. Significant positive correlation exists between the activity of GS, a key N assimilatory enzyme, and the N status of some higher plants. In addition, growth, yield, and/or protein content are sometimes correlated with GS levels in seeds and/or leaves (Hageman, 1979; Srivastava, 1980; Yang et al., 2012).

The use of slow-release fertilizers can regulate N release from soil according to plant demand and can reduce N losses by leaching and/or denitrification (Mosier et al., 1998; Halvorson et al., 2010a,b; Venterea et al., 2011; Drury et al., 2012). This is especially important in regions where the soil moisture is high in the weeks following planting and before the crop is large enough to assimilate large quantities of soil inorganic N (Drury et al., 2012). Slow-release fertilizer products may also be considered by producers when they do not have time to apply N later in the growing season because of other farming operations and/or because of wet soil conditions (Drury et al., 2012). Soon et al. (2011) evaluated regular and polymer-coated urea for N losses. Their results showed that N_2O emissions were between 1.5 and 1.7 times greater from regular urea than polymer-coated urea.

The loss of N can be reduced by adopting techniques that can reduce nitrification. Shoji et al. (2001) reported that in Colorado, N_2O emissions were very high in the first 3 weeks, following urea application to barley. These authors further reported that when the nitrification inhibitor (DCD) or polyolefin-coated urea was applied, the emissions were considerably lower during these 3 weeks and then increased to similar levels as regular urea after this time period. The net effect relative to regular urea was an 81% reduction in N_2O emissions with DCD and a 35% reduction with coated urea. Polymer-coated urea is used to reduce N losses from the soil via leaching or denitrification by delaying urea hydrolysis until later in the growing season thereby reducing the rate of ammonium and nitrate formation until the plants are larger and have bigger root system and consequently greater N uptake capacity (Drury et al., 2012).

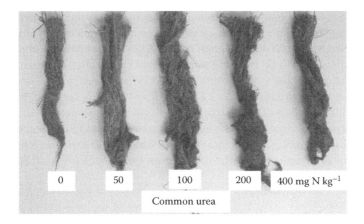

FIGURE 2.2   Root growth of upland rice at different N levels applied by common urea.

Coating urea with NBPT, N-(n-butyl-thiophosphoric triamide), 1.0 g kg^{-10} reduced the cumulative NH$_3$ loss by approximately 66% compared with uncoated urea, although the efficacy of this compound was tied to the size and distribution of precipitation events following fertilization (Engel et al., 2011). Mitigation of NH$_3$ volatilization from urea by NBPT has been attributed to a number of factors, including a moderation of the soil pH rise that results with the production of ammonia bicarbonate (Clay et al., 1990; Christianson et al., 1993) and reduction in the concentration of NH$_4^+$ in the soil solution around the fertilizer placement microsite (Christianson et al., 1993), thereby affecting the NH$_3^+$ hydrolysis. The longevity of NBPT appeared to be greater in the calcareous soil (pH 8.4) than in acidic soil (pH 5.5–6.5) (Engel et al., 2011).

Halvorson and Grosso (2012) reported that N fertilizer source affects the growing season soil N$_2$O emissions from irrigated corn system in Colorado. The use of controlled-release and stabilized N sources reduced N$_2$O emissions under no-till and strip-till corn production systems up to 66% as compared to commonly used urea and up to 43% compared to urea ammonium nitrate. In addition, urease and nitrification inhibitor additions to urea and urea ammonium nitrate resulted in significant reductions in N$_2$O emissions, as did polymer-coated urea. These authors concluded that the choice of N source can be valid management alternatives for reducing N$_2$O emissions to the environment in semiarid areas. They compared the root growth of upland rice at different N rates applied by two N sources (Figures 2.2 and 2.3). The root growth was more vigorous at an adequate N rate in both

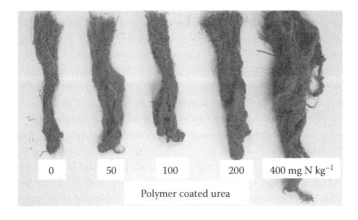

FIGURE 2.3   Root growth of upland rice at different N levels applied by polymer-coated urea.

the sources as compared to the control treatment. Vigorous root system may absorb more water and nutrients, including N and, consequently, the chances are less loss of N from the rhizosphere.

Compared with common urea, polymer-coated urea has been reported to increase apparent N use efficiency in cabbage (*Brassica oleracea* var. *capitata*, Wiedenfeld, 1986), corn (*Zea mays* L., Noellsch et al., 2009), potato (*Solanaum tuberosum* L., Wilson et al., 2009), and barley (*Hordeum vulgare* L., Shoji et al., 2001). Xu et al. (2013) reported that 15–33% of applied N was volatilized as NH_3 under common urea from paddy soils, which was similar to those observed in other studies (Roelcke et al., 2002; Hayashi et al., 2008;). In contrast, only 3–15% of applied N from polymer-coated urea was lost through NH_3 volatilization (Xu et al., 2013). These results are similar to those observed in other studies that showed the application of polymer-coated urea reduced N loss and increased N use efficiency (Chen et al., 2008; Patil et al., 2010; Soon et al., 2011). Polymer-coated urea has been reported to minimize N loss in poorly drained, low-lying areas (Noellsch et al., 2009) and in sandy soils during high rainfall-growing seasons (Zvomuya et al., 2003). Surface applications of polymer-coated urea have also been found to reduce ammonia volatilization loss by 60% as compared to noncoated urea (Rochette et al., 2009a,b).

2.9.4 Adopting Appropriate Cropping System

Adopting the appropriate cropping system is an important strategy in the reduction of gaseous emission by N sources used in crop production (Lemke et al., 1999; Drury et al., 2006; Mosier et al., 2006; Sainju et al., 2012a). The cropping system can affect the quality and quantity of crop residue returned to the soil and can influence the CO_2 and N_2O emissions (Mosier et al., 2006; Dusenbury et al., 2008; Robertson and Vitousek, 2009). The use of the organic farming system is one of the very attractive strategies in improving soil quality and reducing N losses (Teasdale et al., 2007; Lynch et al., 2011). Sainju (2013) studied the effects of tillage, cropping sequence, and the N fertilization rate on N contents in dryland crop biomass, surface residue, and soil at the 0–120 cm depth and also estimated N balance. No-till continuous cropping increased biomass and surface residue of N, but conventional-till crop fallow increased soil available N. Because of increased soil N storage and reduced N requirement to malt barley, NTB-P (no-till malt barley–pea) with 40 kg N ha^{-1} may reduce N loss due to leaching, volatilization, and denitrification compared to other treatments.

Aase and Pikul (1995), Jones and Popham (1997), and Sainju (2013) reported that the traditional farming system using conventional tillage with crop fallow can conserve soil water during fallow, increase N availability due to increased mineralization, control weeds, sustain crop yields, and reduce the risk of crop failure. The system, however, can reduce soil N storage because of increased erosion and mineralization of organic N and reduced plant residue N returned to the soil due to the absence of crops during fallow (Bowman et al., 1999; Halvorson et al., 2002). Enhanced microbial activity due to increased soil temperature and water content during fallow can further reduce N storage (Halvorson et al., 2002). Studies have shown that tillage with crop fallow has reduced soil N storage by 30–50% in the last 50–100 years (Peterson et al., 1998) and reduced annualized crop yields (Aase and Pikul, 1995; Sainju 2013). As a result, the system became unsustainable and uneconomical (Aase and Schaefer, 1996; Dhuyvetter et al., 1996).

2.9.5 Use of Conservation Tillage

Conventional tillage increases nitrification process and hence increases NO_3^- leaching. There are reports in the literature that NO_3^- leaching is lower in the conservation tillage as compared to conventional tillage (Meek et al., 1995). Evans et al. (1996) have reported that greater accumulation of NO_3^- under a grain legume such as peas compared to a cereal, such as barley, indicates greater generation of H^+ ions. The greater NO_3^- concentrations are likely to accelerate NO_3^- leaching, leaving soil acid nonneutralized (Bolan and Hedley, 2003). McKenney et al. (1995) have reported that the addition of residues with high C/N ratio increases the immobilization of N and thereby decreases the nitrate leaching losses. Soil amendment with cereal straw has often found to be effective in reducing NO_3^- concentration under grain legume crops, resulting in reduced acidification through reduced NO_3^- leaching.

2.9.6 USE OF BIOCHAR

One option that has received increasing attention in recent years is the application of charred biomass (biochar) to agricultural soils to protect N loss and leaching from top soil layer (Lehmann et al., 2006; Atkinson et al., 2010; Bruun et al., 2012). *Biochar* is currently the accepted term for pyrolysis-derived charcoal when designated for use as a soil amendment (Sohi et al., 2010). However, *char, charcoal*, and *black carbon*, and other names for this product have also been used (Streubel et al., 2011). Biochar is produced through thermal decomposition (400–600°C) of biomass in the absence of oxygen (pyrolysis) and is defined as "charred organic matter produced and applied to soil in a deliberate manner, with the intent to sequester C and improve soil properties" (Lehmann and Joseph, 2009). The ability of biochar to retain N and other nutrients has been documented in several studies (Ding et al., 2010; Laird et al., 2010). The ammonium that was readily adsorbed to biochar has been reported by Ding et al. (2010) and Dunisch et al. (2007).

The beneficial effect of biochar on N retention has been reported in the field trials (Steiner et al., 2008). Biochar has large surface area and also increases cation-exchange capacity of soil that may be responsible for nutrient retention (Liang et al., 2006; Van Zwieten et al., 2009; Novotny et al. 2009; Lehmann and Joseph, 2009; Roberts et al., 2010). Biochar application in soils has been reported to reduce the nitrification process in the soil by reducing the oxygen and/or substrate availability—for example, by N adsorption to biochar surfaces (Laird et al., 2010) and by microbial N immobilization (Kolb et al., 2009; Novak et al., 2010; Bruun et al., 2011). In addition, the nitrification process inhibited by microbial toxic biochar substances could also explain the lower production of NO_3^- (DeLuca et al., 2006).

Biochars nutrient retention potential is also physically influenced by the biochar particle size. Smaller particles have greater surface area/volume ratios than larger particles and thus, in general, a larger capacity to hold nutrients (Bruun et al., 2012). However, fine biochar particles or components may also be transported downward in soil with the water movement or horizontally by surface water runoff (Leifeld et al., 2007; Major et al., 2010).

Much of the recent interest in biochar as a soil amendment was prompted by studies of Amazonian soils (Terra Preta), where the presence of charcoal was associated with significant improvements in soil quality such as organic matter and nutrient concentrations and increase in crop yields (Glaser et al., 2002; Lehmann, 2007; Lehmann and Joseph, 2009; Novotny et al., 2009). The application of charcoal reportedly increased soil pH, and in acid soils, charcoal decreases the concentration of Al, which often limits crop growth in the tropics (Piccolo et al., 1997). In many tropical and subtropical soils, charcoal increases base saturation, cation-exchange capacity and nutrient availability, decreases soil bulk density, and improves water-holding capacity (Liang et al., 2006; Busscher et al., 2010; Novak et al., 2010). The addition of biochar to soil also increases soil C concentrations that further improve nutrient storage and soil physical properties (Streubel et al., 2011).

An outstanding attribute of biochar is its high sorption affinity and capacity for organic compounds, generally far exceeding those of humic substances and soil organic matter (Zhang et al., 2006; Yang et al., 2009; Chen and Chen, 2009; Graber et al., 2011). At the same time, chars frequently demonstrate sorption hysteresis or hindered desorption kinetics (Braida et al., 2003). The sorption capacity of char may be important to prevent N from the soil–plant system.

2.9.7 USE OF GYPSUM

The use of gypsum can reduce NH_4^+ loss in the runoff (Favaretto et al., 2012). The beneficial effect of gypsum on decreasing NH_4^+ losses in runoff was also observed by Favaretto et al. (2008), due to high leaching as a result of NH_4^+ displacement on the soil exchange complex by Ca^{2+} (Koenig and Pan, 1996). The increase in NH_4^+ concentration in soil solution due to application of gypsum can be beneficial for plants due to the high absorption of NH_4^+ as compared to NO_3^-. The absorption of NH_4^+ requires less energy by plants as compared to absorption of NO_3^- (Fageria et al., 2011). Leaching is the main process by which NO_3^- is transported soil to water (Smith et al., 1990). For NH_4^+, losses by leaching are lower than those for NO_3^- (Piovesan et al., 2009) due to the adsorption of NH_4^+ to the soil particle surfaces (Sparks, 1995).

In addition, gypsum application improves soil structure in heavy-textured soil, so that water infiltration and the ability of roots to penetrate the soil are enhanced (Viator et al., 2002). The improvement in the root growth of upland rice and soybean with the application of gypsum in Brazilian Oxisols has been reported by Fageria (2013). Improved root growth may absorb a higher amount of N that may reflect in increasing crop yields (Fageria, 2013). Improvements in the yield of wheat and sorghum with the addition of gypsum have been reported by Thomas et al. (1995). Gypsum application in irrigation water increased the sugar yield and juice extraction percentage of sugarcane (Kumar et al., 1999). Gypsum also increased the yield in corn and alfalfa up to 50%. This yield response was partially attributed to higher exchangeable Ca and S, and a complementary reduction in exchangeable Al (Toma et al., 1999). Reduction in exchangeable Al may improve root growth and consequently higher N uptake and efficiency of use. For some crops, gypsum is effective in reducing the incidence of soil-borne diseases (Kao and Ko, 1986). The decrease in disease infestation may improve N use efficiency in crop plants.

Subsoil acidity is one of the major yield-limiting factors in acid soils because they restrict root growth (Toma et al., 1999). Surface application of gypsum is an effective technique to ameliorate the effects of subsoil acidity (Ritchey et al., 1980; Hammel et al., 1985; Shainberg et al., 1989; Alcordo and Rechcigl, 1993; Sumner, 1993, 1995; Saigusa et al., 1996; Toma et al., 1999). Vigorous root growth in the soil profile may improve water and nutrient uptake, higher yield, and a higher N efficiency of use. Toma et al. (1999) concluded that the gypsum effect is so long lasting; its use as a subsoil acidity ameliorant becomes highly economic because the initially high cost can be amortized over an extended period of time.

2.9.8 Use of Crop Genotypes Producing High Root Biomass

The use of crop genotypes producing high root biomass is an important strategy in reducing N loss from soil–plant systems. Johnson et al. (2006) summarized the contributions of different plant parts from different plant species to soil organic carbon and gave guidelines, including the contributions of plant roots and rhizodeposition to the total C cycle when analyzing changes in soil organic carbon. The belowground deposition of fixed C in structural root biomass, exudates, mucilage, and sloughed cells may be a major source for soil organic carbon accumulation (Bottner et al., 1999; Allmaras et al., 2004). Benjamin et al. (2010) reported that the contribution of the crop root system to the formation and increase of soil organic carbon is important when considering the selection of a crop rotation in a cropping system. Fageria (2013) discussed the variation in crop species and genotypes within species in root biomass production. Crop plants having C4 photosynthetic pathway (corn, pearl millet, and sorghum) produced more vigorous root system as compared to crop plants with C3 (rice, barley, wheat, and dry bean) photosynthetic pathway (Fageria, 2013). Glass (2003) reported that nutrient uptake by plants is a function of root biomass, root morphology, root age, root/plant growth rates, and root proliferation in regions of abundant nutrients, in addition to the roots physiological capacity for nutrient uptake (Glass, 2003).

The author studied root growth of 12 lowland rice genotypes. The maximum root length was significantly influenced by N rate and genotype treatments (Table 2.2). However, N × G interaction was not significant for this growth parameter, indicating that each of the 12 genotypes reacted similarly to changes in N rates. Maximum root length varied from 21.17 to 29 cm, with an average value of 24.84 cm. Root length was significantly higher at a low N rate as compared to a high N rate. When

TABLE 2.2

Maximum Root Length and Root Dry Weight of Lowland Rice Genotypes as Influenced by N and Genotype Treatments

N Rate/ Genotype	Maximum Root Length (cm)	Root Dry Weight (g plant^{-1})	
		N 0 mg kg^{-1}	N 300 mg kg^{-1}
0 mg N kg^{-1}	26.58a		
300 mg N kg^{-1}	23.61b		
BRS tropical	29.00a	6.04a	5.59d
BRS Jaçanã	24.00ab	4.85abc	2.67e
BRA 02654	25.83ab	5.88ab	8.56bc
BRA 051077	23.83ab	5.59ab	13.10a
BRA 051083	26.00ab	3.09c	6.29d
BRA 051108	27.66ab	6.76a	9.03b
BRA 051126	24.50ab	5.06abc	6.41d
BRA 051129	21.17b	5.85ab	8.88b
BRA 051130	25.67ab	3.73bc	8.64bc
BRA051134	23.50	5.96ab	5.91d
BRA 051135	23.00ab	4.66abc	5.33d
BRA 051250	27.00ab	4.57abc	6.81cd
Average	24.84	5.17b	7.27a
F-test			
N rate (N)	**	**	
Genotype (G)	*	**	
N × G	NS	**	
CV (%) N rate		8.94	
CV (%) Genotype		11.77	

*,**,NS: Significant at the 5% and 1% probability level and nonsignificant, respectively. Means followed by the same letter in the same column (separate for N rate and genotypes) are not significantly different at the 5% probability level by Tukey's test.

there is deficiency of a determined nutrient, roots try to grow longer to take nutrients from the lower soil depth (Fageria and Moreira, 2011). Visually, roots at a lower N rate were having less and smaller hairs as compared to higher N rate.

Root dry weight was significantly influenced by N rate and genotype treatments (Table 2.2). The N × G interaction was also significant for root dry weight, suggesting that genotypes responded differently under two N rates. At a lower N rate, the root dry weight varied from 3.09 g plant^{-1} produced by the genotype BRA 051083 to 6.76 g plant^{-1} produced by genotype BRA051108, with an average value of 5.17 g plant^{-1}. At a higher N rate, the root dry weight varied from 2.67 g plant^{-1} produced by genotype BRS Jaçanã to 13.10 g plant^{-1} produced by genotype BRA 051077, with an average value of 7.27 g plant^{-1}. The application of 300 mg N kg^{-1} produced 41% higher dry weight compared with 0 mg N kg^{-1} of soil. Figures 2.4 and 2.5 show the root growth of the genotypes BRA 02654 and BRA 051108, respectively, at low and high N rates.

2.9.9 INTERCALATION OF UREA IN MONTMORILLONITE

Soil particles could inhibit the volatilization of NH_3 through adsorption reactions, dissolution by adsorbed water, and the physical disturbance of sunlight, heat, airflow, and so on. In fact, a noticeable decrease in n loss could be achieved by simply covering the broadcast urea with soil (Griggs et al., 2007; Kwon et al., 2009), which is rarely put into practice due to the labor cost. Therefore,

FIGURE 2.4   Root growth of lowland rice genotype BRA 02654 at two N levels.

new strategic approaches using natural clay minerals have been developed for efficient practical delivery of fast-acting N fertilizers into deeper soils, through salt occlusion in zeolites (Park and Komarneni, 1998; Park et al., 2005) and urea intercalation in swelling-layered clay minerals (Park et al., 2004). Among the various clay minerals, negatively charged swelling mica-type-layered clay minerals exhibit a high potential for urea delivery because their expandable interlayer space allows

FIGURE 2.5   Root growth of lowland rice genotype BRA 051108 at two N levels.

a considerable amount of urea to be accommodated and released (Kim et al., 2011). Furthermore, their high capacity for water retention and cation adsorption could play an excellent role in suppressing NH_3 emission by exchanging NH_4^+ ions (Kim et al., 2011). Several studies have reported that urea intercalated in montmorillonite leads to a considerable improvement of N use efficiency in soils (Park et al., 2004; Kim et al., 2010, 2011).

2.9.10 ADOPTING CROP–LIVESTOCK INTEGRATION IN CROPPING SYSTEM

Agricultural production systems have trended toward high-input, highly specialized monoculture and simple rotation systems. While these systems have greatly increased productivity, the stress on ecosystems and natural resources has also increased. Adopting crop–livestock integration (CLIS) is an important strategy in reducing N losses from soil–plant system. Crop and livestock activities can compromise the sustainability of an agricultural system (Crusciol et al., 2013). These two activities in isolation can reduce soil fertility, increase soil erosion, decrease soil organic matter and crop yield, and increase the risk of pasture degradation (Crusciol et al., 2010). However, adopting CLIS can decrease the negative effects, being a mixed system of land use characterized by diversification, crop rotation, intercropping systems (cash crop with forage species), and crops and livestock in the same area (Tracy and Zhang, 2008; Crusciol et al., 2013).

Russelle et al. (2007) reviewed numerous studies regarding integrated crop–livestock systems and concluded that the systems may provide economic (Small and McCaughey, 1999) and environmental benefits (Drinkwater et al., 1998). An integrated crop–livestock system has been shown to reduce irrigation demand as compared to a monoculture cotton production system in the Texas High Plains (Allen et al., 2005, 2012; Johnson et al., 2013). Johnson et al. (2013) also reported that the integrated system could be a viable alternative in an area of the region where irrigation is limited due to aquifer depletion and/or pumping regulation. The integrated system had less economic risk associated with the variation in profitability and added ecological diversity that benefited soil health and wildlife populations (Johnson et al., 2013).

In recent years, Brazil has been increasingly using CLIS, which includes, as one of its modalities, the intercropping of grain crops such as corn, soybean, rice, sorghum, and dry bean with perennial tropical forage (mainly palisadegrass—*Brachiaria brizantha* Hochst. Ex A. Rich Stapf) in no-tillage system (Crusciol et al., 2013). This intercropping is an outstanding alternative for forage production during the dry season and provides residue for the no-till system in the next crop season (Crusciol et al., 2010).

Palisadegrass is a perennial forage species originating in Africa, and is well adapted to Brazilian climates. It has a deep root system and can tolerate drought and can produce dry biomass up to 20 Mg ha^{-1} a year (Crusciol et al., 2010). Its addition in the CLIS system can provide adequate residue on the soil surface throughout the year, improves dynamic cycling of nutrients as well as physical characteristics of the soil, and, therefore, proper nutrition and yield of the following crop (Baributsa et al., 2008; Crusciol et al., 2013). Crusciol et al. (2013) reported that the adoption of CLIS in a no-tillage system has been one of the best alternatives in guiding tropical agricultural systems toward sustainability and can result in diversity of production, higher economic returns, and improve crop growing conditions, with emphasis on nutrient cycling and soil quality. Pacheco et al. (2011) reported that palisadegrass growing in the dry season had 142 kg ha^{-1} N, 14 kg ha^{-1} P, 127 kg ha^{-1} K, 91 kg ha^{-1} Ca, and 59 kg ha^{-1} Mg in its tops at the beginning of the subsequent wet season. Crusciol et al. (2010) reported that, after two growing seasons, the soil organic matter in the top 0–5 cm soil layer was 2.5 g m^{-3} in a corn and palisadegrass intercropping system and 2.2 g m^{-3} for corn alone, differing significantly.

2.9.11 ADOPTING ORGANIC FARMING

Several authors have examined the issue of nitrate leaching from arable land or farming systems and concluded that agriculture activities are one of the main sources of NO_3 leaching to aquifers and other natural water sources such as lakes and rivers (Schroder, 1985; Spalding and Exner, 1993; Davies and Sylvester-Bradley, 1995; Krichmann and Bergstrom, 2001). Krichmann and Bergstrom (2001) defined organic farming as a form of agriculture based on the following principles: (i) prohibit the use of synthetic fertilizers and pesticides, and rely upon crop rotation, animal manures, crop residues, and green manure to maintain adequate soil fertility; (ii) enhance and improve the biological conditions for symbiotic N_2 fixation; (iii) emphasize recycling of animal manures; and (iv) create a balance between the number of animals and the cultivated area. Krichmann and Bergstrom (2001) reviewed literature on organic farming and nitrate leaching to lower soil profiles or natural water sources and concluded that lower yields due to lower nutrient inputs in organic cropping systems are often followed by lower N leaching loads than in more intensive conventional systems.

2.9.12 ADOPTING AGROFORESTRY IN CROPPING SYSTEMS

Agroforestry (AGF) is defined as an intensive land management practice that optimizes the benefits (physical, biological, ecological, economical, and social) arising from biophysical interactions created when trees and/or shrubs are deliberately combined with crops and/or livestock (Garrett et al., 1994). Including AGF in cropping systems improves internal drainage, enhances infiltration, reduces runoff, traps sediments and nonpoint source pollution, and creates wildlife habitat and connective travel corridors (Omernick et al., 1981; Schmitt et al., 1999; Udawatta et al., 2002; Qiu, 2003; Schultz et al., 2009; Gold and Garrett, 2009). The inclusion of trees in AGF buffers integrates long-term environmental benefits of forest ecosystems such as improved soil physical properties, moisture storage capacity, soil quality, and carbon sequestration into agricultural land (Seobi et al., 2005; Kumar et al., 2008; Jose, 2009; Anderson et al., 2009; Senavirante et al., 2012).

Adopting AGF system is an important strategy in reducing soil degradation, improving crop yields, and N use efficiency (Dossa et al., 2012). AGF systems that combine trees or shrubs with crops can potentially provide organic inputs and improve nutrient recovery and use efficiency (Mafongoya et al., 2006). The mechanisms include nutrient recovery from subsoil layers not exposed by the crop that are deposited on the surface layer through litter input and root turnover, reduced nutrient loss (Young, 1989; Gathumbi et al., 2004), and improved soil, physical, chemical, and biological properties (Buresh and Tian, 1998). Hartemink et al. (2000) in Kenya showed that *Sesbania sesban* during fallow periods can retrieve considerable subsoil inorganic N in highly weathered Alfisol and Oxisol. This N was shown to be effectively recycled for subsequent crops.

More than 95% of farmland in sub-Saharan Africa is rain-fed, and crop yields are generally low and variable as a consequence of variable rainfall, drought, and land degradation (Wani et al., 2009). Corn accounts for >50% of the cropped area and the calories consumed in many countries in sub-Saharan Africa (Sileshi et al., 2010). Growing corn in association with legume trees in AGF arrangements has been shown to increase yields in many parts of sub-Saharan Africa (Sileshi et al., 2012). Sileshi et al. (2011) also reported that integration of legume trees into cropping systems is one option to mitigating land degradation because they add considerable amounts of organic matter and N to the soil (Snapp et al., 1998; Mafongoya et al., 2006; Akinnifesi et al., 2007; Beedy et al., 2010). More than two decades of AGF research in southern Africa has shown that organic matter added to the system increases the structural stability of the soil, resistance to rainfall impact, infiltration rates, and faunal and microbial activities (Sileshi and Mafongoya, 2006; Beedy et al., 2010).

Dossa et al. (2012) reported that the indigenous shrub, *Guiera senegalensis*, coexists with crops to varying degrees in farmer fields throughout the drier Sahel (Africa), as an integral part of the cropping systems. The presence of this shrub in the cropping systems improved crop yields as compared to plots with no shrub (Dossa et al., 2012). These authors further reported that the improvement in crop yields in association with *Guiera senegalensis* shrub was related to improved nutrient availability and higher soil quality measured in terms of particulate organic matter content. Dossa et al. (2012) further reported that the positive crop response in the presence of shrubs in a dry year suggests that shrubs were not significantly competing with the crop for water. This was indeed found by Kizito et al. (2007) who showed no competition for water between shrubs and millet and a significantly higher soil moisture profile in millet–shrub intercrop plots than in sole millet plots on the same experimental plots (Dossa et al., 2012). Additionally, *Guiera senegalensis* could have been providing water by performing hydraulic lift (Richards and Caldwell, 1987) of water from the wet subsoil to the dry surface soil as shown by Kizito et al. (2007).

2.9.13 USE OF NATURAL ZEOLITE

Among the effective solutions for minimizing N losses from soil–plant systems, the use of natural minerals such as zeolite has been recommended (Gholamhoseini et al., 2012). Zeolites are hydrated aluminosilicates, characterized by a three-dimensional network of SIO_4 and AlO_4, which are linked by shared oxygen atoms. Partial substitution of silicon and aluminum results in an excessive negative charge, which is compensated by cations (Gholamhoseini et al., 2012). These cations are located with water molecules in the cavities and channels inside the aluminosilicate framework. Water and cations can be removed or replaced by other cations (Rehakova et al., 2004; Murphy et al., 2005). Selective absorption and controlled release of cations by zeolite enhances nutrient availability and improves plant growth and development (Gholamhoseini et al., 2012).

Zeolite has certain unique features such as high cation-exchange capacity (200–300 $cmol_c\ kg^{-1}$) (Leggo et al., 2006), selective absorption, slow release of ammonium (He et al., 2002), and structure stability over the long term (Huang and Petrovic, 1994). Regarding zeolite availability, Mumpton (1999) reported that extensive deposits of zeolite have been found in western United States, Bulgaria, Hungary, Japan, Australia, China, and Iran. In Iran, the cost of zeolite is approximately 2.5 cents (U.S.)/kg; so, the application of zeolite would be economical (Gholamhoseini et al., 2012). Many studies have reported that the use of zeolite with organic fertilizers improved their efficiencies in crop production (Daryaei et al., 2011; Khodaei et al., 2012). Hence, it is expected that the application of zeolite with chemical fertilizers, especially N fertilizers, can increase efficiency of use and optimize chemical fertilizer use (Gholamhoseini et al., 2012). Gholamhoseini et al. (2012) reported that a combined application of zeolite and chemical N for canola (*Brassica napus* L.) production in a poor sandy soil is recommended to ensure an acceptable forage yield and soil protection from excess N leaching loss.

2.9.14 CALCIUM STIMULATES AMMONIUM ABSORPTION

Calcium ion in soil solution in an adequate amount was reported to increase NH_4^+ absorption by plants (Horst et al., 1985; Taylor et al., 1985; Fenn et al., 1987, 1994). Strong substantiation for the use of Ca^{2+} to increase plant growth, in the presence of NH_4^+, was reported by Sung and Lo (1990). They showed that tillering and seed weight of rice substantially increased with increasing Ca^{2+} concentration. These authors concluded that the increase in plant growth was related to an increase in photosynthetic activity. Robinson and Baysdorfer (1991) also reported an increase in photosynthetic activity with the addition of NH_4^+ along with calcium in soybean. Plant yields were increased in calcareous soils where Ca^{2+} was applied with urea, compared with yield from plots fertilized with urea alone, NH_4NO_3, or other common N fertilizers (Horst et al., 1985; Taylor et al., 1985; Fenn, 1986; Fenn et al., 1987). The Ca^{2+} concentrations needed to enhance absorption in calcareous soils were above those normally considered necessary for adequate plant nutrition (Fenn et al., 1994). Fenn et al. (1991, 1994) and Fenn and Miyamoto (1981) reported that calcium might stimulate metabolite transfers within plants and enhance photosynthesis. The higher uptake of NH_4^+ in the presence of Ca^{2+} may reduce the conversion of ammonium into NO_3^-; hence, less leaching or loss of N from soil–plant system.

2.9.15 ADOPTING APPROPRIATE INTERCROPPING

Adopting the appropriate intercropping in the cropping system is an important strategy in improving N losses from soil–plant system and improving their efficiency. Intercropping is an agricultural production system in which two or more species develop simultaneously during part or the entire growing season and compete with each other for available resources (Fukai and Trenbath, 1993). Such type of a cropping system improves the efficiency of resources such as water and nutrients in comparison with their sole crop counterparts (Andrews and Kassam, 1976; Ofori and Stern, 1987; Caviglia et al., 2004; Echarte et al., 2011; Coll et al., 2012; Andrade et al., 2012). In addition to the improvement in resource use efficiency, other mechanisms that can contribute to the increased yield observed in intercropping systems include increased ecological stability and resilience (Reddy and Willey, 1981; Tilman, 1996; Trenbath, 1999; Szumigalski and Van Acker, 2008). Indeed, there are many examples of intercropping systems that have demonstrated greater grain or forage yield as compared to monoculture systems on an equivalent land area basis (Ikeorgu et al., 1989; Chen et al., 2004; Agegnehu et al., 2006; Ghosh et al., 2006; Wortman et al., 2012).

Sunflower and soybean are two species that can be intercropped (Calvino and Monzon, 2009). Andrade et al. (2012) reported higher yields of sunflower–soybean intercrops in the southern Pampas of Argentina compared to sole crops. The greater intercrop yield was associated with an increase in capturing the resource use efficiency (Andrade et al., 2012). Intercropped sunflower and soybean complement each other in the use of the resource because critical periods for yield determination occur at different times during a period of low resource demand by the other component (Coll et al., 2012). The critical period for yield determination of a crop is defined as the stage where a reduction in resources availability (water, nutrients, and solar radiation) determines the greater grain yield lost (Cantagallo et al., 2004; Egli and Bruening, 2005). In the southern Pampas of Argentina, sunflower sown in mid-October has its critical period for yield determination when soybean (sown in mid-November) is still in its vegetative stages, and soybean critical period takes place close to sunflower maturity (Andrade et al., 2012).

Intercropping soybean and palisadegrass (*Bracharia brizantha* Hochst. Ex A. Rich) favors soybean production as it reduces weeds and breaks the pest disease cycle (Silva et al., 2006). In addition, palisadegrass has an aggressive root system that favors nutrient cycling, increased soil biological activity, and increased organic matter content (Dabney et al., 2001; Nacent and Crusciol, 2012). Furthermore, palisadegrass produces a high amount of dry matter, which provides good ground cover in no-tillage system, with greater persistence in the soil surface (Crusciol et al., 2012). Borghi et al. (2007) also reported that soybean can supply N to the palisadegrass from the biological fixation of soybean.

2.10 CONCLUSIONS

N is the most essential plant nutrient for crop production. Soils cultivated with crops are always deficient in N and the use of chemical N fertilizers is a prerequisite to obtaining a higher yield. N cycle in soil–plant system is very dynamic and influenced by soil, plant, and climatic factors. In modern agriculture, the major source of N is chemical fertilizers. The major part of N applied to plants is lost through leaching, volatilization, denitrification, surface erosion, and loss of NH_3 from plant foliage. Among these processes of N loss, the major part of N loss occurs through leaching and denitrification. These two processes can constitute about 50% of the total N loss from the soil–plant system. N loss through ammonia volatilization can be significant if N is not incorporated into the soil and no irrigation and/or precipitation occurred after N fertilization was applied to the soil surface. Factors affecting N losses from soil–plant system are precipitation, temperature, soil pH, soil texture, methods of N application, N source and particle size, and plant factors. A discussion on these factors is provided in this chapter.

Some practices that can be adopted to reduce N losses are irrigation immediately after N fertilizer application, incorporation of applied N fertilizers into the soil, and the use of a slow-release source of N fertilizer such as polymer-coated or sulfur-coated urea. The use of adequate rate, source, and timing of N application can also reduce N losses. In addition, the use of gypsum, use of biochar, and intercalation of urea in montmorillonite can also reduce N losses and improve N use efficiency in crop plants. Management practices should aim at the optimum utilization of soil N and applied fertilizer N, and water, through the increased production of plant root biomass, and the retention of N in the root zone.

REFERENCES

Aase, J. K. and J. L. Pikul. Jr. 1995. Crop and soil response to long-term tillage practices in the northern Great Plains. *Agron. J.* 87:652–656.

Aase, J. K. and G. M. Schaefer. 1996. Economics of tillage practices and spring wheat and barley crop sequence in northern Great Plains. *J. Soil Water Conserv.* 51:167–170.

Agegnehu, G., A. Ghizaw, and W. Sinebo. 2006. Yield performance and land-use efficiency of barley and faba bean mixed cropping in Ethiopian highlands. *Eur. J. Agron.* 25:202–207.

Akinnifesi, F. K., W. Makumba, G. Sileshi, O. C. Ajayi, and D. Mweta. 2007. Synergistic effect of inorganic N and P fertilizers and organic inputs from *Gliricidia sepium* on productivity of intercropped maize in southern Malawi. *Plant Soil* 294:203–217.

Alcordo, I. S. and J. E. Rechcigl. 1993. Phosphogypsum in agriculture: A review. *Adv. Agron.* 49:55–118.

Allen, V. G., P. Brown, R. Kellison, P. Green, C. J. Zilverberg, and P. Johnson. 2012. Integrating cotton and beef production in the Texas southern High Plains: I. Water use and measurement of productivity. *Agron. J.* 104:1625–1642.

Allen, V. G., P. Brown, R. Kellison, E. Segarra, T. Wheeler, and P. A. Dotry. 2005. Integrating cotton and beef production to reduce water withdrawal from the Ogallala aquifer in the southern High Plains. *Agron. J.* 97:556–567.

Allmaras, R. R., D. R. Linden, and C. E. Clapp. 2004. Corn-residue transformation into root and soil carbon as related to nitrogen, tillage, and stover management. *Soil Sci. Soc. Am. J.* 68:1366–1375.

Amer, F. M. and M. V. Bartholomew. 1951. Influence of oxygen concentration in soil air on nitrification. *Soil Sci.* 71:215–219.

Anderson, S. H., R. P. Udawatta, T. Seobi, and H. E. Garrett. 2009. Soil water content and infiltration in agroforestry buffer strips. *Agrofor. Syst.* 75:5–16.

Andrade, J. F., A. Cerrudo, R. H. Rizzalli, and J. P. Monzon. 2012. Sunflower–soybean intercrop productivity under different water conditions and sowing managements. *Agron. J.* 104:1049–1055.

Andrews, D. J. and A. H. Kassam. 1976. The importance of multiple cropping in increasing world food supplies. In: *Multiple Cropping*, eds., R. I. Papendic, A. Sanchez, and G. B. Triplett, pp. 1–10. Madison, Wisconsin: ASA.

Aneja, V. P., D. Nelson, P. Roelle, and J. Walker. 2003. Agricultural ammonia emissions and ammonium concentrations associated with aerosols and precipitation in the southeast United States. *J. Geophys. Res.* 108 (D4):12-1–12-11.

Angers, D. A., N. Samson, and A. Legere. 1993. Early changes in water stable aggregation induced by rotation and tillage in a soil under barley production. *Can. J. Soil Sci.* 73:51–59.

Angle, J. S., C. M. Gross, R. L. Hill, and M. S. Mcintosh. 1993. Soil nitrate concentrations under corn as affected by tillage, manure and fertilizer applications. *J. Environ. Qual.* 22:141–147.

Arriaga, H., G. Salcedo, L. Martinez-Suller, S. Calsamigli, and P. Merino. 2010. Effect of dietary crude protein modification on ammonia and nitrous oxide concentration on a tie-stall dairy barn floor. *J. Dairy Sci.* 93:3158–3165.

Asman, W. A. H., M. A. Sutton, and J. K. Schjorring. 1998. Ammonia: Emission, atmospheric transport and deposition. *New Phytol.* 139:27–48.

Atkinson, C. J., J. D. Fitzgerald, and N. A. Hipps. 2010. Potential mechanisms for achieving agricultural benefits from biochar application to temperate soils: A review. *Plant Soil* 337:1–18.

Aulakh, M. S., J. W. Doran, and A. R. Mosier. 1992. Soil denitrification—Significance, measurement, and effects of management. *Ad. Soil Sci.* 18:1–57.

Aulakh, M. S., J. W. Doran, D. T. Walters, and J. F. Power. 1991. Legume residues and soil water effects on denitrification in soils of different textures. *Soil Biol. Biochem.* 23:1161–1167.

Aulakh, M. S., D. A. Rennie, and E. A. Paul. 1983. The effect of various clover management practices on gases N losses and mineral N accumulation. *Can. J. Soil Sci.* 63:593–605.

Aulakh, M. S., D. A. Rennie, and E. A. Paul. 1984. Gases nitrogen losses from soils under zero till as compared with conventional-till management systems. *J. Environ. Qual.* 13:130–136.

Bailey, L. D. 1976. Effects of temperature and root on denitrification in a soil. *Can. J. Soil Sci.* 56:79–87.

Baker, J. L. and H. P. Johnson. 1981. Nitrate–nitrogen in tile drainage as affected by fertilization. *J. Environ. Qual.* 10:519–522.

Baributsa, D. N., E. F. F. Foster, K. D. Thelen, A. N. Kravchenko, D. R. Mutch, and M. Ngouajio. 2008. Corn and cover crop response to corn density in an interseeding system. *Agron. J.* 100:981–987.

Bauer, S., D. Lloyd, B. P. Horgan, and D. J. Soldat. 2012. Agronomic and physiological responses of cool-season turfgrass to fall-applied nitrogen. *Crop Sci.* 52:1–10.

Beedy, T. L., S. S. Snapp, F. K. Akinnfesi, and G. W. Sileshi. 2010. Impact of *Gliricidia sepium* intercropping on soil organic matter fractions in a maize-based cropping system. *Agric. Ecosyst. Environ.* 138:139–146.

Benjamin, J. G., A. D. Halvorson, D. C. Nielsen, and M. M. Mikha. 2010. Crop management effects on crop residue production and changes in soil organic carbon in the central Great Plains. *Agron. J.* 102:990–997.

Beyrouty, C. A., L. E. Sommers, and D. W. Nelson. 1988. Ammonium volatilization from surface-applied urea as affected by several phosphoroamide compounds. *Soil Sci. Soc. Am. J.* 52:1173–1178.

Black, C. A. 1968. *Soil–Plant Relationships*, 2nd edition. New York: John Wiley & Sons.

Black, A. S., R. R. Sherlock, and N. P. Smith. 1987. Effect of timing of simulated rainfall on ammonia volatilization from urea, applied to soil of varying moisture content. *J. Soil Sci.* 38:679–687.

Black, A. S. and S. A. Waring. 1972. Ammonium fixation and availability in some cereal producing soils in Queensland. *Aust. J. Soil Res.* 10:197–207.

Blackmer, A. M. 2000. Bioavailability of nitrogen. In: *Handbook of Soil Science*, ed., M. E. Sumner, pp. 1–17. Boca Raton, Florida: CRC Press.

Blankenau, K., H. W. Olfs, and H. Kuhlman. 2000. Strategies to improve the use efficiency of mineral fertilizer nitrogen applied to winter wheat. *J. Agron. Crop Sci.* 188:146–154.

Bolan, N. S. and M. J. Hedley. 2003. Role of carbon, nitrogen, and sulfur cycles in soil acidification. In: *Handbook of Soil Acidity*, ed., Z. Rengel, pp. 29–56. New York: Marcel Dekker.

Borghi, E., N. V. Costa, C. A. C. Crusciol, and G. P. Matheus. 2007. Influence of the spatial distribution of maize and *Brachiaria brizantha* intercropping on the weed population under no-tillage. *Planta Daninha* 26:559–568.

Boswell, F. C., J. J. Meisinger, and N. L. Case. 1985. Production, marketing, and use of nitrogen fertilizers. In: *Fertilizer Technology and Use*, ed., O. P. Engelstad, pp. 229–292. Madison, Wisconsin: SSSA.

Bottcher, J., D. Weymann, R. Well, C. V. Heide, A. Schwen, H. Flessa, and W. H. M. Duijnisveld. 2011. Emission of groundwater-derived nitrous oxide into the atmosphere: Model simulations based on a ^{15}N field experiment. *Eur. J. Soil Sci.* 62:216–225.

Bottner, P., M. Pansu, and Z. Sallih. 1999. Modeling the effect of active roots on soil organic matter turnover. *Plant Soil* 216:15–25.

Bouwman, A. F., L. J. M. Boumans, and N. H. Batjes. 2002. Estimation of global NH_3 volatilization loss from synthetic fertilizers and animal manure applied to arable lands and grasslands. *Global Biogeochem. Cycle* 16:8–13.

Bouwmeester, R. J. B., P. L. G. Vlek, and J. M. Stumpe. 1985. Effect of environmental factors on ammonia volatilization from a urea fertilized soil. *Soil Sci. Soc. Am. J.* 49:376–381.

Bouwmeester, R. J. B., P. L. G. Vlek, and J. M. Stumpe. 1985. Effect of environmental factors on ammonia volatilization from urea fertilized soil. *Soil Sci. Soc. Am. J.* 49:376–381.

Bowman, D. C., J. L. Paul, W. B. Davis, and S. H. Nelson. 1987. Reducing ammonium volatilization from Kentucky bluegrass turf by irrigation. *HortScience* 22:84–87.

Bowman, R. A., M. F. Vigil, D. C. Nielsen, and R. L. Anderson. 1999. Soil organic matter changes in intensively cropped dryland systems. *Soil Sci. Soc. Am. J.* 63:186–191.

Boyer, E. W., C. L. Goodale, N. A. Jaworski, and R. W. Howarth. 2002. Anthropogenic nitrogen sources and relationships to riverine nitrogen export in the northeastern USA. *Biogeochemistry* 57/58:137–169.

Brady, N. C. and R. E. Weil. 2002. *The Nature and Properties of Soils*, 13th edition. Upper Saddle River, New Jersey: Prentice-Hall.

Braida, W. J., J. J. Pignatello, Y. F. Lu, P. I. Ravikovitch, A. V. Neimark, and B. S. Xing. 2003. Sorption hysteresis of benzene in charcoal particles. *Environ. Sci. Technol.* 37:409–417.

Broderick, G. A. and M. K. Clayton. 1997. A statistical evaluation of animal and nutritional factors influencing concentrations of milk urea nitrogen. *J. Dairy Sci.* 80:2964–2971.

Bruun, E. W., D. Miller-Stover, P. Ambus, and H. Hauggaard-Nielsen. 2011. Biochar soil application and N_2O emissions—Potential effects of blending biochar with anaerobically digested slurry. *Eur. J. Soil Sci.* 62:581–589.

Bruun, E. W., C. Peterson, B. W. Strobel, and H. H. Nielsen. 2012. Nitrogen and carbon leaching in repacked sandy soil with added fine particulate biochar. *Soil Sci. Soc. Am. J.* 76:1142–1148.

Brye, K. R., J. M. Norman, L. G. Bundy, and S. T. Gower. 2001. Nitrogen and carbon leaching in agroecosystems and their role in denitrification potential. *J. Environ. Qual.* 30(58):58–70.

Buchkina, N., E. Balashov, E. Rizhiya, and K. Smith. 2010. Nitrous oxide emissions from a light-textured arable soil of north-western Russia: Effects of crops, fertilizers, manure and climate parameters. *Nutr. Cycl. Agroecosyst.* 87:429–442.

Buresh, R. J. and G. Tian. 1998. Soil improvement by trees in sub-Saharan Africa. *Agrofor. Syst.* 38:51–76.

Busscher, W. J., J. M. Novak, D. E. Evans, D. W. Watts, M. A. S. Niandou, and M. Ahmedna. 2010. Influence of pecan biochar on physical properties of a Norfolk loamy sand. *Soil Sci.* 175:10–14.

Byrnes, B. H. 1990. Environmental effects of n fertilizer use: An overview. *Fert. Res.* 26:209–215.

Calvino, P. and J. P. Monzon. 2009. Farming systems of Argentina: Yield constraints and risk management. In: *Crop Physiology: Applications for Genetic Improvement and Agronomy*, eds., V. Sadras and D. Calderini, pp. 55–70. Amsterdam: Academic Press.

Campbell, D. H., J. S. Baron, K. A. Tonnessen, P. D. Brooks, and P. F. Schuster. 2000. Controls on nitrogen flux in alpine/subalpine watersheds of Colorado. *Water Resour. Res.* 36:37–47.

Cantagallo, J. E., D. Medan, and A. J. Hall. 2004. Grain number in sunflower as affected by shading during floret growth, anthesis and grain setting. *Field Crops Res.* 85:191–202.

Cao, Y., X. Fan, S. Sun, G. Xu, and Q. Shen. 2008. Effect of nitrate on activities and transcript levels of nitrate reductase and glutamine synthetase in rice. *Pedosphere* 18:664–673.

Caviglia, O. P., V. O. Sadras, and F. H. Andrade. 2004. Intensification of agriculture in the south eastern Pampas. I. Capture and efficiency in the use of water and radiation in double-cropped wheat–soybean. *Field Crops Res.* 87:117–129.

Chen, B. L. and Z. M. Chen. 2009. Sorption of naphthalene and 1-naphthol by biochar of orange peels with different pyrolytic temperatures. *Chemosphere* 76:127–133.

Chen, C., M. Westcott, K. Neill, D. Wichman, and M. Knox. 2004. Row configuration and nitrogen application for barley pea intercropping in Montana. *Agron. J.* 96:1730–1738.

Chen, D., J. R. Freney, I. Rochester, G. A. Constable, A. R. Mosier, and P. M. Chalk. 2008. Evaluation of a polyolefin coated urea as a fertilizer for irrigated cotton. *Nutr. Cycl Agroecosyst.* 81:245–254.

Chighladze, G., A. Kaleita, S. Birrell, and S. Logsdon. 2012. Estimating soil solution nitrate concentration from dielectric spectra using partial least squares analysis. *Soil Sci. Soc. Am.* 76:1536–1547.

Christianson, C. B., W. E. Baethgen, G. Carmona, and R. G. Howard. 1993. Microsite reactions of urea–nBTPT fertilizer on the soil surface. *Soil Biol. Biochem.* 25:1107–1117.

Cisse, M. and P. L. G. Vlek. 2003. Conservation of urea N by immobilization–remobilization in a rice-*Azolla* intercrop. *Plant Soil* 250:95–104.

Clay, D. E., G. L. Malzer, and J. L. Anderson. 1990. Ammonia volatilization from urea as influenced by soil temperature, soil water content, and nitrification and hydrolysis inhibitors. *Soil Sci. Soc. Am. J.* 54:263–266.

Cole, C. V., J. Duxbury, J. Freney, O. Heinemeyer, K. Minami, A. Mosier, K. Paustian, N. Rosenberg, N. Sampson, D. Sauerbeck, and Q. Zhao. 1997. Global estimates of potential mitigation of greenhouse gas emissions by agriculture. *Nutr. Cycl. Agroecosyst.* 49:221–228.

Coll, L., A. Cerrudo, R. H. Rizzalli, J. P. Monzon, and F. H. Andrade. 2012. Capture and use of water and radiation in summer intercrops in the southern Pampas of Argentina. *Field Crops Res.*134:105–113.

Collins, H. P., A. K. Alva, J. D. Streubel, S. F. Fransen, C. Frear, S. Chen, C. Kruger, and D. Granatstein. 2011. Greenhouse gas emissions from an irrigated silt loam soil amended with anaerobically digested dairy manure. *Soil Sci. Soc. Am. J.* 75:2206–2216.

Crews, T. E. and M. B. Peoples. 2004. Legume versus fertilizer sources of nitrogen: Ecological tradeoff and human needs. *Agric. Ecosyst. Environ.* 102:279–297.

Crusciol, C. A. C., G. P. Mateus, A. S. Nascente, P. O. Martins, E. Borghi, and C. M. Pariz. 2012. An innovative crop–forage intercrop system: Early cycle soybean cultivars and palisadegrass. *Agron. J.* 104:1085–1095.

Crusciol, C. A. C., A. S. Nascente, G. P. Mateus, E. Borghi, E. P. Leles, and N. C. B. Santos. 2013. Effect of intercropping on yields of corn with different relative maturities and palisadegrass. *Agron. J.* 105:599–606.

Crusciol, C. A. C., R. P. Soratto, E. Borghi, and G. P. Matheus. 2010. Benefits of integrating crops and tropical pastures as systems of production. *Better Crops* 94:14–16.

Curtin, D., M. H. Beare, and G. Hernandez-Ramirez. 2012. Temperature and moisture effects on microbial biomass and soil organic matter mineralization. *Soil Sci. Soc. Am. J.* 76:2055–2067.

Dabney, S. M., J. A. Delgado, and D. W. Reeves. 2001. Use of winter cover crops to improved soil and water quality. *Commun. Soil Sci. Plant Anal.* 32:1221–1250.

Dalal, R. C., W. Wang, D. E. Allen, S. Reeves, and N. W. Menzies. 2011. Soil nitrogen and nitrogen use efficiency under long term no-till practice. *Soil Sci. Soc. Am. J.* 75:2251–2261.

Daryaei, F., A. Ghalavand, A. Sorooshzadeh, M. R. Chaichian, and M. Aqaalikhani. 2011. Effect of different fertilizing systems using green manure and zeoponix on qualitative and quantitative yield of forage rape in sequential cropping system. *Int. Res. J. Appl. Basic Sci.* 2:20–27.

Davies, D. B. and R. Sylvester-Bradley. 1995. The contribution of fertilizer nitrogen to leachable nitrogen in the vUK: A review. *J. Sci. Food and Agric.* 68:399–406.

De Boer, W. and G. A. Kowalchuk. 2001. Nitrification in acid soils: Microorganisms and mechanisms. *Soil Biol. Biochem.* 33:853–866.

Decau, M. L., J. C. Simon, and A. Jacquer. 2004. Nitrate leaching under grassland as affected by mineral nitrogen fertilization and cattle urine. *J. Environ. Qual.* 33:637–644.

DeLuca, T. H., M. D. MacKenzie, M. J. Gundale, and W. E. Holben. 2006. Wildfire-produced charcoal directly influences nitrogen cycling in ponderosa pine forests. *Soil Sci. Soc. Am. J.* 70:448–453.

Delwiche, C. C. and B. A. Bryan. 1976. Denitrification. *Annu. Rev. Microbiol.* 30:241–262.

Dentener, F. J. and P. J. Crutzen. 1994. A three dimensional model of the global ammonia cycle. *J. Atmos. Chem.* 19:331–369.

Dhuyvetter, K. C., C. R. Thompson, C. A. Norwood, and A. D. Halvorson. 1996. Economics of dryland cropping systems in the Great Plains. A review. *J. Prod. Agric.* 9:216–222.

Di, H. J. and K. C. Cameron. 2002. Nitrate leaching in temperate agroecosystems: Sources, factors and mitigating strategies. *Nutr. Cycl. Agroecosyst.* 64:237–256.

Dillon, K. A., T. W. Walker, D. L. Harrell, L. J. Krutz, J. J. Varco, C. H. Koger, and M. S. Cox. 2012. Nitrogen sources and timing effects on nitrogen loss and uptake in delayed fold rice. *Agron. J.* 104:466–472.

Ding, Y., Y. Liu, W. Wu, D. Shi, M. Yang, and Z. Zhong. 2010. Evaluation of biochar effects on nitrogen retention and leaching in multi-layered soil columns. *Water Air Soil Pollut.* 213:47–55.

Dinnes, D. L., D. L. Karlen, D. B. Jaynes, T. C. Kaspar, J. L. Hatfield, T. S. Colvin, and C. A. Cambardella. 2002. Nitrogen management strategies to reduce nitrate leaching in tile-drained Midwestern soils. *Agron. J.* 94:153–171.

Dobermann, A. and K. G. Cassman. 2002. Plant nutrient management for enhanced productivity in intensive grain production system of the United States and Asia. *Plant Soil* 247:153–175.

Dossa, E. L., I. Diedhiou, M. Khouma, M. Sene, A. Lufafa, F. Kizito, S. A. N. Samba, A. N. Badiane, S. Diedhiou, and R. P. Dick. 2012. Crop productivity and nutrient dynamics in a shrub (*Guiera senegalensis*)-based farming system of the Sahel. *Agron. J.* 104:1255–1264.

Drinkwater, L. E., P. Wagoner, and M. Sarrantonio. 1998. Legume-based cropping systems have reduced carbon and nitrogen losses. *Nature* 396:262–265.

Drury, C. F. and E. G. Beauchamp. 1991. Ammonium fixation, releases nitrification, and immobilization in high and low fixing soils. *Soil Sci. Soc. Am. J.* 55:125–129.

Drury, C. F., E. G. Beauchamp, and L. J. Evans. 1989. Fixation and immobilization of recently applied $^{15}NH_4$ in selected Ontario and Quebec soils. *Can. J. Soil Sci.* 69:391–400.

Drury, C. F., W. D. Reynolds, C. S. Tan, T. W. Welacky, W. Calder, and N. B. McLaughlin. 2006. Emissions of nitrous oxide and carbon dioxide: Influence of tillage type and nitrogen placement depth. *Soil Sci. Soc. Am. J.* 70:570–581.

Drury, C. F., W. D. Reynolds, X. M. Yang, N. B. McLaughlin, T. W. Welacky, W. Calder, and C. A. Grant. 2012. Nitrogen source, application time, and tillage effects on soil nitrous oxide emissions and corn grain yields. *Soil Sci. Soc. Am. J.* 76:1268–1279.

Duan, Z. and H. Xiao. 2000. Effects of soil properties on ammonia volatilization. *Soil Sci. Plant Nutr.* 46:845–852.

Dunisch, O., V. C. Lima, G. Seehann, J. Donath, V. R. Montoia, and T. Schwarz. 2007. Retention properties of wood residues and their potential for soil amelioration. *Wood Sci. Technol.* 41:169–189.

Dusenbury, M. P., R. F. Reynolds, P. R. Miller, R. L. Lemke, and R. Wallander. 2008. Nitrous oxide emissions from a northern Great Plains soils as influenced by nitrogen management and cropping systems. *J. Environ. Qual.* 37:542–550.

Echarte L., A. D. Maggiora, D. Cerrudo, V. H. Gonzales, P. Abbate, A. Cerrudo, V. O. Sadras, and P. Calvino. 2011. Yield response to plant density of maize and sunflower intercropped with soybean. *Field Crops Res.* 121:423–429.

Economic Research Service. 2010. Fertilizer use and price. Available at http://www.ers.usda.gov/data/fertilizeruse (verified 3 September 2011). USDA, Economic Research Service, Washington, DC.

Eerden, L. V. D., W. D. Vries, and H. V. Dobben. 1998. Effect of ammonia deposition on forest in the Netherlands. *Atmos. Environ.* 32:525–532.

Egli, D. B. and W. P. Bruening. 2005. Shade and temporal distribution of pod production and pod set in soybean. *Crop Sci.* 45:1764–1769.

Emmett, B. A. 2007. Nitrogen saturation of terrestrial ecosystem: Some recent findings and their implications for our conceptual framework. *Water Air Soil Pollut.* 7:99–109.

Engel, R., C. Jones, and R. Wallander. 2011. Ammonia volatilization from urea and mitigation by NBPT following surface application to cold soils. *Soil Sci. Soc. Am. J.* 75:2348–2357.

Evans, J., N. A. Fettell, and G. E. O'Connor. 1996. Nitrate accumulation under pea cropping and the effects of crop establishment methods: A sustainability issue. *Aust. J. Exp. Agric.* 36:581–586.

Fageria, N. K. 2013. *The Role of Plant Roots in Crop Production.* Boca Raton, Florida: CRC Press.

Fageria, N. K. 2014. *Mineral Nutrition of Rice.* Boca Raton, Florida: CRC Press.

Fageria, N. K. and V. C. Baligar. 2005. Enhancing nitrogen use efficiency in crop plants. *Adv. Agron.* 88:97–185.

Fageria, N. K., V. C. Baligar, and C. A. Jones. 2011. *Growth and Mineral Nutrition of Field Crops,* 3rd edition. Boca Raton, Florida: CRC Press.

Fageria, N. K. and H. R. Gheyi. 1999. *Efficient Crop Production.* Campina Grande, Paraiba, Brazil: Federal University of Paraiba, Brazil.

Fageria, N. K. and A. M. Knupp. 2013. Upland rice phenology and nutrient uptake in tropical climate. *J. Plant Nutr.* 36:1–14.

Fageria, N. K. and A. Moreira. 2011. The role of mineral nutrition on root growth of crop plants. *Adv. Agron.* 110:251–331.

Fageria, N. K., A. B. Santos, and M. F. Moraes. 2010. Influence of urea and ammonium sulfate on soil acidity indices in lowland rice production. *Commun. Soil Sci. Plant Anal.* 41:1565–1575.

FAO. 2006. Plant nutrition for food security: A guide for integrated nutrient management. *Fertilizer Plant Nutr. Bull.* 16. Food and Agricultural Organization of the United Nations. ftp://ftp.fao.org/agl/agll/docs/fpnb16.pdf (accessed 18 August 2012).

FAO. 2008. *The State of Food and Agriculture. Biofuels: Prospects, Risks and Opportunities.* Rome: FAO.

FAO/IFA. 2001. Global estimates of gases emissions of NH_3, NO, and N_2O from agricultural land. Food and Agricultural Organization of the United Nations (FAO) and International Fertilizer Industry Association, Rome. ftp://ftp.fao.org/agl/agll/docs/globest.pdf (accessed 5 July 2012).

Farquhr, G. D., P. M. Firth, R. Wetselaar, and B. Weir. 1980. On the gases exchange of ammonia between leaves and the environment: Determination of the ammonia compensation point. *Plant Physiol.* 66:710–714.

Favaretto, N., L. D. Norton, B. C. Joern, and S. M. Brouder. 2008. Gypsum amendment and calcium and magnesium effects on plant nutrition under conditions of intensive nutrient extraction. *Soil Sci.* 173:108–118.

Favaretto, N., L. D. Norton, C. T. Johnston, J. Bigham, and M. Sperrin. 2012. Nitrogen and phosphorus leaching as affected by gypsum amendment and exchangeable calcium and magnesium. *Soil Sci. Soc. Am. J.* 76:575–585.

Feng, G., B. Sharratt, and L. Wendling. 2011. Fine particle emission potential from loam soils in a semiarid region. *Soil Sci. Soc. Am. J.* 75:2262–2270.

Fenn, L. B. and L. R. Hossner. 1985. Ammonia volatilization from ammonium or ammonium forming nitrogen fertilizers. *Adv. Soil Sci.* 1:123–169.

Fenn, L. B. and D. E. Kissel. 1986. Effects of soil drying on ammonia volatilization from surface applied urea. *Soil Sci. Soc. Am. J.* 50:485–490.

Fenn, L. B. and S. Miyamoto. 1981. Ammonia loss and associated reaction of urea in calcareous soils. *Soil Sci. Soc. Am. J.* 45:537–540.

Fenn, L. B., R. M. Taylor, and G. L. Horst. 1987. *Phaseolus vulgaris* growth in an ammonium based nutrient solution with variable calcium. *Agron. J.* 79:89–91.

Fenn, L. B., R. Taylor, M. L. Binzel, and C. M. Burks. 1991. Calcium stimulation of ammonium absorption in onions. *Agron. J.* 83:840–843.

Fenn, L. B., R. M. Taylor, and C. M. Burks. 1994. Calcium stimulation of ammonium absorption and growth by beet. *Agron. J.* 86:916–920.

Fenn, M. E., M. A. Poth, J. D. Aber, J. S. Baron, B. T. Bormann, D. W. Johnson, A. D. Lemly, S. G. McNulty, D. F. Ryan, and R. Stottlemyer. 1998. Nitrogen excess in North American ecosystems: Predisposing factors, ecosystem responses, and management strategies. *Ecol. Appl.* 8:706–733.

Ferguson, R. B. and D. E. Kissel. 1986. Effects of soil drying on ammonia volatilization from surface applied urea. *Soil Sci. Soc. Am. J.* 50:485–490.

Ferm, M. 1998. Atmospheric ammonia and ammonium transport in Europe and critical loads: A review. *Nutr. Cycl Agroecosyst.* 51:5–17.

Filella, I., J. Penuelas, L. Llorens, and M. Estiarte. 2004. Reflectance assessment of seasonal and annual changes in biomass and CO_2 uptake of a Mediterranean shrubland submitted to experimental warming and drought. *Remote Sens. Environ.* 90:308–318.

Fillery, I. R. P. and P. L. G. Vlex. 1986. Reappraisal of the significance of ammonia volatilization as an N loss mechanism in flooded rice fields. *Fert. Res.* 9:79–98.

Finlayson-Pitts, B. J. and J. N. Pitts. Jr. 2000. Chemistry of inorganic nitrogen compounds. In: *Chemistry of the Upper and Lower Atmosphere*, eds., B. J. Finlayson-Pitts and J. N. Pitts. Jr., pp. 265–293. San Diego, California: Academic Press.

Follet, R. F. and D. J. Walker. 1989. Ground water concerns about nitrogen. In: *Nitrogen Management and Ground Water*, ed., R. F. Follet, pp. 1–22. Amsterdam: Elsevier.

Foulkes, M. J., R. Sylvester-Bradley, and R. K. Scott. 1998. Evidence for differences between winter wheat cultivars in acquisition of soil mineral nitrogen and uptake and utilization of applied fertilizer nitrogen. *J. Agric. Sci.* 130:29–44.

Fox, R. H. and L. D. Hoffman. 1981. The effect of N fertilizer source on grain yield, N uptake, soil pH and time requirements in no-till corn. *Agron. J.* 73:891–895.

Francis, D. D., J. S. Schepers, and M. F. Vigil. 1997. Post-anthesis nitrogen loss from corn. *Agron. J.* 85:659–663.

Franzluebbers, A. J. 2007. Integrated crop–livestock systems in the southeastern USA. *Agron. J.* 99:361–372.

Freney, J. R., O. T. Denmead, and J. R. Simpson. 1978. Soil as a source or sink for atmospheric nitrous oxide. *Nature* 273:530–532.

Fukai, S. and B. R. Trenbath. 1993. Processes determining intercrop productivity and yields of component crops. *Field Crops Res.* 34:247–271.

Galloway, J. N., E. B. Cowling, S. P. Seitzinger, and R. H. Socolow. 2002. Reactive nitrogen: Too much of a good thing? *Ambio* 31:60–63.

Garrett, H. E., W. Kurtz, L. F. Buck, J. P. Lassoie, M. A. Gold, H. A. Pearson, L. H. Hardesty, and J. P. Slusher. 1994. *Agroforestry: An Integrated Land Use Management System for Production and Farmland Conservation*. Resource conservation act appraisal document. Washington, DC: NRCS.

Gathumbi, S. M., G. Cadisch, and K. E. Giller. 2004. Improved fallows: Effects of species interaction on growth and productivity in monoculture and mixed stands. *For. Ecol. Manage.* 187:267–280.

Gay, S. W. and K. F. Knowlton. 2005. *Ammonia Emission and Animal Agriculture*. Publication 442-110. Virginia Coop. Ext./Biological Systems Engineering, Blackburg. p. 1–5.

Geisseler, D., P. A. Lazicki, G. S. Pettygrove, B. Ludwig, P. A. M. Bachand, and W. R. Horwath. 2012. Nitrogen dynamics in irrigated forage systems fertilized with liquid dairy manure. *Agron. J.* 104:897–907.

Gerik, T. J., D. M. Oosterhuis, and H. A. Torbert. 1998. Managing cotton nitrogen supply. *Adv. Agron.* 64:115–147.

Gholamhoseini, M., M. A. Alikhani, A. Dolatabadian, A. Khodaei-Joghan, and H. Zakikhani. 2012. Decreasing nitrogen leaching and increasing canola forage yield in a sandy soil by application of natural zeolite. *Agron. J.* 104:1467–1475.

Ghosh, P. K., M. C. Manna, K. K. Bandyopadhyaay, A. A. K. Triathi, R. H. Wanjari, K. M. Hati, A. K. Misra, C. L. Acharya, and A. Subba Rao. 2006. Interspecific interaction and nutrient use in soybean/sorghum intercropping system. *Agron. J.* 98:1097–1108.

Ghoshal, N. 2002. Available pool and mineralization rate of soil N in a dryland agroecosystems: Effect of organic soil amendments and chemical fertilizer. *Trop. Ecol.* 43:363–366.

Giambalvo, D., P. Ruisi, G. D. Miceli, A. S. Frenda, and G. Amato. 2010. Nitrogen use efficiency and nitrogen fertilizer recovery of durum wheat genotypes as affected by interspecific competition. *Agron. J.* 102:707–715.

Gioacchini, P., A. Nastri, C. Marzadori, C. Giovannini, L. V. Antisari, and C. Gessa. 2002. Influence of urease and nitrification inhibitors on N losses from soils fertilized with urea. *Biol. Fertil. Soils* 36:129–135.

Glaser, B., J. Lehmann, and W. Zech. 2002. Ameliorating physical and chemical properties of highly weathered soils in the tropics with charcoal: A review. *Biol. Fertil. Soils* 35:219–230.

Glass, A. D. M. 2003. Nitrogen use efficiency of crop plants: Physiological constraints upon nitrogen absorption. *Crit. Rev. Plant Sci.* 22:453–470.

Glover, J. D., S. W. Culman, S. T. DuPont, W. Broussard, L. Young, M. E. Mangan, J. G. Mai et al. 2010. Harvested perennial grasslands provide ecological benchmark for agricultural sustainability. *Agric. Ecosyst. Environ.* 137:3–12.

Gold, M. A. and H. E. Garrett. 2009. Agroforestry nomenclature, concepts, and practices. In: *North American Agroforestry: An Integrated Science and Practice*, ed., H. E. Garrett, pp. 45–55. Madison, Wisconsin: ASA.

Goodchild, R. G. 1998. EC policies for reduction of nitrogen in water: The example of the nitrates directives. In: *First International Conference*, eds., V. D. Hoek and W. Klass, pp. 737–740. Oxford, UK: Elsevier.

Graber, E. R., L. Tsechansky, J. Khanukov, and Y. Oka. 2011. Sorption, volatilization, and efficacy of the fumigant 1,3-dichloropropene in a biochar-amended soil. *Soil Sci. Soc. Am. J.* 75:1365–1373.

Grant, C. A., S. Jia, K. R. Brown, and L. D. Bailey. 1996. Volatile losses of NH_3 from surface applied urea and urea ammonium nitrate with and without the urease inhibitors NBPT or ammonium thiosulphate. *Can. J. Soil Sci.* 76:417–419.

Gray, S. B., A. T. Classen, P. Kardol, Z. Yermakov, and R. M. Miller. 2011. Multiple climate change factors interact to alter soil microbial community structure in an old-field ecosystem. *Soil Sci. Soc. Am. J.* 75:2217–2226.

Greenhouse Gas Working Group. 2010. Agricultures role in greenhouse gas emissions and capture. Greenhouse Gas Working Group Report, Madison, Wisconsin: ASA, CSSA, and SSSA.

Griggs, B. R., R. J. Norman, C. E. Wilson, Jr., and N. A. Slaton. 2007. Ammonia volatilization and nitrogen uptake for conventional and conservation tilled dry-seeded, delayed-flood rice. *Soil Sci. Soc. Am. J.* 71:745–751.

Gruber, N. and J. N. Galloway. 2008. An earth-system perspective of the global nitrogen cycle. *Nature* 451:293–296.

Gupta, S. K., R. C. Gupta, A. B. Gupta, A. K. Seth, J. K. Bassin, and A. Gupta. 2000. Recurrent acute respiratory tract infections in areas with high nitrate concentrations in drinking water. *Environ. Health Perspect.* 108:363–366.

Hageman, R. H. 1979. Integration of nitrogen assimilation in relation to yield. In: *Nitrogen Assimilation of Plants*, eds., E. J. Hewitt and C. V. Cutting, pp. 591–611. London: Academic Press.

Halvorson, A. D. and S. J. Del Grosso. 2012. Nitrogen source and placement affect soil nitrous oxide emissions from irrigated corn in Colorado. *Better Crops Plant Food* 96:7–9.

Halvorson, A. D., S. J. Del Grosso, and F. Alluvione. 2010a. Nitrogen source effects on nitrous oxide emissions from irrigate no-till corn. *J. Environ. Qual.* 39:1554–1562.

Halvorson, A. D., S. J. Del Grosso, and F. Alluvione. 2010b. Tillage and inorganic nitrogen source effects on nitrous oxide emissions from irrigated cropping systems. *Soil Sci. Soc. Am. J.* 74:436–445.

Halvorson, A. D., B. J. Wienhold, and A. L. Black. 2002. Tillage, nitrogen, and cropping system effects on soil carbon sequestration. *Soil Sci. Soc. Am. J.* 66:906–912.

Hammel, J. E., M. E. Sumner, and H. Shahandeh. 1985. Effect of physical and chemical profile modification on soybean and corn production. *Soil Sci. Soc. Am. J.* 49:1508–1511.

Harper, L. A., V. R. Catchpoole, R. Davis, and K. L. Weir. 1983. Ammonia volatilization: Soil, plant and microclimate effects on diurnal and seasonal fluctuations. *Agron. J.* 75:212–218.

Harper, L. A. and R. R. Sharpe. 1995. Nitrogen dynamics in irrigated corn: Soil–plant nitrogen and atmospheric ammonia transport. *Agron. J.* 87:669–675.

Harrison, R. and J. Webb. 2001. Effects of N fertilizer types on gases emission. *Adv. Agron.* 73:67–110.

Hartemink, A. E., R. J. Buresh, P. M. Van Bogedom, A. R. Braun, B. Jama, and B. H. Janssen. 2000. Inorganic nitrogen dynamics in fallows and maize on an Oxisol and Alfisol in Kenya. *Geoderma* 98:11–33.

Hauck, R. D. and H. F. Stephenson. 1965. Nitrification of nitrogen fertilizers: Effect of nitrogen source, size, and pH of the granule, and concentrations. *J. Agric. Food Chem.* 13:486–492.

Hauck, R. D. and K. K. Tanji. 1982. Nitrogen transfers and mass balances. In: *Nitrogen in Agricultural Soils*, ed., F. J. Stevenson, pp. 891–925. Madison, Wisconsin: American Society of Agronomy.

Hayashi, K., N. Koga, and N. Fueki. 2011. Limited ammonia volatilization loss from upland fields of Andosols following fertilizer applications. *Agric. Ecosyst. Environ.* 140:534–538.

Hayashi, K., S. Nishimura, and K. Yagi. 2008. Ammonia volatilization from a paddy field following application of urea: Rice plants are both an absorber and an emitter for atmospheric ammonia. *Sci. Total Environ.* 390:485–495.

He, X. T., R. L. Mulvaney, F. J. Stevenson, and R. M. Vanden Heuvel. 1990. Characterization of chemically fixed liquid anhydrous ammonia in an Illinois drummer soil. *Soil Sci. Soc. Am. J.* 54:775–780.

He, Z. L., D. V. Calvert, A. K. Alva, Y. C. Li, and D. J. Banks. 2002. Clinoptilolite zeolite and cellulose amendments to reduce ammonia volatilization in a calcareous sandy soil. *Plant Soil* 247:253–260.

Herrero, M., P. K. Thorton, A. M. Notenbaert, S. Wood, S. Msangi, and H. A. Freeman. 2010. Smart investments in sustainable food productions: Revisiting mixed crop–livestock systems. *Science* 327:822–825.

Hinman, W. C. 1966. Ammonium fixation in relation to exchangeable K and organic matter content of two Saskatchewan soils. *Can. J. Soil Sci.* 46:223–225.

Holcomb, J. C., D. M. Sullivan, D. A. Horneck, and G. H. Clough. 2011. Effect of irrigation rate on ammonia volatilization. *Soil Sci. Soc. Am. J.* 75:2341–2347.

Hooda, P. S., A. C. Edwards, H. A. Anderson, and A. Miller. 2000. A review of water quality concerns in livestock farming areas. *Sci. Total Environ.* 250:143–147.

Horst, G. L., L. B. Fenn, and N. Beadle. 1985. Bermudagrass turf responses to nitrogen sources. *J. Am. Soc. Hort. Sci.* 166:393–396.

Huang, W. Y., W. McBride, and U. Vasavada. 2008. Recent volatility in U. S. fertilizer prices. *Amber Waves* 7:28–31.

Huang, Z. T. and A. M. Petrovic. 1994. Clinoptilolite zeolite influence on nitrate leaching and nitrogen use efficiency in simulated sand based golf greens. *J. Environ. Qual.* 23:1190–1194.

Huckaby, E. C. K., C. W. Wood, and E. A. Guertal. 2012. Nitrogen source effects on ammonia volatilization from warm-season sod. *Crop Sci.* 52:1379–1384.

Hull, R. J. and H. Liu. 2005. Turfgrass nitrogen: Physiology and environmental impacts. *Int. Turfgrass Soc. Res. J.* 10:962–975.

Ikeorgu, J. E. G., H. C. Ezumah, and T. A. T. Wahua. 1989. Productivity of species in cassava/maize/okra/egusi melon complex mixtures in Nigeria. *Field Crops Res.* 21:1–7.

Jabro, J. D., A. D. Jabro, and S. L. Fales. 2012. Models performance and robustness for simulating drainage and nitrate nitrogen fluxes without recalibration. *Soil Sci. Soc. Am. J.* 76:1957–1964.

Jansson, S. L. and J. Persson. 1982. Mineralization and immobilization of soil nitrogen. In: *Nitrogen in Agricultural Soils*, ed., F. J. Stevenson, pp. 229–252. Madison, Wisconsin: ASA, CSSA, and SSSA.

Jantalia, C. P., A. D. Halvorson, R. F. Follett, B. J. Rodrigues Alves, J. C. Polidoro, and S. Urquiaga. 2012. Nitrogen source effects on ammonia volatilization as measured with semi-static chambers. *Agron. J.* 104:1595–1603.

Jarvis, S. C. and D. J. Hatch. 1994. Potential for denitrification at depth below long-term grass swards. *Soil Biol. Biochem.* 26:1629–1636.

Jayaweera, G. R. and D. S. Mikkelsen. 1990. Ammonia volatilization from flooded soil systems: A computer model I. Theoretical aspects. *Soil Sci. Soc. Am. J.* 54:1447–1455.

Johnson, J. M. F., R. R. Allmaras, and D. C. Reicosky. 2006. Estimating source carbon from crop residues roots and rhizodeposits using the national grain-yield database. *Agron. J.* 98:622–636.

Johnson, J. M. F., S. L. Weyers, D. W. Archer, and N. W. Barbour. 2012. Nitrous oxide, methane emission, and yield-scaled emission from organically and conventional management systems. *Soil Sci. Soc. Am. J.* 76:1347–1357.

Johnson, P., C. J. Zilverberg, V. G. Allen, J. Weinheimer, P. Brown, R. Kellison, and E. Segarra. 2013. Integrating cotton and beef production in the Texas southern High Plains: III. An economic evaluation. *Agron. J.* 105:929–937.

Jones, O. R. and T. W. Popham. 1997. Cropping and tillage systems for dryland grain production in southern High Plains. *Agron. J.* 89:222–232.

Jonker, J. S., R. A. Kohn, and J. High. 2002. Dairy herd management practices that impact nitrogen utilization efficiency. *J. Dairy Sci.* 85:1218–1226.

Joosten, L. T. A., S. T. Buijze, and D. M. Jansen. 1998. Nitrate in sources of drinking water: Dutch drinking water companies aim at prevention. In: *First International Conference*, eds., W. Van der Hoek and W. Klass, pp. 487–491. Oxford, UK: Elsevier.

Jose, S. 2009. Agroforestry for ecosystem services and environmental benefits: An overview. *Agrofor. Syst.* 76:1–10.

Juma, N. G. and E. A. Paul. 1983. Effect of a nitrification inhibitor on N immobilization and release of ^{15}N from nonexchangeable ammonium and microbial biomass. *Can. J. Soil Sci.* 63:167–175.

Jurgens-Gschwind, S. 1989. Ground water nitrates in other developed countries (Europe)—Relationships to land use patterns. In: *Nitrogen Management and Ground Water*, ed., R. F. Follet, pp. 75–138. Amsterdam: Elsevier.

Kao, C. W. and W. H. Ko. 1986. The role of calcium and microorganisms in suppression of cucumber damping-off caused by *Pythum splendens* in a Hawaiian soil. *Phytopathology* 76:221–225.

Keerthisinghe, G., K. Mengel, and S. K. De Datta. 1984. The release of nonexchangeable ammonium (^{15}N labeled) in wetland rice soils. *Soil Sci. Soc. Am. J.* 48:291–294.

Khodaei, J., A. A. Ghalavand, M. Aghaalikhani, M. Gholamhoseini, and A. Dolatabadian. 2012. How organic and chemical nitrogen fertilizers, zeolite, and comnations influence wheat yield and grain mineral content. *J. Crop Improv.* 26:116–129.

Kim, K. S., M. Park, C. L. Choi, D. H. Lee, Y. J. Seo, C. Y. Kim, J. S. Kim, S. I. Yun, H. M. Rao, and S. Komarneni. 2010. Suppression of NH_3 and N_2O emissions by massive urea intercalation in montmorillonite. *J. Soils Sediments* 11:416–422.

Kim, K. S., M. Park, W. T. Lim, and S. Komarneni. 2011. Massive intercalation of urea in montmorillonite. *Soil Sci. Soc. Am. J.* 75:2361–2366.

Kissel, D. E. and M. L. Cabrera. 1988. Ammonia volatilization from urea and an experimental triazon fertilizer. *HortScience* 23:1087.

Kissel, D. E., M. L. Cabrera, and R. B. Ferguson. 1988. Reactions of ammonia and urea hydrolysis products with soil. *Soil Sci. Soc. Am. J.* 52:1793–1796.

Kissel, D. E., M. L. Cabrera, N. Vaio, J. R. Craig, J. A. Rema, and L. A. Morris. 2004. Rainfall timing and ammonia loss from urea in a loblolly pine plantation. *Soil Sci. Soc. Am. J.* 68:1744–1750.

Kissel, D. E., M. L. Cabrera, N. Vaio, J. R. Craig, J. A. Rema, and L. A. Morris. 2009. Forest floor composition and ammonia loss from urea in a loblolly pine plantation. *Soil Sci. Soc. Am. J.* 73:630–637.

Kizito, F., M. Sene, M. I. Dragila, A. Lufafa, I. Diedhiou, E. Dossa, R. Cuenca, J. Selker, and R. P. Dick. 2007. Soil water balance of annual crop native shrub systems in Senegal's peanut basin: The missing link. *Agric. Water Manage.* 90:137–148.

Kjelgaard, J. F., D. G. Chandler, and K. E. Saxton. 2004. Evidence for direct suspension of loessial soils on the Columbia Plateau. *Earth Surf. Processes Landf.* 29:221–236.

Klemedtsson, L., B. H. Svensson, T. Lindberg, and T. Rosswall. 1977. The ise of acetylene inhibition of nitrous oxide reductase in quantifying denitrification in soils. *Swed. J. Agric. Res.* 7:179–185.

Knight, E. C., E. A. Guertal, and C. W. Wood. 2007. Mowing and nitrogen source effects on ammonia volatilization from turfgrass. *Crop Sci.* 47:1628–1634.

Koenig, R. T. and W. L. Pan. 1996. Calcium effects on quantity–intensity relationships and plant availability of ammonium. *Soil Sci. Soc. Am. J.* 60:492–497.

Kolb, S., K. Fermanich, and M. Dornbush. 2009. Effect of charcoal quantity on microbial biomass and activity in temperate soils. *Soil Sci. Soc. Am. J.* 73:1173–1181.

Kirchmann, H. and L. Bergstrom. 2001. Do organic farming practices reduce nitrate leaching? *Commun. Soil Sci. Plant Anal.* 32:997–1028.

Kumar, S., S. H. Anderson, L. G. Bricknell, and R. P. Udawatta. 2008. Soil hydraulic properties influenced by agroforestry and grass buffers for grazed pasture systems. *J. Soil Water Conserv.* 63:224–232.

Kumar, V., S. Singh, and H. D. Yadav. 1999. Performance of sugarcane growth under soil and water conditions. *Agric. Water Manage.* 41:1–9.

Kwon, H. Y., J. M. R. Hudson, and R. L. Mulvaney. 2009. Characterization of the organic nitrogen fraction determined by the Illinois soil nitrogen test. *Soil Sci. Soc. Am. J.* 73:1033–1043.

Laird, D., P. Fleming, B. Wang, R. Horton, and D. Karlen. 2010. Biochar impact on nutrient leaching from a Midwestern agricultural soil. *Geoderma* 158:436–442.

Lal, R. 1999. Soil quality and food security: The global perspective. In: *Soil Quality and Soil Erosion,* ed., R. Lal, pp. 3–16. Ankeny, IA: Soil and Water Conservation Society.

Lal, R., T. J. Logan, D. J. Eckert, W. Dick, and M. J. Shipitalo. 1994. Conservation tillage in the corn belt of United States. In: *Conservation Tillage in Temperate Agroecosystems: Development and Adaptation to Soil, Climate, and Biological Constraints*, pp. 73–114. Boca Raton, Florida: CRC Press.

Larson, S. and D. Gunary. 1962. Ammonia loss from ammonical fertilizers applied to calcareous soils. *J. Sci. Food Agr.* 13:566–572.

Leggo, P. J., B. L. Sert, and C. Graham. 2006. The role of clinoptilolite in organo-zeolitic soil systems used for phytoremediation. *Sci. Total Environ.* 363:1–10.

Lehmann, J. 2007. Bio-energy in the black. *Front. Ecol. Environ.* 5:381–387.

Lehmann, J. and S. Joseph. 2009. Biochar for environmental management: An introduction. In: *Biochar for Environmental Management*, eds., J. Lehmann and S. Joseph, pp. 1–12. Earthscan, London: Science and Technology.

Lehmann, J., J. Gaunt, and M. Rondon. 2006. Biochar sequestration in terrestrial ecosystems: A review. *Mitigat. Adaptat. Strateg. Global Change* 11:403–427.

Leifeld, J., S. Fenner, and M. Muller. 2007. Mobility of black carbon in drained peatland soils. *Biogeosciences* 4:425–432.

Lehrsch, G. A., R. E. Sojka, and D. T. Westermann. 2001. Furrow irrigation and N management strategies to protect water quality. *Commun. Soil Sci. Plant Anal.* 32:1029–1050.

Lemke, R. L., R. C. Izaurralde, M. Nyborg, and E. D. Solberg. 1999. Tillage and nitrogen source influence soil-emitted nitrous oxide in the Alberta Parkland region. *Can. J. Soil Sci.* 79:15–24.

Liang, B., J. Lehmann, D. Solomon, J. Kinyangi, J. Grossman, B. O'Neill, J. O. Skjemstad et al. 2006. Black carbon increases cation exchange capacity in soils. *Soil Sci. Soc. Am. J.* 70:1719–1730.

Lightner, J. W., D. B. Mengel, and C. L. Rhyderd. 1990. Ammonia volatilization from nitrogen fertilizer surface applied to orchardgrass sod. *Soil Sci. Soc. Am. J.* 54:1478–1482.

Limaux, F., S. Recous, J. M. Meynard, and A. Gukert. 1999. Relationship between rate of crop growth at date of fertilizer nitrogen application and fate of fertilizer nitrogen applied to winter wheat. *Plant Soil* 214:49–59.

Linn, D. M. and J. W. Doran. 1984. Aerobic and anaerobic microbial populations in no-till and plowed soils. *Soil Sci. Soc. Am. J.* 48:794–799.

Livingston, M. L. and D. C. Cory. 1998. Agricultural nitrate contamination of ground water: An evaluation of environmental policy. *J. Am. Water Resour. Assoc.* 34:1311–1317.

Lowrance, R., G. Vellidis, and R. K. Hubbard. 1995. Denitrification in a restored riparian forest wetland. *J. Environ. Qual.* 24:808–815.

Lynch, D., R. MacRae, and R. Martin. 2011. The carbon and global warming impacts of organic farming: Does have a significant role in an energy constrained world? *Sustainability* 3:322–362.

Ma, B. L., T. Y. Wu, N. Tremblay, W. Deen, N. B. McLaughlin, M. J. Morrison, and G. Stewart. 2010a. On farm assessment of the amount and timing of nitrogen fertilizer on ammonia volatilization. *Agron. J.* 102:134–144.

Ma, L., W. Q. Ma, G. L. Velthof, F. H. Wang, F. Quin, F. S. Zhang, and O. Oenema. 2010b. Modeling nutrient flows in the food chains of China. *J. Environ. Qual.* 39:1279–1289.

Mafongoya, P. L., E. Kuntushulu, and G. Sileshi. 2006. Managing soil fertility and nutrient cycles through fertilizer trees in southern Africa. In: *Biological Approach to Sustainable Soil Systems*, eds., N. Uphoff, A. A. Ball, E. Fernandes, H. Herren, O. Husson, M. Laing, C. Palm, J. Pretty, P. Sanchez, N. Sanginga, and J. Thais, pp. 273–289. Boca Raton, Florida: CRC Press.

Mahendrappa, M. K., R. L. Smith, and A. T. Christianson. 1966. Nitrifying organisms affected by climatic region in western United States. *Soil Sci. Soc. Am. Proc.* 30:60–62.

Major, J., J. Lehmann, M. Rondon, and C. Goodale. 2010. Fate of soil applied black carbon: Downward migration, leaching and soil respiration. *Glob. Change Biol.* 16:1366–1379.

Mchunu, C. N., S. Lorentz, G. Jewitt, A. Manson, and V. Chaplot. 2011. No-till impact on soil and soil organic carbon erosion under crop residue scarcity in Africa. *Soil Sci. Soc. Am. J.* 75:1503–1512.

McKenney, D. J., S. W. Wang, C. F. Drury, and W. I. Findlay. 1995. Denitrification, immobilization, and mineralization in nitrate limited and nonlimited residue amended soil. *Soil Sci. Soc. Am. J.* 59:118–124.

Meek, B. D., D. L. Carter, D. T. Westerman, J. L. Wright, and R. E. Peckenpaugh. 1995. Nitrate leaching under furrow irrigation as affected by crop sequence and tillage. *Soil Sci. Soc. Am. J.* 59:204–210.

Mikkelsen, D. S. 1987. Nitrogen budgets in flooded soils used for rice production. *Plant Soil* 100:71–97.

Minamikawa, K., S. Nishimura, Y. Nakajima, K. I. Osaka, and K. Yagi. 2011. Upward diffusion of nitrous oxide produced by denitrification near shallow groundwater table in the summer: A lysimeter experiment. *Soil Sci. Plant Nutr.* 57:719–732.

Moller, K. and W. Stinner. 2009. Effects of different manuring systems with and without biogas digestion on soil mineral nitrogen content and on gases nitrogen losses (ammonia, nitrous oxide). *Eur. J. Agron.* 30:1–16.

Molodovskaya, M., O. Singurindy, B. K. Richards, J. Warland, M. S. Johnson, and T. S. Steenhuis. 2012. Temporal variability of nitrous oxide from fertilized croplands: Hot moment analysis. *Soil Sci. Soc. Am. J.* 76:1728–1740.

Mosier, A. R., J. M. Duxbury, J. R. Freney, O. Heinemeyer, and K. Minani. 1998. Assessing and mitigating N_2O emissions from agricultural soils. *Clim. Change* 40:7–38.

Mosier, A. R., A. D. Halvorson, C. A. Reule, and X. J. Liu. 2006. Net global warming potential and greenhouse gas intensity in irrigated cropping systems in northeastern Colorado. *J. Environ. Qual.* 35:1584–1598.

Mosier, A. R., D. S. Schimel, D. W. Valentine, K. F. Bronson, and W. J. Parton. 1991. Methane and nitrous oxide fluxes in native, fertilized and cultivated grasslands. *Nature* 350:330–332.

Mugasha, A. G. and D. J. Pluth. 1995. Ammonium loss following surface application of urea fertilizer to undrained and drained forest minerotrophic peatland sites in central Alberta, Canada. *For. Ecol. Manage.* 78:139–145.

Muller, C., R. J. Stevens, R. J. Laughlin, and H. J. Jager. 2004. Microbial processes and the site of N_2O production in a temperate grassland soil. *Soil Biol. Biochem.* 36:453–461.

Muller, M. M., V. Sundman, and J. J. Skujins. 1980. Denitrification in low pH spodosols and peats determined with the acetylene inhibition method. *Appl. Environ. Microbiol.* 40:235–239.

Mumpton, F. A. 1999. La roca magica: Uses of natural zeolite in agriculture and industry. *Proc. Natl. Acad. Sci. USA* 96:3463–3470.

Murphy, J. A., H. Samaranayake, J. A. Honig, T. J. Lawson, and S. L. Murphy. 2005. Creeping bent grass establishment on amended sand zones in two micro environments. *Crop Sci.* 45:1511–1520.

Murray, P. J., D. J. Hatch, E. R. Dixon, R. J. Stevens, R. J. Laughlin, and S. C. Jarvis. 2004. Denitrification potential in a grassland subsoil: Effects of carbon substrates. *Soil Biol. Biochem.* 36:545–547.

Neeteson, J. J., D. J. Greenwood, and E. J. M. H. Habets. 1986. Dependence of soil mineral N on N-fertilizer application. *Plant Soil* 91:417–420.

Nett, L., S. Ruppel, J. Ruehlmann, E. George, and M. Fink. 2012. Influence of soil amendment history on decomposition of recently applied organic amendments. *Soil Sci. Soc. Am. J.* 76:1290–1300.

Nieder, R. and D. K. Benbi. 2008. *Carbon and Nitrogen in the Terrestrial Environments*. New York: Springer.

Noellsch, A. J., P. P. Motavalli, K. A. Nelson, and N. R. Kitchen. 2009. Corn response to conventional and slow release nitrogen fertilizers across a claypan landscape. *Agron. J.* 101:607–614.

Nommik, H. 1957. Fixation and defixation of ammonium in soils. *Acta Agric. Scandanavian* 7:395–439.

Nommik, H. and K. Vahtras. 1982. Retention and fixation of ammonium and ammonia in soils In: *Nitrogen in Agricultural Soils*, ed., F. J. Stevenson, pp. 123–171. Madison, Wisconsin: ASA, CSSA, and SSSA.

Norman, R. J., C. E. Wilson Jr., N. A. Slaton, B. R. Griggs, J. T. Bushong, and E. E. Gbur. 2009. Nitrogen fertilizer sources and timing before flooding dry-seeded, delayed-flood rice. *Soil Sci. Soc. Am. J.* 73:2184–2190.

Nousiainen, J., K. J. Shingfield, and P. Huhtanen. 2004. Evaluation of milk urea nitrogen as diagnostic of protein feeding. *J. Dairy Sci.* 87:386–398.

Novak, J. M., W. J. Busscher, D. W. Watts, D. A. Laird, M. A. Ahmedna, and M. A. S. Niandou. 2010. Short term CO_2 mineralization after additions of biochar and switchgrass to a typic Kandiudult. *Geoderma* 154:281–288.

Novotny, E. H., M. H. B. Hayes, B. E. Madri, T. J. Bonagamba, E. R. Azevedo, A. A. Souza, G. Song, C. M. Nogueira, and A. S. Mangrich. 2009. Lessons from the Terra Preta de Indios of the Amazon region for the utilization of charcoal for soil amendment. *J. Braz. Chem. Soc.* 20:1003–1010.

NRCS. 1997. Water quality and agriculture: Status, conditions, and trends. Working paper 16. U. S. Govt. Print. Office, Washington, DC.

Ofori, F. and W. R. Stern. 1987. Relative sowing time and density of component crops in a maize/cowpea intercrop system. *Exp. Agric.* 23:41–52.

Olmos Colmenero, J. J. and G. A. Broderick. 2006. Effect of dietary crude protein concentration on milk production and nitrogen utilization in lactating dairy cows. *J. Dairy Sci.* 89:1704–1712.

Olson, R. A. and L. T. Kurtz. 1982. Crop nitrogen requirements, utilization, and fertilization. In: *Nitrogen in Agricultural Soils*, ed., F. J. Stevenson, pp. 567–604. Madison, Wisconsin: ASA, CSSA, and SSSA.

Omernick, J. M., A. R. Abernathy, and L. M. Male. 1981. Stream nutrient levels and proximity of agricultural and forest lands to streams: Some relationships. *J. Soil Water Conserv.* 36:227–231.

Owens, L. B. 1994. Impacts of soil N management on the quality of surface and subsurface water. *Adv. Soil Sci.* 10:137–162.

Pacheco, L. P., W. M. Leandro, P. L. O. A. Machdo, R. L. Assis, and T. Cobucci. 2011. Biomass production and nutrient accumulation and release by cover crops in the off season. *Pesq. Agropec. Bras.* 46:17–25.

Pacholski, A., G. Cai, R. Nieder, J. Richter, X. Fan, Z. Zhu, and M. Roelcke. 2006. Calibration of a simple method for determining ammonia volatilization in the field—Comparative measurements in Henan Province, China. *Nutr. Cycl. Agroecosyst.* 74:259–273.

Park, M., C. Y. Kim, D. H. Lee, C. L. Choi, J. Choi, S. R. Lee, and J. H. Choi. 2004. Intercalation of magnesium urea complex into swelling clay. *J. Phys. Chem. Solids* 65:409–412.

Park, M., J. S. Kim, C. L. Choi, J. E. Kim, N. H. Heo, S. Komarneni, and J. Choi. 2005. Characteristics of nitrogen release from synthetic zeolite Na-P1 occluding NH_4NO_3. *J. Control. Release* 106:44–50.

Park, M. and S. Komarneni. 1998. Ammonium nitrate occlusion vs. nitrate ion exchange in natural zeolites. *Soil Sci. Soc. Am. J.* 62:1455–1459.

Parkin, T. B. and J. J. Meisinger. 1989. Denitrification below the crop rooting zone as influenced by surface tillage. *J. Environ. Qual.* 18:12–16.

Patil, M. D., B. S. Das, E. Barak, P. B. S. Bhadoria, and A. Polak. 2010. Performance of polymer coated urea in transplanted rice: Effect of mixing ratio and water input on nitrogen use efficiency. *Paddy Water Environ.* 8:189–198.

Pendall, E., S. Bridgham, P. J. Hanson, B. Hungate, D. W. Kicklighter, D. W. Johnson, B. E. Law et al. 2004. Below ground process responses to elevated CO_2 and temperature: A discussion of observations, measurement methods, and models. *New Phytol.* 162:311–322.

Perakis, S. S. and L. O. Hedin. 2002. Nitrogen loss from unpolluted South American forest mainly via dissolved organic compounds. *Nature* 415:416–419.

Pesek, J., G. Stanford, and N. L. Case. 1971. Nitrogen production and use. In: *Fertilizer Technology and Use*, 2nd edition, ed., R. C. Dinauer, pp. 217–269. Madison, Wisconsin: SSSA.

Peterson, G. A., A. D. Halvorson, J. L. Havlin, O. R. Jones, D. G. Lyon, and D. L. Tanaka. 1998. Reduced tillage and increasing cropping intensity in the Great Plains conserve soil carbon. *Soil Tillage Res.* 47:207–218.

Phillips, R. L. 2007. Organic agriculture and nitrous oxide emissions at sub-zero soil temperatures. *J. Environ. Qual.* 36:23–30.

Piccolo, A., G. Pietramellara, and J. S. C. Mbagwu. 1997. Reduction in soil loss from erosion susceptible soils amended with humic substances from oxidized coal. *Soil Technol.* 10:235–245.

Piovesan, R. P., N. Favaretto, V. Pauletri, A. C. V. Motta, and C. B. Reissmann. 2009. Nutrients loss from subsurface in soil column under mineral and organic fertilization. *Rev. Bras. Ci. Solo* 33:757–766.

Powell, J. M. and G. A. Broderick. 2011. Transdisciplinary soil science research: Impacts of dairy nutrition on manure chemistry and environment. *Soil Sci. Soc. Am. J.* 75:2071–2078.

Powell, J. M., G. A. Broderick, and T. H. Misselbrook. 2008. Seasonal diet affects ammonia emissions from tie-stall dairy barns. *J. Dairy Sci.* 91:857–869.

Pye, K. 1987. *Aeolian Dust and Dust Deposits*. New York: Academic Press.

Qiu, Z. 2003. A VSA-based strategy for placing conservation buffers in agricultural watersheds. *J. Environ. Manage.* 32:299–311.

Raun, W. R. and V. G. Johnson. 1999. Improving nitrogen use efficiency for cereal production. *Agron. J.* 1:357–363.

Reddy, M. S. and R. W. Willey. 1981. Growth and resource use studies in an intercrop of pearl millet/groundnut. *Field Crops Res.* 4:13–24.

Rehakova, M., S. Cuvanova, M. Dzivak, J. Rimaran, and Z. Gavalova. 2004. Agricultural and agrochemical uses of natural zeolite of the clinoptilolite. *Curr. Opin. Solid State Mater. Sci.* 8:397–404.

Richards, J. H. and M. M. Caldwell. 1987. Hydraulic lift substantial nocturnal water transport between soil layers by *Artemisia tridentate* roots. *Oecologia* 73:486–489.

Ritchey, K. D., D. M. G. Souza, and U. F. Costa. 1980. Calcium leaching to increase rooting depth in a Brazilian savanna Oxisol. *Agron. J.* 72:40–44.

Roberts, K. G., B. A. Gloy, S. Joseph, N. R. Scott, and J. Lehmann. 2010. Life cycle assessment of biochar systems: Estimating the energetic, economic, and climate change potential. *Environ. Sci. Technol.* 44:827–833.

Robertson, G. P. and P. M. Groffman. 2007. Nitrogen transformations. In: *Soil Microbiology, Ecology, and Biochemistry*, 3rd edition, ed., E. A. Paul, pp. 341–364. New York: Academic Press.

Robertson, G. P. and P. M. Vitousek. 2009. Nitrogen in agriculture: Balancing the cost of an essential resource. *Annu. Rev. Environ. Resour.* 34:97–125.

Robinson, C. A., M. Ghaffarzadeh, and R. M. Cruse. 1996. Vegetative filter strip effects on sediment concentration in cropland runoff. *J. Soil Water Conserv.* 51:227–230.

Robinson, J. M. and C. Baysdorfer. 1991. Interrelationship between photosynthetic carbon and nitrogen metabolism in manure soybean leaves and isolated leaf mesophyll cells. In: *Regulation of Carbon Partitioning in Photosynthetic Tissue, Proceedings of Annual Symposium of Plant Physiology*, 8th edition, eds., R. L. Hath and J. Preiss, Riverside, California: University of California.

Rochette, P., D. A. Angers, G. Belanger, M. H. Chantigny, D. Prevost, and G. Levesque. 2004. Emissions of N_2O from alfalfa and soybean crops in eastern Canada. *Soil Sci. Soc. Am. J.* 68:493–506.

Rochette, P., D. A. Angers, M. H. Chantigny, J. D. MacDonald, M. Gasser, and N. Bertrand. 2009a. Reducing ammonia volatilization in a no-till soil by incorporating urea and pig slurry in shallow bands. *Nutr. Cycl. Agroecosyst.* 84:71–80.

Rochette, P., D. Guilmetre, M. H. Chantigny, D. A. Angers, J. D. MacDonald, and N. Bertrand. 2008. Ammonia volatilization following application of pig slurry increase with slurry interception by grass foliage. *Can. J. Soil Sci.* 88:585–593.

Rochette, P., J. D. MacDonald, D. A. Angers, M. H. Chantigny, M. Gasser, and N. Bertrand. 2009b. Banding of urea increased ammonia volatilization in a dry acidic soil. *J. Environ. Qual.* 38:1383–1390.

Rodriguez, M. B., A. Godeas, and R. S. Lavado. 2008. Soil acidity changes in bulk soil and maize rhizosphere in response to nitrogen fertilization. *Commun. Soil Sci. Plant Anal.* 39:2597–2607.

Roelcke, M., S. X. Li, X. H. Tian, Y. J. Gao, and J. Richter. 2002. *In situ* comparisons of ammonia volatilization from N fertilizers in Chinese loess soils. *Nutr. Cycl. Agroecosyst.* 62:73–88.

Ross, G. H., E. M. J. Temminghoff, and E. Hoffland. 2011. Nitrogen mineralization: A review and meta-analysis of the predictive value of soil tests. *Eur. J. Soil Sci.* 62:162–173.

Russelle, M. P., M. H. Entz, and A. J. Franzluebbers. 2007. Reconsidering integrated crop–livestock systems in North America. *Agron. J.* 99:325–334.

Ryden, J. C. and D. R. Lockyer. 1985. Evaluation of a system of wind tunnels for field studies of ammonia loss from grassland through volatilization. *J. Sci. Food Agric.* 36:781–788.

Ryden, J. C. and L. J. Lund. 1980. Nature and extent of directly measured denitrification losses from some irrigated vegetable crop production units. *Soil Sci. Soc. Am. J.* 44:505–511.

Sahrawat, K. L. 1982. Assay of nitrogen supplying capacity of tropical rice soils. *Plant Soil* 65:111–121.

Sahrawat, K. L. 2008. Factors affecting nitrification in soils. *Commun. Soil Sci. Plant Anal.* 39:1436–1446.

Saigusa, M., M. Toma, and M. Nanzyo. 1996. Alleviating of subsoil acidity in nonallophanic Andosolos by phosphogypsum application in topsoil. *Soil Sci. Plant Nutr.* 42:221–227.

Sainju, U. M. 2013. Tillage, cropping sequence, and nitrogen fertilization influence dryland soil nitrogen. *Agron. J.* 105:1253–1263.

Sainju, U. M., T. Caesar-Tonthat, A. W. Lenssen, and J. L. Barsotti. 2012a. Dryland soil greenhouse gas emissions affected by cropping sequence and nitrogen fertilization. *Soil Sci. Soc. Am. J.* 76:1741–1757.

Sainju, U. M., A. W. Lenssen, T. Caesar-Ton That, J. D. Jabro, R. T. Lartey, R. G. Evans, and B. L. Allen. 2012b. Dryland soil nitrogen cycling influenced by tillage, crop rotation, and cultural practices. *Nutr. Cycl. Agroecosyst.* 93:309–322.

Scherer, H. W. and K. Mengel. 1986. Importance of soil type on the release of nonexchangeable NH_4^+ and availability of fertilizer NH_4^+ and fertilizer NO_3^-. *Fert. Res.* 8:249–258.

Schlegel, A. J., K. C. Dhuyvetter, and J. L. Havlin. 1996. Economic and environmental impacts of long term nitrogen and phosphorus fertilization. *J. Prod. Agric.* 9:114–118.

Schlesinger, W. H. 1997. *Biogeochemistry: An Analysis of Global Change*, 2nd edition. San Diego, California: Academic Press.

Schmitt, T. J., M. G. Dosskey, and K. D. Hoagland. 1999. Filter strip performance and processes for different vegetation, widths, and contaminants. *J. Environ. Qual.* 28:1479–1489.

Schroder, H. 1985. Nitrogen losses from Danish agriculture—Trends and consequences. *Agric. Ecosyst. Environ.* 14:279–289.

Schultz, R. C., T. M. Isenhart, J. P. Colletti, W. W. Simpkins, R. P. Udawatta, and P. L. Schultz. 2009. Riparian and upland buffer practices. In: *North American Agroforestry: An Integrated Science and Practice*, ed., H. E. Garrett, pp. 163–218. Madison, Wisconsin: ASA.

Sehy, U., R. Ruser, and J. C. Munch. 2003. Nitrous oxide fluxes from maize fields: Relationship to yield, site specific fertilization, and soil conditions. *Agric. Ecosyst. Environ.* 99:97–111.

Seiler, W. 1986. Other greenhouse gases and aerosol; nitrous oxide. In: *The Greenhouse Effect, Climatic Change and Ecosystem,* eds., B. Bolin, B. R. Doos, J. Jager, and R. A. Warrick, pp.170–174. New York: John Wiley &Sons.

Senavirante, G. M. M. M., R. P. Udawatta, K. A. Nelson, K. Shanon, and S. Jose. 2012. Temporal and spatial influence of perennial upland buffers on corn and soybean yields. *Agron. J.* 104:1356–1362.

Seobi, T., S. H. Anderson, R. P. Udawatta, and C. J. Gantzer. 2005. Influence of grass and agroforestry buffer strips on soil hydraulic properties for an Albaqualf. *Soil Sci. Soc. Am. J.* 69:893–901.

Shah, S. B., M. L. Wolfe, and J. T. Borggaard. 2004. Simulating the fate of subsurface banded urea. *Nutr. Cycl. Agroecosyst.* 70:47–66.

Shainberg, I., M. E. Sumner, W. P. Miller, M. P. W. Farnaa, M. A. Pavan, and M. V. Fey. 1989. Use of gypsum on soils: A review. *Adv. Soil Sci.* 9:1–111.

Sharma, M., S. Kishore, S. N. Tripathi, and S. N. Behera. 2007. Role of atmospheric ammonia in the formation of inorganic secondary particulate matter: A study at Kanpur, India. *J. Atmos. Chem.* 58:1–17.

Sharpe, R. R. and L. A. Harper. 1997. Apparent atmospheric nitrogen loss from hydroponically grown corn. *Agron. J.* 89:605–609.

Sharratt, B. 2011. Size distribution of windblown sediment emitted from agricultural fields in the Columbia Plateau. *Soil Sci. Soc. Am. J.* 75:1054–1060.

Shaviv, A. 2001. Advances in controlled release fertilizers. *Adv. Agron.* 71:1–49.

Shoji, S. 2005. Innovative use of controlled viability fertilizers with high performance for intensive agriculture and environmental conservation. *Sci. Chin. C Life Sci.* 48:912–920.

Shoji, S., J. Delgado, A. Mosier, and Y. Miura. 2001. Use of controlled release fertilizers and nitrification inhibitors to increase nitrogen use efficiency and to conserve air and water quality. *Commun. Soil Sci. Plant Anal.* 32:1051–1070.

Sileshi, G. W., F. K. Akinnifesi, O. C. Ajayi, and B. Muys. 2011. Integration of legume trees in maize based cropping systems improves rainfall use efficiency and crop yield stability. *Agric. Water Manage.* 98:1364–1372.

Sileshi, G., F. K. Akinifesi, L. K. Bebusho, T. Beedy, O. C. Ajayi, and S. Mogomba. 2010. Variation in maize yield gaps with plant nutrient inputs, soil type and climate across sub-Saharan Africa. *Field Crops Res.* 116:1–13.

Sileshi, G. W., L. K. Debusho, and F. K. Akinnifesi. 2012. Can integration of legume trees increase yield stability in rainfed maize cropping systems in southern Africa. *Agron. J.* 104:1392–1398.

Sileshi, G. and P. L. Mafongoya. 2006. Long-term effect of legume improved fallows on soil invertebrate and maize yield in eastern Zambia. *Agric. Ecosyst. Environ.* 115:69–78.

Silgram, M. and M. A. Shepherd. 1999. The effects of cultivation on soil nitrogen mineralization. *Adv. Agron.* 65:267–311.

Silva, A. C., J. E. S. Carneiro, L. R. Ferreira, and P. R. Cecon. 2006. Bean intercropped with *Bracharia brizantha* under reduced graminicide doses. *Planta Daninha* 24:71–76.

Singh, B., V. Singh, H. S. Thind, A. Kumar, and R. K. Gupta. 2012a. Fixed time adjustment does site-specific fertilizer nitrogen management in transplanted irrigated rice in South Asia. *Field Crops Res.* 126:63–69.

Singh, P., R. P. Singh, and N. Ghoshal. 2011. Influence of herbicide and soil amendments on soil nitrogen dynamics, microbial biomass, and crop yield in tropical dryland agroecosystems. *Soil Sci. Soc. Am. J.* 76:2208–2220.

Singh, U., J. Sanabria, E. R. Austin, and S. Agyin-Birikorang. 2012b. Nitrogen transformation, ammonia volatilization loss, and nitrate leaching in organically enhanced nitrogen fertilizers relative to urea. *Soil Sci. Soc. Am. J.* 76:1842–1854.

Sistani, K. R., J. G. Warren, N. Lovanh, S. Higgins, and S. Shearer. 2010. Greenhouse gas emissions from swine effluent applied to soil by different methods. *Soil Sci. Soc. Am. J.* 74:429–435.

Small, J. A. and W. P. McCaughey. 1999. Beef cattle management in Manitoba. *Can. J. Anim. Sci.* 79:539–545.

Smith, S. J., J. S. Schepers, and L. K. Porter. 1990. Assessing and managing agricultural nitrogen losses to the environment. *Adv. Soil Sci.* 14:1–43.

Snapp, S. S., P. L. Mafongoya, and S. R. Waddington. 1998. Organic matter technologies for integrated nutrient management in smallholder cropping systems of southern Africa. *Agric. Ecosyst. Environ.* 71:185–2000.

Snyder, C. S. 2008. *Fertilizer Nitrogen BMPs to Limit Losses That Contribute to Global Warming.* Norcross, Georgia: International Plant Nutrition Institute.

Sohi, S., E. Krull, E. Lopez-Capel, and R. Bol. 2010. A review of biochar and its use and function in soil. *Adv. Agron.* 105:47–82.

Soil Science Society of America. 2008. *Glossary of Soil Science Terms.* Madison, Wisconsin: Soil Science Society of America.

Sommer, S. G. and B. T. Christensen. 1992. Ammonia volatilization after injection of anhydrous ammonia into arable soils of different moisture levels. *Plant Soil* 142:143–146.

Sommer, S. G. and A. K. Ersboll. 1996. Effect of air flow rate, lime amendments, and chemical soil properties on the volatilization of ammonium from fertilizer applied to sandy soils. *Biol. Fertil. Soils.* 21:53–60.

Sommer, S. G. and J. E. Olesen. 1991. Effects of dry matter content and temperature on ammonia loss from surface applied cattle slurry. *J. Environ. Qual.* 20:679–683.

Soon, Y. K., S. S. Malhi, R. L. Lemke, N. Z. Lupwayi, and C. A. Grant. 2011. Effect of polymer-coated urea and tillage on the dynamics of available N and nitrous oxide emissions from gray Luvisols. *Nutr. Cycle. Agroecosyst.* 90:267–279.

Sorensen, P. and G. H. Rubek. 2012. Leaching of nitrate and phosphorus after autumn and spring application of separated solid animal manures to winter wheat. *Soil Use. Manage.* 28:1–11.

Spalding, R. F. and M. E. Exner. 1993. Occurrence of nitrate in groundwater: A review. *J. Environ. Qual.* 22:392–402.

Sparks, D. L. 1995. *Environmental Soil Chemistry.* San Diego, California: Academic Press.

Srivastava, H. S. 1980. Regulation of nitrate reductase activity in higher plants. *Phytochemistry* 19:725–733.

Stanford, G., S. Dzienia, and R. A. Vanderpol. 1975. Effects of temperature on denitrification rate in soils. *Soil Sci. Soc. Am. Proc.* 39:867–870.

Stehouwer, R. C. and J. W. Johnson. 1991. Soil adsorption interactions of band-injected anhydrous ammonia and potassium chloride fertilizers. *Soil Sci. Soc. Am. J.* 55:1374–1381.

Stefanson, R. C. and D. G. Greenland. 1970. Measurement of nitrogen and nitrous oxide evolution from soil–plant systems using sealed growth chambers. *Soil Sci.* 109:203–206.

Steiner, C., B. Glaser, W. G. Teixeira, J. Lehmann, W. E. H. Blum, and W. Zech. 2008. Nitrogen retention and plant uptake on a highly weathered central Amazonian Ferresol amended with compost and charcoal. *J. Plant Nutr. Soil Sci.* 171:893–899.

Stevens, C., N. B. Dise, J. O. Mountford, and D. J. Gowing. 2004. Impact of nitrogen deposition on the species richness of grasslands. *Science* 303:1876–1879.

Stevenson, F. J. 1982. Origin and distribution of nitrogen in soils. In: *Nitrogen in Agricultural Soils*, ed. F. J. Stevenson, pp. 1–41. Madison, Wisconsin: ASA, CSSSA, and SSSA.

Stevenson, F. J. 1986. *Cycles of Soil.* New York: Wiley Interscience.

Stewart, W. M. 2008. Fertilizer sources for irrigated corn. In: *Fertilizing for Irrigated Corn: Guide to Best Management Practices*, eds., W. M. Stewart and W. B. Gordon, pp. 2.1–2.6. Norcross, Georgia: International Plant Nutrition Institute.

Stout, W. L. 2003. Effect of urine volume on nitrate leaching in the northeast USA. *Nutr. Cycl. Agroecosyst.* 67:197–203.

Stout, W. L., S. L. Fales, L. D. Muller, R. R. Schnabel, and S. R. Weaver. 2000. Water quality implications of nitrate leaching from intensively grazed pasture swards in the northeast US. *Agric. Ecosyst. Environ.* 77:203–210.

Stout, W. L., W. J. Ghurek, R. R. Schnabel, G. J. Folmer, and S. R. Weaver. 1998. Soil climate effects on nitrate leaching from cattle excreta. *J. Environ. Qual.* 27:992–998.

Strebel, O., W. H. M. Duynisveld, and J. Bottcher. 1989. Nitrate pollution of groundwater in western Europe. *Agric. Ecosyst. Environ.* 26:189–214.

Streubel, J. D., H. P. Collins, M. Garcia-Perez, J. Tarara, D. Granatstein, and C. E. Kruger. 2011. Influence of contrasting biochar types on five soils at increasing rates of application. *Soil Sci. Soc. Am. J.* 75:1402–1413.

Sumner, M. E. 1993. Gypsum and acid soils: The world scene. *Adv. Agron.* 51:1–32.

Sumner, M. E. 1995. Ameliorating of subsoil acidity with minimum disturbance. In: *Subsoil Management Techniques*, eds., N. S. Jayawardane, and B. A. Stewart, pp. 147–185. Boca Raton, Florida: CRC Press.

Sun, Y., Y. Sun, X. Li, X. Guo, and J. Ma. 2009. Relationship of nitrogen utilization and activities of key enzymes involved in nitrogen metabolism in rice under water–nitrogen interaction. *Acta Agron. Sin.* 35:2055–2063.

Sung, F. J. M. and W. S. Lo. 1990. Growth responses of rice in ammonium based nutrient solution with variable calcium supply. *Plant Soil* 125:239–244.

Szumigalski, A. R. and R. C. Van Acker. 2008. Land equivalent ratios, light interception, and water use in annual intercrops in the presence or absence of in-crop herbicides. *Agron. J.* 100:1145–1154.

Taylor, R. M., L. B. Fenn, and C. Pety. 1985. The influence of calcium on growth of selected vegetable species in the presence of ammonium nitrogen. *J. Plant Nutr.* 8:1013–1023.

Teasdale, J. R., C. B. Coffman, and R. W. Mangum. 2007. Potential long term benefits of no-tillage and organic cropping systems for grain production and soil improvement. *Agron. J.* 99:1297–1305.

Tegen, I. and A. A. Lacis. 1996. Modeling of particle size distribution and its influence on the radiative properties of mineral dust aerosol. *J. Geophys. Res.* 101:19237–19244.

Terman, G. L. and C. M. Hunt. 1964. Volatilization losses of nitrogen from surface-applied fertilizers, as measured by crop response. *Soil Sci. Soc. Am. Proc.* 23:667–673.

Thomas, G. A., G. Gibson, R. G. H. Nielsen, W. D. Martin, and B. J. Radford. 1995. Effects of tillage, stubble, gypsum, and nitrogen fertilizer on cereal cropping on a red brown earth in southwest Queensland. *Aust. J. Exp. Agric.* 35:997–1008.

Thomas, S., H. Waterland, R. Dann, M. Close, G. Francis, and F. Cook. 2012. Nitrous oxide dynamics in a deep-alluvial gravel vadose zone following nitrate leaching. *Soil Sci. Soc. Am. J.* 76:1333–1346.

Tilman, D. 1996. Biodiversity: Population versus ecosystem stability. *Ecology* 77:350–363.

Titko, S., J. R. Street, and T. J. Logan. 1987. Volatilization of ammonia from granular and dissolved urea applied to turfgrass. *Agron. J.* 79:535–540.

Toma, M., M. E. Sumner, G. Weeks, and M. Saigusa. 1999. Long-term effects of gypsum on crop yield and subsoil chemical properties. *Soil Sci. Soc. Am. J.* 63:891–895.

Torello, W. A. and D. J. Wehner. 1983. Urease activity in a Kentucky bluegrass turf. *Agron. J.* 75:654–656.

Torello, W. A., D. J. Wehner, and A. J. Turgeon. 1983. Ammonia volatilization from fertilized turfgrass stands. *Agron. J.* 75:454–457.

Tracy, B. F. and Y. Zhang. 2008. Soil compaction, corn yield response, and soil nutrient pool dynamics within an integrated crop–livestock system in Illinois. *Crop Sci.* 48:1211–1218.

Trenbath, B. R. 1999. Multispecies cropping systems in India: Predictions of their productivity, stability resilience and ecological sustainability. *Agroforest. Syst.* 45:81–107.

Turpin, J. E., J. P. Thompson, S. A. Waring, and J. Mackenzie. 1998. Nitrate and chloride leaching in vertisols for different tillage and stubble practices in fallow–grain cropping. *Aust. J. Soil Res.* 36:31–44.

Udawatta, R. P., J. J. Krstansky, G. S. Henderson, and H. E. Garrett. 2002. Agroforestry practices, runoff, and nutrient loss: A paired watershed study. *J. Environ. Qual.* 31:1214–1225.

Urban, W. J., W. L. Hargrove, B. R. Bock, and R. A. Raunikar. 1987. Evaluation of urea–urea phosphate as a nitrogen source for no-tillage production. *Soil Sci. Soc. Am. J.* 51:242–246.

USEPA. 2011. Inventory of U. S. greenhouse gas emissions and sinks. 1990–2009. EPA 430-R-11-605, USEPA, Office of Atmospheric Programs, Washington, DC. http://www.epa.gov/climatechange/emissions.usinventoryreport.html (accessed 26 July 2011).

U. S. Environmental Protection Agency. 1990. *Clean Air Act Amendments of 1990, Detailed Summary of Titles, Title 1-Provisions for Attainment and Maintenance of National Ambient Air Quality Standards—Ozone, Carbon Monoxide and PM10 Nonattainment Provisions*. pp. 24–28. Washington, DC: USEPA.

Vaio, N., M. L. Cabrera, D. E. Kissel, J. A. Rema, J. F. Newsome, and V. H. Calvert. 2008. Ammonia volatilization from urea-based fertilizers applied to tall fescue pastures in Georgia, USA. *Soil Sci. Soc. Am. J.* 72:1665–1671.

Van Cleemput, O. 1998. Subsoils chemo and biological denitrification, N_2O and N_2 emissions. *Nutr. Cycl. Agroecosyst.* 52:187–194.

Van Groenigen, J. W., P. J. Georgius, C. V. Kessel, E. W. J. Hummelink, G. L. Velthof, and K. B. Zwart. 2005. Subsoil N-15 fertilizer. *Nutr. Cycl. Agroecosyst.* 72:13–25.

Van Zwieten, L., S. Kimber, S. Morris, K. Y. Chan, A. Downie, J. Rust, S. Joseph, and A. Cowie. 2009. Effects of biochar from slow pyrolysis of papermill waste on agronomic performance and soil fertility. *Plant Soil* 327:235–246.

Venterea, R. T., M. Burger, and K. A. Spokas. 2005. Nitrogen oxide and methane emission under varying tillage and fertilizer management. *J. Environ. Qual.* 34:1467–1477.

Venterea, R. T., B. Maharjan, and M. S. Dolan. 2011. Fertilizer source and tillage effects on yield-scaled nitrous oxide emissions in a corn cropping system. *J. Environ. Qual.* 40:1521–1531.

Veum, K. S., K. W. Goyne, R. Kremer, and P. P. Motavalli. 2012. Relationships among water stable aggregates and organic matter fractions under conservation management. *Soil Sci. Soc. Am. J.* 76:2143–2153.

Veum, K. S., K. W. Goyne, P. P. Motavalli, and R. P. Udawatta. 2009. Runoff and dissolved organic carbon loss from a paired-watershed study of three adjacent agricultural watersheds. *Agric. Ecosyst. Environ.* 130:115–122.

Viator, R. P., J. L. Kovar, and W. B. Hallmark. 2002. Gypsum and compost effects on sugarcane root growth, yield and plant nutrients. *Agron. J.* 94:1332–1336.

Von Lutzow, M. and I. Kogel-Knabner. 2009. Temperature sensitivity of sol organic matter decomposition. What do we know? *Biol. Fertil. Soils* 46:1–15.

Wang, W. H., B. Kohler, F. Q. Cao, and L. H. Liu. 2008. Molecular and physiological aspects of urea transport in higher plants. *Plant Sci.* 175:467–477.

Wang, Z. H., X. J. Liu, X. T. Ju, F. S. Zhang, and S. S. Malhi. 2004. Ammonia volatilization loss from surface-broadcast urea: Comparison of vented and closed chamber methods and loss in winter wheat summer maize rotation in north China plain. *Commun. Soil Sci. Plant Anal.* 35:2917–2939.

Wani, S. P., P. Singh, K. Boomiraj, and K. L. Sahrawat. 2009. Climate change and sustainable rainfed agriculture: Challenges and opportunities. *Agric. Situation India* 66:221–239.

Wells, R., J. Augustin, J. Davis, S. M. Griffith, K. Meyer, and D. D. Myrold. 2001. Production and transport of denitrification gases in shallow ground water. *Nutr. Cycl. Agroecosyst.* 60:65–75.

Wendland, F., M. Bach, and R. Kunkel. 1998. The influence of nitrate reduction strategies on the temporal development of nitrate pollution of soil and ground water throughout Germany: A regionally differentiated case study. *Nut. Cycl. Agroecosyst.* 50:167–179.

Wiebe, K. and N. Gollehon. 2006. *Agricultural Resources and Environmental Indicators.* Hauppaugr, New York: Nova Science Publishers Inc.

Wiedenfeld, R. P. 1986. Rate, timing, and slow-release nitrogen fertilizers on cabbage and onions. *HortScience* 21:236–238.

Wilson, M. L., C. J. Rosen, and J. F. Moncrief. 2009. Potato response to polymer coated urea on an irrigated, coarse textured soil. *Agron. J.* 101:897–905.

Wortman, S. E., C. A. Francis, and J. L. Lindquist. 2012. Cover crop mixtures for the western corn belt: Opportunities for increased productivity and stability. *Agron. J.* 104:699–705.

Wu, Y., L. Lu, X. Lin, J. Zhu, Z. Cai, X. Yan, and Z. Jia. 2011. Long term field fertilization significantly alters community structure of ammonia-oxidizing bacteria rather than archaea in a paddy soil. *Soil Sci. Soc. Am. J.* 75:1431–1439.

Xing, G. X. and Z. L. Zhu. 2000. An assessment of N loss from agricultural fields to the environment in China. *Nutr. Cycl. Agroecosyst.* 57:67–73.

Xu, J., X. Ma, S. D. Ogsden, and R. Horton. 2011. Short, multineedle frequency domain reflectometry sensor suitable for measuring soil water. *Soil Sci. Soc. Am. J.* 76:1929–1937.

Xu, M., D. Li, J. Li, D. Qin, Y. Hosen, H. Shen, R. Cong, and X. He. 2013. Polyolefin-coated urea decreases ammonia volatilization in a double rice system of southern China. *Agron. J.* 105:277–284.

Yang, Y., W. Hunter, S. Tao, and J. Gan. 2009. Effects of black carbon on pyrethroid availability in sediment. *J. Agric. Food Chem.* 57:232–238.

Yang, Y., M. Zhang, Y. C. Li, X. Fan, and Y. Geng. 2012. Controlled release urea improved nitrogen use efficiency, activities of leaf enzymes, and rice yield. *Soil Sci. Soc. Am J.* 76:2307–2317.

Yeomans, J. C., J. M. Bremner, and G. W. McCarty. 1992. Denitrification capacity and denitrification potential of subsurface soils. *Commun. Soil Sci. Plant Anal.* 23:919–927.

Young, A. 1989. *Agroforestry for Soil Conservation.* Wallingford, UK: CAB International.

Young, J. L. and R. A. Cattani. 1962. Mineral fixation of anhydrous ammonia by air-dry soils. *Soil Sci. Soc. Am. Proc.* 26:147–152.

Zhang, J. S., F. P. Zhang, J. H. Yang, J. P. Wang, M. L. Cai, C. F. Li, and G. C. Cou. 2011. Emissions of N_2O and NH_3, and nitrogen leaching from direct seeded rice under different tillage practices in central China. *Agric. Ecosyst. Envion.* 140:164–173.

Zhang, P., G. Y. Sheng, Y. H. Feng, and D. M. Miller. 2006. Predominance of char sorption over substrate concentration and soil pH in influencing biodegradation of benzonitrile. *Biodegradation* 17:1–8.

Zobeck, T. M. and D. W. Fryrear. 1986. Chemical and physical characteristics of windblown sediment II. Chemical characteristics and total soil and nutrient discharge. *Trans. ASAE* 29:1037–1041.

Zvomuya, F., C. J. Rosen, M. P. Russelle, and S. C. Gupta. 2003. Nitrate leaching and nitrogen recovery following application of polyolefin-coated urea to potato. *J. Environ. Qual.* 32:480–489.

3 Diagnostic Techniques for Nitrogen Requirements in Crop Plants

3.1 INTRODUCTION

The grain yield of cereals and legumes must increase continuously to feed the increasing world population to avoid malnutrition and bring stability to social order. A long-term goal for any nation must be food security, the basis of a stable society, and the foundation of national security (Allen et al., 2012). In modern agriculture, the main objectives are to maximize the yield or production per unit area and minimize the cost of production. Furthermore, preserving natural resources (land and water) and maintaining a clean environment have become important components of the crop production system around the world. A sustainable and secure food and fiber production and an economically and ecological viable agriculture cannot deplete the resources or destroy the environment on which they depend (Allen et al., 2012). In conclusion, there are two main components or objectives of crop production in modern agriculture. These are maximizing crop yields without degrading natural resources (air, water, and soil).

To achieve these objectives, the use of intensive crop production techniques or practices is necessary. These techniques include the use of higher-yield potential crop cultivars, supply of adequate moisture to a crop during its growth cycle, and control of insects, weeds, and diseases. Adequate preparation of land and adoption of appropriate spacing and seed rate are also required to produce higher crop yields (Fageria, 1992). In addition, the use of essential plant nutrients in adequate amounts and proportions is fundamental not only to produce a higher yield but also to maintain sustainability of the cropping systems (Fageria, 2009, 2013, 2014).

There are 17 essential nutrients for higher plants. These nutrients are carbon (C), hydrogen (H), oxygen (O), nitrogen (N), phosphorus (P), potassium (K), calcium (Ca), magnesium (Mg), sulfur (S), zinc (Zn), copper (Cu), manganese (Mn), iron (Fe), boron (B), molybdenum (Mo), chlorine (Cl), and nickel (Ni). Among these nutrients, the first three (C, H, and O) are supplied to plants by air and water. Hence, the remaining 14 elements have to be supplied to plants through organic and/or inorganic fertilizers if the soil is deficient in certain nutrients. The essential plant nutrients are divided into two groups such as macro- and micronutrients. The macronutrients are N, P, K, Ca, Mg, and S and the micronutrients are Zn, Cu, Mn, Fe, B, Mo, Cl, and Ni. Figure 3.1 shows the supply of C, H, and O by air and rain water and division of macro- and micronutrients for crop production. All the essential nutrients are equally important for crop production. The division between macro- and micronutrients is only associated with the quantity required. Macronutrients are required in higher amounts compared to micronutrients. Most of the micronutrients are responsible for the activation of enzymes in metabolic processes of the plant and hence they are required in small amounts by plants.

There are three criteria for the essentiality of a nutrient for crop plants. These criteria are: (i) a plant cannot complete its life cycle (germination to grain production) if a determined nutrient is deficient in the growth medium, (ii) an essential nutrient cannot be substituted by another nutrient, and (iii) a nutrient is directly involved in the metabolic process of the plant. The essential nutrients required by higher plants are exclusively inorganic, a feature distinguishing these organisms from

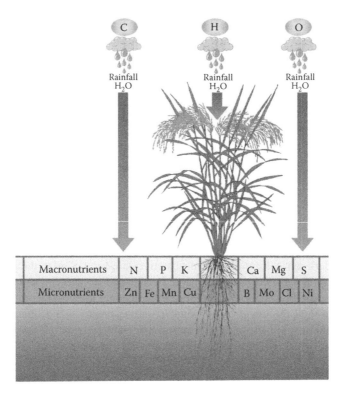

FIGURE 3.1 (**See color insert.**) Essential nutrients for plant growth and development. (Adapted from Fageria, N. K. 1989b. *Tropical Soils and Physiological Aspects of Crops.* EMBRAPA: Brasilia, Brazil.)

man, animals, and many species of microorganism, which additionally need organic foodstuffs to provide energy (Mengel et al., 2001).

Among essential nutrients, N is the most yield limiting in crop production around the world. The deficiency of N in crop production is widely reported in the literature (Marschner, 1995; Fageria et al., 2001; Brady and Weil, 2002; Fageria and Baligar, 2005b; Fageria, 2009, 2013). More research has been done with N than any other mineral nutrient, because the world's ecosystems are probably influenced more by deficiency or excess of N than by those of any other essential element (Brady and Weil, 2002). However, still there is a gap of information for the best management of N in the sustainable cropping systems. Hence, more information is required to know the quantity of N required to increase its availability in the soil to achieve crop yields to a designated level in different cropping systems. The objective of this chapter is to discuss diagnostic techniques for nitrogen requirements in crop plants. This information will be helpful in identifying nitrogen disorder in crop plants and consequently in improving the management of nitrogen fertilization for maximizing crop yields.

3.2 NITROGEN DEFICIENCY, SUFFICIENCY, AND TOXICITY DIAGNOSTIC TECHNIQUES

Diagnostic techniques for nutrient disorders, including N, refer to the methods for identifying nutrient deficiencies, toxicities, or imbalances in the soil–plant system (Fageria and Baligar, 2005a). Nutritional deficiency can occur when there is insufficient nutrient in the medium or when it cannot be absorbed and utilized by plants as the result of unfavorable environmental conditions. Nutritional deficiencies are very common in almost all field crops worldwide. The magnitude varies from crop to crop and from region to region. Even some cultivars are more susceptible to nutritional deficiencies than others within a crop species.

Nitrogen deficiency, sufficiency, and toxicity in crop plants can be identified with the help of observations on the growth and development of crop plants, including visual deficiency and toxicity symptoms on the plant parts, especially on the leaves. Other criteria of nutritional disorder diagnostic in crop plants are soil testing, plant tissue testing, and plant response to applied nutrients under controlled and field conditions (Fageria and Baligar, 2005).

3.2.1 DEFICIENCY AND TOXICITY SYMPTOMS

Nutrient deficiency and toxicity symptoms are important diagnostic techniques of nutritional status of crop plants. It is one of the cheapest nutritional disorder diagnostic techniques in plants. However, it requires a lot of experience on the part of the researcher or extension workers who identify the disorders, because it may be confused with plant abnormality caused by other factors. These factors are herbicide damage, drought, soil salinity problem, insect or disease problem, and low and high temperatures. A symptom typical of nutrient deficiency might even result from a hereditary condition (Bennett, 1993).

3.2.1.1 Deficiency Symptoms

Nutrient deficiency is defined as a low concentration of an essential element or nutrient that reduces plant growth and prevents the completion of the normal plant life cycle (Soil Science Society of America, 2008). Nutrient deficiency symptoms are an indication that the plant cannot take up sufficient nutrients or metabolize sufficient amounts of a nutrient (Voss, 1993). Seeds contain sufficient essential nutrients for germination and seedling emergence in the absence of supplemental nutrition (Voss, 1993). Hence, abnormal plant growth in the early stage of plant growth is generally not related to nutritional problems. Bennett (1993) reported that deficiency symptoms in plants can be grouped into five categories: (i) chlorosis, which is a yellowing of the leaves, which is related to limited chlorophyll synthesis, (ii) necrosis or death of plant tissue, (iii) lack of new growth or terminal growth resulting in resetting, (iv) accumulation of anthocyanin and appearance of a reddish color, and (v) stunting or reduced growth with either a normal or dark green color or yellowing. Further, the most important key to identifying nutrient deficiency symptoms in plants is their location in older or younger leaves. In case of mobile nutrients, deficiency symptoms first start in the older leaves. The mobile nutrients are nitrogen, phosphorus, potassium, and magnesium. In case of immobile nutrients, deficiency symptoms first start in the younger leaves and the older leaves appear normal. The immobile nutrients are calcium, sulfur, boron, copper, iron, manganese, molybdenum, and zinc (Voss, 1993).

In addition to the visual color symptoms, N deficiency symptoms can also be measured in terms of reducing plant height, leaf size, tillering in cereals, and branching in legumes. A reduction in the number of panicles or heads in cereals and pods in legumes is indicative of nutrient deficiency if other environmental factors are at an adequate level. A decrease in the root growth of cereals and legumes has been reported by Fageria and Moreira (2011) and Fageria (2013) when N was deficient in the growth medium. Nitrogen deficiency is favored by climatic, soil, and plant factors. Climatic factors that favor N deficiency in crop plants are low temperature, which inhibits the mineralization of organic N. Soil factors that contribute to N deficiency are soil acidity, crop residue with a low N content or high C/N ratio, sandy soil, soil with low organic matter content, large amount of leaching due to excess rain or flooding, and ponded area when the temperature is warm (Voss, 1993).

In soybean, N deficiency is also reported when roots are infected by cyst nematodes or other nematodes, root or Mo deficiency, all of which interfere with nodule development (Sinclair, 1993). In peanut N deficiency, chlorosis can result from lack of nodulation associated with an inadequate amount of soil bacteria, or from an insufficient N reduction attributable to Mo deficiency, a condition usually associated with extreme soil acidity. Chlorosis may also result from translocation of a limited supply of N to developing fruit late in the season, or from water-logged conditions that limit root respiration and inhibit N fixation (Smith et al., 1993).

A summary of N deficiency symptoms in important food crops is presented in Table 3.1. Figures 3.2 through 3.5 show the growth of tops of cove crops grown at 0 and 200 mg N kg^{-1}. A reduction

TABLE 3.1

Summary of N Deficiency Symptoms in Principal Field Crop Plants

Crop Species	Description of Deficiency Symptoms	Reference
Rice	Reduced tillering, leaf area index and panicle density, and yellowing of older leaves	Fageria et al. (2011a)
Wheat	Stunted growth, reduced tillering, and yellowing of older leaves	Wiese (1993)
Corn	Whole plant to be pale, yellowish green, and have spindly stalks. Yellowing begins on the older, lower leaves and progresses up the plant if the deficiency persists. V-shaped yellowing on the tips of the leaves appears later	Voss (1993)
Sorghum	Plants deficient in N are usually stunted, spindly, and pale green to pale yellow. N deficiency symptoms first appear on the older or lower leaves and advance to younger (upper) leaves. A fairly uniform pale or deep yellow color develops near the tips and margins and progresses toward the base and midrib of the leaf. Dark brown necrotic spots often develop when severe N deficiency occurs	Clark (1993)
Sugarcane	Older leaves die back. Leaf blades of N-deficient plants turn uniformly light green to yellow. Stalks become short and slender, and the vegetative growth rate is reduced. The tips and margins of older leaves become necrotic prematurely	Gascho et al. (1993)
Sugar beet	Overall yellowing of the leaves occurs when a plant first becomes N deficient. Yellowing continues as the plant ages, accompanied by wilting and an accelerated death rate of the older leaves. Newly formed leaves in the center of the plant are much smaller and narrower than older leaves and turn an intense green. Leaves often lie nearly parallel to the soil surface, with the petioles curved slightly upward	Ulrich et al. (1993)
Soybean	N-deficient plants become pale green. Later, the leaves turn distinctly and uniformly yellow. Symptoms first appear on the basal leaves and quickly spread to the upper parts. The plants eventually defoliate and often are spindly and stunted	Sinclair (1993)
Peanut	Nitrogen deficiency in peanut plants is characterized by varying degrees of foliar chlorosis. Young plants not yet adequately colonized by bradyrhizobia usually appear lighter green than normal. In severe cases, the entire leaf becomes a uniform, pale yellow, and stems may be slender and elongated. As the plant develops, lower older leaves are most affected because soluble N from older leaves moves to new younger leaves. The stem may appear reddish because of the accumulation of anthocyanin pigments	Smith et al. (1993)
Cotton	Early and midseason N deficiency symptoms include a yellowish green leaf color, which first appears on older leaves, and reduced size of younger leaves. Plant height is reduced, few vegetative branches develop, fruiting branches are short, and many bolls are shed in the first 10–12 days after flowering. When N deficiency occurs late in the season on plants with a moderate load of maturing bolls, foliar symptoms appear as a reddening in the middle of the canopy, and few bolls are retained at late fruiting positions	Cassman (1993)
Dry bean	Deficiency symptoms appear as a uniformly pale green to yellow discoloration of older leaves. Growth is reduced, and few flowers develop or pods fill poorly	Hall and Schwartz (1993)

continued

TABLE 3.1   (continued)
Summary of N Deficiency Symptoms in Principal Field Crop Plants

Crop Species	Description of Deficiency Symptoms	Reference
Potato	The entire potato plant turns a light green to pale yellow when it first becomes N deficient. As the deficiency increases in severity, the leaves at the apex become smaller and curl upward. The lower, older leaves become deep yellow to light brown to necrotic, and when dry they separate from the stem easily	Ulrich (1993)
Tomato	Nitrogen-deficient plants grow slowly. Leaves are small and light green to yellowish green to pale yellow. Leaves near the top will be yellow-green with purple veins. Stems are thick and hard. Flower buds turn yellow and drop off. Fruits may be small and pale green before ripening. Yields are reduced	Wilcox (1993)
Cucumber, muskmelon, and watermelon	Inadequate N supply to these cucurbit plants reduced vegetative growth and a yellowing of the leaves due to a loss in chlorophyll will be observed. The loss of the green color occurs first on the more mature leaves and last on the younger leaves because of the degradation of N-containing compounds in the older leaves and movement of N to the young tissue	Locascio (1993)
Onions	Deficienct of N in onions can lead to small plants and bulbs and early maturity. The N deficiency is characterized by a uniformly light green foliage that becomes increasingly yellow as the deficiency progresses. The symptoms first appear on the older outer leaves and move progressively toward the newly emerging leaves in the center of the plant as N is translocated toward the stronger sinks near the growing point	Bender (1993)

Source: Bender, D. A. 1993. In: *Nutrient Deficiencies and Toxicities in Crop Plants*, ed., W. F. Bennett, pp. 131–135. St. Paul, Minnesota: The American Phytopathological Society; Cassman, K. G. 1993. Cotton. In: *Nutrient Deficiencies and Toxicities in Crop Plants*, ed., W. F. Bennett, pp. 111–119. St. Paul, Minnesota: The American Phytopathological Society; Clark, R. B. 1993. Sorghum. In: *Nutrient Deficiencies and Toxicities in Crop Plants*, ed., W. F. Bennett, pp. 21–26. St. Paul, Minnesota: The American Phytopathological Society; Fageria, N. K., V. C. Baligar, and C. A. Jones. 2011a. *Growth and Mineral Nutrition of Field Crops*, 3rd edition. Boca Raton, Florida: CRC Press; Gascho, G. J., D. L. Anderson, and J. E. Bowen. 1993. Sugarcane. In: *Nutrient Deficiencies and Toxicities in Crop Plants*, ed., W. F. Bennett, pp. 37–42. St. Paul, Minnesota: The American Phytopathological Society; Hall, R., and H. F. Schwartz. 1993. Common bean. In: *Nutrient Deficiencies and Toxicities in Crop Plants*, ed., W. F. Bennett, pp. 143–147. St. Paul, Minnesota: The American Phytopathological Society; Locascio, S. J. 1993. Cucurbits: Cucumber, muskmelon, and watermelon. In: *Nutrient Deficiencies and Toxicities in Crop Plants*, ed., W. F. Bennett, pp. 123–130. St. Paul, Minnesota: The American Phytopathological Society; Sinclair, J. B. 1993. Soybeans. In: *Nutrient Deficiencies and Toxicities in Crop Plants*, ed., W. F. Bennett, pp. 99–103. St. Paul, Minnesota: The American Phytopathological Society; Smith, D. H., M. A. Wells, D. M. Porter, and F. R. Cox. 1993. Peanuts. In: *Nutrient Deficiencies and Toxicities in Crop Plants*, ed., W. F. Bennett, pp. 105–110. St. Paul, Minnesota: The American Phytopathological Society; Ulrich, A. 1993. Potato. In: *Nutrient Deficiencies and Toxicities in Crop Plants*, ed., W. F. Bennett, pp. 149–156. St. Paul, Minnesota: The American Phytopathological Society; Ulrich, A., J. T. Moraghan, and E. D. Whitney. 1993. Sugar beet. In: *Nutrient Deficiencies and Toxicities in Crop Plants*, ed., W. F. Bennett, pp. 91–98. St. Paul, Minnesota: The American Phytopathological Society; Voss, R. D. 1993. Corn. In: *Nutrient Deficiencies and Toxicities in Crop Plants*, ed., W. F. Bennett, pp. 11–19. St. Paul, Minnesota: The American Phytopathological Society; Wiese, M. V. 1993. Wheat. In: *Nutrient Deficiencies and Toxicities in Crop Plants*, ed., W. F. Bennett, pp. 27–33. St. Paul, Minnesota: The American Phytopathological Society; Wilcox, G. E. 1993. Tomato. In: *Nutrient Deficiencies and Toxicities in Crop Plants*, ed., W. F. Bennett, pp. 137–141. St. Paul, Minnesota: The American Phytopathological Society.

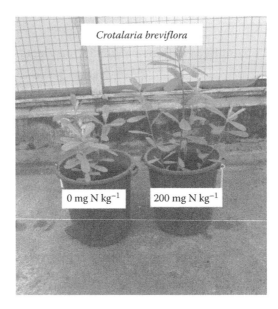

FIGURE 3.2 (**See color insert.**) Growth of cover crop *Crotalaria breviflora* at low and high N levels.

in growth of these cover crops at 0 mg N kg^{-1} compared to 200 mg N kg^{-1} is visible. Similarly, root growth of cover crops at low and high N levels is shown in Figures 3.6 through 3.8. Root growth at 200 mg N kg^{-1} was much more vigorous compared to 0 mg N kg^{-1}. Figure 3.9 shows a plot with N deficiency symptoms and a plot without N deficiency symptoms grown on a Brazilian Oxisol. Similarly, Figure 3.10 shows N deficiency symptoms in dry bean and Figure 3.11 shows dry bean plants growth without N and with N application. Similarly, deficiency nitrogen deficiency symptoms are very clear in plants that did not receive N application. Figure 3.12 shows the growth of dry bean and corn at low (control) and high (200 mg N kg^{-1}). Both the species show yellowing of the leaves at low N levels. Figures 3.13 and 3.14 show the growth of upland rice at two N levels. Nitrogen deficiency is very clear in the plants that did not receive N fertilization. Similarly, root growth of upland rice as influenced by N fertilization is shown in Figures 3.15 through 3.17. Root growth of these upland rice genotypes was much more vigorous at 300 mg N kg^{-1} compared to the control treatment.

FIGURE 3.3 (**See color insert.**) Growth of cover crop *Crotalaria mucronata* at low and high N levels.

FIGURE 3.4 **(See color insert.)** Growth of cover crop *Crotalaria spectabilis* at low and high N levels.

FIGURE 3.5 **(See color insert.)** Growth of cover crop *Lablab purpureus* at low and high N levels.

FIGURE 3.6   Root growth of cover crop lablab at low and high N levels.

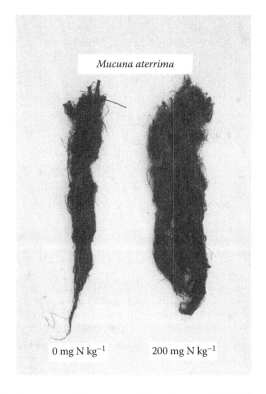

FIGURE 3.7   Root growth of cover crop *Mucuna aterrima* at low and high N levels.

FIGURE 3.8   Root growth of cover crop *Mucuna cinereum* at low and high N levels.

The influence of N on the growth of lowland rice tops and roots is presented in Figures 3.18 and 3.19. Nitrogen deficiency in the tops of lowland rice genotypes is shown by a yellowing of the foliage of plants that did not receive N. However, plants that received 300 mg N kg⁻¹ were green and having larger foliage compared to plants that did not receive N. The growth of the roots of lowland rice genotypes was also affected by N treatments (Figures 3.20 and 3.21). It was higher in the treatments that received N compared to those that did not receive N.

Nitrogen deficiency has many adverse effects on a plant's physiological and biochemical processes. In N-deficient plants, photosynthetic activity is decreased. Equally important to declining

FIGURE 3.9   (**See color insert.**) Upland rice plot without N at the left and with N at the right grown on a Brazilian Oxisol.

FIGURE 3.10 (**See color insert.**) Dry bean leaves and plants with nitrogen deficiency symptoms.

photosynthesis is a reduction in leaf expansion and leaf area and increased sensitivity to water stress when N deficiency occurs. Physiological responses of N-stressed cotton were similar to those encountered with water stress (Radin and Mauney, 1986). Similar to water stress, N stress decreases stomatal and mesophyll conductance of CO_2 (Radin and Ackerson, 1981), decreases hydraulic conductivity, that is, water uptake and transport in the plant (Radin and Parker, 1979; Radin and Boyer,

FIGURE 3.11 (**See color insert.**) Dry bean plants with N deficiency symptoms (left) and without N deficiency symptoms (right).

FIGURE 3.12 **(See color insert.)** Growth of dry bean and corn at two N levels grown on a Brazilian Oxisol. Half of the N was applied at sowing and the remaining half at 35 days after sowing.

1982), reduces leaf expansion and leaf area (Radin and Matthews, 1989), increases starch and soluble carbohydrates in roots (Radin et al., 1978), and decreases leaf osmotic and turgor potential (Radin and Parker, 1979).

3.2.1.2 Toxicity Symptoms

When essential plant nutrients are absorbed in excess of plant requirement, they create adverse effects on plant growth and development. Nitrogen is rarely toxic to crop plants when absorbed in excess amount. However, when it is absorbed in higher amount than the required amount, it may creates an imbalance with other essential nutrients. Bennett (1993) while describing general guidelines for critical,

FIGURE 3.13 **(See color insert.)** Growth of upland rice genotype BRA 02535 at low and high N levels. Half of the N was applied at sowing and the remaining half at the active tillering growth stage.

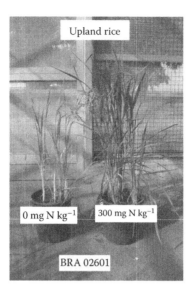

FIGURE 3.14 **(See color insert.)** Growth of upland rice genotype BRA 02601 at low and high N levels. Half of the N was applied at sowing and the remaining half at the active tillering growth stage.

sufficient, and toxic levels of plant nutrients reported that N is nontoxic element. As a general rule, micronutrients are more toxic to plants when absorbed in excess than macronutrients. Nutrient toxicity symptoms first appear on the older leaves and spread to other parts of the plant if excess absorption persists for a longer duration. Nutrient toxicity symptoms in plants can appear easily in solution culture when applied in excess. However, under field conditions, it rarely happens, especially with macronutrients, including N. Excess nitrate-N reduced root growth, especially secondary roots, and increased fibrous root hair growth in sorghum (Clark, 1993). Excess N in plants may cause excessive vegetative growth, flowering may be delayed, and the grain set and yield are reduced.

Seedling injury has been reported for seed-placed urea at rates as low as 40 kg N ha^{-1} for barley (Grant and Bailey, 1999; O'Donovan et al., 2008). Furthermore, seed-placed rates above 28 kg

FIGURE 3.15 Root growth of upland rice genotype BRA 02535 at low and high N levels. Half of the N was applied at sowing and the remaining half at the active tillering growth stage.

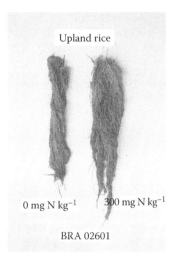

FIGURE 3.16 Root growth of upland rice genotype BRA 02601 at low and high N levels. Half of the N was applied at sowing and the remaining half at the active tillering growth stage.

N ha^{-1} were shown to reduce canola and durum wheat seedling emergence (Malhi et al., 2003). McKenzie et al. (2006, 2007) reported a substantial reduction in the winter wheat stand and in mustard at rates as low as 30 kg N ha^{-1}. Figures 3.22 through 3.25 show N toxicity in dry bean, upland rice, corn, and wheat. Toxicity in all the crops was observed (reductions in growth) at 400 mg N kg^{-1}, except corn. In corn, N toxicity was observed at 600 mg N kg^{-1}. Toxicity symptoms in all the crops were observed in lower or older leaves in the beginning.

3.3 SOIL TEST

Soil testing is the most common criterion to determine the nutrient and lime requirements of crop plants. In modern agriculture, soil testing is considered as an indispensable tool essential for the assessment of the fertility status of soils. A good soil testing program is essential to sound fertilizer

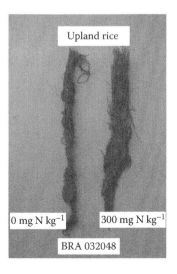

FIGURE 3.17 Root growth of upland rice genotype BRA 032048 at low and high N levels. Half of the N was applied at sowing and the remaining half at the active tillering growth stage.

FIGURE 3.18 **(See color insert.)** Growth of lowland rice genotype BRA 02654 at low and high N levels. Half of the N was applied at sowing and the remaining half at the active tillering growth stage.

use. If the value of a soil test can be treated or expressed as an independent variable, fewer personal judgments are required and more accurate fertilizer recommendations will be made (Melsted and Peck, 1973). Melsted and Peck (1973) reported that a sound soil testing program is best and perhaps the only way to determining what constitutes adequate but not excessive fertilizer use for high and efficient crop production.

A soil test is defined as a physical, chemical, or biological procedure that estimates the suitability of the soil to support plant growth (Soil Science Society of America, 2008). Soil test calibration is defined as the process of determining the crop nutrient requirement at different soil

FIGURE 3.19 **(See color insert.)** Growth of lowland rice cultivar BRA 051077 at low and high N levels. Half of the N was applied at sowing and the remaining half at the active tillering growth stage.

Lowland rice

0 mg N kg^{-1} 300 mg N kg^{-1}

BRA 051077

FIGURE 3.20 Root growth of lowland rice genotype BRA 051077 at low and high N levels. Half of the N was applied at sowing and the remaining half at the active tillering growth stage.

test values (Soil Science Society of America, 2008). The basic terminologies related to soil test calibration are soil test correlation, soil test critical concentration, and soil test interpretation. According to the Soil Science Society of America (2008), these terms are defined as follows: Soil test correlation is defined as the process of determining the relationship between plant nutrient uptake or yield and the amount of nutrient extracted by a particular soil test method. Soil test critical concentration is defined as the concentration of an extractable nutrient above which a crop response to the added nutrient would not be expected. Soil test interpretation is defined as the process of developing nutrient application recommendations from soil test concentration and other soil, crop, economic, environmental, and climatic information. Based on the above discussion, a

Lowland rice

300 mg N kg^{-1}

0 mg N kg^{-1} BRA 051135

FIGURE 3.21 Root growth of lowland rice genotype BRA 051135 at low and high N levels. Half of the N was applied at sowing and the remaining half at the active tillering growth stage.

FIGURE 3.22 **(See color insert.)** Response of dry bean to nitrogen fertilization to a Brazilian Oxisol. Half of the N was applied at sowing and the remaining half as topdressing at 35 days after sowing.

soil test has four components. These components are collecting the soil sample, extracting and determining the available nutrients, interpreting the analytical results, and making the fertilizer recommendations.

Traditionally, soil testing has involved the evaluation of nutrient deficiencies in soils; today, with the increased emphasis on environment quality, soil tests are a logical tool to determine areas where adequate or excess fertilization has taken place. In the soil testing program, the most important point is to get a representative soil sample of an area to determine the fertility status of the soil and make fertilizer recommendations. There is always a variation in the field in relation to soil fertility. This is especially true when fertilizers are band or furrow applied. Hence, a large number of soil samples are required at randomly selected points in a field. Fageria and Baligar (2005a)

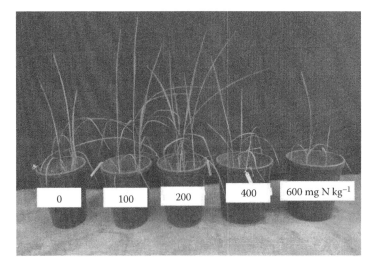

FIGURE 3.23 **(See color insert.)** Response of dry bean to nitrogen fertilization to a Brazilian Oxisol. Half of the N was applied at sowing and the remaining half as topdressing at 35 days after sowing.

FIGURE 3.24 **(See color insert.)** Response of dry bean to nitrogen fertilization to a Brazilian Oxisol. Half of the N was applied at sowing and the remaining half as topdressing at 35 days after sowing.

recommended taking a minimum of one composite sample per 12–15 hectares for lime and fertilizer recommendations. A representative soil sample is composed of 15–20 subsamples from a uniform field with no major variations in the slope, drainage, or past fertilizer history. Any of these listed factors, if changed, will have an effect on the number of samples and unit area from which the sample is obtained.

Generally, soil samples are collected from a 0 to 20 cm soil depth for making fertilizer recommendations for field crops, because a major part of the roots of field crops is located in this layer (Fageria, 2013). However, Fageria and Baligar (2005a) reported that the depth of sampling for mobile nutrients such as nitrogen should be 60 cm, and for immobile nutrients such as P, K, Ca, and Mg, a 15–20 cm sampling depth can give satisfactory results. For pasture crops, a sampling depth

FIGURE 3.25 **(See color insert.)** Growth of wheat at different N levels. Half of the N was applied at sowing and the remaining half as topdressing at 35 days after sowing.

of 0–10 cm is normally sufficient to evaluate the nutrient status and to make liming and fertilizer recommendations.

The inorganic combined nitrogen in most soils is in the form of ammonium (NH_4^+) and nitrate (NO_3^-). Nitrogen is also taken up by crop plants in the form of both NO_3–N and NH_4^+–N. Hence, the soil test values of nitrogen (NO_3^- and NH_4^+) should be calibrated against the straw yield or grain yield of crops. However, in the author's opinion, as yet there does not seem to be any well-accepted method for testing soils for available N. This is a reflection, in part, on the fact that 97–99% of the N in the soil is present in very complex organic compounds that are not available to plants (Dahnke and Vasey, 1973). These authors further reported that developing a test for available soil N is further complicated by the fact that the mineralization process of N in the soil is very complex and influenced by several factors such as temperature, moisture, aeration, type of organic matter, pH, and other factors. In addition, the soil test values of NO_3^- and NH_4^+ are not stable during crop growth because they change very rapidly due to nitrification and losses by volatilization, leaching, and denitrification. In addition, the NO_3^- ion is very soluble and generally does not form insoluble compounds with any of the soil constituents. Another complicating factor in the study of NO_3–N is that it can be rapidly immobilized by soil microorganisms if a suitable source of energy is present, only to reappear after an indefinite period of time as the microorganisms reduce the C/N ratio of the energy source (Dahnke and Vasey, 1973).

Meisinger (1984) concluded, in his review of methods to predict soil N availability to crops, that since scientists have been unsuccessful in their attempts, over the past 75 years, to develop a quick test to predict N mineralization in the field, it is unlikely that, such a test will be developed in the near future. Meisinger et al. (2008) also reported that, in contrast to P, N that is not used by a crop cannot reliably be banked in the soil from year to year and both the spatial and temporal variabilities in soil N provide limitations to the potential accuracy of fertilizer N recommendations. Similarly, Anthony et al. (2012) also reported that the challenges of site-specific N management are different from those involving P, primarily because of the nature of the transient nature of soil N. Hence, soil test calibration for rapid movable or loosing form of N (NO_3^- and NH_4^+) has doubt about its applicability under field conditions.

Soil test calibration values are recommended for immobile nutrients such as P and K. However, data regarding NO_3–N (which is a major form of N in the oxidized soils) interpretation for crop production are available in the literature (Table 3.2). The NO_3–N values in Table 3.2 can serve only as reference points. Hence, these values of the NO_3–N test should be used with caution for different crop species and soil types in crop production processes. In the state of Iowa, the initial interpretation of the NO_3 test for corn was that if the test result is above a critical value of 25 mg kg^{-1}, no N is recommended. However, if the test result is below 25 mg kg^{-1}, 9 kg N ha^{-1} is recommended for each 1 mg kg^{-1} below the critical level (Blackmer et al., 1993). This interpretation has been modified a little bit in 1997 (Blackmer et al., 1997). The critical level dropped by 3–5 mg kg^{-1} when the spring rainfall was more than 20% above normal.

TABLE 3.2
Interpretation of NO_3^- Soil Test Values for Making Nitrogen Recommendations for Crop Plants

NO_3 Values (mg kg^{-1})	Interpretation
<10	Low
10–30	Medium
>30	High

Source: Adapted from Keeney, D. R. and D. W. Nelson. 1982. *Methods of Soil Analysis.* Part 2, 2nd edition, eds., A. L. Page et al., pp. 643–687. Madison, Wisconsin: ASA and SSSA.

The current N fertilizer recommendations for spring wheat are mainly based on the determination of the soil NO_3–N content (0–60 cm) at sowing (Calvo et al., 2013). To use it, different N availability thresholds (soil + fertilizer) have been suggested, which vary according to the area, farming systems, and crop yield objective. A recent study conducted with different wheat genotypes has shown that critical levels of N availability for applications at sowing and at tillering were 174 and 152 kg ha^{-1}, respectively (Barbieri et al., 2012). These kinds of simplified models do not explicitly take into account the N supply through mineralization, however, which represents one of the main N sources for crops (Campbell et al., 1993, 2008), especially in soils with high organic matter content (Calvo et al., 2013). Under medium- to high-yielding conditions, mineralization of N from soil organic pools during the growing season meets 30% of wheat (Gonzalez Montaner et al., 1997) and 60% of corn (Steinbach et al., 2004) demands for N. It is worth mentioning that soil potential N mineralization is affected by cropping history, soil management practices, and weather conditions (Calvo et al., 2013).

Soil analyses are direct measurements of the soil N status. Several approaches have been adopted to directly assess soil N (Stanford, 1982). Gerik et al. (1998) reported that in the western states of the United States, where arid conditions prevail, soil NO_3^- has been successfully used to determine the existing N levels and to adjust N application rates. Other locations have adopted a preseason soil NO_3^- test for adjusting fertilizer N rates based on the existing NO_3^- levels. As with fertilizer tests, soil analysis should be confined to the general location and soil type where testing was performed (Gerik et al., 1998).

3.4 PLANT TISSUE TEST

Plant tissue analyses were developed to overcome variations inherent in fertilizer tests and soil analyses. They play an increasingly important role in modern crop production. A plant tissue test or plant analysis is one of the most important diagnostic techniques in determining nutrient deficiency, sufficiency, or toxic levels in plants. The concept of using a plant tissue test or plant analysis as a guide to identify the nutritional status of the plant is relatively old. However, there is renewed interest in recent years in using this technique due to the increase in the number of spectrographs or the analytical technique. In addition, large quantities of data have been accumulated in the last few decades to facilitate the interpretation of tissue analysis results or reference standards are available. The third reason is that farmers and those who assist them in crop growing techniques are becoming increasingly sophisticated in the technology they demand and are capable of testing. The fourth reason is that fertilization has contributed significantly in increasing crop yields in the last few decades of the twentieth century of important food crops, which permitted the economical use of tissue tests as a diagnostic technique. Smith et al. (1990) reported that to date tissue tests have found application mainly as a diagnostic tool to indicate the N deficiency/sufficiency for high-value crops, including cotton and corn.

A plant tissue test or plant analysis is defined as the determination of the nutrient concentration in plants or plant parts with analytical procedures (Soil Science Society of America, 2008). According to Ulrich and Hills (1973), plant analysis, in its simplest terms, is a study of the relationship of the nutrient content of the plant to its growth. Reference concentrations of mineral nutrients in specific plant parts are determined and used as a guide to indicate how well plants are supplied with essential plant nutrients at a certain time of sampling. Such reference concentrations provide a tool to assist the agronomist in evaluating nutrient disorders and in improving fertilization in the present or succeeding crops (Ulrich and Hills, 1973). Ulrich and Hills (1973) further reported that a basic concept is that the concentration of a nutrient within the plant at any particular moment is an integrated value of all the factors that have influenced the nutrient concentration up to the time of sampling. Rauschkolb et al. (1984) stated that an attractive feature of a tissue test is that the plant root system tends to integrate the spatial variability of soil N supplying power over a relatively large field volume.

By modern chemical methods, it is possible to analyze for the nutrient element in a plant sample with great accuracy. In plant analysis, concentration is the nutrient content per unit of dry plant material. Plant analysis results (concentration) are expressed in percentage (%), g kg⁻¹, or mg kg⁻¹.

3.4.1 Nitrogen Concentration in Plants

This approach is used to determine the *critical nutrient level* or optimum concentration of a nutrient, including N in the plant tissue of a crop. Critical nutrient level is defined as the concentration of a nutrient in a plant tissue at 90% or 95% relative yield. Ulrich and Hills (1973) reported that the critical nutrient level in a plant tissue can be calculated by using a 10% reduction in yield. They also defined critical nutrient concentration as that concentration at which the growth rate of the plant begins to decline significantly. It involves adding increasing amounts of a deficient element to others present in adequate amounts, and relating the growth or yield to the nutrient concentration. The critical concentration is estimated best through the use of solution culture and, to a lesser extent, by the soil culture or field experiment technique (Ulrich, 1950). The reason for this is that in solution culture it is easy to create a large variation in nutrient concentration and get a good crop response curve.

Ulrich and Hills (1973) further stressed that plant response to nutrient addition is largely independent of their source, and consequently it makes very little difference to the plant whether its roots are in a culture solution or in a soil. The symptoms and nutrient concentration of the affected leaf blades are, for all practical purposes, identical in both growth media. It follows, therefore, that the critical values have a nearly universal application once they are properly calibrated (Ulrich and Hills, 1973). Varietal differences have not been observed in critical concentrations of sugar beet (Ulrich and Hills, 1973). Similarly, the authors did not observe significant differences in critical levels of N among upland and lowland rice cultivars having similar yield levels. However, Barker and Bryson (2007) reported that N concentrations in plants vary within species and with varieties within species.

In the author's opinion, yield level is more important in defining adequate or critical N concentration rather than variety. Nutrient concentration at a higher yield level will be relatively lower compared with that at a lower yield level due to dilution effects, especially in the vegetative growth stage. During the grain filling stage, nutrient concentration is more or less constant in the crop species. A hypothetical relationship between nutrient concentration in plant tissue and relative yield is presented in Figure 3.26. According to Ulrich and Hills (1973), the most useful calibration curve is one in which the transition zone is sharp, that is, there is a narrow range in nutrient concentration between plants that are deficient and those that are well supplied with the nutrient in question. Barker and Bryson (2007) reported that herbaceous crops from fertilized fields commonly have concentrations of N that exceed 3% (30 g kg⁻¹) of the dry mass of mature leaves. Leaves of grasses (Gramineae, Poaceae) (1.5–3.5% N) are typically lower in total N concentration than those of legumes (Leguminosae, Fabaceae) (>3%). Leaves of trees and woody ornamentals may have <1.5% N in mature leaves. The choice of tissue for plant analysis is important in N disorder diagnosis (Table 3.3).

However, Novoa and Loomis (1981) reported that the N concentration 1.4–1.6% (14–16 g kg⁻¹) in mature biomass reflects the final balance, which is achieved between the grain and straw fractions. These authors reported that the demand for N is determined by the growth rate and the N composition of the new tissues. In the field, both the growth rate and tissue composition will vary with the nitrogen and water supply, plant competition, and other environmental factors. The maximum demand for N will be achieved under nonlimiting conditions for photosynthesis when the growth rate approaches its genetic potential (Novoa and Loomis, 1981).

In the literature, several different terminologies have been used in classifying nutrient concentrations in plant tissue. On the basis of Figure 3.26, nutrient concentrations can be classified as deficient, marginal, excess, and toxic. When nutrients are in the deficiency range, the plant growth

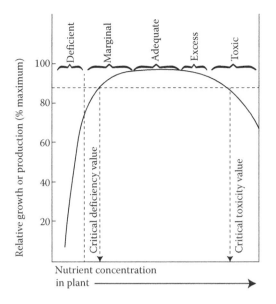

FIGURE 3.26 Relationship between nutrient concentration in plant tissue and relative growth or yield. (From Fageria, N. K., V. C. Baligar, and C. A. Jones. 2011a. *Growth and Mineral Nutrition of Field Crops*, 3rd edition. Boca Raton, Florida: CRC Press. With permission.)

and yield are significantly reduced and foliar deficiency symptoms appear. In this range, application of the nutrient results in a sharp increase in growth with very little change in nutrient concentration in the plant. In the marginal range, the growth or yield is reduced but plants do not show deficiency symptoms, and both nutrient concentrations and growth increase as more nutrients are absorbed. Sometimes the marginal range is also called the transition zone (Ulrich and Hills, 1973). Within the marginal or transition zone lies the critical level or concentration.

The third range or zone is the adequate zone, in which there is no increase in growth but the nutrient concentration increases. This classification is also known as satisfactory, normal, or sufficient concentration. The high classification range represents the range of concentrations between the adequate and toxic ranges. Fertilizer use on crops with values in this range should be reduced until the nutritional status of plants lies in the adequate range (Reuter and Robinson, 1986). The toxic range is the range of nutrients in which there is a reduction in growth and yield but the concentration of nutrients continues to increase. In this range, plants start showing toxicity symptoms. The critical toxicity value lies in this range (Figure 3.26).

TABLE 3.3

Average Total N Concentration in Different Plant Parts

Plant Part Analyzed	N Concentration Range (g kg⁻¹)	Optimum N Concentration (g kg⁻¹)
Leaves (blades)	10–60	>30
Stems	10–40	>20
Roots	10–30	>10
Fruits	10–60	>30
Seeds	29–70	>20

Source: Adapted from Barker, A. V. and G. M. Bryson. 2007. Nitrogen. In: *Handbook of Plant Nutrition*, eds., A. V. Barker, and D. J. Pilbeam, pp. 21–50. Boca Raton, Florida: CRC Press.

Generally, nutrient concentration in most tissues of crop plants decreased with increasing plant age. A relationship between plant age and N concentration in the tissue of corn, upland rice, soybean, and dry bean is shown in Figure 3.27. Nitrogen concentration in four crops decreased with increasing plant age. The decreasing trend was different between cereals and legumes. The decrease in N concentration with plant age was expected because, with increasing plant age, more dry matter was produced, diluting the concentration of nutrients accumulated (Fageria, 2009). Barker and Bryson (2007) also reported that the concentrations of N in leaves, stems, and roots change during the growing season. In the early stages of growth, concentrations will be high throughout the plant. As plants mature, the concentrations of N in these organs fall and are usually independent of the initial external supply of N. If the development of a plant is restricted by low levels of external factors, such as other nutrients, water, or temperature, the internal concentration of N may rise (Barker and Bryson, 2007). Table 3.4 shows the average adequate concentrations for essential nutrients in crop plants. No doubt these values vary with the soil, climate, crop, and management practices, and it is very difficult to make generalizations. Generalized values do give the reader some idea about what levels of nutrients are adequate in crop plants.

Various studies have indicated that plant N concentration declines as plants grow (Greenwood et al., 1986; Lemaire and Gastal, 2009; Yue et al., 2012a). This decline in N is described by a negative power function called the dilution curve: $N = aW^{-b}$, where W is the aboveground biomass or dry matter (DM), expressed in Mg DM ha^{-1}, N is the N_c in aboveground DM, expressed in % DM, a represents plant N concentration in percent when the crop mass is 1 Mg DM ha^{-1}, and b represents the dilution coefficient (Greenwood et al., 1990). Greenwood et al. (1990) defined two general critical N dilution curves when W is >1 Mg DM ha^{-1}. One is for C3 crop species, that is, $N_c = 5.7 W^{-0.5}$. This curve value is valid for tall fescue (*Festuca arundinacea* Schreb.), Lucerne (*Medicago sativa* L.), potato (*Solanum tuberosum* L.), wheat (*Triticum aestivum* L.), and rape

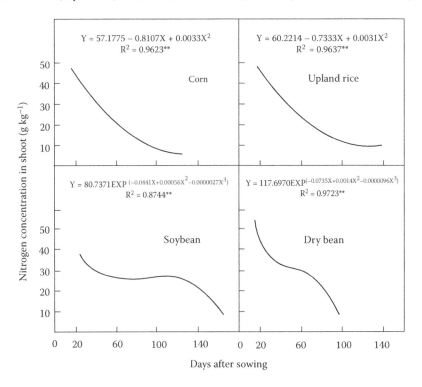

FIGURE 3.27 Relationship between plant age in days and nitrogen concentration in plant tissue of corn, upland rice, soybean, and dry bean. (Adapted from Fageria, N. K. 2004. *Commun. Soil Sci. Plant Anal.* 35:961–974.

TABLE 3.4
Adequate Total Nitrogen Concentration in the Plant Tissue of Principal Food Crops

Crop Species	Growth Stage	Plant Part Analyzed	N Concentration (g kg^{-1})
Rice[1]	Heading	Uppermost mature leaves	26–42
Wheat	Tillering	Leaf blade	43–52
Wheat	Shooting	Leaf blade	36–44
Wheat	Heading	Leaf blade	21–30
Wheat	Flowering	Leaf blade	27–30
Barley	Tillering	Leaves	47–51
Barley	Shooting	Leaves	45–47
Barley	Heading	Whole tops	20–30
Barley	Flowering	Leaves	29–35
Corn	30–45 DAE	Whole tops	35–50
Corn	Before tasseling	LB below the whorl	30–35
Corn	Silking	BOAC	>32
Corn	Silking	Ear LB	28–35
Sorghum	Seedling	Whole tops	35–51
Sorghum	Early vegetative	Whole tops	30–40
Sorghum	Vegetative	YMB	32–42
Sorghum	Bloom	3BBP	33–40
Soybean	Prior to pod set	UFDTU	45–55
Dry bean	Early flowering	UMB	52–54
Dry bean	Peak harvest	Pods	31
Cowpea	30 DAS	Whole tops	28
Cowpea	Early flowering	PUMB	11–17
Peanut	Early pegging	USL	35–45
Sugarcane	3 month plants	TVD	24–25
Sugarcane	6 month plants	TVD	19
Sugarcane	4–5 month ratoon	TVD	19
Cassava	Vegetative	UMB	50–60
Potato	42 DAE	UMB + P	40–50
Potato	Early flowering	UMB + P	55–65
Potato	Tubers half grown	UMB + P	30–50
Cotton	45 DAS	LB	>50
Cotton	First flowering	YMB	38–45

Source: Compiled from Fageria, N. K., V. C. Baligar, and C. A. Jones. 2011a. *Growth and Mineral Nutrition of Field Crops*, 3rd edition. Boca Raton, Florida: CRC Press. With permission.

Note: DAE = days after emergence; LB = leaf blade; BOAC = blade opposite and above cob; YMB = younger upper most mature leaf blade; UFDT = upper fully developed trifoliate; UMB = uppermost mature leaf blade; DAS = days after sowing; PUMB = petiole of uppermost mature leaf blade; USL = upper stems and leaves; TVD = top visible dewlap, which is approximately the third leaf from the shoot apex; UMB + P = uppermost mature leaf blade + petiole.

(*Brassica oleracea* L.). The curve for C4 crop species is $N_c = 4.1\ W^{-0.5}$, which is mainly for crops such as corn, sorghum, pearl millet, and sugarcane. Justes et al. (1994) estimated the critical N dilution curve for winter wheat in France as $N_c = 5.35\ W^{-0.442}$ when the aboveground biomass was between 1.55 and 12 Mg DM ha^{-1}. When the aboveground biomass was <1.55 mg DM ha^{-1}, the constant critical value N = 4.4% DM was applied, which was independent of the aboveground biomass. This critical N dilution curve is used for wheat production worldwide (Stockle and Debaeke, 1997). However, some investigators have reported variations in the critical N curve between regions, species, and even genotypes within species (Belanger et al., 2001; Tei et al., 2002; Ziadi et al., 2008; Yue et al., 2012a,b).

The author studied nitrogen concentration in the shoots and grains of 19 upland rice genotypes (Table 3.5). Nitrogen concentration in the shoot and N uptake in the shoot (N concentration in the shoot × shoot dry weight) was significantly influenced by N and genotype treatments (Table 3.5). However, N concentration in the grain was not influenced either by N or by genotype treatments. But N uptake in the grain was only influenced by N treatment. Nitrogen concentration in the shoot varied from 4.8 to 8.9 g kg^{-1} with an average value of 6.4 g kg^{-1} at low N levels. All the genotypes showed visual N deficiency symptoms at low N levels. This means that an average N concentration of 6.4 g kg^{-1} in the shoot can be considered as a deficient level in upland rice. At higher N levels, N concentration in the shoot varied from 5.7 to 12.8 g kg^{-1} with an average value of 9.5 g kg^{-1}. Visual N deficiency symptoms were not observed in any genotype at a higher N level. Hence, the average value of 9.5 g N kg^{-1} can be taken as an adequate concentration in the shoots of upland rice at harvest. Fageria (2014) reported an adequate concentration for maximum yield of about 8.7 g kg^{-1} in the shoots of upland rice under field conditions at harvest. The slightly higher value of N concentration in shoots in the present study may be due to the use of different genotypes. Nitrogen concentration

TABLE 3.5
Concentration of N in the Shoot and Grain of Upland Rice Genotypes at Two N Levels

Genotype	N Concentration in Shoot (g kg^{-1})		N Concentration in Grain (g kg^{-1})
	0 mg N kg^{-1}	400 mg N kg^{-1}	Across Two N levels
CRO 97505	6.3abc	8.7ab	10.9a
CNAs 8993	6.3abc	7.0ab	11.3a
CNAs 8812	6.9abc	9.2ab	13.3a
CNAs 8938	5.7bc	11.3ab	11.9a
CNAs 8960	6.0bc	10.4ab	12.4a
CNAs 8989	6.3abc	12.8a	11.7a
CNAs 8824	4.9c	9.8ab	12.6a
CNAs 8957	5.4bc	10.8ab	12.8a
CRO 97422	6.7abc	9.8ab	12.5a
CNAs 8817	5.6bc	10.5ab	12.1a
CNAs 8934	8.9a	10.5ab	11.5a
CNAs 9852	7.2abc	10.5ab	11.3a
CNAs 8950	7.6ab	9.1ab	12.6a
CNA 8540	4.8c	9.8ab	12.3a
CNA 8711	6.6abc	10.0ab	12.5a
CNA 8170	5.7bc	8.7ab	11.1a
BRS Primavera	6.0bc	7.2ab	12.5a
BRS Canastra	7.6ab	5.7b	12.1a
BRS Carisma	6.7abc	8.3ab	12.8a
Average	6.4	9.5	12.1
F-test			
N level (N)	*		NS
Genotype (G)	**		NS
N × G	**		NS
CV (%)	16		14

Source: From Fageria, N. K., O. P. Morais, and A. B. Santos. 2010. *J. Plant Nutr*. 33:1696–1711. With permission.

*,**, NS: Significant at the 5% and 1% probability levels and nonsignificant, respectively. Means followed by the same letter in the same column are not significantly different at the 5% probability level by Tukey's test.

in shoots was having a significant quadratic association with grain yield (Y = −38.3532 + 126.1739X 41.0378X^2, R^2 = 0.34*). This means that improving the N concentration in upland rice genotypes can improve grain yield.

N concentration in grains was 11.3 g kg^{-1} at low N levels and 12.8 g kg^{-1} at high N levels across the 19 upland rice genotypes (Fageria, 2014). These concentrations can be considered deficient and sufficient for upland rice, respectively. Cassman et al. (2002) reported that N concentration of about 12 g kg^{-1} is desired in grains of rice for optimal cooking and eating quality. Nitrogen concentration was higher in grains compared to shoots. Nitrogen concentration is always higher in grains of rice than in stover (Kiniry et al., 2001). However, there was less variation in N concentration in grains among genotypes compared to N concentration in shoots. Nitrogen concentration in grains was having a highly significant positive association with grain yield (r = 0.31*). Hence, increasing the N concentration in grains can increase the upland rice grain yield.

The author also studied N concentration in the shoots and grains of dry bean genotypes (Table 3.6). The average value of N concentration in shoots was 13.3 g kg^{-1} at the zero N level and

TABLE 3.6
Nitrogen Concentration in the Shoot and Grain of Dry Bean Genotypes under Two N Rates

Genotype	N Concentration in Shoot (g kg^{-1})		N Concentration in Grain (g kg^{-1})	
	N0	N400	N0	N400
Pérola	13.7ab	26.7abc	55.0a	54.0abcd
BRS Valente	21.0a	28.0ab	49.0abc	44.3bcde
CNFM 6911	11.3b	14.0cd	45.0abcd	48.0bcde
CNFR 7552	13.3ab	12.0d	42.0abcd	32.7e
BRS Radiante	12.3b	17.0bcd	45.33abcd	40.0cde
Jalo Precoce	14.0ab	15.0cd	53.7ab	53.0abcd
Diamante Negro	11.0b	18.0bcd	43.7abcd	53.3abcd
CNFP 7624	15.0ab	10.3d	47.3abcd	36.7de
CNFR 7847	9.0b	9.3d	45.3abcd	43.3cde
CNFR 7866	15.0ab	36.0a	43.0abcd	72.7a
CNFR 7865	16.3ab	18.3bcd	48.7abc	57.7abc
CNFM 7875	15.3ab	9.0d	48.7abc	42.0cde
CNFM 7886	14.7ab	12.0d	39.7abcd	42.0cde
CNFC 7813	21.0a	9.3d	41.0abcd	43.0cde
CNFC 7827	10.7b	8.7d	44.0abcd	37.0de
CNFC 7806	9.7b	16.0bcd	35.0cd	50.0bcde
CNFP 7677	10.0b	8.7d	40.7abcd	41.3cde
CNFP`7775	9.7b	8.3d	39.0bcd	38.0cde
CNFP 7777	8.7b	11.7d	33.7cd	46.0bcde
CNFP 7792	14.0ab	25.0abc	33.0d	63.3ab
Average	13.3	15.7	43.6	46.9
F-test				
N rate (N)	**		**	
Genotype (G)	**		**	
N × G	**		**	

Source: From Fageria, N. K. L. C. Melo, and. P. de Oliveira. 2013. *J. Plant Nutr.* 36:2179–2190. With permission.

**Significant at the 1% probability level. Within the same column, means followed by the same letter do not differ significantly at the 5% probability level by Tukey's test.

15.7 g kg^{-1} at the 400 mg N level. There was an 18% increase in N concentration in shoots at a higher N level compared to a low N level. Similarly, the average N concentration in grains was 43.6 g kg^{-1} at a low N level and 46.9 g kg^{-1} at a high N level. The increase in N concentration at a high N level was 7.6% compared to a low N level across 20 genotypes. The average nitrogen concentration value in the shoot (15.7 g kg^{-1}) and the grain (46.9 g kg^{-1}) at a high N level can be considered as an adequate level for the bean crop. Similarly, adequate values of N concentration in bean shoots have been reported by Fageria (1989a) and Piggott (1986). Nitrogen concentration in the grain was about three times more compared with N concentration in the shoot at low as well as high N rates. Similar results have been reported by Fageria (1989) and Fageria et al. (2011a).

3.4.2 Nitrogen Uptake in Plants

Nitrogen uptake is defined as the quantity of N taken up from the growth medium by plants at a determined growth stage. The uptake of nutrient in crop plants is determined by the analysis of straw and grain separately at harvest and expressed in kg ha^{-1}. It is determined by the multiplication of concentration with dry matter, which includes both grain and straw. Nutrient uptake, including N, increased with increasing plant age of crop plants in a quadratic fashion, Figure 3.28. The most appropriate growth stage to determine N uptake is at the harvest of crop plants because straw and grain yield are maximum at this growth stage. Nitrogen uptake data serve as a measurement of soil fertility depletion. They can also be used to determine how much N a crop requires to produce a determined yield. Data related to the uptake of nutrients, including N by upland rice, dry bean, corn, and soybean, are presented in Table 3.7. In addition, information on the uptake of macro- and micro-nutrients to produce 1 metric ton of grain is also presented. It is clear from these data that N uptake by all crops is significant and substantial except in the cases of upland and lowland rice. Similarly,

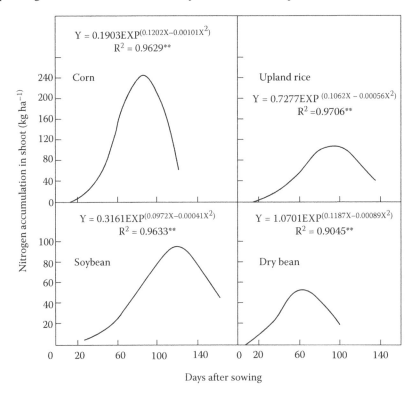

FIGURE 3.28 Nitrogen accumulation in the straw of four crop species. (Adapted from Fageria, N. K. 2004. *Commun. Soil Sci. Plant Anal.* 35:961–974.)

TABLE 3.7
Uptake of Macro- and Micronutrients by Upland Rice, Lowland Rice, Dry Bean, Corn, and Soybean Grown on Brazilian Oxisols

Nutrient	Straw	Grain	Total	Required to Produce 1 t Grain[b]
Upland rice	(6343 kg ha^{-1})[a]	(4568 kg ha^{-1})[a]		
Nitrogen (kg ha^{-1})	56	70	126	28
Phosphorus (kg ha^{-1})	3	9	12	3
Potassium (kg ha^{-1})	150	56	206	45
Calcium (kg ha^{-1})	23	4	27	6
Magnesium (kg ha^{-1})	14	5	19	4
Zinc (g ha^{-1})	161	138	299	65
Copper (g ha^{-1})	35	57	92	20
Iron (g ha^{-1})	654	117	771	169
Manganese (g ha^{-1})	1319	284	1603	351
Boron (g ha^{-1})	53	30	83	18
Lowland rice	(9423 kg ha^{-1})[a]	(6389 kg ha^{-1})[a]		
Nitrogen (kg ha^{-1})	65	86	151	24
Phosphorus (kg ha^{-1})	15	15	30	5
Potassium (kg ha^{-1})	156	20	176	28
Calcium (kg ha^{-1})	26	5	31	5
Magnesium (kg ha^{-1})	15	7	22	3
Zinc (g ha^{-1})	546	224	770	121
Copper (g ha^{-1})	77	102	179	28
Iron (g ha^{-1})	2553	505	3058	479
Manganese (g ha^{-1})	4724	369	5093	797
Boron (g ha^{-1})	69	33	102	16
Dry bean	(2200 kg ha^{-1})	(3409 kg ha^{-1})[a]		
Nitrogen (kg ha^{-1})	16.9	124.1	141	36.5
Phosphorus (kg ha^{-1})	1.6	14.6	16.2	4.2
Potassium (kg ha^{-1})	41.2	63.8	105	27.2
Calcium (kg ha^{-1})	22.0	8.5	30.5	7.9
Magnesium (kg ha^{-1})	8.9	6.2	15.1	3.9
Zinc (g ha^{-1})	61.7	123.3	185	48.0
Copper (g ha^{-1})	9.2	34.6	43.8	11.4
Iron (g ha^{-1})	1010.3	274.9	1285.2	333.1
Manganese (g ha^{-1})	31.1	48.7	79.8	20.7
Boron (g ha^{-1})	20	14	34	
Corn	(11,873 kg ha^{-1})[a]	(8501 kg^{-1})[a]		
Nitrogen (kg ha^{-1})	72	127	199	24
Phosphorus (kg ha^{-1})	4	17	21	3
Potassium (kg ha^{-1})	153	34	187	23
Calcium (kg ha^{-1})	33	8	41	5
Magnesium (kg ha^{-1})	20	9	29	4
Zinc (g ha^{-1})	184	192	376	46
Copper (g ha^{-1})	53	14	67	8
Iron (g ha^{-1})	2048	206	2254	274
Manganese (g ha^{-1})	452	82	534	65
Boron (g ha^{-1})	103	43	146	18

continued

TABLE 3.7   (continued)

Uptake of Macro- and Micronutrients by Upland Rice, Lowland Rice, Dry Bean, Corn, and Soybean Grown on Brazilian Oxisols

Nutrient	Straw	Grain	Total	Required to Produce 1 t Grain[b]
Soybean	(3518 kg ha^{-1})[a]	(4003 kg ha^{-1})[a]		
Nitrogen (kg ha^{-1})	37.5	280.0	317.5	79
Phosphorus (kg ha^{-1})	1.76	14.3	16.1	4
Potassium (kg ha^{-1})	57.5	77.5	135.0	34
Calcium (kg ha^{-1})	31.0	13.4	44.4	11
Magnesium (kg ha^{-1})	20.3	10.2	30.5	8
Zinc (g ha^{-1})	29.3	169.3	198.6	50
Copper (g ha^{-1})	32.8	60.3	93.1	23
Iron (g ha^{-1})	187.3	373.0	560.3	140
Manganese (g ha^{-1})	117.4	120.1	237.5	59
Boron (g ha^{-1})	22	21	43	

Sources: Adapted from Fageria, N. K. 2001b. *Rev. Bras. Eng. Agric. Ambiental.* 5:416–424; Fageria, N. K., V. C. Baligar, and R. W. Zobel. 2007. *Commun. Soil Sci. Plant Anal.* 38:1637–1653; Fageria, N. K., V. C. Baligar, and C. A. Jones. 2011a. *Growth and Mineral Nutrition of Field Crops*, 3rd edition. Boca Raton, Florida: CRC Press.

[a] Straw and grain yield.

[b] Accumulation in straw and grain (macronutrients in kg and micronutrients in g).

it is also required in large amounts to produce 1 metric ton of grain of all the five crops. In addition, a large amount of N is accumulated in grains that cannot be recycled back to the soil. Hence, the application of a large amount of N fertilizers is required to maintain the sustainability of cropping systems or succeeding crops in rotation with these crops. Thompson et al. (1975) reported that each ton of wheat grain contains 23 kg N. Bhatia and Rabson (1976) reported that, in rice, 10 kg N is required to produce 1 ton of rice grain, whereas in wheat and rye, 16 kg N is required to produce 1 ton of grain. In an examination of 16 winter wheat varieties in the United Kingdom and France, the N requirement per Mg grain ranged from 14.4 to 31.1 kg under low and high N treatments, respectively (Gaju et al., 2011). Similarly, in an investigation of 10 winter wheat varieties in Mexico, the requirement per Mg grain ranged from 23 to 37 kg (Ortiz-Monasterio et al., 1997). Liu et al. (2006) reported that across widely differing environments in China, the average N requirement of winter wheat was 25 kg for yields that ranged from 0.35 to 8.73 Mg ha^{-1} (Liu et al., 2006). Yue et al. (2012b) reported that in winter wheat, for the optimum N fertilizer treatment, the average N requirement was 24.3 kg per Mg grain yield and it declined with increasing grain yield. These authors further reported that for the yield ranges between 6.0 and 7.5 Mg ha^{-1}, N requirement decreased from 27.1 to 24.5 kg due to the increasing harvest index (HI, from 0.39 to 0.46) and decreasing grain N concentration (from 2.41% to 2.21%).

Historically, N fertilizer recommendations for corn in the United States have been based on a simplified empirical formula called the yield goal (Hoeft et al., 2000; Kyveryga et al., 2013). The major premise of yield goal recommendations is that corn N requirements or optimum N rates are proportional to corn yields, with a constant multiplier of 21.4 kg N Mg^{-1} corn grain. These calculations were based on the assumption of a constant supply of N from the soil under a wide range of soil and weather conditions (Kyveryga et al., 2013). While the yield goal recommendations were based on a mass balance approach (N rates should approximate N removed by grain plus adjustments for N losses and N supplied by the soil). However, several studies have shown a low correlation between corn yields and optimal N rates (Vanotti and Bundy, 1994; Scharf et al., 2006). The low correlation is often attributed to the large variability in the N supply from the soil

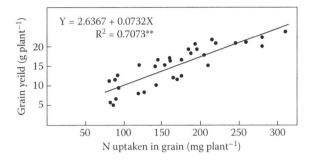

$$Y = 2.6367 + 0.0732X$$
$$R^2 = 0.7073**$$

FIGURE 3.29 Relationship between N uptake in grain and grain yield of lowland rice. (From Fageria, N. K. 2009. *The Use of Nutrients in Crop Plants*. Boca Raton, Florida: CRC Press. With permission.)

and variable N losses by different mechanisms such as leaching, volatilization, or denitrification (Kyveryga et al., 2013).

Nitrogen uptake is positively related to the grain yield in crops and understanding the N uptake–yield relationship and quantifying N requirements would be of great benefit for optimizing N fertilization for annual crops (Yue et al., 2012b). Nitrogen uptake–yield relationship for winter wheat has been reported in various studies (Ortiz-Monasterio et al., 1997; Pathak et al., 2003; Liu et al., 2006; Lopez-Bellido et al., 2008; Barraclough et al., 2010; Gaju et al., 2011; Giuliani et al., 2011; Pask et al., 2012; Yue et al., 2012b). A higher amount of N uptake or accumulation in the grain of crop plants is important because crop yield is significantly and linearly associated with N accumulated in the grain (Figure 3.29). Generally, N uptake in the grain has significant positive associations with the grain yield (Fageria and Baligar, 2001, López-Bellido et al., 2003). Hence, improving N uptake in grain may lead to improved grain yield. Higher N concentration in shoots is also desirable because if the shoot has a higher N concentration during crop growth, it is translocated to the grain during higher plant demands and the yield is improved. Figure 3.30 shows that N uptake in the shoot is significantly associated with grain yield in a quadratic fashion.

The author studied N uptake in the shoots and grains of upland rice at two N levels (Table 3.8). Nitrogen uptake in the shoot varied from 72.1 to 142.3 mg pot^{-1} at a low N level with an average value of 98.9 mg pot^{-1}. At a higher N level, N uptake in the shoot varied from 352.2 to 635.1 mg pot^{-1} with an average value of 511.2 mg pot^{-1}. The uptake of N in the grain varied from 298.1 to 513.5 mg pot^{-1} across two N levels. The uptake of N in the shoot and grain followed the dry matter yield of these two plant parameters. The uptake of N in the shoot (r = 0.85*) as well as in the grain (0.93*) was having highly significant associations with grain yield. This indicates that increasing N

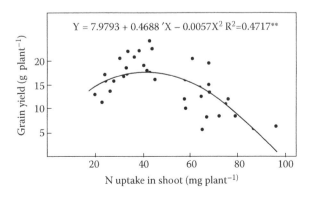

$$Y = 7.9793 + 0.4688\,'X - 0.0057X^2 \quad R^2=0.4717**$$

FIGURE 3.30 Relationship between N uptake in shoot and grain yield of lowland rice. (From Fageria, N. K. 2009. *The Use of Nutrients in Crop Plants*. Boca Raton, Florida: CRC Press. With permission.)

TABLE 3.8

Uptake of Nitrogen in the Shoot and Grain of 19 Upland Rice Genotypes at Two N Levels

Genotype	N Uptake in Shoot (mg Pot⁻¹)		N Uptake in Grain (mg Pot⁻¹)
	0 mg N kg^{-1}	400 mg N kg^{-1}	Across Two N Levels
CRO 97505	100.0abc	473.1a	409.0a
CNAs 8993	93.5abc	363.4a	513.5a
CNAs 8812	117.7abc	574.5a	510.3a
CNAs 8938	79.2bc	588.1a	386.3a
CNAs 8960	89.2bc	478.3a	429.7a
CNAs 8989	91.2bc	635.1a	470.9a
CNAs 8824	68.2c	456.9a	417.4a
CNAs 8957	100.5abc	481.1a	465.5a
CRO 97422	109.1abc	491.1a	387.3a
CNAs 8817	84.9bc	581.9a	444.0a
CNAs 8934	142.3a	620.2a	370.5a
CNAs 9852	100.7abc	473.9a	385.3a
CNAs 8950	115.4abc	434.6a	422.0a
CNA 8540	72.1c	578.0a	405.3a
CNA 8711	111.8abc	541.8a	414.9a
CNA 8170	99.0abc	654.3a	298.1a
BRS Primavera	87.1bc	352.2a	484.8a
BRS Canastra	123.3abc	382.1a	381.5a
BRS Carisma	94.4abc	552.4a	439.4a
Average	98.9	511.2	422.9
F-test			
N level (N)	**		NS
Genotype (G)	**		NS
N × G	**		NS
CV (%)	21		25

Source: Adapted from Fageria, N. K., O. P. Morais, and A. B. Santos. 2010a. *J. Plant Nutr.* 33:1696–1711.
**, NS: Significant at the 1% probability level and nonsignificant, respectively. Means followed by the same letter in the same column are not significantly different at the 5% probability level by Tukey's test.

accumulation in the shoot as well as the grain can improve upland rice yield. However, the influence of N accumulation in the grain is higher compared to that in the shoot.

The author also studied N uptake in the shoots and grains of dry bean genotypes (Table 3.9). Nitrogen uptake in the shoot varied from 121.5 to 296.6 mg pot⁻¹ across two N rates with an average value of 195.3 mg pot⁻¹. Similarly, N uptake in the grain varied from 267 to 838.3 mg pot⁻¹, with an average value of 613 mg N pot⁻¹. The N uptake in the shoot and grain varied according to the variation in the shoot and grain yield among different genotypes. The N uptake in the grain was 214% higher compared with the N uptake in the shoot. The higher N uptake in the grain is related to a higher grain yield and higher N concentration in the grain compared to the shoot yield and N concentration in the shoot. A higher uptake of N in the bean grain compared to the shoot has been reported by Fageria (1989a). Nitrogen uptake in the shoot was having a significant quadratic association with the grain yield ($Y = 3.4063 + 0.0613X - 0.000026X^2$, $R^2 = 0.4153^*$). Similarly, N uptake in the grain was having a significant linear association with grain yield ($Y = 0.4282 + 0.0216X$, $R^2 = 0.9101^*$). This means that the variation in grain yield was higher due to N accumulation in the grain compared to N accumulation in the shoot.

TABLE 3.9
Nitrogen Uptake in the Shoot or Straw and Grain of Dry Bean Genotypes

Genotype	N Uptake in Shoot (mg Pot^{-1})	N Uptake in Grain (mg Pot^{-1})
Pérola	292.8a	838.3a
BRS Valente	242.4ab	267.0c
CNFM 6911	181.1ab	727.6ab
CNFR 7552	202.3ab	405.8bc
BRS Radiante	296.6a	581.9ab
Jalo Precoce	224.1ab	608.1ab
Diamante Negro	237.8ab	610.5ab
CNFP 7624	191.0ab	564.4ab
CNFR 7847	124.5b	712.5ab
CNFR 7866	271.0ab	648.9ab
CNFR 7865	263.5ab	723.0ab
CNFM 7875	138.0ab	615.3ab
CNFM 7886	193.4ab	594.9ab
CNFC 7813	136.5ab	616.6ab
CNFC 7827	127.9b	660.8ab
CNFC 7806	157.9ab	569.0ab
CNFP 7677	138.2ab	711.0ab
CNFP'7775	121.5b	634.6ab
CNFP 7777	195.5ab	759.8a
CNFP 7792	170.2ab	410.4bc
Average	195.3	613.0
F-test		
N rate (N)	**	**
Genotype (G)	*	*
N × G	NS	NS

Source: From Fageria, N. K., L. C. Melo, and J. P. Oliveira. 2013a. *J. Plant Nutr.* 36:2179–2190. With permission.
Note: Values are across two N rates. N rates were 0 and 400 mg kg^{-1}.
*,**, NS: Significant at the 5% and 1% probability levels and nonsignificant, respectively. Within the same column, means followed by the same letter do not differ significantly at the 5% probability level by Tukey's test.

The author studied nutrient uptake, including N in the straw and grain of soybean, under different lime rates. Liming in the shoot of soybean significantly influenced the uptake of P, K, Ca, and Mg (Table 3.10). Similarly, in grains, lime had a significant influence on the uptake of N, P, K, and Mg. In shoots, the uptake of N at 6 Mg lime ha^{-1} (rate of maximum economic yield) was 9% higher, the uptake of P was 14% higher, the uptake of K was 24% higher, the uptake of Ca was 63% higher, and the uptake of Mg was 27% higher compared to the control treatment. Similarly, in grains, there was a 55% increase in the uptake of N, a 56% increase in the uptake of P, a 57% increase in the uptake of K, a 61% increase in the uptake of Ca, and a 62% increase in the uptake of Mg at 6 Mg ha^{-1} lime rate compared to the control treatment. The relationship between lime rate and nutrient uptake was determined in shoots and grains (Table 3.11). There was a significant and quadratic increase in the uptake of all the macronutrients in shoots and grains, except N uptake in the shoot. The increase in the uptake of these macronutrients with increasing lime rate was associated with the increase in shoot dry matter and grain yield by liming.

The accumulation of macronutrients in the soybean plant (shoot and grain) was in the order of N > K > Ca > Mg > P. Fageria (2001b) reported a similar accumulation pattern in soybean plant grown on Brazilian Oxisol. Overall, macronutrient use efficiency (kg grain/kg nutrient

TABLE 3.10

Macronutrient Uptake in the Shoot and Grain of Soybean under Different Lime Rates

Lime Rate (Mg ha⁻¹)	N (kg ha⁻¹)	P (kg ha⁻¹)	K (kg ha⁻¹)	Ca (kg ha⁻¹)	Mg (kg ha⁻¹)
			Shoot		
0	34.4	1.54	46.2	19.9	12.6
3	35.4	1.77	58.8	27.6	16.0
6	37.5	1.76	57.5	31.0	20.3
12	32.3	1.79	63.3	31.6	18.0
18	45.7	2.22	76.0	40.6	23.7
Average	37.1	1.82	60.4	30.1	18.1
F-test					
Lime	NS	*	*	**	**
CV (%)	20.0	14.9	19.1	18.5	20.1
			Grain		
0	180.8	9.17	49.3	8.3	6.3
3	256.3	13.18	71.2	11.8	9.5
6	280.0	14.29	77.5	13.4	10.2
12	278.5	15.37	78.3	12.6	10.7
18	296.5	17.09	80.2	12.0	10.6
Average	258.4	13.82	71.3	11.6	9.5
F-test					
Lime	**	**	**	NS	**
CV (%)	7.0	8.5	7.7	19.5	7.5

Source: From Fageria, N.K. et al. 2013b. *Commun. Soil Sci. Plant Anal.* 44:2941–2951. With permission.

*Significant at the 5% probability level.

**, NS: Significant at the 1% probability level and nonsignificant, respectively.

accumulated in the grain) was 11 kg for N, 200 kg for P, 39 kg for K, 238 kg for Ca, and 292 kg for Mg. This means Mg was having maximum efficiency in grain production and N was having minimum efficiency. Similar results were reported by Fageria (2001a) for soybean grown on Brazilian Oxisol.

The uptake of all the macronutrients in the grain was having a highly significant ($P < 0.01$) quadratic association with grain yield (Table 3.12). Grain yield was having 89% variability due to N uptake, 91% variability due to P uptake, 86% variability due to K uptake, 61% variability due to Ca uptake, and 79% variability due to Mg uptake. This means that the N and P uptakes were having maximum influence on grain yield compared to the K, Ca, and Mg uptakes. Nitrogen and P improved the number of pods per plant or per unit area in legumes in Oxisols, which might have been responsible for the higher variation in grain yield due to these nutrients (Fageria et al., 2006).

A large part of the N accumulation in crop species during the growth cycle is translocated from the vegetative to the reproductive plant parts. Mae (1997) reported that the amount of N absorbed by the plant during the grain filling period is much smaller than the amount of N accumulated in mature grain, and a large part of grain N is translocated from vegetative organs. Nitrogen distribution studies showed that 30–80% of the N accumulated in the rice grain originated from translocation from vegetative tissue after heading (Guindo et al., 1992; Ntanos and Koutroubas, 2002). Xiong et al. (2013) also reported that, in rice preheading, N accumulation was having a highly significant correlation ($r = 94*$) to N translocated to grain.

TABLE 3.11

Relationship between Lime Rate (X) and Nutrient Uptake (Y) in Soybean Shoot and Grain

| | Regression Equation | |
Nutrient	Shoot	R^2
N (kg ha^{-1})	$Y = 36.2499 - 0.8188X + 0.0701X^2$	0.2216[NS]
P (kg ha^{-1})	$Y = 1.6261 + 0.0069X - 0.0013 X^2$	0.4067*
K (kg ha^{-1})	$Y = 49.6413 + 1.2491X - 0.0076 X^2$	0.4241*
Ca (kg ha^{-1})	$Y = 21.8085 + 1.3067X - 0.0183X^2$	0.5869**
Mg (kg ha^{-1})	$Y = 13.5985 + 0.7763X - 0.0148X^2$	0.4531**
	Grain	
N (kg ha^{-1})	$Y = 195.6751 + 15.8686X - 0.5951X^2$	0.7462**
P (kg ha^{-1})	$Y = 9.8974 + 0.8371X - 0.0253X^2$	0.8073**
K (kg ha^{-1})	$Y = 53.2387 + 4.8377X - 0.1917X^2$	0.7491**
Ca (kg ha^{-1})	$Y = 8.8891 + 0.8883X - 0.0408X^2$	0.3768**
Mg (kg ha^{-1})	$Y = 6.8257 + 0.7184X - 0.0290X^2$	0.7985**

Source: From Fageria, N. K. et al. 2013b. *Commun. Soil Sci. Plant Anal.* 44:2941–2951. With permission.

*,**, NS: Significant at the 5% and 1% probability levels and nonsignificant, respectively.

3.4.3 Forms of Nitrogen Uptake by Plants

Nitrogen is mainly absorbed as NO_3^- and NH_4^+ by the roots. In oxidized soils, NO_3^- is the dominant form of N uptake, since ammonium is quickly transformed in the soil solution to nitrate through nitrification when typical weather conditions for crop growth prevail. Crop plants absorb nitrate through their roots and transport it directly to the leaves in the transpiration stream. Once in the leaf, nitrate is reduced to ammonium and combines with organic acids to form amino acids and proteins (Gerik et al., 1998). These processes require considerable energy in the form of reductants, such as nicotinamide adenine dinucleotide (NADH), and a ready supply of organic acids from carbon assimilation. Up to 55% of the net carbon assimilated in some tissues is committed to N metabolism (Huppe and Turpin, 1994).

TABLE 3.12

Relationship between Nutrient Uptake (X) in the Grain and Yield (Y) of Soybean

Nutrient	Regression Equation	R^2
Nitrogen	$Y = -1281.9460 + 23.9024X - 0.0308X^2$	0.8901**
Phosphorus	$Y = -1688.9170 + 547.4529X - 15.5759X^2$	0.9126**
Potassium	$Y = -1588.2320 + 97.2274X - 0.4910X^2$	0.8630**
Calcium	$Y = -1290.9050 + 597.0891X - 20.2308X^2$	0.6131**
Magnesium	$Y = 3208.7790 - 385.0645X + 34.6338X^2$	0.7921**

Source: From Fageria, N. K. et al. 2013b. *Commun. Soil Sci. Plant Anal.* 44:2941–2951. With permission.

**Significant at the 1% probability level.

In reduced soil conditions, such as flooded rice, NH_4^+ may predominate in the absorption process. The topic of NH_4^+ versus NO_3^- nutrition of plants has been extensively reviewed (Hayes and Goh, 1978; Hageman, 1984; Mengel et al., 2001; Fageria and Baligar, 2005b). It has been proven that most annual crops grow best when supplied mixtures of NH_4^+ and NO_3^- under controlled conditions (Fageria and Baligar, 2005b).

When NH4+ uptake exceeds NO_3^- uptake, soil solution pH decreases and when NO_3^- uptake exceeds NH_4^+ uptake, soil solution pH increases. Furthermore, nearly 70% of the cations or anions taken up by the plants are ammonium or nitrate (Van Beusichem et al., 1988). The NH_4^+/NO_3^- ion uptake can change the rhizosphere pH up to two units higher or lower compared with the bulk soil (Mengel et al., 2001). This change in pH may influence the uptake of other essential nutrients from soil solution by plants. In consequence of the lowering of rhizosphere pH by ammonium, an increased uptake has been reported of phosphate (Riley and Barber, 1971) and also of micronutrients (Schug, 1985) as compared with nitrate treatment.

Legumes are exceptions in that they acidify the rhizosphere even after the supply of nitrate (Marschner and Romheld, 1983). In legumes, the uptake rate of cations is greater than that of anions, since they acquire nitrogen from the atmosphere through nitrogen fixation rather than by nitrate uptake (Aguilars and Van Diest, 1981; Marschner and Romheld, 1983; Jarvis and Hatch, 1985). However, some workers have reported that the acidification of the legume rhizosphere is associated with specific properties of legumes, which normally acidify the rhizosphere even under nitrate-fed conditions (Hinsinger, 1998; Marschner, 1995). The pH also changes with the excretion of organic acids by roots and by microorganism activities in the rhizosphere. Further, the CO_2 produced by roots and microorganism respiration can dissolve in soil solution and may form carbonic acid and lower the pH. Soil buffering capacity (clay and organic matter content) and initial pH are the main parameters that determine changes in soil pH.

In addition, the absorption of NH_4^+ occurs faster than the absorption of NO_3^- (Gaudin and Dupuy, 1999). In addition, the assimilation of NO_3^- required energy equivalents up to 20 adenosine triphosphate (ATP) mol^{-1} NO_3^-, whereas NH_4^+ assimilation required only 5 APT mol^{-1} NH_4^+ (Salsac et al., 1987). Similarly, Bloom et al. (1992, 2003) reported that root absorption and assimilation of 1 mol NH_4^+ requires or consumes 0.31 mol O_2, whereas 1 mol of NO_3^- consumes 1.5 mol O_2. This means that NO_3^- uptake consumes about five times more energy compared to NH_4^+ ion uptake. While NH_4^+ can be assimilated directly into amino acids, NO_3^- must first be reduced to NO_2^- and then NH_4^+ via nitrate reductase and nitrite reductase, a process that implies an additional energetic cost (Hopkins, 1999). Potential energy savings for yield could be obtained if plants were supplied only NH_4^+ (Huffman, 1989). This concept has not been consistently observed, nor is it easy to conduct experiments on this given the nature of the N-cycle dynamic (Raun and Johnson, 1999; Fageria et al., 2006).

The effectiveness of the two N forms on the growth and N uptake varies with the type of cultivar and NH_4^+/NO_3^- ratio. For example, supplying N in solution entirely as NH_4^+ or NO_3^- has been shown to inhibit plant growth when compared to plant growth in solution containing 25% or 50% of either N form (Nittler and Kenny, 1976). Gashaw and Mugwira (1981) reported that triticale, wheat, and rye produced higher dry matter with combinations of 25/75; 50/50, and 75/25 $NH_4^+ - N/NO_3 - N$ ratios than with either N source alone. Warncke and Barber (1973) found that there were no differences between the relative rates of NH_3^+ and NO_3^- absorption by corn but increasing each N form reduced the uptake of the other N form. It has been shown that corn, wheat, and oats prefer 50% of N as NO_3^- for maximum growth (Diest, 1976). Barber et al. (1992) reported that, overall, manipulation of soil NH_4^+/NO_3^- ratios had few effects on corn development or yield under field conditions.

Plants grown on NO_3^- maintain a relative homeostasis of their N concentration and internal NO_3^- over a wide range of external concentrations, which suggests an efficient mechanism to control NO_3^- uptake (Glass et al., 2002). In contrast, NH_4^- uptake seems poorly regulated (Britto and Kronzucker, 2002). The imbalance between NH_4^+ uptake and assimilation rates depends on a variety of factors, including plant species and carbohydrate availability (Schjoerring et al., 2002). Although the

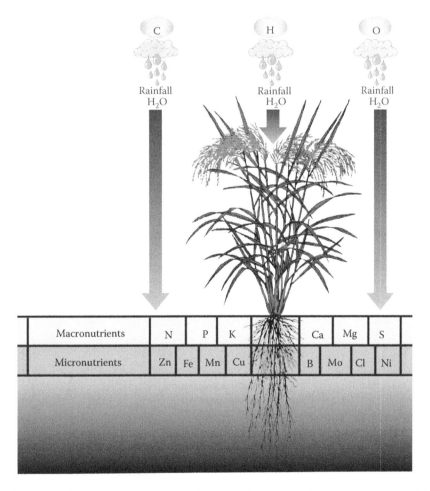

FIGURE 3.1 Essential nutrients for plant growth and development. (Adapted from Fageria, N. K. 1989b. *Tropical Soils and Physiological Aspects of Crops*. EMBRAPA: Brasilia, Brazil.)

FIGURE 3.2 Growth of cover crop *Crotalaria breviflora* at low and high N levels.

FIGURE 3.3 Growth of cover crop *Crotalaria mucronata* at low and high N levels.

FIGURE 3.4 Growth of cover crop *Crotalaria spectabilis* at low and high N levels.

FIGURE 3.5 Growth of cover crop *Lablab purpureus* at low and high N levels.

Without N

With N

FIGURE 3.9 Upland rice plot without N at the left and with N at the right grown on a Brazilian Oxisol.

FIGURE 3.10 Dry bean leaves and plants with nitrogen deficiency symptoms.

FIGURE 3.11 Dry bean plants with N deficiency symptoms (left) and without N deficiency symptoms (right).

FIGURE 3.12 Growth of dry bean and corn at two N levels grown on a Brazilian Oxisol. Half of the N was applied at sowing and the remaining half at 35 days after sowing.

FIGURE 3.13 Growth of upland rice genotype BRA 02535 at low and high N levels. Half of the N was applied at sowing and the remaining half at the active tillering growth stage.

FIGURE 3.14 Growth of upland rice genotype BRA 02601 at low and high N levels. Half of the N was applied at sowing and the remaining half at the active tillering growth stage.

FIGURE 3.18 Growth of lowland rice genotype BRA 02654 at low and high N levels. Half of the N was applied at sowing and the remaining half at the active tillering growth stage.

FIGURE 3.19 Growth of lowland rice cultivar BRA 051077 at low and high N levels. Half of the N was applied at sowing and the remaining half at the active tillering growth stage.

FIGURE 3.22 Response of dry bean to nitrogen fertilization to a Brazilian Oxisol. Half of the N was applied at sowing and the remaining half as topdressing at 35 days after sowing.

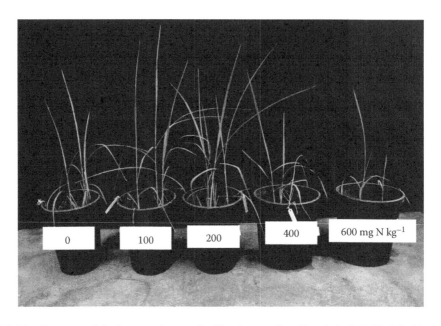

FIGURE 3.23 Response of dry bean to nitrogen fertilization to a Brazilian Oxisol. Half of the N was applied at sowing and the remaining half as topdressing at 35 days after sowing.

FIGURE 3.24 Response of dry bean to nitrogen fertilization to a Brazilian Oxisol. Half of the N was applied at sowing and the remaining half as topdressing at 35 days after sowing.

FIGURE 3.25 Growth of wheat at different N levels. Half of the N was applied at sowing and the remaining half as topdressing at 35 days after sowing.

FIGURE 3.33 Response of lowland rice applied with ammonium sulfate. Half of the N was applied at sowing and the remaining half at the active tillering growth stage.

FIGURE 3.34 Response of lowland rice to N fertilization applied with urea. Half of the N was applied at sowing and the remaining half at the active tillering growth stage.

FIGURE 3.35 Response of lowland rice to nitrogen fertilization applied with polymer-coated urea. Half of the N was applied at sowing and the remaining half at the active tillering growth stage.

mechanisms of toxicity have not been elucidated completely, NH_4^+ accumulation in leaves has been found to depress photosynthesis and leaf growth (Raab and Terry, 1994). Also, NO_3^- is an important osmoticum that intervenes in the expansion of plant cells, and a reduction of its absorption may depress growth (McIntyre, 1997). The reduction of the absorption of cations such as K^+, Ca^{2+}, and Mg^{2+} by NH_4^+ may also limit plant growth (Salsac et al., 1987).

Numerous studies have demonstrated that the presence of NH_4^+ in the growth medium reduces NO_3^- uptake into roots of crop species and into microorganisms within minutes of exposure (Lee and Drew, 1989; Glass, 2003). The mechanism responsible for this effect is still unclear. Based upon their demonstration that NH_4^+ reduced NO_3^- uptake within 3 min in barley roots, Lee and Drew (1989) and Ayling (1993) proposed that the effect might be based upon the direct effects of NH_4^+ on membrane potential, since NH_4^+ typically depolarizes the membrane electrical potential difference. Because NO_3^- traverses the plasma membrane as a cation (through its association with at least two H^+), membrane depolarization would reduce the proton motive force driving the uphill transport of NO_3^- and concomitantly reduce NO_3^- uptake (Glass, 2003).

Uptake of N (NO_3^- and NH_4^+) is regulated by genes in higher plants (Glass, 2003). Glass (2003) reported that there are many genes involved fin NO_3^- and NH_4^+ absorption and transport in the higher plants. Higher plants may possess genes encoding as many as 11 nitrate transporters and at least 6 high-affinity ammonium transporters, and the systems for regulating N fluxes are complex and highly integrated (Glass et al., 2001, 2002). Nevertheless, there are clear indications that only a limited number of the NO_3^- transporter genes are responsible for nitrate absorption from soils; the remainder probably encode transporters that participate in internal redistribution. A similar situation applied with respect to NH_4^+ absorption (Glass, 2003).

3.5 USE OF CHLOROPHYLL METER

Traditionally, nutrient deficiency/toxicity symptoms, soil testing, plant tissue analysis, and greenhouse and field trials have been used to assess N availability for crops (Kitchen and Goulding, 2001; Fageria, 2013, 2014). However, since the early 1990s, handheld chlorophyll meters have been available to monitor plant N status by measuring the transmittance of radiation through a leaf in two wavelength bands centered near 60 and 940 nm (Blackmer et al., 1994; Wood et al., 1993; Souza et al., 2010). Previous research has shown that corn reflectance of green and near-infrared (NIR) light measured with a radiometer is sensitive to N status (Bausch and Duke, 1996) and can be used to predict the amount of N fertilizer needed by the crop (Dellinger et al., 2008; Scharf and Lory, 2009). Walburg et al. (1982) confirmed that corn spectral properties associated with N deficiency are likely to be apparent by the V12 growth stage, when the crop still has the potential for large yield responses to added N (Russelle et al., 1983; Scharf et al., 2002). Blackmer et al. (1996) measured the reflected radiation form from R5 growth-stage corn canopies using reference areas with nonlimiting N to calculate the relative reflectance. They concluded that the reflected radiation around 550 and 710 nm provided the best detection of N deficiency in the 400–1000 nm spectral range.

The use of a chlorophyll meter in topdressing N during a crop growth cycle is an important aid in correcting N deficiency. After visual deficiency symptoms, it is one of the cheapest techniques to identify N status in growing crop plants. Instantaneous and nondestructive chlorophyll meter reading represents an alternative to traditional tissue analysis for diagnosing crop N status, and this approach has been used in barley (Wienhold and Krupinsky, 1999), corn (Schepers et al., 1992), rice (Peng et al., 1993), and wheat (Follett et al., 1992; Ziadi et al., 2010).

A chlorophyll meter (Minolta SPAD–soil–plant analysis development) measures the greenness intensity of plant leaves. The greenness intensity is associated with chlorophyll content, which in turn is associated with leaf N concentration (Wolfe et al., 1988). In most crop species (rice, wheat, corn, potato), the relationship between SPAD readings and crop yield was poor at the early crop development stages, but it improved at later stages because of the greater N deficiency expression as crop requirements increased (Blackmer and Schepers, 1995; Pagani et al., 2009). A normalization of absolute SPAD readings is necessary because greenness intensity can vary up to 10% among corn hybrids at a fixed N level (Schepers, et al., 1992). Moreover, interactions between hybrids and seeding dates have been reported in greenness intensity (Jemison and Litle, 1996). To avoid this interference, a strip without N deficiency is necessary in each field to provide a reference of the maximum GI (greenness index), which is used to normalize the absolute crop greenness intensity through nitrogen sufficiency index (NSI). It has been shown that a threshold value of 0.95 separates sites with and without response to N applications (Fox et al., 1992).

3.6   NITROGEN HARVEST INDEX

Nitrogen harvest index (NHI) is defined as the amount of N accumulated in the grain divided by the amount of N accumulated in the grain plus straw. It is an index and hence has no unit. However, sometimes it is expressed as a percentage. A higher NHI in crop plants or genotypes is desirable because it has a positive association with grain yield (Figure 3.31). Figure 3.31 shows that a 71% variability in the grain yield of lowland rice grown on Brazilian Inceptisol was due to NHI. The NHI values varied from crop species to crop species and among genotypes of the same species. Figure 3.32 shows that the NHI of five lowland rice genotypes varied from 0.53 to 0.64. Genotype CNA 8569 was having the lowest NHI and genotype BRS Bigua was having the highest NHI. This may be related to the larger remobilization of stored N in the grain during the grain filling period. The author studied the influence of two N rates on the NHI of upland rice genotypes (Table 3.13). There was a significant difference among genotypes in NHI at a higher N rate (400 mg N ha^{-1}). Overall, NHI was higher with the application of N rate compared to the control treatment.

Novoa and Loomis (1981) reported that the high harvest index of modern wheat varieties implies that a larger fraction of the stored N is remobilized during grain filling. McNeal et al. (1971) also observed that wheat varieties differ in their ability to remobilize N, and Huffaker and Rains (1978) have underlined the importance of remobilization for its influence on the protein content of wheat grain. Remobilization is a major determinant of NUE by the whole crop. The extreme remobilization common to soybean, lentil, and pea that led Sinclair and De Wit (1975) to their self-destruction hypothesis may simply represent an adaption for high efficiency in the use of scarce supplies of N. Leopold (1961) reported that remobilization of N is so efficient in oats that the plant can acquire enough N during vegetative phase for the entire life cycle. Novoa and Loomis (1981) reported that it may be possible to select cereal and legume cultivars that are highly efficient in remobilization under conditions of low water supply.

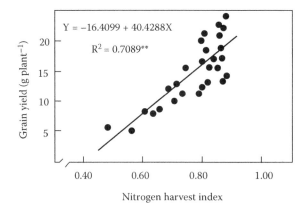

FIGURE 3.31   Relationship between NHI and grain yield of lowland rice. (From Fageria, N. K. 2009. *The Use of Nutrients in Crop Plants*. Boca Raton, Florida: CRC Press. With permission.)

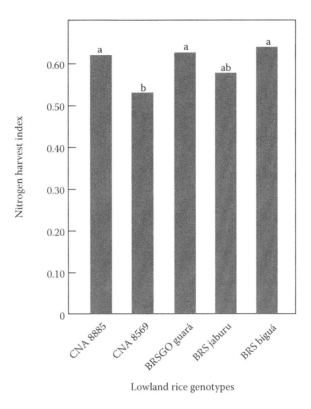

FIGURE 3.32 NHI of five lowland rice genotypes. (From Fageria, N. K. 2007a. Yield physiology of rice. *J. Plant Nutr.* 30:843–879. With permission.)

The author studied the NHI of dry bean genotypes (Table 3.14). The N rate genotype significantly influenced NHI and the N × G interaction was significant for this trait (Table 3.14). The average N harvest index at a low N rate was 0.63 and at a high N rate it was 0.75. This means that overall 63% of the total N accumulated in the bean plant (grain plus straw) was translocated to the grain at a low N rate. At a higher N rate, the translocation was 75% of the total N uptake. NHI is a measure of N partitioning in bean, which provides an indication of how efficiently the plant utilized the acquired N for grain production. Harvest index was having significant linear relationship with grain yield ($Y = -18.3691 + 46.0572X$, $R^2 = 0.4044^*$). This is proved by the highest-yielding genotypes, that is, CNFC 7827, CNFP 7677, and CNFP 7775 at a higher N rate were also having higher N harvest indices. Genetic variability for NHI has been reported in crop plants (Rattunde and Frey, 1986). High NHI is associated with efficient utilization of N (Welch and Yong, 1980). Selecting crop genotypes for high NHI may simultaneously improve grain yield (Loffler and Busch, 1982). Generally, NHI values are higher in legumes compared to cereals (Fageria et al., 2006). Bender et al. (2013) reported that the NHI of corn was 58% and Karlen (1988) reported that it was 60%.

3.7 PLANT RESPONSE TO APPLIED NITROGEN

Soil and plant analyses are the common practices for identifying nutritional deficiencies in crop production. The best criterion, however, for diagnosing nutritional deficiencies in annual crops is through evaluating crop responses to applied nutrients. If a given crop responds to an applied nutrient in a given soil, this means that the nutrient is deficient for that crop. The relative decrease in yield in the absence of a nutrient, as compared to an adequate soil fertility level, can give an idea of the

TABLE 3.13
Nitrogen Harvest Index of Upland Rice Genotypes at Two
N Levels

Genotype	Nitrogen Harvest Index	
	0 mg N kg^{-1}	400 mg N kg^{-1}
CRO 97505	0.82a	0.73ab
CNAs 8993	0.44a	0.53b
CNAs 8812	0.80a	0.79ab
CNAs 8938	0.57a	0.68ab
CNAs 8960	0.67a	0.69ab
CNAs 8989	0.67a	0.75ab
CNAs 8824	0.65a	0.71ab
CNAs 8957	0.43a	0.78ab
CRO 97422	0.80a	0.85a
CNAs 8817	0.47a	0.74ab
CNAs 8934	0.71a	0.69ab
CNAs 9852	0.45a	0.85a
CNAs 8950	0.44a	0.78ab
CNA 8540	0.47a	0.85a
CNA 8711	0.71a	0.85a
CNA 8170	0.69a	0.78ab
BRS Primavera	0.71a	0.85a
BRS Canastra	0.70a	0.85a
BRS Carisma	0.79a	0.88ab
Average	0.62b	0.77a
F-test		
N level (N)	**	
Genotype (G)	**	
N × G	**	

**Significant at the 1% probability level. Within the same column, means followed by the same letter do not differ significantly at the 5% probability level by Tukey's test. Average values were compared in the same line.

magnitude of nutrient deficiency (Fageria and Baligar, 2005a). Crop response to the applied nutrient can be evaluated under both greenhouse and field conditions. Greenhouse or controlled condition experiments are those of short duration conducted to develop and understand the basic principles of soil fertility and plant nutrition. On the other hand, field experiments are of relatively longer duration and are conducted to understand the applied part of crop production. Field experiments are more costly compared to greenhouse experiments.

In the author's opinion, controlled condition experimental results cannot be extrapolated to field conditions due to the large variation in environmental factors under field conditions. In general, however, properly conducted greenhouse and laboratory studies often give useful evidence as to the factors that merit investigation under field conditions. For example, if greenhouse experiments indicate rather marked differences in response among the fertilizers, the consideration of field experiments is justified. Pot experiments are also valuable in relating the observed laboratory measurement to actual plant response and they are important in establishing the basic principles of soil–fertilizer–plant relationships. Both types of experimentation are important and should be complementary in solving the problem of crop production in general. Hence, these two techniques

TABLE 3.14
Nitrogen Harvest Index of Dry Bean Genotypes at Two N Levels

Genotype	0 mg N kg⁻¹	400 mg N kg⁻¹
Pérola	0.82	0.73ab
BRS Valente	0.44	0.53b
CNFM 6911	0.80	0.79ab
CNFR 7552	0.57	0.68ab
BRS Radiante	0.67	0.69ab
Jalo Precoce	0.67	0.75ab
Diamante Negro	0.65	0.71ab
CNFP 7624	0.43	0.78ab
CNFR 7847	0.80	0.85a
CNFR 7866	0.47	0.74ab
CNFR 7865	0.71	0.69ab
CNFM 7875	0.45	0.85a
CNFM 7886	0.44	0.78ab
CNFC 7813	0.47	0.85a
CNFC 7827	0.71	0.85a
CNFC 7806	0.69	0.78ab
CNFP 7677	0.71	0.85a
CNFP`7775	0.70	0.85a
CNFP 7777	0.79	0.88ab
CNFP 7792	0.62	0.72ab
Average	0.63	0.75
F-test		
N rate (N)	**	
Genotype (G)	**	
N × G	**	

Source: From Fageria, N. K., L. C. Melo, and J. P. Oliveira. 2013a. *J. Plant Nutr.* 36:2179–2190.
With permission.

**Significant at the 1% probability level. Within the same column, means followed by the same letter do not differ significantly at the 5% probability level by Tukey's test.

should be used in combination rather than in isolation. The basic principles of conducting greenhouse and field experiments are discussed in the succeeding section.

3.7.1 GREENHOUSE OR CONTROLLED CONDITION EXPERIMENTS

Research is the foundation of technological improvements. The standard of living of a country is correlated to the use of technology (Fageria, 2005). The low yields of crops in some parts of the world or countries are the result of actions and interactions of many factors and there are no simple, easily implementable solutions. A better understanding of the biological, climatic, edaphic, and management factors through research and the development of production technologies in the appropriate sociopolitical–economic climate can help increase crop production in such regions. Further, in the twenty-first century research, one of the key factors guiding research will be the need to develop low-cost technology components that do not require a heavy application of purchase inputs with minimum degradation of natural resources (Fageria, 2005).

In agriculture science, soil fertility and plant nutrition are important subjects whose contribution in increasing crop yields is well known. Borlaug and Dowswell (1994) reported that as much as 50% of the increase in crop yields worldwide during the twentieth century was due to the use of chemical fertilizers. In the twenty-first century, the importance of chemical fertilizers in improving crop yields will continue and is expected to be still higher due to the necessity of increase in yields per unit land area rather than increasing land areas. Further, judicious use of chemical fertilizers along with other complementary methods, such as the use of organic manures, and exploiting the genetic potential of crop species and cultivars within species in nutrient utilization will be extremely useful and necessary (Fageria, 2005).

The main objectives of controlled condition experiments are to understand the basic principles. In the case of soil fertility and plant nutrition, such experiments are mainly conducted to understand nutrient movements, absorption, and utilization processes in soil plant systems. In addition, nutrient/elemental deficiency/toxicity symptoms and adequate and toxic concentrations in plant tissue are also determined under controlled conditions. Further, for example, pot experiments with different types of soils can show the degree of response that may be anticipated at different soil test levels and serve as excellent checks on the ratings being used. Since such tests provide no measure of the cumulative effects of treatments on yield or soil buildup or depletion, they have limited value in determining the rates of fertilizer that should be recommended for sustained productivity. Greenhouse pot studies, in which plants are used for estimating the relative availability of nutrients, can also provide useful indices of the relative availability of a standard fertilizer source in different soils and indices of different fertilizer sources. A detailed discussion regarding the basic principles and methodology in conducting controlled condition experiments is given by Fageria (2005).

3.7.1.1 Observation and Data Transformation in Greenhouse Experiments

In greenhouse experiments, important observations that should be recorded and that will be helpful in the analysis and interpretation of experimental results are growth, yield, and yield components. For example, in cereals, important measurements or observations are plant height, dry weight of straw, grain yield, panicle number, grain sterility, and thousand grain weight. Similarly, in legumes, the straw yield, grain yield number of pods, seeds per pod, and hundred grain weight are generally recorded. In addition, GHI is also calculated based on the straw and grain yield data to know how photosynthetic products were translocated to economic parts (grain or seeds) of the plant. Greenhouse experimental data should be transformed per plant basis (generally 3–4 plants per pot are used for annual crops) for statistical analysis. The following equations are used for data transformation for statistical analysis in the greenhouse experiments:

$$\text{Straw yield (g plant}^{-1}) = \frac{\text{straw yield in g pot}^{-1}}{\text{number of plants pot}^{-1}}$$

$$\text{Grain yield (g plant}^{-1}) = \frac{\text{grain yield in g pot}^{-1}}{\text{number of plants pot}^{-1}}$$

$$\text{Number of panicles (plant}^{-1}) = \frac{\text{number of panicles pot}^{-1}}{\text{number of plants pot}^{-1}}$$

$$\text{Spikelet sterility (\%)} = \frac{\text{unfilled spikelets}}{\text{unfilled and filled spikelets}} \times 100$$

$$\text{Thousand grain weight (g)} = \frac{\text{grain weight pot}^{-1} \text{ in g}}{\text{number of grains pot}^{-1}} \times 1000$$

$$\text{Number of pods (plant}^{-1}) = \frac{\text{number of pods pot}^{-1}}{\text{number of plants pot}^{-1}}$$

$$\text{Grain harvest index (GHI)} = \frac{\text{grain yield in g}}{\text{grain + straw yield in g}}$$

$$\text{Hundred grain weight (g)} = \frac{\text{grain weight pot}^{-1} \text{ in g}}{\text{number of grains pot}^{-1}} \times 100$$

$$\text{Grain per pod} = \frac{\text{pods pot}^{-1}}{\text{number of grains pot}^{-1}}$$

3.7.1.2 Experimental Results

The author conducted an experiment to study the influence of three sources of nitrogen and different N rates on the growth and yield of lowland rice (Table 3.15). The plant height, straw yield, grain yield, and panicle density increased significantly with the addition of N fertilizer from the three N sources (Table 3.15). The increase in plant height was quadratic with the addition of N in the range of 0–400 mg kg^{-1} from the three sources (Table 3.16). The variation in plant height with the addition of N rate was 71% for ammonium sulfate, 84% for common urea, and 74% for polymer-coated urea. Based on the regression equation, maximum plant height was achieved with the addition of 407 mg N kg^{-1} by ammonium sulfate, 674 mg N kg^{-1} by common urea, and 232 mg N kg^{-1} by polymer-coated urea. Improvement in plant height with the addition of N has been

TABLE 3.15

Plant Height, Straw Yield, Grain Yield, and Panicle Density of Lowland Rice as Influenced by N Rate and Sources

Treatments[a]	Plant Height (cm)	Straw Yield (g Plant^{-1})	Grain Yield (g Plant^{-1})	Panicle Density (Plant^{-1})
0 (control)	83.75d	5.91e	1.31ef	2.31bc
AS 100	95.50bcd	10.82d	5.25cd	4.56a
CU 100	93.75cd	11.08d	3.64de	3.31ab
PU 100	95.00bcd	13.25cd	3.12def	3.37ab
AS 200	101.75abc	14.08cd	7.84b	4.62a
CU 200	99.25abc	14.73c	7.00bc	4.37a
PU 200	100.25abc	15.52c	7.72b	4.43a
AS 400	108.50ab	20.86b	13.59a	5.12a
CU 400	110.50a	20.04b	7.29bc	3.43ab
PU 400	92.00cd	26.30a	0.77f	1.25c
Average	98.02	15.26	5.75	1.97
F-test	**	**	**	**
CV (%)	5.91	8.87	17.08	22.17

[a] Treatments were: AS = ammonium sulfate, CU = common urea, PU = polymer-coated urea, and N rates were 0, 100, 200, and 400 mg kg^{-1} of soil.

**Significant at the 1% probability level. Means followed by the same letter in the same column are not significant at the 5% probability level by Tukey's test.

TABLE 3.16

Relationship between the Nitrogen Rate and Plant Height, Straw Yield, Grain Yield, and Panicle Density of Lowland Rice

Variable[a]	Regression Equation	R^2
N versus PH (AS)	$Y = 84.04 + 0.1220X - 0.00015X^2$	0.71**
N versus PH (UC)	$Y = 84.12 + 0.0931X - 0.000069X^2$	0.84**
N versus pH (UP)	$Y = 83.73 + 0.1441X - 0.00031X^2$	0.74**
N versus SY (AS)	$Y = 6.51 + 0.0365X$	0.95**
N versus SY (UC)	$Y = 5.95 + 0.0549X - 0.000049X^2$	0.93**
N versus SY (UP)	$Y = 6.69 + 0.0488X$	0.98**
N versus GY (AS)	$Y = 1.73 + 0.0301X$	0.96**
N versus GY (UC)	$Y = 1.05 + 0.0385X - 0.000056X^2$	0.85**
N versus GY (UP)	$Y = 0.64 + 0.0565X - 0.00014X^2$	0.78**
N versus PD (AS)	$Y = 2.50 + 0.0186X - 0.000031X^2$	0.55**
N versus PD (UC)	$Y = 2.22 + 0.0166X - 0.000034X^2$	0.66**
N versus PD (UP)	$Y = 2.16 + 0.0219X - 0.000060X^2$	0.78**
N versus RDW (AS)	$Y = 1.39 + 0.0134X - 0.000024X^2$	0.46*
N versus RDW (UC)	$Y = 1.66 + 0.0038X$	0.51**
N versus RDW (UP)	$Y = 1.52 + 0.0083X$	0.70**

[a] PH = plant height, SY = straw yield, GY = grain yield, and PD = panicle density. AS = ammonium sulfate, UC = urea common, and UP = urea polymer coated.

*,**Significant at the 5% and 1% probability levels, respectively.

reported by Fageria and Barbosa Filho (2001) and Fageria et al. (2011a) in lowland rice grown on Brazilian Inceptisol.

Plant height varied from 83.75 to 110.50 cm, with an average value of 98.02 cm. At 100 and 200 mg N kg^{-1}, maximum plant height was produced by ammonium sulfate and polymer-coated urea and minimum plant height was produced by common urea. At 400 mg N kg^{-1}, maximum plant height was produced by common urea, followed by ammonium sulfate and polymer-coated urea. A significant variation in plant height due to N sources has been reported by Fageria et al. (2011). Plant height is an important trait in determining plant lodging and response to N fertilization (Fageria, 2007a). It is also an important trait due to the positive significant relationship with grain yield ($Y = -61.36 + 1.07X - 0.0040X^2$, $R^2 = 0.47*$). A positive significant relationship between plant height and grain yield of lowland rice has been reported by Fageria and Baligar (2001), Fageria (2007a), and Fageria (2001b). Plant height is genetically controlled and also influenced by environmental factors such as essential plant nutrients at an adequate level (Fageria, 2007a). During the 1960s, rice breeders made excellent progress in the development of dwarf cultivars that responded to heavy application of N (Jennings et al., 1979).

Straw yield increased significantly with the addition of N fertilizer by three N sources. The response of plant height to N fertilizer was linear for ammonium sulfate and polymer-coated urea and quadratic with the addition of N in the form of common urea (Table 3.16). The variation in straw yield was 95% with the addition of N in the form of ammonium sulfate, 93% in the form of common urea, and 98% when N was added as polymer-coated urea. Maximum straw yield was obtained with the addition of 400 mg N kg^{-1} compared to the lower N rate at three N sources. Among the N sources, maximum straw yield was produced by polymer-coated urea, followed by the remaining two sources. Improvement in straw yield with the addition of N has been reported by Fageria and Baligar (2001), Fageria et al. (2009), and Fageria et al. (2011a). Straw yield was having a significantly positive relationship with grain yield ($Y = -10.52 + 2.08X - 0.058X^2$, $R^2 = 0.42*$). The positive association between straw yield and grain yield of lowland rice has been reported by Fageria and Barbosa Filho (2001), Fageria et al. (2008), Fageria et al. (2009), and Fageria et al. (2001lb).

Grain yield increased in a linear fashion with the addition of N in the form of ammonium sulfate and in a quadratic fashion with the addition of N in the form of common urea and polymer-coated urea (Table 3.16). Based on a regression equation, maximum grain yield was obtained with the addition of 340 mg N kg^{-1} by common urea and 203 mg N kg^{-1} by polymer-coated urea. Among the three N sources, the highest grain yield was obtained with the addition of ammonium sulfate at the three N rates. However, at the intermediate N rate (200 mg N kg^{-1}), three N sources produced an almost equal grain yield. At the highest N rate of ammonium sulfate, the increase in grain yield was about 10 times compared to control treatments. Fageria et al. (2011b) reported that, at a lower as well at a higher N rate, ammonium sulfate produced a higher grain yield in upland rice compared to urea fertilizer under greenhouse conditions. Similarly, Fageria et al. (2010b) also reported a higher grain yield of lowland rice in a field experiment compared to urea fertilization. The higher yield of rice with ammonium sulfate may be related to the higher acidity produced by this fertilizer in the rhizosphere because rice is an acid-tolerant plant (Fageria, 2014). The higher acidity might have favored balanced nutrition for rice plants due to the acidity tolerance characteristics that may be responsible for higher grain yield.

Panicle density was significantly increased with the addition of N fertilizer with the three nitrogen sources. However, values varied from source to source and from rate to rate. Panicle density increased in a quadratic fashion with the increasing N rate in the range of 0–400 mg kg^{-1} from three N sources (Table 3.16). The variation in panicle density with the addition of N was 55% by ammonium sulfate, 66% by common urea, and 78% by polymer-coated urea. Maximum panicle density was obtained with the addition of ammonium sulfate at a lower as well as at a higher N rate. This might be one of the factors responsible for the higher grain yield with ammonium sulfate compared to two other sources, since panicle density was having a significant exponential quadratic relationship with grain yield ($Y = 0.22 \exp(1.14X - 0.082X^2)$, $R^2 = 0.73*$). Fageria and Barbosa Filho (2001), Fageria et al. (2001), and Fageria et al. (2001la) reported a significant positive association between panicle density and grain yield of lowland rice. Fageria et al. (2011a) also reported that ammonium sulfate produced a higher panicle density compared to urea in lowland rice grown on Brazilian Inceptisol. Figures 3.33 through 3.35 show the growth of lowland rice with the addition of N from three different sources of nitrogen.

The author also conducted another greenhouse experiment to evaluate the response of lowland rice genotypes to N fertilization (Table 3.17). Grain yield was significantly ($P < 0.01$) influenced by N rate, genotype, and N × genotype interaction. A significant N × G interaction indicates different responses of genotypes to N rates. Grain yield varied from 9.12 g plant^{-1} produced by genotype BRA 051135 to 15.16 g plant^{-1} produced by genotype BRS Jaçanã, with an average value of 12.92 at a low N rate. Similarly, at a high N rate, grain yield varied from 17.05 g plant^{-1} produced by genotype BRA 051135 to 24.43 g plant^{-1} produced by genotype BRS Tropical, with an average value of 20.91 g plant^{-1}. The increase in grain yield at a high N rate was 67% compared with a low N rate. An increase in grain yield of lowland rice with the addition of N fertilizer has been reported by Fageria and Barbosa Filho (2001) under greenhouse conditions.

Panicle density varied from 5.00 per plant produced by genotype BRA 051134 to 8.08 per plant produced by genotype BRA 051083, with an average value of 6.58 at a low N rate. At a high N rate, panicle density varied from 10.00 per plant produced by genotypes BRA 051134 and BRA 051135, with an average value of 12.85 per plant. The increase in panicle density was almost double at a high N rate compared with a low N rate. The panicle density or panicle number was significantly and linearly related with grain yield ($Y = 4.6694 + 1.2607X$, $R^2 = 0.9350*$). Fageria and Barbosa Filho (2001) reported a highly significant correlation ($r = 96*$) between grain yield and panicle density in lowland rice grown in Brazilian Inceptisol.

FIGURE 3.33 (**See color insert.**) Response of lowland rice applied with ammonium sulfate. Half of the N was applied at sowing and the remaining half at the active tillering growth stage.

FIGURE 3.34 (**See color insert.**) Response of lowland rice to N fertilization applied with urea. Half of the N was applied at sowing and the remaining half at the active tillering growth stage.

FIGURE 3.35 (**See color insert.**) Response of lowland rice to nitrogen fertilization applied with polymer-coated urea. Half of the N was applied at sowing and the remaining half at the active tillering growth stage.

TABLE 3.17

Grain Yield and Panicle Number of 12 Lowland Rice Genotypes as Influenced by N Rate and Genotype Treatments

	Grain Yield (g Plant⁻¹)		Panicle Number (Plant⁻¹)	
Genotype	0 mg N kg⁻¹	300 mg N kg⁻¹	0 mg N kg⁻¹	300 mg N kg⁻¹
BRS Tropical	12.53abc	24.43a	7.25abcd	15.00a
BRS Jaçanã	15.16a	20.47cd	7.08abcd	13.50ab
BRA 02654	14.27ab	22.63ab	7.75ab	13.91ab
BRA 051077	13.54abc	21.91bc	6.33bcde	14.00ab
BRA 051083	13.17abc	22.34bc	8.08a	14.33a
BRA 051108	14.80ab	22.07bc	7.50abc	14.00ab
BRA 051126	12.45abc	19.61de	6.42bcde	12.48bc
BRA 051129	11.09cd	18.10ef	5.25e	11.33cd
BRA 051130	13.00abc	22.37bc	5.83de	13.33ab
BRA051134	12.28bc	18.02ef	5.00e	10.00d
BRA 051135	9.12d	17.05f	6.00cde	10.00d
BRA 051250	13.61abc	21.91bc	6.50abcde	12.25bc
Average	12.92b	20.91a	6.58b	12.85a
F-test				
N rate (N)	**		**	
Genotype (G)	**		**	
N × G	**		**	
CV (%) N rate	4.23		7.28	
CV (%) genotype	4.91		5.84	

**Significant at the 1% probability level. Means followed by the same letter in the same column are not significantly different at the 5% probability level by Tukey's test.

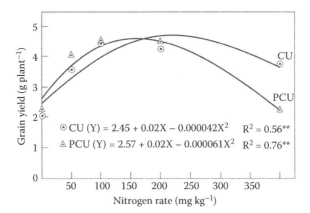

FIGURE 3.36 Grain yield of upland rice as influenced by N rate and sources. The symbol CU means common urea and PCU means polymer-coated urea.

In another greenhouse experiment, the author studied the effects of two N sources (common urea and polymer-coated urea) on the grain yield and panicle density of upland rice. The grain yield of upland rice increased in a quadratic fashion by both sources when N was added in the range of 0–400 mg kg^{-1} (Figure 3.36). Maximum grain yield was obtained with the addition of 167 mg N kg^{-1} by polymer-coated urea. Similarly, maximum grain yield was obtained with the addition of 238 mg N kg^{-1} by common urea. Fageria et al. (2011a) reported that maximum grain yield of upland rice was obtained with the addition of 271 mg N kg^{-1} by urea as a source of N fertilizer. Grain yield was very similar by both sources of N. However, at the highest N rate, grain yield was higher with the application of common urea compared to polymer-coated urea. But at the lower rate of N (up to 200 mg N kg^{-1}), it was inverse. The increase in grain yield of upland rice with the addition of N was expected because Oxisols are deficient in N (Fageria and Baligar, 2005b; Fageria, 2013).

Panicle density was also increased in a quadratic fashion by both sources of N fertilization (Figure 3.37). Maximum panicle density was obtained with the addition of 233 mg N kg^{-1} by polymer-coated

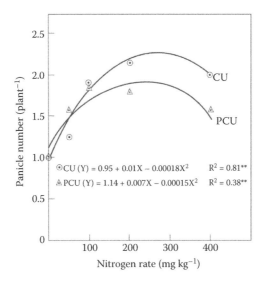

FIGURE 3.37 Influence of two nitrogen sources on panicle number in upland rice. The symbol CU means common urea and PCU means polymer-coated urea.

urea. Similarly, maximum panicle density was achieved with the addition of 278 mg N kg^{-1}. Fageria et al. (2011b) reported that maximum panicle density of upland rice was obtained with the addition of 270 mg N kg^{-1} by common urea in a Brazilian Oxisol. The results of the present study fall within this range. Fageria et al. (2011b) also reported a significant increase in the panicle density of 20 upland rice with the addition of 400 mg N kg^{-1} compared to the control treatment. Similarly, Fageria (2009) also reported an increase in yield and yield components of upland rice with the application of N in Brazilian Oxisols.

3.7.2 Field Experiments

The era of field experimentation, which began in 1834 when J.B. Boussingault, a French chemist, set up the first field experiments at Bechelbonn, Alsace (France), was placed on a modern scientific basis by Liebig's report of 1840 (Collis-George and Davey, 1960). The first field experiments in the form used today were established by Lawes and Gilbert at Rothamsted in 1843. Since then, field experiments have sought for and confirmed the importance of the essential elements in influencing the production of field crops. However, a great deal of the evidence for the discovery of the essentiality of nutrients has been in the laboratory experiments in nutrient solution and not from field experiments (Collis-George and Davey, 1960).

Soil fertilizer tests are the oldest and most widely used method of determining fertilizer N requirements and developing recommendations for crop plants. Yield response trials to develop N fertilizer recommendations for crops were developed in the late 1950s (Kyveryga et al., 2013). These tests indirectly account for the mineralization potential of soil, N leaching, denitrification, N immobilization, availability of fertilizer N, and climatic variability on N uptake and yield (Gerik et al., 1998). They involve the measurement of yield response to increasing levels of applied N fertilizer from experiments conducted at specific locations over several years. Generally, a quadratic equation is used to calculate the N rate for maximum yield (Gerik et al., 1998).

The application of field trial results led to a large increase in agricultural production around the world. Research in agriculture is a complex process and demands constant efforts and experimentation due to change in weather conditions, soil heterogeneity, and release of new cultivars (Barley, 1964). These changes are sometimes so significant that all management practices in use to produce good yields of crops need reevaluation and adjustments to changed situations. For example, when a new cultivar of a crop is released, its nutritional requirements are different from those under cultivation due to the difference in the yield potential, diseases, insect's resistance, and change in architecture. Therefore, field experiments are a basic need in modern agriculture to evaluate nutritional requirements under different agroecological regions. It is very hard to transfer the experimental results of one region to another due to the differences in soil properties, climatic differences, and socioeconomic conditions of farmers (Fageria, 2007b). All these factors determine the technological development and its adaptation by farmers. In conducting field experimentation, certain basic principles should be followed to arrive at meaningful conclusions. Some of these important principles or considerations in field experimentation are discussed in this section. The principles discussed here will help agricultural scientists in the planning and execution of their research trials. Of course, the discussion is mainly concerned with the field of soil fertility and plant nutrition but some basic principles are applicable to other disciplines of agricultural science too. These principles are applicable everywhere with a slight modification according to the circumstances of a particular situation. Most of the points discussed are the outcome of the author's practical experience of more than 40 years in the field of agriculture, in general, and soil fertility and plant nutrition, in particular.

Soil fertility is one of the important factors in determining crop yields. Further, maintaining soil fertility at an appropriate level is also vital for sustainable agriculture and in reducing environmental pollution. To achieve these objectives, research data are required for different agroecological

regions for different crops and cropping systems. A good research project involving experimentation should have appropriate planning to get meaningful results. The planning includes well-defined
objectives based on the priority of problems and to achieve these objectives, experimental methodology should be adequate. Statistical analysis and interpretation of experimental data are as important as planning and executing the experiments. A detailed discussion regarding the basic principles
and methodology in conducting field experiments is given by Fageria (2007b).

3.7.2.1 Observation and Data Transformation in Field Experiments

As in the greenhouse experiments, in field experiments too, growth, yield, and yield components are
measured and data are transformed for statistical analysis. The following equations are used to transform field experimental data for statistical analysis and presentation of results into technical articles.

In the field experiments, at the time of harvest, plant height and panicle density from a 1 m row
should be determined. A one meter row should also be harvested from each plot to determine the
straw dry weight. In addition, 10–20 panicles should be harvested from each plot to determine
spikelet sterility and 1000 grain weight in the case of cereals such as rice. Grain and straw yield
ha^{-1}, panicle density m^{-2}, spikelet sterility, 1000 grain weight and GHI can be determined by using
the following equations:

$$\text{Straw yield (kg ha}^{-1}) = \frac{\text{dry weight of straw m}^{-1}\text{ row in g}}{\text{spacing between row in m}} \times 10$$

$$\text{Grain yield (kg ha}^{-1}) = \frac{\text{grain yield plot}^{-1}\text{ in g}}{\text{area harvested plot}^{-1}\text{ in m}^2} \times 10$$

$$\text{Panicle density m}^{-2} = \frac{\text{panicle number m}^{-1}\text{ row}}{\text{row spacing in m}}$$

$$\text{Spikelet sterility (\%)} = \frac{\text{number of unfilled spikelets in 10 panicles}}{\text{number of filled and unfilled spikelets in 10 panicles}} \times 100$$

$$\text{Thousand grain wt. (g)} = \frac{\text{filled grain weight in 10 panicles in g}}{\text{number of filled grain in 10 panicles}}$$

$$\text{Grain harvest index (GHI)} = \frac{\text{grain yield}}{\text{grain plus straw yield}}$$

In addition to the determination of plant parameters, soil samples should also be taken from each
plot or treatment and the following soil chemical indices should be determined to evaluate fertilizer
treatment effects and correlate these soil indices with grain yield:

$$\text{CEC (cmol}_c\text{ kg}^{-1}) = \sum(\text{Ca, Mg, K, H, Al})$$

where Ca, Mg, K, H, and Al are in $cmol_c$ kg^{-1}, H + Al determined at pH 7.

$$\text{Base saturation (\%)} = \frac{\sum(\text{Ca, Mg, K})}{\text{CEC at pH 7}} \times 100$$

$$\text{Acidity saturation} (\%) = \frac{H + Al}{CEC} \times 100$$

$$\text{Saturation of Ca, Mg, or K} (\%) = \frac{Ca}{CEC} \times 100, \ \frac{Mg}{CEC} \times 100, \text{ or } \frac{K}{CEC} \times 100$$

$$\text{Ca, Mg, or K ratios} = \frac{Ca}{Mg}, \ \frac{Ca}{K}, \text{ or } \frac{Mg}{K}$$

Analysis of variance should be used for data analysis and the quadratic regression model and is generally used to describe the yield and yield component responses to fertilizer or nutrient rates and soil chemical properties or indices. The quadratic response function is the most common functional form to evaluate the yield response to applied nutrient rates and soil chemical properties or indices. The quadratic model is a second-order polynomial function written as

$$Y = a + bx + cx^2$$

where Y is the estimated yield, x is the application rate of the nutrients, and soil chemical properties or indices a, b, and c, are coefficients estimated by fitting the model to the data. The quadratic function assumes that crop yield will increase at the decreasing rate as the nutrient application rate increases until the maximum yield is achieved at

$$N(Y_{max}) = b/2c$$

where $N(Y_{max})$ is the level of applied nutrient that achieves maximum yield; past this point, the yield decreases.

In addition to the above observations related to the greenhouse and field experiments, data related to yield should be presented in metric units. Similarly, nutrient concentration in soil and plants should be expressed in mol m^{-3} or mmol.

3.7.2.2 Experimental Results

The author studied the response of lowland rice to N fertilization under field conditions. The significance of F values derived from the analysis of variance showed significant responses of rice grain yield and yield components to N rates and years of cultivation, but the year × nitrogen rate (Y × N) interactions were significant only for grain yield (Fageria and Baligar, 2001) (Figure 3.38). Therefore, the grain yield data for 3 years as well as the average values of 3 years are presented. Grain yield increased with N fertilization and showed significant ($P < 0.01$) quadratic responses in the 3 year experimentation (Figure 3.38). Based on the regression equations, in the first year, maximum grain yield (6937 kg ha^{-1}) was obtained at 209 kg N ha^{-1}; in the second year, maximum grain yield (6958 kg ha^{-1}) was obtained at 163 kg N ha^{-1}; and in the third year, maximum grain yield of 5682 kg ha^{-1} was obtained at 149 kg N ha^{-1}.

The average data for three years showed that the maximum grain yield of 6465 kg ha^{-1} was obtained with the application of 171 N ha^{-1}. Singh et al. (1998) reported that the maximum average grain yield of 7700 kg ha^{-1} of 20 lowland rice genotypes was obtained at 150–200 kg N ha^{-1} at the International Rice Research Institute in the Philippines. Our results fall more or less in the same range. In our fertilizer experimentations, however, 90% of maximum yield is considered as an economical rate (Fageria et al., 2011a); in the first year, it was 6298 kg kg^{-1} achieved at 120 kg N ha^{-1}. In the second and third years, 90% of the maximum grain yields (6345 and 5203 kg ha^{-1}) was achieved at 90 and 78 kg N ha^{-1}, respectively. The average of 3 year data showed that 90% of

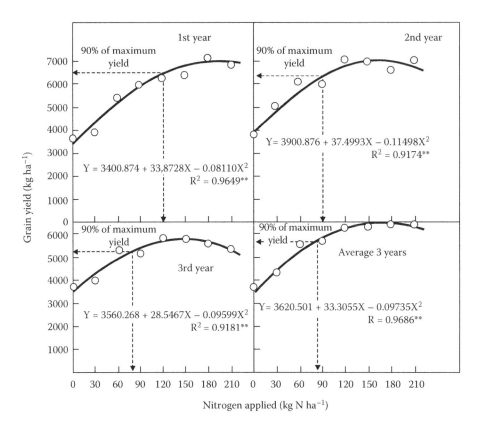

FIGURE 3.38 Response of lowland rice to nitrogen fertilization. (From Fageria, N. K. and V. C. Baligar. 2001. *Commun. Soil Sci. Plant Anal.* 32:1405–1429. With permission.)

the maximum grain yield (5731 kg ha^{-1}) was obtained at 84 kg N ha^{-1}. This means that there was a residual effect of N application in lowland rice grown on an Inceptisol. The increase in grain yield of lowland rice at an economical rate (120 kg N ha^{-1}) in the first year was 76% as compared to the control N treatment. Similarly, the increase in grain yield in the second and third years at economical N rates (90 and 78 kg ha^{-1}) was 69% and 41%, respectively. The average increase of grain yield across the 3 years was 56% at the economical N rate of 84 kg ha^{-1}. At the zero N level, the grain yield was 3579, 3754, and 3702 kg ha^{-1} in the first, second, and third years, respectively. The average value of grain yield across the 3 years was 3678 kg ha^{-1} at the zero N rate. This means that rice grain yield under the control treatment (no N application) was quite good during 3 years of experimentation. In the control N treatment, rice yields increased during the second and third years of cultivation as compared to the first year of cultivation. Fageria and Baligar (1996) also reported significant increases in grain yields of lowland rice grown on an Inceptisol in the central part of Brazil. These authors reported that an average yield of 3 years (5523 kg ha^{-1}) of lowland rice was achieved with the application of 100 kg N ha^{-1} and that grain yields at low fertility levels increased with succeeding cropping years.

Increases in corn yield with the application of N are widely documented in different parts of the world (Azeez et al., 2006; Riedell et al., 2009; Abbasi et al., 2012). The application of 250 kg N ha^{-1} has been reported as an optimum rate of N for corn in the semiarid environment of Pakistan (Hammad et al., 2011). Torbert et al. (2001) reported that corn grain yields in Texas increased with N fertilizer up to 168 kg N ha^{-1} in a year with sufficient rainfall. Similarly, Ma and Subedi (2005) observed that the yield of corn increased with up to 120 kg N ha^{-1} in Ontario, Canada. Abbasi et al.

(2012) reported that the application of 180 kg N ha⁻¹ in split application (half at sowing and the remaining half at the V6 stage—six-leaf stage) is a successful and sustainable management strategy for corn production in rainfed mountainous ecosystems. A significant linear and quadratic increase in the yield of corn hybrids has been reported by Costa et al. (2002). The quadratic equation was $(GY = 8654.03 + 44.49N - 0.14N^2, R^2 = 0.85)$. Based on a regression equation, the maximum grain yield was obtained with the application of 156 kg N ha⁻¹. This rate is within the local limits (Ottawa region) recommended for the Canadian corn range of 120–170 kg N ha⁻¹ (Costa et al., 2002). Both linear (Oberle and Keeney, 1990) and quadratic (Oberle and Keeney, 1990; Stecker et al., 1995) models have been reported for N fertilization versus grain yield relationships for conventional corn hybrids.

The author also studied the response of four upland rice genotypes to nitrogen fertilization. There was a significant quadratic increase in grain yield of all the four genotypes (Figure 3.39). However, the magnitude of response varied. Yue et al. (2012b) reported that the average winter wheat yield was 7.2 Mg ha⁻¹ with the addition of 130 kg N ha⁻¹ in China. The N application timing was 55 kg ha⁻¹ at sowing and the remaining 75 kg N ha⁻¹ was topdressed at the stem elongation stage. Berzsenyi and Tokatlidis (2012) reported that in corn maximum grain yield was obtained with the application of 200 kg N ha⁻¹. Two-thirds of this N was applied at sowing and the remaining as topdressed. Similarly, Blumenthal et al. (2003) also reported that maximum corn yield was obtained with the addition of 2002 kg N ha⁻¹. Boomsma et al. (2009) did not find significant differences in the grain yield of corn with N rates ranging from 165 to 330 kg ha⁻¹. Sindelar et al. (2012) determined the optimum rate of N for corn, which ranged from 168 to 233 kg N ha⁻¹ depending on the locations. Half of the N was applied at sowing and the remaining half as topdressed when corn plants were 20–30 cm tall.

Studies of dryland wheat production in the western Prairies of Canada have demonstrated that grain yield is optimized within the range of 40–90 kg N ha⁻¹ (Beres et al., 2008, 2012a,b; Karamanos et al., 2005; Selles and Zentner, 1993). Irrigated production would increase supplemental N requirements to at least 75 kg N ha⁻¹ (Beres et al., 2008), and as high as 180 kg N ha⁻¹ (Mckenzie et al., 2008).

The author determined the soil acidity indices of a Brazilian Oxisol under different liming treatments after harvest of three soybean crops. Analysis of variance showed that the year × lime rate interactions were significant for all the soil chemical properties analyzed (Table 3.18.). Hence, 3 year values of these chemical properties are presented (Table 3.18). In the first year, pH increases from 5.3 to 7.3 with a lime rate of 0–18 Mg ha⁻¹. In the second year, the pH increase was 5.2–7.2 in the same lime rate range, and in the third year, the increase in pH was 4.7–7.7. Overall, the pH

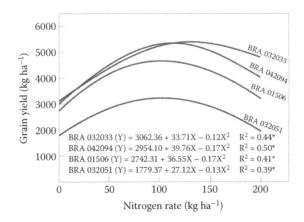

FIGURE 3.39 Upland rice genotype responses to nitrogen fertilization.

TABLE 3.18

Selected Soil Chemical Properties after Harvest of Three Soybean Crops as Influenced by Liming Treatments

Soil Property	Lime Rate (Mg ha⁻¹)				
	0	**3**	**6**	**12**	**18**
			1st Crop		
pH	5.3	6.0	6.5	7.0	7.3
Base saturation (%)	17.1	54.1	68.2	80.6	91.1
H + Al (cmol$_c$ kg⁻¹)	3.90	2.53	1.85	1.23	0.63
Al (cmol$_c$ kg⁻¹)	3.75	0	0	0	0
Ca (cmol$_c$ kg⁻¹)	0.47	1.78	2.84	3.80	4.77
Mg (cmol$_c$ kg⁻¹)	0.25	1.11	1.06	1.26	1.41
CEC (cmol$_c$ kg⁻¹)	4.71	5.51	5.84	6.39	6.90
			2nd Crop		
pH	5.2	5.7	6.4	6.9	7.2
Base saturation (%)	16.9	37.2	56.8	72.8	83.3
H + Al (cmol$_c$ kg⁻¹)	3.85	2.75	1.99	1.22	0.67
Al (cmol$_c$ kg⁻¹)	0.38	0.1	0	0	0
Ca (cmol$_c$ kg⁻¹)	0.49	1.03	1.64	2.05	2.13
Mg (cmol$_c$ kg⁻¹)	0.13	0.44	0.85	1.09	1.09
CEC (cmol$_c$ kg⁻¹)	4.64	4.38	4.64	4.53	4.06
			3rd Crop		
pH	4.7	6.1	6.7	7.5	7.7
Base saturation (%)	15.2	52.5	70.4	95.5	100
H + Al (cmol$_c$ kg⁻¹)	3.88	2.04	1.26	0.19	0
Al (cmol$_c$ kg⁻¹)	0.43	0	0	0	0
Ca (cmol$_c$ kg⁻¹)	0.38	1.33	1.69	2.34	2.50
Mg (cmol$_c$ kg⁻¹)	0.18	0.76	1.10	1.41	1.46
CEC (cmol$_c$ kg⁻¹)	4.57	4.30	4.22	4.1	4.08
			Average of Three Crops		
pH	5.1	5.9	6.5	7.1	7.4
BS (%)	16.4	47.9	65.1	83.0	91.5
H + Al (cmol$_c$ kg⁻¹)	3.88	2.44	1.70	0.88	0.43
Al (cmol$_c$ kg⁻¹)	1.52	0.03	0	0	0
Ca (cmol$_c$ kg⁻¹)	0.45	1.38	2.06	2.73	3.14
Mg (cmol$_c$ kg⁻¹)	0.19	0.77	1.00	1.25	1.32
CEC (cmol$_c$ kg⁻¹)	4.64	4.73	4.9	5.01	5.01
			Statistical Analysis		
Year (Y)		**			
Lime rate (L)		**			
Y × L		**			

Source: From Fageria, N. K., L. C. Melo, and J. P. de Oliveira. 2013a. *J. Plant Nutr.* 36: 2179–2190. With permission.
**Significant at the 1% probability level.

increase was 5.1–7.4 in the lime rate range of 0–18 Mg ha^{-1}. The increase in pH with lime application was associated with a neutralization of Al + H ions and an increase in Ca and Mg concentration in the soil solution. Fageria and Stone (2004) and Fageria (2006) reported a similar increase in pH with the application of lime in the range of 0–24 Mg ha^{-1} in the Brazilian Oxisol. Overall, the increase in base saturation was 16.4–91.5%, H + Al decrease was 3.88–0.43 cmol$_c$ kg^{-1}, Al decrease was 1.52–0 cmol$_c$ kg^{-1}, Ca increase was 0.45–3.14 cmol$_c$ kg^{-1}, Mg increase was 0.19–1.32 cmol$_c$ kg^{-1}, and cation exchange capacity increase was 4.64–5.01 cmol$_c$ kg^{-1} with the application of 0–18 Mg lime ha^{-1}. Fageria and Stone (2004) reported a similar increase or decrease in the acidity indices of Brazilian Oxisol with the application of lime in the range of 0–24 Mg ha^{-1}. Fageria (2001a) also reported similar increases in Ca and Mg concentration Brazilian Oxisol with the application of lime in the range of 0–20 Mg ha^{-1}.

An interesting feature of these results is that, with the application of 3 Mg lime ha^{-1}, practically all the Al^{3+} ions were neutralized. This means that acidity at a higher lime rate was represented by H$^+$ ions in the soil solution. Fageria and Morais (1987) reported similar results with the application of lime in the Brazilian Oxisol. Soil acidity indices (pH, Ca, Mg, base saturation, H + Al, acidity saturation, Ca/K, and Mg/K) were having a significant quadratic association with grain yield (Table 3.19). The variability in grain yield was 93% due to soil pH, 96% due to the soil Ca content, 94% due to the soil Mg content, 97% due to base saturation, 91% due to the H + Al content, 94% due to acidity saturation, 89% due to the Ca/Mg ratio, 90% due to the Ca/K ratio, and 91% due to the Mg/K ratio (Table 3.19). This means that the importance of acidity indices in increasing soybean yield was in the order of base saturation > Ca > Mg > acidity saturation > pH > Mg/K > H + Al > Ca/K > Ca/Mg. Fageria (2001b) reported a more or less similar importance of increasing soybean grain yield in Brazilian Oxisol.

Values for maximum grain yield (3100 kg ha^{-1}) calculated by quadratic regression equations were 7.1 for pH, 2.7 comol$_c$ kg^{-1} for Ca, 1.6 comol$_c$ kg^{-1} for Mg, 88% for base saturation, 0.49 comol$_c$ kg^{-1} for H + Al, 5.2 comol$_c$ kg^{-1} for CEC, 1.92 Ca/Mg ratio, 9.5 Ca/K ratio, and 5.4 Mg/K ratio. Fageria (2001b) reported that the maximum grain yield of soybean in Brazilian Oxisol was obtained with 63% base saturation and at a pH of 6.8. The Ca and Mg values for maximum grain yield of soybean

---

TABLE 3.19

Relationship between Soil Chemical Property (X) and Soybean Grain Yield

Soil Property	Regression Equation	R^2	VMY[a]	VMEY[b]
pH in H$_2$O	$Y = -9884.7040 + 3636.8190X - 254.6528X^2$	0.9260**	7.1	6.0
Ca (comol$_c$ kg^{-1})	$Y = 1484.3560 + 1189.5530X - 216.6682X^2$	0.9577**	2.7	1.6
Mg (comol$_c$ kg^{-1})	$Y = 1650.7640 + 1881.7360X - 584.0436X^2$	0.9362**	1.6	0.9
Base saturation (%)	$Y = 1397.4520 + 38.7096X - 0.2203X^2$	0.9713**	88	51.0
H + Al (comol$_c$ kg^{-1})	$Y = 3080.3400 + 93.4309X - 95.7709X^2$	0.9076**	0.49	0
Acidity saturation (%)	$Y = 3041.1380 + 11.3545X - 0.5417X^2$	0.9409**	10.5	0
CEC (comol$_c$ kg^{-1})	$Y = -42,520.15 + 17,455.66X - 1670.3430X^2$	0.5101**	5.2	4.8
Ca/Mg ratio	$Y = 5359.008 - 2288.174X + 281.131X^2$	0.8903**	1.92	1.9
Ca/K ratio	$Y = 1277.9740 + 397.1924X - 20.9609X^2$	0.9006**	9.5	5.6
Mg/K	$Y = 1599.9570 + 573.1361X - 52.9977X^2$	0.9124**	5.4	3.0

Source: From Fageria, N. K. et al. 2013b. *Commun. Soil Sci. Plant Anal.* 44: 2941–2951. With permission.

Note: Values are averages of three crops.

[a] VMY = value of maximum yield was calculated by a quadratic regression equation.

[b] VMEY = value of maximum economic yield was calculated by a regression equation on the basis of 90% of the maximum yield.

**Significant at the 1% probability level.

were 4 comol$_c$ kg^{-1} and 1.4 comol$_c$ kg^{-1}. According to EMBRAPA (1995), base saturation for soybean grain yield in the central region of Brazil should be near 70%. Variations in the results of acidity indices for maximum and economic yields in the present study compared with those reported earlier in the literature may be due to the use of different cultivars, variation in yield level, and other soil and crop management practices.

3.8   CONCLUSIONS

There are four main criteria or techniques to diagnose nutrient deficiency, sufficiency, or toxicity in soil and/or plants. These techniques are visual deficiency or toxicity symptoms, soil testing, plant tissue testing, and plant or crop response to the applied nutrient. Among these techniques, visual symptoms are the cheapest and a plant tissue test is the most expensive technique. However, care should be taken in identifying N deficiency disorder through visual symptoms because sometimes it may be confused with biotic and other abiotic stresses. Nitrogen deficiency is first observed in the older leaves of plants. Because N is highly mobile in plants, when it is limited, it translocates from the older to the younger part of the plants. Typical N deficiency symptoms are yellowing of the older leaves and when deficiency persists for a longer period, the leaves become dry and almost dead. Nitrogen toxicity rarely occurs in crop plants but it may create nutrient imbalance in plants when present in excess. This nutrient imbalance may be detected by reducing the growth of plant organs such as leaves, height and tillering in cereals, and branches in legumes.

A plant tissue test is an important N disorder diagnostic technique in crop plants. To compare the tissue analysis results for N, preestablished critical or sufficiency level data are required for each crop species under each agroclimatic region. In addition, the nutrient concentration in plant tissue changes significantly with the advancement of plant age, and sufficiency level values should be established at different growth stages during the crop growth cycle. Analytical results of plant tissue should be compared with the same physiological age to obtain comparable results. Values of tissue analysis of N are stable across a wide range of climatic conditions. There exist differences among crop species for critical or sufficiency levels. However, varietal differences in an N critical or adequate level are minimal for most crop species. There is no sound or authentic soil test for N to make fertilizer recommendations, because N is absorbed mainly in the form of NO_3^- and NH_4^+ by plants and their concentration changes significantly with time and space due to mineralization, immobilization, and leaching from the soil–plant system.

Plant response to applied N in the soil is the most important diagnostic technique for N deficiency and sufficiency. It can be done under greenhouse/controlled conditions and field conditions. However, for making an N fertilizer recommendation to a given crop species, field experiments are required. These experiments should be conducted at several locations. The ideal period to repeat these experiments is 3 years but sometimes 2 year data can also be used to make N recommendations provided there is no significant variation from 1 year to another. Generally, average values are used to make N recommendations for a crop species under an agroecological region. These N disorder diagnostic techniques should be used in combination rather than in isolation to get good results.

REFERENCES

Abbasi, M. K., M. M. Tahir, A. Sadiq, M. Iqbal, and M. Zafar. 2012. Yield and nitrogen use efficiency of rainfed maize response to splitting and nitrogen rates in Kashmir, Pakistan. *Agron. J.* 104:448–457.

Allen, V. G., C. P. Brown, R. Kellison, P. Green, C. J. Zilverberg, P. Johnson, J. Weinheimer et al. 2012. Integrating cotton and beef production in the Texas southern high plains: I. Water use and measures of productivity. *Agron. J.* 104:1625–1642.

Aguilar, S. A. and van Diest, A. 1981. Root-phosphate mobilization induced by the alkaline uptake pattern of legume utilizing symbiotically fixed nitrogen. *Plant Soil* 61:27–42.

Anthony, P., G. Malzer, M. Zhang, and S. Sparrow. 2012. Soil nitrogen and phosphorus behavior in a long-term fertilization experiment. *Agron. J.* 104:1223–1237.

Ayling, S. M. 1993. The effect of ammonium ion on membrane potential and anion flux in roots of barley and tomato. *Plant Cell Environ.* 16:297–303.

Azeez, J. O., M. T. Adetunji, and S. T. O. Lagoke. 2006. Response of low nitrogen tolerant maize genotypes to nitrogen application in a tropical Alsisols in northern Nigeria. *Soil Till. Res.* 9:181–185.

Barraclough, P. B., J. R. Howarth, J. Jones, R. Lopez-Bellido, S. Parmar, C. E. Shepherd, and M. J. Hawkesford. 2010. Nitrogen efficiency of wheat: Genotypic and environmental variation and prospects for improvement. *Eur. J. Agron.* 33:1–11.

Barber, K. L., L. D. Maddux, D. E. Kissel, and G. M. Pierzynski. 1992. Corn responses to ammonium and nitrate-nitrogen fertilization. *Soil Sci. Soc. Am. J.* 56:1166–1171.

Barbieri, P. A., H. E. Echeverria, and H. R. Sainz Rozas. 2012. Alternatives for nitrogen diagnosis for wheat with different yield potentials in the humid pampas of Argentina. *Commun. Soil Sci. Plant Anal.* 43:1512–1522.

Barker, A. V. and G. M. Bryson. 2007. Nitrogen. In: *Handbook of Plant Nutrition*, eds., A. V. Barker, and D. J. Pilbeam, pp. 21–50. Boca Raton, Florida: CRC Press.

Barley, K. P. 1964. The utility of field experiments. *Soils Fertilizers* 27:267–269.

Bausch, W. C. and H. R. Duke. 1996. Remote sensing of plant nitrogen status in corn. *Trans. ASAE* 39:1869–1875.

Blackmer, T. M. and J. S. Schepers. 1995. Use of chlorophyll meter to monitor nitrogen status and schedule fertigation for corn. *J. Prod. Agric.* 8:56–60.

Blackmer, T. M., J. S. Schepers, and G. E. Varvel. 1994. Light reflectance compared with other nitrogen stress measurements in corn leaves. *Agron. J.* 86:934–938.

Blackmer, T. M., J. S. Schepers, G. E. Valvel, and E. A. Walter-Shea. 1996. Nitrogen deficiency detection using reflected shortwave radiation from irrigated corn canopies. *Agron. J.* 88:1–5.

Belanger, G., J. Walsh, J. Richards, P. Milburn, and N. Ziadi. 2001. Critical nitrogen curve and nitrogen nutrition index for potato in eastern Canada. *Am. J. Potato Res.* 78:355–364.

Bender, D. A. 1993. Onions. In: *Nutrient Deficiencies and Toxicities in Crop Plants*, ed., W. F. Bennett, pp. 131–135. St. Paul, Minnesota: The American Phytopathological Society.

Bender, R. R., J. W. Haegele, M. L. Ruffo, and F. E. Below. 2013. Nutrient uptake, portioning, and remobilization in modern, transgenic insect protected maize hybrids. *Agron. J.* 105:161–170.

Bennett, W. F. 1993. Plant nutrient utilization and diagnostic plant symptoms. In: *Nutrient Deficiencies and Toxicities in Crop Plants*, ed., W. F. Bennett, pp. 1–7. St. Paul, Minnesota: The American Phytopathological Society.

Beres, B. L., R. S. Sadasivaiah, R. H. McKenzie, R. J. Graf, and R. J. Dyck. 2008. The influence of moisture regime, N management and cultivar on the agronomic performance and yield stability of soft white spring wheat. *Can. J. Plant Sci.* 88:859–872.

Beres, B. L., R. H. McKenzie, R. E. Dowbenko, C. V. Badea, and D. M. Spaner. 2012. Does handling physically alter the coating integrity of ESN urea fertilizers? *Agron. J.* 104:1149–1159.

Beres, B. L., R. H. McKenzie, H. A. Carcamo, L. M. Dosdall, M. L. Evenden, R. C. Yang, and D. M. Spaner. 2012. Influence of seeding rate, nitrogen management and micronutrient blend applications on pith expression in solid stemmed spring wheat. *Crop Sci.* 52:1316–1329.

Berzsenyi, Z. and I. S. Tokatlidis. 2012. Density dependence rather than maturity determines hybrid selection in dryland maize production. *Agron. J.* 104:331–336.

Bhatia, C. R. and R. Rabson. 1976. Bioenergetic considerations in cereal breeding for protein improvement. *Science* 194:1418–1421.

Blackmer, A. M., R. D. Voss, and A. P. Mallarino. 1997. *Nitrogen Fertilizer Recommendations for Corn in Iowa.* Iowa State University Extension/Leopold Center, Ames, Iowa, Publ. Pm-1714.

Blackmer, A. M., T. F. Morris, B. G. Meese, and A. P. Mallarino. 1993. *Soil Testing to Optimize Nitrogen Management for Corn.* Iowa State University Extension/Leopold Center, Ames, Iowa, Publ. Pm 1521.

Bloom, A. J., P. A. Meyerhoff, A. R. Taylor, and T. L. Rost. 2003. Root development and absorption of ammonium and nitrate from the rhizosphere. *J. Plant Growth Regulation* 21:416–431.

Bloom, A. J., S. S. Sukrapanna, and R. L. Warner. 1992. Root respiration associated with ammonium and nitrate absorption and assimilation by barley. *Plant Physiol.* 99:1294–1301.

Blumenthal, J. M., D. J. Lyon, and W. W. Stroup. 2003. Optimal plant population and nitrogen fertility for dryland corn in western Nebrasks. *Agron. J.* 95:878–883.

Boomsma, C. R., J. B. Santini, M. Tollenaar, and T. J. Vyn. 2009. Maize morphological responses to intense crowding at low nitrogen availability: An analysis and review. *Agron. J.* 101:1426–1452.

Borlaug, N. E. and C. R. Dowswell. 1994. Feeding a human population that increasingly crowds a fragile planet. Paper presented at the 15th World Congress of Soil Science, July 10–16, Acapulco, Mexico.

Brady, N. C. and R. R. Weil. 2002. *The Nature and Properties of Soils*, 13th edition. Upper Saddle River, New Jersey: Prentice Hall.

Britto, D. T. and H. J. Kronzucker. 2002. NH_4^+ toxicity in higher plants: A critical review. *J. Plant Physiol.* 159:567–584.

Calvo, N. I. R., H. Sainz Rozas, H. Echeverria, and A. Berardo. 2013. Contribution of anaerobically incubated nitrogen to the diagnosis of nitrogen status in spring wheat. *Agron. J.* 105:321–328.

Campbell, C., R. Zentner, P. Basnyat, R. Dejong, R. Lemke, R. Desjardins, and M. Reiter. 2008. Nitrogen mineralization under summer fallow and continuous wheat in the semiarid Canadian prairie. *Can. J. Soil Sci.* 88:681–696.

Campbell, C., R. Zentner, F. Sells, B. McConkey, and F. Dyck. 1993. Nitrogen management for spring grown annually on zero tillage: Yield and nitrogen use efficiency. *Agron. J.* 85:107–114.

Cassman, K. G. 1993. Cotton. In: *Nutrient Deficiencies and Toxicities in Crop Plants*, ed., W. F. Bennett, pp. 111–119. St. Paul, Minnesota: The American Phytopathological Society.

Cassman, K. G., A. Bobermann, and D. T. Walters. 2002. Agroecosystems, nitrogen use efficiency, and nitrogen management. *Ambio* 31:132–140.

Clark, R. B. 1993. Sorghum. In: *Nutrient Deficiencies and Toxicities in Crop Plants*, ed., W. F. Bennett, pp. 21–26. St. Paul, Minnesota: The American Phytopathological Society.

Collis-George, N. and B. G. Davey. 1960. The doubtful utility of present day field experimentation and other determinations involving soil-plant interactions. *Soil Fertilizers* 23:307–310.

Costa, C., L. M. Dwyer, D. W. Stewart, and D. L. Smith. 2002. Nitrogen effects on grain yield and yield components of leafy and nonleafy maize genotypes. *Crop Sci.* 42:1556–1563.

Dahnke, W. C. and E. H. Vasey. 1973. Testing soils for nitrogen. In: *Soil Testing and Plant Analysis*, ed., R. C. Dinauer, pp. 97–114. Madison, Wisconsin: SSSA.

Dellinger, A. E., J. P. Schmidt, and D. B. Beegle. 2008. Developing nitrogen fertilizer recommendations for corn using an active sensor. *Agron. J.* 100:1546–1552.

Diest, A. B. 1976. Ammonium and nitrate nutrition of crops. *Stikstof 7* 83/84:389–394.

EMBRAPA (Empresa Brasileira de Pesquisa Agropecuaria). 1995. *Technical Recommendations for Soybean Production in the Central Region of Brazil*. Londrina, Brazil: EMBRAPA-National Soybean Research Center, Document 88.

Fageria, N. K. 1989a. Effects of phosphorus on growth, yield and nutrient accumulation in the common bean. *Trop. Agric.* 66:249–255.

Fageria, N. K. 1989b. *Tropical Soils and Physiological Aspects of Crops*. Brasilia, Brazil: EMBRAPA.

Fageria, N. K. 1992. *Maximizing Crop Yields*. New York: Marcel Dekker.

Fageria, N. K. 2001a. Effect of liming on upland rice, common bean, corn, and soybean production in cerrado soil. *Pesq. Agropec. Bras.* 36:1419–1424.

Fageria, N. K. 2001b. Response of upland rice, dry bean, corn, and soybean to base saturation in cerrado soil. *Rev. Bras. Eng. Agric. Ambiental* 5:416–424.

Fageria, N. K. 2004. Dry matter yield and shoot nutrient concentrations of upland rice, common bean, corn and soybean in rotation on an Oxisol. *Commun. Soil Sci. Plant Anal.* 35:961–974.

Fageria, N. K. 2005. Soil fertility and plant nutrition research under controlled conditions: Basic principles and methodology. *J. Plant Nutr.* 28:1975–1999.

Fageria, N. K. 2006. Liming and copper fertilization in dry bean production on an Oxisol in no-tillage system. *J. Plant Nutr.* 29:1–10.

Fageria, N. K. 2007a. Yield physiology of rice. *J. Plant Nutr.* 30:843–879.

Fageria, N. K. 2007b. Soil fertility and plant nutrition research under field conditions: Basic principles and methodology. *J. Plant Nutr.* 30:203–223.

Fageria, N. K. 2009. *The Use of Nutrients in Crop Plants*. Boca Raton, Florida: CRC Press.

Fageria, N. K. 2013. *The Role of Plant Roots in Crop Production*. Boca Raton, Florida: CRC Press.

Fageria, N. K. 2014. *Mineral Nutrition of Rice*. Boca Raton, Florida: CRC Press.

Fageria, N. K. and V.C. Baligar. 1996. Response of lowland rice and common bean grown in rotation to soil fertility levels on a varzea soil. *Fertilizer Res.* 45:13–20.

Fageria, N. K. and V. C. Baligar. 2001. Lowland rice response to nitrogen fertilization. *Commun. Soil Sci. Plant Anal.* 32:1405–1429.

Fageria, N. K. and V. C. Baligar. 2005a. Nutrient availability. In: *Encyclopedia of Soils in the Environment*, ed., D. Hillel, pp. 63–71. San Diego, CA: Elsevier.

Fageria, N. K. and V. C. Baligar. 2005b. Enhancing nitrogen use efficiency in crop plants. *Adv. Agron.* 88:97–185.

Fageria, N. K., V. C. Baligar, and R. B. Clark. 2006. *Physiology of Crop Production.* New York: The Haworth Press.

Fageria, N. K., V. C. Baligar, and C. A. Jones. 2011a. *Growth and Mineral Nutrition of Field Crops*, 3rd edition. Boca Raton, Florida: CRC Press.

Fageria, N. K., V. C. Baligar, and R. W. Zobel. 2007. Yield, nutrient uptake, and soil chemical properties as influenced by liming and boron application in common bean in a no-till system. *Commun. Soil Sci. Plant Anal.* 38:1637–1653.

Fageria, N. K. and M. P. Barbosa Filho. 2001. Nitrogen use efficiency in lowland rice genotypes. *Commun. Soil Sci. Plant Anal.* 32:2079–2090.

Fageria, N. K., L. C. Melo, and J. P. de Oliveira. 2013a. Nitrogen use efficiency in dry bean genotypes. *J. Plant Nutr.* 36:2179–2190.

Fageria, N. K. and O. P. Morais. 1987. Evaluation of rice cultivars for utilization of calcium and magnesium in the cerrado soil. *Pesq. Agropec. Bras.* 22:667–672.

Fageria, N. K., O. P. Morais, and A. B. Santos. 2010a. Nitrogen use efficiency in upland rice genotypes. *J. Plant Nutr.* 33:1696–1711.

Fageria, N. K. and A. Moreira. 2011. The role of mineral nutrition on root growth of crop plants. *Adv. Agron.* 110:251–331.

Fageria, N. K., A. Moreira, and M. A. Coelho. 2011b. Yield and yield components of upland rice as influenced by nitrogen sources. *J. Plant Nutr.* 34:361–370.

Fageria, N. K., A. Moreira, C. Castro, and M. F. Moraes. 2013b. Optimal acidity índices for soybean production in Brazilian Oxisol. *Commun. Soil Sci. Plant Anal.* 44:2941–2951.

Fageria, N. K., A. B. Santos, and M. F. Moraes. 2010b. Influence of urea and ammonium sulfate on soil acidity indices in lowland rice production. *Commun. Soil Sci. Plant Anal.* 41:1565–1575.

Fageria, N. K., A. B. Santos, and V. A. Cutrim. 2008. Dry matter and yield of lowland rice genotypes as influenced by nitrogen fertilization. *J. Plant Nutr.* 31:788–795.

Fageria, N. K., A. B. Santos, and V. A. Cutrim. 2009. Nitrogen uptake and its association with grain yield in lowland rice genotypes. *J. Plant Nutr.* 32:1965–1974.

Fageria, N. K., A. B. Santos, and J. P. Oliveira. 2013. Nitrogen use efficiency in lowland rice genotypes under field conditions. *Commun. Soil Sci. Plant Anal.* 44:2497–2506.

Fageria, N. K. and L. F. Stone. 2004. Yield of dry bean in no-tillage system with application of lime and zinc. *Pesq. Agropec. Bras.* 39:73–78.

Follett, R. H., R. F. Follett, and A. D. Halvorson. 1992. Use of chlorophyll meter to evaluate the nitrogen status of dryland winter wheat. *Commun. Soil Sci. Plant Anal.* 23:687–697.

Fox, R. H., J. J. Meisinger, J. Y. Sims, and W. P. Piekielek. 1992. Predicting N fertilizer needs for corn in humid regions: Advances in the Mid-Atlantic States. In: *Predicting N Fertilizer Need for Corn in Humid Regions*, eds., B. R. Block, and K. R. Kelley, pp. 43–56. Muscle Shoals, Alabama: National Fertilizer Environmental Research Center, TVA.

Gaju, O., V. Allard, P. Martre, J. W. Snape, E. Heume, and J. Legouis. Identification of traits to improve the nitrogen use efficiency of wheat genotypes. *Field Crops Res.* 123:139–152.

Gascho, G. J., D. L. Anderson, and J. E. Bowen. 1993. Sugarcane. In: *Nutrient Deficiencies and Toxicities in Crop Plants*, ed., W. F. Bennett, pp. 37–42. St. Paul, Minnesota: The American Phytopathological Society.

Gashaw, L. and L. M. Mugwira. 1981. Ammonium-N and nitrate-N effects on the growth and mineral composition of triticale, wheat, and rye. *Agron. J.* 73:47–52.

Gaudin, R. and J. Dupuy. 1999. Ammonical nutrition of transplanted rice fertilizer with large urea granules. *Agron. J.* 91:33–36.

Gerik, T. J., D. M. Oosterhuis, and H. A. Torbert. 1998. Managing cotton nitrogen supply. *Adv. Agron.* 64:115–147.

Giuliani, M. M., L. Giuzio, A. De Caro, and Z. Flagella. 2011. Relationships between nitrogen utilization and grain technological quality in durum wheat: I. Nitrogen translocation and nitrogen use efficiency for protein. *Agron. J.* 103:1487–1494.

Glass, D. M. 2003. Nitrogen use efficiency of crop plants: Physiological constraints upon nitrogen absorption. *Crit. Rev. Plant Sci.* 22:453–470.

Glass, A. D. M., D. T. Britto, B. N. Kaiser, H. J. Kronzucker, A. Kumar, M. Okamoto, S. R. Rawat, M. Y. Siddiqi, S. M. Slim, J. J. Vidmar, and D. Zhou. 2001. Nitrogen transport in plants, with emphasis on the regulation of fluxes to match plant demand. *Pflanzenernahr. Bodenkd.* 164:199–207.

Glass, A. D. M., D. T. Britto, B. N. Kaiser, J. R. Kinghorn, H. J. Kronzucker, A. Kumar, M. Okamoto et al. 2002. The regulations of nitrate and ammonium transport systems in plants. *J. Exp. Botany* 53:855–864.

Gonzalez Montaner, J. H., G. A. Maddonni, and M. R. DiNapoli. 1997. Modeling grain yield and grain response to nitrogen in spring wheat crops in the Argentinean Southern Pampa. *Field Crops Res.* 51:241–252.

Grant, C. A. and L. D. Bailey. 1999. Effect of seed placed urea fertilizer and *N*-(*n*-buty)thiophosphoric triamide (NBPT) on emergence and grain yield of barley. *Can. J. Plant Sci.* 79:491–496.

Greenwood, D. J., G. Lemaire, G. Gosse, P. Cruz, A. Draycott, and J. J. Neeteson. 1990. Decline in percentage n of C3 and C4 crops with increasing plant mass. *Ann. Bot.* 66:425–436.

Greenwood, D. J., J. J. Neeteson, and A. Draycott. 1986. Quantative relationships for the dependence of growth rate of arable crops on their nitrogen content, dry weight and aerial environment. *Plant Soil* 91:281–201.

Guindo, D., B. R. Wells, C. E. Wilson, and R. J. Norman. 1992. Seasonal accumulation and partitioning of nitrogen-15 in rice. *Soil Sci. Soc. Am. J.* 56:1521–1527.

Hageman, R. H. 1984. Ammonium versus nitrate nutrition of higher plants. In: *Nitrogen in Crop Production,* ed., R. D. Hauck, pp. 67–85. Madison, Wisconsin: ASA, CSSA, and SSSA.

Hall, R. and H. F. Schwartz. 1993. Common bean. In: *Nutrient Deficiencies and Toxicities in Crop Plants*, ed., W. F. Bennett, pp. 143–147. St. Paul, Minnesota: The American Phytopathological Society.

Hammad, H. M., A. Ahmad, T. Khaliq, W. Farhad, and M. Mubeen. 2011. Optimizing rate of nitrogen application for higher yield and quality in maize under semiarid environment. *Crop Environ.* 2:38–41.

Hayes, R. J. and K. M. Goh. 1978. Ammonium and nitrate nutrition of plants. *Biol. Rev.* 53:465–510.

Hinsinger, P. 1998. How do plant roots acquire mineral nutrients? Chemical processes involved in the rhizosphere. *Adv. Agron.* 64:225–265.

Hoeft, R. G., E. D. Nafziger, R. R. Johnson, and S. R. Aldrich. 2000. *Modern Corn and Soybean Production.* Champaign, Illinois: MCSP Publication.

Hopkins, W. G. 1999. *Introduction to Plant Physiology*, 2nd edition. New York: John Wiley.

Huffaker, R. C. and D. W. Rains. 1978. Factors influencing nitrate acquisition by plants; assimilation and fate of reduced nitrogen. In: *Nitrogen in the Environment*, eds., D. R. Nelson, and J. G. McDonald, pp. 1–43. New York: Academic Press.

Huffman, J. R. 1989. Effects of enhanced ammonium nitrogen availability for corn. *J. Agron. Edu.* 18:325–339.

Huppe, H. C. and D. H. Turpin. 1994. Integration of carbon and nitrogen metabolism in plant and algal cells. *Annu. Rev. Plant Physiol. Plant Mol. Biol.* 45:577–607.

Jarvis, S. C. and D. J. Hatch. 1985. Rates of hydrogen ion efflux by nodulated legumes grown in flowing solution culture with continuous pH monitoring and adjustment. *Anna. Bot.* 55:41–51.

Jemison, J. M. and D. E. Litle. 1996. Field evaluation of two nitrogen testing methods in Maine. *J. Prod. Agric.* 9:108–113.

Jennings, P. R., W. R. Coffman, and H. E. Kauffman. 1979. *Rice Improvement.* Los Bānos, the Philippines: International Rice Research Institute.

Justes, E., B. Mary, J. M. Meynard, J. M. Macher, and L. Thelier-Huche. 1994. Determination of a critical nitrogen dilution curve for winter wheat crops. *Ann. Bot.* 74:397–407.

Karamanos, R. E., N. A. Flore, and J. T. Harapiak. 2005. Effect of post-emergence nitrogen application on the yield and protein content of wheat. *Can. J. Plant Sci.* 85:327–342.

Karlen, D. L., R. L. Flannery, and E. J. Sadler. 1988. Aerial accumulation and partitioning of nutrients by corn. *Agron. J.* 80:232–242.

Keeney, D. R. and D. W. Nelson. 1982. Nitrogen-inorganic forms. In: *Methods of Soil Analysis.* Part 2, 2nd edition, eds., A. L. Page et al., pp. 643–687. Madison, Wisconsin: ASA and SSSA.

Kiniry, J. R., G. McCauley, Y. Xie, and J. G. Arnold. 2001. Rice parameters describing crop performance of four U. S. cultivars. *Agron. J.* 93:1354–1361.

Kitchen, N. R. and K. W. Goulding. 2001. On-farm technologies and practices to improve nitrogen use efficiency. In: *Nitrogen in the Environment; Sources, Problems, and Management*, eds., R. F. Follett, and J. L. Hatfield, pp. 335–369. Amsterdam: Elsevier.

Kyveryga, P. M., P. C. Caragea, M. S. Kaiser, and T. M. Blackmer. 2013. Predicting risk from reducing nitrogen fertilization using hierarchical models and on-farm data. *Agron. J.* 105:85–94.

Lee, R. B. and M. C. Drew. 1989. Rapid, reversible inhibition of nitrate influx in barley by ammonium. *J. Exp. Bot.* 40:741–752.

Lemaire, G. and F. F. Gastal. 2009. Quantifying crop response to nitrogen deficiency and avenues to improve nitrogen use efficiency. In: *Crop Physiology: Applications for Genetic Improvement and Agronomy*, eds., V. Sadras, and D. Calderini, pp. 171–211: San Diego, California: Academic Press.

Leopold, A. C. 1961. Senescence in plant development. *Science* 134:1727–1732.

Liu, M., Z. Yu, Y. Liu, and N. Konijn. 2006. Fertilizer requirements for wheat and maize in China: The QUEFTS approach. *Nutr. Cycling Agroecosyst.* 74:245–258.

Locascio, S. J. 1993. Cucurbits: Cucumber, muskmelon, and watermelon. In: *Nutrient Deficiencies and Toxicities in Crop Plants*, ed., W. F. Bennett, pp. 123–130. St. Paul, Minnesota: The American Phytopathological Society.

Loffler, C. M. and R. H. Busch. 1982. Selection for grain protein, grain yield, and nitrogen partitioning efficiency in hard red spring wheat. *Crop Sci.* 22:591–595.

López-Bellido, R., L. Lopez-Bello, F. J. Lopez-Bello, and J. Castillo. 2003. Faba bean (*Vicia faba* L.) response to tillage and soil residual nitrogen in a continuous rotation with wheat (*Triticum aestivum* L.) under rainfed Mediterranean conditions. *Agron. J.* 95:1253–1261.

Lopez-Bellido, R., J. Castilo, and L. Lopez-Bellido. 2008. Comparative response of bred and durum wheat cultivars to nitrogen fertilizer in a rainfed Mediterranean environment: Soil nitrate and N uptake and efficiency. *Nutr. Cycling Agroecosyst.* 80:121–130.

Ma, B. L. and K. D. Subedi. 2005. Development, yield, grain moisture and nitrogen uptake of Bt corn hybrids and their conventional near-isoline. *Field Crops Res.* 93:199–211.

Mae, T. 1997. Physiological nitrogen efficiency in rice: Nitrogen utilization, photosynthesis, and yield potential. *Plant Soil* 196:201–210.

Malhi, S. S., E. Oliver, G. Mayerle, G. Kruger, and K. S. Gill. 2003. Improving effectiveness of seedrow-placed urea with urease inhibitor and polymer coating for durum wheat and canola. *Commun. Soil Sci. Plant Anal.* 34:1709–1727.

Marschner, H. 1995. *Mineral Nutrition of Higher Plants*, 2nd edition. New York: Academic Press.

Marschner, H. and V. Romheld. 1983. *In vivo* measurement of root-induced pH changes at soil-root interface. Effect of plant species and nitrogen source. *Z. Pflanzenphysiol.* 111:241–251.

McIntyre, G. I. 1997. The role of nitrate in the osmotic and nutritional control of plant development. *Aust. J. Plant Physiol.* 24:103–118.

McKenzie, R. H., A. B. Middleton, R. Dunn, R. S. Sadasivaiah, B. Bers, and E. Bremer. 2008. Response of irrigated soft white spring wheat to seeding date, seeding rate and fertilization. *Can. J. Plant Sci.* 88:291–298.

McKenzie, R. H., E. Bremer, A. B. Middleton, P. G. Pfiffner, R. E. Dunn, and B. L. Bers. 2007. Efficacy of high seedling rates to increase grain yield of winter wheat and winter triticale in southern Alberta. *Can. J. Plant Sci.* 87:503–507.

McKenzie, R. H., A. B. Middleton, and E. Bremer. 2006. Response of mustard to fertilization, seeding date, and seeding rate in southern Alberta. *Can. J. Plant Sci.* 86:353–362.

McNeal, F. N., M. A. Berg, P. L. Brown, and C. F. McGuire. 1971. Productivity and quality response of five spring wheat genotypes, *Triticum aestivum* L., to nitrogen fertilizer. *Agron. J.* 63:908–910.

Meisinger, J. J. 1984. Evaluating plant available nitrogen in soil-crop systems. In: *Nitrogen in Crop Production,* ed. R. D. Hauck, pp. 391–414. Madison, Wisconsin. ASA, CSSSA and SSSA.

Meisinger, J. J., J. S. Schepers, and W. R. Raun. 2008. Crop nitrogen requirement and fertilization. In: *Nitrogen in Agricultural Systems*, eds., J. S. Schepers, and W. R. Raun, pp. 563–612. Madison, Wisconsin: ASA, CSSA, and SSSA.

Melsted, S. W. and T. R. Peck. 1973. The principles of soil testing. In: *Soil Testing and Plant Analysis*, ed., R. Dinauer, pp. 13–21. Madison, Wisconsin: SSSA.

Mengel, K., E. A. Kirkby, H. Kosegarten, and T. Appel. 2001. *Principles of Plant Nutrition*, 5th edition. Dordrecht: Kluwer Academic Publishers.

Nittler, L. W. and J. J. Kenny. 1976. Effect of ammonium to nitrate ratio on growth and anthocyanin development of perennial ryegrass cultivars. *Agron. J.* 68:680–682.

Novoa, R. and R. S. Loomis. 1981. Nitrogen and plant production. *Plant Soil* 58:177–204.

Ntanos, D. A. and S. D. Koutroubas. 2002. Dry matter and n accumulation and translocation for Indica and Japonica rice under Mediterranean conditions. *Field Crops Res.* 74:93–101.

Oberle, S. L. and D. R. Keeney. 1990. Soil type, precipitation, and fertilizer N effects on corn yields. *J. Prod. Agric.* 3:522–527.

O'Donovan, J. T., G. W. Clayton, C. A. Grant, K. N. Harker, T. K. Turkington, and N. Z. Lupwayi. 2008. Effect of nitrogen rate and placement and seeding rate on barley productivity and wild oat fecundity in a zero tillage system. *Crop Sci.* 48:1569–1574.

Ortiz-Monasterio, J. I., K. D. Sayre, S. Rajaram, and M. McMahon. 1997. Genetic progress in wheat yield and nitrogen use efficiency under four nitrogen rates. *Crop Sci.* 37:898–904.

Pagani, A., H. E. Echeverria, F. H. Andrade, and H. R. Sainz Rozas. 2009. Characterization of corn nitrogen status with a greenness index under different availability of sulfur. *Agron. J.* 101:315–322.

Pask, A. J. D., R. Sylvester-Bradley, P. D. Jamieson, and M. J. Foulkes. 2012. Quantifying how winter wheat crops accumulate and use reserves during growth. *Field Crops Res.* 126:104–118.

Pathak, H., P. K. Aggarwal, R. Roetter, N. Kalra, S. K. Bandyopadhaya, S. Prasad, and H. Van Keulen. 2003. Modeling the quantitative evaluation of soil nutrient supply, nutrient use efficiency, and fertilizer requirements of wheat in India. *Nutr. Cycling Agroecosyst.* 65:105–113.

Peng, S., F. V. Garcia, R. C. Laza, and K. G. Cassman. 1993. Adjustment for specific leaf weight improves chlorophyll meters estimate of rice leaf nitrogen concentration. *Agron. J.* 85:987–990.

Piggott, T. J. 1986. Vegetable crops. In: *Plant Analysis: An Interpretation Manual*, eds., D. J. Reuter, and J. B. Robinson, pp. 148–187. Melbourne: Inkata Press.

Raab, T. K. and N. Terry. 1994. Nitrogen source regulation of growth and photosynthesis in *Beta vulgaris* L. *Plant Physiol.* 105:1159–1166.

Radin, J. W. and R. C. Ackerson. 1981. Water relations of cotton plants under nitrogen deficiency. III. Stomatal conductance, photosynthesis, and abscisic acid accumulation during drought. *Plant Physiol.* 67:115–119.

Radin, J. W. and J. S. Boyer. 1982. Control of leaf expansion by nitrogen nutrition in sunflower plants: Role of hydraulic conductivity and turgor. *Plant Physiol.* 69:771–775.

Radin, J. W. and M. A. Matthews. 1989. Water transport properties of cortical cells in roots of nitrogen and phosphorus deficient cotton seedlings. *Plant Physiol.* 89:264–268.

Radin, J. W. and J. R. Mauney. 1986. The nitrogen stress syndrome. In: *Cotton Physiology*, eds., J. R. Mauney, and J. M. Stewart, pp. 91–105. Memphis, TN: The Cotton Foundation.

Radin, J. W. and L. L. Parker. 1979. Water relations of cotton plants under nitrogen deficiency. I. Dependence upon leaf structure. *Plant Physiol. Plant Physiol.* 64:495–498.

Radin, J. W., L. L. Parker, and C. R. Sell. 1978. Partitioning of sugar between growth and nitrate reduction in cotton roots. *Plant Physiol.* 62:550–553.

Rattunde, H. F. and K. J. Frey, K. J. 1986. Nitrogen harvest index in oats: Its repeatability and association with adaptation. *Crop Sci.* 26:606–610.

Raun, W. and G. V. Johnson. 1999. Improving nitrogen use efficiency for cereal production. *Agron. J.* 91:357–363.

Rauschkolb, R. S., T. L. Jackson, and A. I. Dow. 1984. Management of nitrogen in the Pacific States. In: *Nitrogen in Crop Production*, ed., R. D. Hauck, pp. 765–777. Madison, Wisconsin: ASA.

Reuter, D. J. and J. B. Robinson. 1986. *Plant Analysis: An Interpretation Manual.* Melbourne: Inkata Press.

Riedell, W. E., J. L. Pikul, Jr., A. A. Jaradat, and T. E. Schumacher. 2009. Crop rotation and nitrogen input effects on soil fertility, maize mineral nutrition, yield, and seed composition. *Agron. J.* 101:870–879.

Riley, D. and S. A. Barber. 1971. Effect of ammonium and nitrate fertilization on phosphorus uptake as related to root induced pH changes at the root-soil interface. *Soil Sci. Soc. Am. Proc.* 35:301–306.

Russelle, M. P., R. D. Hauck, and R. A. Ilson. 1983. Nitrogen accumulation rates of irrigated corn. *Agron. J.* 75:593–598.

Salsac, L., S. Chaillou, J. F. Morot-Gaudry, C. Lesaint, and E. Jolivoe. 1987. Nitrate and ammonium nutrition in plants. *Plant Physiol. Biochem.* 25:805–812.

Scharf, P. C., N. R. Kitchen, K. A. Sudduth, and J. G. Davis. 2006. Spatially variable corn yield is a weak predictor of optimal nitrogen rate. *Soil Sci. Soc. Am. J.* 70:2154–2160.

Scharf, P. C. and J. A. Lory. 2009. Calibrating reflectance measurements to predict optimal sidedress nitrogen rate for corn. *Agron. J.* 101:615–625.

Scharf, P. C., W. J. Wiebold, and J. A. Lory. 2002. Corn yield response to nitrogen fertilizer timing and deficiency level. *Agron. J.* 94:435–441.

Schepers, J. S., D. D. Francis, M. Vigil, and F. E. Below. 1992. Comparison of corn leaf nitrogen concentration and chlorophyll meter readings. *Commun Soil Sci. Plant Anal.* 23:2173–2187.

Schjoerring, J. K., S. Husted, G. Mack, and M. Mattson. 2002. The regulation of ammonium translocation in plants. *J. Exp. Bot.* 53:883–890.

Schug, E. 1985. Mikronahrstoff-mangel-ein stussfaktor im ertragreichen pflanzenbau. *Kali-Briefe* 17:419–430.

Selles, E. and R. P. Zentner. 1993. Spring wheat yield trends in long term fertility trials. *Can. J. Plant Sci.* 73:83–92.

Sinclair, J. B. 1993. Soybeans. In: *Nutrient Deficiencies and Toxicities in Crop Plants*, ed., W. F. Bennett, pp. 99–103. St. Paul, Minnesota: The American Phytopathological Society.

Sinclair, T. R. and C. T. De Wit. 1975. Comparative analysis of photosynthate and nitrogen requirements in the production of seed by various crops. *Science* 89:565–567.

Sindelar, A. K., J. A. Lamb, C. C. Sheaffer, H. G. Jung, and C. J. Rosen. 2012. Response of corn grain, cellulosic biomass, an ethanol yields to nitrogen fertilization. *Agron. J.* 104:363–370.

Singh, U., J. K. Ladha, E. G. Castillo, G. Punzalan, A. Tirol-Padre, and M. Duqueza. 1998. Genotypic variation in nitrogen use efficiency in medium and long duration rice. *Field Crops Res.* 58: 35–53.

Smith, S. J., J. S. Schepers, and L. K. Porter. 1990. Assessing and managing agricultural nitrogen losses to the environment. *Adv. Soil Sci.* 14:1–43.

Smith, D. H., M. A. Wells, D. M. Porter, and F. R. Cox. 1993. Peanuts. In: *Nutrient Deficiencies and Toxicities in Crop Plants*, ed., W. F. Bennett, pp. 105–110. St. Paul, Minnesota: The American Phytopathological Society.

Soil Science Society of America. 2008. *Glossary of Soil Science Terms*. Madison, Wisconsin: SSSA.

Souza, E. G., P. C. Scharf, and K. A. Sudduth. 2010. Sun position and cloud effects on reflectance and vegetation indices of corn. *Agron. J.* 102:734–744.

Stanford, G. 1982. Assessment of soil nitrogen availability. In: *Nitrogen in Agricultural Soils*, ed., F. J. Stevenson, pp. 651–688. Madison, Wisconsin: ASA and SSSA.

Stecker, J. A., D. D. Buchhplz, R. G. Hanson, N. C. Wollenhaupt, and K. A. McVay. 1995. Tillage and rotation effects on corn yield response to fertilizer nitrogen on Aqualf soils. *Agron. J.* 87:409–415.

Steinbach, H. S., R. Alvarez, and C. Valente. 2004. Balance between mineralization and immobilization of nitrogen as affected by soil mineral nitrogen level. *Agrochima* 48:204–212.

Stockle, C. O. and P. Debaeke. 1997. Modeling crop nitrogen requirements: A critical analysis. *Eur. J. Agron.* 7:161–169.

Tei, F., P. Benincasa, and M. Guiducci. 2002. Critical nitrogen concentration in processing tomato. *Eur. J. Agron.* 18:45–55.

Thompson, R. K., E. B. Jackson, and J. R. Gebert. 1975. Irrigated wheat production to water and nitrogen fertilizer. *Univ. Arizona Agric. Exp. Stn. Tech Bull* 229:16.

Torbert, H. A., K. N. Porter, and J. E. Morrison. 2001. Tillage system, fertilizer nitrogen rate, and timing effect on corn yields in the Texas Blackland Prairie. *Agron. J.* 93:1119–1124.

Ulrich, A. 1950. Critical nitrate levels of sugarbeets estimated from analysis of petioles and blades with special reference to yield and sucrose concentrations. *Soil Sci.* 69:291–309.

Ulrich, A. 1993. Potato. In: *Nutrient Deficiencies and Toxicities in Crop Plants*, ed., W. F. Bennett, pp. 149–156. St. Paul, Minnesota: The American Phytopathological Society.

Ulrich, A. and F. J. Hills. 1973. Plant analysis as an aid in fertilizing sugar crops. Part 1. Sugarbeet, In: *Soil Testing and Plant Analysis*, eds., L. M. Walsh, and J. D. Beaton, pp. 271 288. Madison, Wisconsin: Soil Sci. Soc. Am.

Ulrich, A., J. T. Moraghan, and E. D. Whitney. 1993. Sugar beet. In: *Nutrient Deficiencies and Toxicities in Crop Plants*, ed., W. F. Bennett, pp. 91–98. St. Paul, Minnesota: The American Phytopathological Society.

Van Beusichem, M. L., E. A. Kirkby, and R. Baas. 1988. Influence of nitrate and ammonium nutrition and the uptake, assimilation, and distribution of nutrients in *Ricinus communis*. *Plant Physiol.* 86:914–921.

Vanotti, M. B. and L. G. Bundy. 1994. Corn nitrogen recommendations based on yield response data. *J. Prod. Agric.* 7:249–260.

Voss, R. D. 1993. Corn. In: *Nutrient Deficiencies and Toxicities in Crop Plants*, ed., W. F. Bennett, pp. 11–19. St. Paul, Minnesota: The American Phytopathological Society.

Walburg, G., M. E. Bauer, C. S. T. Daughtry, and T. L. Housley. 1982. Effects of nitrogen nutrition on the growth, yield, and reflectance characteristics of corn canopy. *Agron. J.* 74:677–683.

Warncke, D. D. and S. A. Barber. 1973. Ammonium and nitrate uptake by corn as influenced by nitrogen concentration and NH_4^+/No_3^- ratio. *Agron. J.* 65:950–953.

Welch, R. W. and Y. Y. Yong. 1980. The effects of variety and nitrogen fertilization on protein production in oats. *J. Sci. Food Agric.* 31:541–548.

Wienhold, B. J. and J. M. Krupinsky. 1999. Chlorophyll meter as a nitrogen management tool in malting barley. *Commun. Soil Sci. Plant Anal.* 30:2551–2562.

Wiese, M. V. 1993. Wheat. In: *Nutrient Deficiencies and Toxicities in Crop Plants*, ed., W. F. Bennett, pp. 27–33. St. Paul, Minnesota: The American Phytopathological Society.

Wilcox, G. E. 1993. Tomato. In: *Nutrient Deficiencies and Toxicities in Crop Plants*, ed., W. F. Bennett, pp. 137–141. St. Paul, Minnesota: The American Phytopathological Society.

Wood, C. W., D. W. Reeves, and D. G. Himelrik. 1993. Relationships between chlorophyll meter readings and leaf chlorophyll concentration, N status, and corn yield: A review. *Proc. Agron. Soc. New Zealand* 23:1–9.

Wolfe, D. W., D. W. Henderson, T. C. Hsiao, and A. Alvino. 1988. Interactive water and nitrogen effects on senescence of maize. II. Photosynthetic decline and longevity of individual leaves. *Agron. J.* 80:865–870.

Xiong, J., C. Q. Ding, G. B. Wei, Y. F. Ding, and S. H. Wang. 2013. Characteristic of dry matter accumulation and nitrogen uptake of super-high yielding early rice in China. *Agron. J.* 105:1142–1150.

Yue, R.,Q. F. Meng, R. F. Zhao, F. Li, X. P. Chen, F. S. Zhang, and Z. L. Cui. 2012a. Critical nitrogen dilution curve for optimizing N management of winter wheat production in the North China Plain. *Agron. J.* 104:523–529.

Yue, S., Q. Meng, R. Zhao, Y. Ye, F. Zhang, Z. Cui, and X. Chen. 2012b. Change in nitrogen requirement with increasing grain yield for winter wheat. *Agron. J.* 104:1687–1693.

Ziadi, N., M. Brassard, G. Belanger, A. N. Cambouris, N. Tremblay, M. C. Nolin, A. Claessens, and L. Parent. 2008. Critical nitrogen curve and nitrogen nutrition index for corn in eastern Canada. *Agron. J.* 100:271–276.

Ziadi, N., G. Belanger, A. Claessens, L. Lefebvre, N. Tremblay, A. N. Cambouris, M. C. Nolin, and L. F. Parent. 2010. Plant based diagnostic tools for evaluating wheat N status. *Crop Sci.* 50:2580–2590.

4 Management of Soil Organic Matter

4.1 INTRODUCTION

Soil organic matter (SOM) has long been recognized as an important indicator of soil productivity (Haynes, 2005). Organic matter (OM) refers to the solid, nonmineral portions of the soil, originating from plant and animal residues (Aust and Lea, 1991). According to the Soil Science Society of America (2008), SOM can be defined as the organic fraction of the soil exclusive of undecayed plant and animal residues. Hayes and Swift (1983) defined OM as the term used to refer more specifically to the nonliving components, which are a heterogeneous mixture, largely composed of products resulting from microbial and chemical transformations of organic debris. This transformation, collectively known as the humification process, gives rise to humus, a mixture of substances, which has a degree of resistance to further microbial attack. An adequate amount of OM in the soil plays an important role in improving the soil physical, chemical, and biological properties, and consequently improves or maintains the sustainability of cropping systems. In agricultural systems, maintenance of SOM has long been recognized as a strategy to reduce soil degradation (Mikha and Rice, 2004; Baldock and Broos, 2012).

The major role that OM plays in soil is to stabilize soil aggregates, make soil easier to cultivate, increase the soil water-holding and buffering capacity, and release plant nutrients upon mineralization (Carter and Stewart, 1996; Fageria, 2012). There is no critical level of OM established for different cropping systems below which soil quality decreases markedly or irreversibly; decreasing SOM is still of concern since it might adversely affect some or all of the above properties (Webb et al., 2003). OM also adsorbs heavy metals, which may be toxic to plants or may contaminate soils and reduce their quality. Wander et al. (1996) reported that SOM characteristics are potentially the single best integrator of inherent soil productivity and should be developed as an index of soil quality. Maintenance of soil quality, which is the capacity of soils to sustain productivity, maintain environmental quality, and promote plant and animal health (Doran and Parkin, 1994), is the key to agricultural sustainability (Wander et al., 1996).

SOM consists of a heterogeneous mixture of components with hydrophilic and hydrophobic functional groups (Jenkinson, 1988; Ellerbrock et al., 2005). The SOM formation is a consequence of a feedback relationship between organic carbon input and decomposition (Hsieh, 1996). Hence, the amount of OM in a soil that has been under a given system of cropping and management for a long time depends on how much OM enters the soil each year and how fast this OM decomposes in the soil (Jenkinson and Ayanaba, 1977). The turnover of SOM represents energy (C) and nutrient flows of a soil and, therefore, is closely related to intrinsic soil productivity (Hsieh, 1996). Improving the SOM content is difficult in arable lands due to the rapid decomposition rate of added organic materials. In cultivated soils, fertility management practices may not change the SOM contents by more than 10% during time periods of 0–10 years (Paustian et al. 1992; Wander and Traina, 1996). The small magnitude of C change may easily be overshadowed by natural soil C heterogeneity. This may be the reason why, even though it is well recognized that SOM should be maintained to sustain soil productivity, SOM contents are generally not effectively used within sites to assay management practice impacts on soil productivity or fertility (Wander and Traina, 1996). Although by addition of organic materials the total SOM content may not improve, there may be beneficial changes in the microbial biomass and/or SOM characteristics (Doran et al., 1987; Liebhardt et al., 1989; Wander et al., 1994).

The importance of OM in maintaining the sustainability of cropping systems is *Indiscutivel.* Allison (1973), Campbell (1978), and Haynes (2005) reported that SOM has long been suggested as the single most important indicator of soil productivity. This is because OM greatly affects the chemical, physical, and biological properties and processes in soils. Several workers have tabulated and discussed these effects in detail (Stevenson, 1994; Baldock and Nelson, 2000; Haynes, 2005).

The ability to increase pools of soil organic carbon (SOC) in agricultural ecosystems is of interest both for sequestering atmospheric CO_2 and for restoring OM pools important to soil health (Hooker et al., 2005). The decline in SOC following cultivation and the detrimental effects of decreased SOC have been well documented (Follett, 2001; Mann et al. 2002; Hooker et al., 2005). The continuous crop production potential of soils has a direct relationship to its OM content (Lal, 1998; Mann et al., 2002; Wilhelm et al., 2004). Within limits, productivity is positively related to the SOM content (Reicosky and Forcella, 1988). Many of the characteristics of productive soils are associated with the organic fraction of the soil (Doran et al. 1998; Doran, 2002; Wilhelm et al., 2004).

SOM is a reactive, dynamic component of soils. It is a major component of biogeochemical cycles of major nutrients, including N and the quantity and quality of SOM, both of which reflect and control crop productivity (Burke et al., 1989). SOM, a major source of system stability in agrosystems, is controlled by many factors that have complex interactions. The SOM content depends on the soil type (Schimel et al., 1994; Webb et al., 2003), frequency and type of cultivation (Heenan et al., 1995), cropping and residue management (Grace et al., 1995), fertilizer N input (Bhogal et al., 1997), and climatic conditions (Webb et al., 2003). Understanding the process that controls SOM dynamics and its response to management is essential for the appropriate use of agricultural land (Burke et al., 1989). OM is the main component that supplies N to crop plants. The objective of this chapter is to provide updated information on the role of OM in improving the crop productivity and sustainability of cropping systems and suggesting management practices to improve/maintain the OM content of the soils.

4.2   SOM-RELATED TERMS

Providing definitions of SOM-related terms in the beginning of the chapter may contribute to an understanding of the concepts and discussion on OM dynamics in soil–plant systems. The Soil Science Society of America (2008) and Fageria (2012) provide the following definitions of the most common terms generally used in the OM contest.

1. *Organic farming:* Crop production system that reduces, avoids, or largely excludes the use of synthetically compound fertilizers, pesticides, growth regulators, and livestock feed additives.
2. *Organic fertilizer:* A by-product from the processing of animals or vegetable substances that contains sufficient plant nutrients to be of value as fertilizers.
3. *Organic soils:* A soil in which the sum of the thickness of layers containing organic soil materials is generally greater than the sum of the thickness of mineral layers.
4. *Organic soil materials:* Soil materials that are saturated with water and have $174 \ g \ kg^{-1}$ or more organic carbon if the mineral fraction has $500 \ g \ kg^{-1}$ or more clay, or $116 \ g \ kg^{-1}$ organic carbon if the mineral fraction has no clay, or has proportional intermediate contents, or, if never saturated with water, has $203 \ g \ kg^{-1}$ or more organic carbon.
5. *Organotroph:* An organism able to derive carbon and energy for growth and cell synthesis by utilizing organic compounds.
6. *Humic substances:* A series of relatively high-molecular-weight, yellow-to-black-colored organic substances formed by secondary synthesis reactions in soils.

7. *Humification:* The process whereby the carbon of organic residues is transformed and converted into humic substances through biochemical and abiotic processes.

8. *Humin:* The fraction of SOM that cannot be extracted from the soil with dilute alkali and is most resistant to microbial oxidation.

9. *Humus:* The total of the organic compounds in soil exclusive of undecayed plant and animal tissues, their partial decomposition products, and the soil biomass. The term is often synonymously used with SOM.

10. *Humus form:* A group of soil horizons, located at or near the surface of a pedon, which have formed from organic residues either separated from or intermixed with mineral material.

11. *Humic acid (HA):* The dark-colored organic material that can be extracted from the soil with dilute alkali and other reagents and that is precipitated by acidification to pH 1.0–2.0. HA is intermediate in resistance to microbial attack.

12. *Fulvic acid (FA):* The pigmented organic material that remains in solution after the removal of HA by acidification. It is separated from the FA fraction by adsorption on a hydrophobic resin at low pH values. FA is most susceptible to microbial degradation but is relatively more stable to microbial attack than most freshly applied plant residues.

13. *Fluvic acid fraction:* A fraction of SOM that is soluble both in alkali and dilute acid.

14. *Organan:* A cutan composed of a concentration of OM.

15. *Parent material:* The unconsolidated and more or less chemically weathered mineral or OM from which the solum of soils is developed by pedogenic processes.

16. *Soil organic residues:* Animal and vegetative materials added to the soil of recognizable origin.

4.3   SOM FRACTIONS

SOM is a heterogeneous, dynamic substance that varies in C and N content, molecular structure, decomposition rate, and turnover time (Oades, 1988; McLauchlan and Hobbie, 2004). However, SOM can be conceptually defined as a series of fractions that comprise a continuum based on the decomposition rate (Stanford and Smith, 1972; Paul and Clark, 1996). The various fractions of SOM varied in the degree of decomposition, recalcitrance, and turnover rate and management practices may affect these fractions differently (Schimel et al., 1985; Echeverria et al., 2004). Tirol-Padre and Ladha (2004) and McLauchlan (2004), however, reported that SOM is mainly divided into two main groups or fractions. These fractions are *labile*, that is, smaller in size and most rapidly decomposable. The larger pool with a slow turnover is termed *recalcitrant*. The labile fraction of OM may decompose in a few weeks or months. The labile fraction of OM constitutes plant litter, macroorganic matter or light fraction, the living component or biomass, and nonhumic substances that are not bound to soil minerals (Theng et al. 1989; Tirol-Padre and Ladha, 2004). The most common components of rapidly decomposable OM fractions are carbohydrates, amino acids, peptides, amino sugars and lipids, cellulose, hemicellulose, waxes, fats, resins, and lignin. Labile SOM fractions are highly responsive to changes in C inputs to the soil and will provide a measurable change before any such change in the total OM (Gregorich and Janzen, 1996; Tirol-Padre and Ladha, 2004).

HAs, formed as a result of the humification process, are recalcitrant fractions of SOM (Dodla et al., 2012). They play an important role in controlling C cycles and the biogeochemistry of soils. HAs enhance soil C sequestration through hydrophobic protection (Spaccini et al., 2002) and stabilizing soil aggregates (Hayes and Edward, 2001). The stabilized aggregates can, in turn, protect easily degradable C such as polysaccharides (Bronick and Lai, 2005). Humic substances affect the transport and retention of contaminants. Their amphipathic nature enables them to interact with a wide variety of inorganic and organic pollutants, including heavy metals and charged organic pollutants, via chemical bonding and with nonpolar organic pollutants through nonspecific physical interactions (Li et al., 2003).

The stable fraction of OM may persist in the soil for years or even decades. Stable organic constituents in the soil include humic substances and other organic macromolecules that are highly resistant to microbial decomposition, or that are physically protected by adsorption on mineral surfaces or entrapment within clay and mineral aggregates (Theng et al., 1989; Tirol-Padre and Ladha, 2004). A stable fraction of OM is probably a more appropriate and representative fraction for C sequestration characterization (Cheng and Kimble, 2001; Tirol-Padre and Ladha, 2004). SOM can also be divided into functional pools based on turnover rates (Tiro-Padre and Ladha, 2004). A small pool (1–5%) with a rapid turnover that may take weeks to years and two large pools that are designated as a slow turnover that may take decades and very slow turnover that may take centuries (Scholes and Scholes, 1995).

4.4   OM CONTENT OF THE SOIL

The SOM content is generally measured as organic carbon (C) and/or total N content (Haynes, 2005). Although the organic fraction of soils typically accounts for a small, but variable, proportion (typically 5–10%) of soil mass, it exerts far-reaching effects on soil properties (Haynes, 2005). The input of C to the soil occurs mainly as aboveground plant litter, turnover of root material, and exudation of carbonaceous material from the roots (Cadisch and Giller, 1997; Paustian et al., 1997; Haynes, 2005). This C originates from atmospheric CO_2 that has been photosynthetically fixed and incorporated into organic compounds in plants. Once the organic residues are added to the soil, they are decomposed by the combined actions of soil fauna and microorganisms (Haynes, 2005). During this process, the bulk of the residue C (about 70%) is returned to the atmosphere as CO_2 through faunal and microbial respiration (Jenkinson et al., 1991). The remainder of added C, including that incorporated into the microbial biomass, undergoes further transformations with the eventual formation of relative recalcitrant humic substances (Haynes, 2005).

Soils having widely different OM content are often found even within the same climatic zone. Such differences in the OM content of soils are normally attributed to the effects of vegetation, microbial population, temperature and moisture content, and management practices adopted in crop production. Natural processes leading to the development of soils having variable OM contents are related to the so-called factor of soil formation (Stevenson, 1982).

$$OM = f(\text{time, climate, vegetation, parent material, topography,...})$$

where f stands for "depends" or "function of" and the dots indicate that other factors may be involved.

Carbon is the chief element of SOM that is readily measured quantitatively. Hence, estimates of OM are frequently based on organic-C that is mainly determined by two methods: (i) those based on quantitative combustion procedures wherein C is determined as CO_2 and (ii) those based on the reduction of the $Cr_2O_7^{2-}$ ion by OM, wherein the unreduced $Cr_2O_7^{2-}$ is measured by titration (Allison, 1965). Values for the organic-C content of soils may be expressed as such or may be reported as total OM by multiplying the figure for organic-C by the conventional "Van Bemmelen factor" of 1.72. The use of this factor is based on the assumption that SOM contains 58% C. Organic nitrogen may also be estimated from organic carbon values by dividing by 12 for most soils (Brady and Weil, 2002). Many studies have been reported on the other factor, with highly variable results indicating that it is at best only an approximation (Allison, 1965). However, Broadbent (1953) reported that the factor for converting organic-C into OM in surface soils is approximately 1.9, and that the factors for subsoils are about 2.5. Generally, Histosols and Cambisols have a higher C content as compared to other soil groups.

A typical prairie grassland soil (Mollisol) may contain 5–6% OM in the top 15 cm and a sandy soil, <1%. Poorly drained soils (Aquepts) often have OM content approaching 10%. Tropical Oxisols are known for their low content of OM (Stevenson, 1982). Fageria and Breseghello (2004) determined the OM content of 43 sites, covering 33 rural properties in the State of Mato Grosso, Brazil. The soils of these sites were classified as Oxisols and the OM content varied from 8 to 31 g kg^{-1} (0.8–3.1%) with an average value of 21.5 g kg^{-1} (2.2%). These authors reported that the average OM content of these soils was good but the mineralogy of highly weathered tropical soils such as Oxisols is rich in oxides and hydroxides of Fe and Al. These oxides and hydroxides did not permit adequate mineralization of OM and a major part of OM remains in an inactive form (Lopes, 1983).

From their OM content, soil can be classified as mineral soils (<l00 g OM kg^{-1} soil), organic soils (100–250 g OM kg^{-1} soil), or peat soils (>250 g OM kg^{-1} soil) (Bailey et al., 1991). In Brazil, there are about 35 million hectares of lowland, locally known as "varzea." These areas are distributed throughout the country and generally have favorable climatic conditions for crop production. Owing to poor drainage, lowland rice is a suitable crop to plant on these soils during the rainy season. During dry periods, other crops can be planted in rotation, provided there is adequate drainage. Currently, <2 million hectares of the varzea land is under rice cultivation. Fageria et al. (1997) also determined the OM content of varzea soils of the Mato Grosso and Mato Grosso do Sul States of Brazil (Table 4.1). Data in Table 4.1 show that the OM content in these soils varied from 5 to 257 g kg^{-1} (0.5–25.7%) in the top 0–20 cm soil layer, 1 to 37 g kg^{-1} (0.1–3.7%) in the 20–40 cm soil depth, 21 to 148 g kg^{-1} (2.1–14.8%) in the 40–60 cm soil depth, and 31 to 50 g kg^{-1} (3.1–5.0%) in the 60–80 cm soil depth. The averaged values of the SOM content of the two states were 46 g kg^{-1} at the 0–20 cm soil depth, 12 g kg^{-1} at the 20–40 cm soil depth, 17 g kg^{-1} at the 40–60 cm soil depth, and 7 g kg^{-1} at the 60–80 cm soil depth. Overall, the OM content in the top layer (0–20 cm) was higher compared to the lower soil depths. The turnover of organic carbon depends on environmental conditions, such as aeration, temperature, and water content, and for this reason, the concentration of organic C is higher in the topsoil layer (Mengel et al., 2001). Additionally, most of the plant residues are added on the soil surface and their decomposition can increase the top layer OM content. Cochrane et al. (1985) reported the OM levels in soils from the lowlands of tropical America. These authors classified soils as high, medium, and low in OM if they contained >45, 45-15, and <15 g kg^{-1}, respectively. According to this classification, the average OM content in the soils of the two states (Mato Grosso and Mato Grosso do Sul, Brazil) in the 0–20 cm depth, 46 g kg^{-1} was high. In the lower soil depth, the values of OM content were low. These results showed that there is great variability in the OM content of varzea soils. Fageria et al. (1991) also reported ample variability in the OM content of varzea soils of the Goias State of Brazil.

Some soils such as Histosols, humic Oxisols and Ultisols, and Andisols contain very high OM and create problems for crop production (Sanchez and Miller, 1986). Vast areas of deep organic soils classified as Histosols cover 32 million hectares of Southeast Asia (Driessen, 1978). Many of these soils have more than 20% C to a depth >50 cm and attempts to cultivate them have met with very limited success (Dent, 1980). The main constraints are poor foothold for roots, H toxicity, and

TABLE 4.1

OM Content (g kg^{-1}) of the Varzea Soils of the States of Mato Grosso and Mato Grosso do Sul, Brazil

Soil Depth (cm)	Minimum	Maximum	Average
0–20	5	257	46
20–40	1	37	12
40–60	21	148	17
60–80	31	50	7

Source: Adapted from Fageria, N. K. et al. 1997. *Commun. Soil Sci. Plant Anal.* 28:37–47.

severe Cu deficiency. Definitely too much of a good thing! Another case of too much SOM occurs in certain well-drained soils of the tropics classified as Humox or Humults. These are Oxisols or Ultisols with more than 16 kg C m^{-2} to a depth of 1 m, exclusive of surface litter (Sanchez and Miller, 1986). Their topsoils have organic C contents ranging from 4% to 8% and show a strong black color. Although initially attractive to farmers, the areas of these soils in Brazil are under extensive cattle grazing, subsistence farming, or are not farmed at all, while adjacent nonhumic Oxisols and Ultisols are commercially farmed (Sanchez and Miller, 1986). Two factors are thought to account for the low fertility of these high OM content soils: manganese toxicity and high lime requirement. Their high SOM contents promote the reduction of manganese compounds to Mn^{2+} that can become toxic to many plants (Pavan and Miyazawa, 1984). To overcome Mn toxicity, these soils generally have to be limed to a pH of 6.2. The amount of lime required to raise the pH from 4.6 to 6.2 is astronomical and certainly not economical. A third example relates to the better known group of soils derived from volcanic ash that have high topsoil SOM contents of 10–20% C. Plants grown on these soils exhibit N and P deficiencies that are completely out of proportion with the soils' total N and P contents. The dominant allophane minerals of Andisols are intimately mixed with SOM and fix large quantities of P, which inhibit N mineralization (Bornemisza and Pineda, 1969).

4.5 OM VERSUS SOIL PHYSICAL PROPERTIES

Physical properties of soils are those characteristics, processes, or reactions of a soil that are caused by physical forces and that can be described by, or expressed in, physical terms or equations (Soil Science Society of America, 2008). Examples of physical properties are soil texture, structure or porosity, bulk density, and water-holding capacity. The soil physical properties mainly influence air–water relations in the soil, which in turn affect the growth of plants. The addition of OM to soil improves these physical properties. With the improvement of the soil physical properties, there is an improvement in soil quality that consequently improves crop productivity (Stevenson, 1982; Bauer and Black, 1994). The main soil physical, chemical, and biological properties that are influenced or are having a positive correlation with SOM are summarized in Table 4.2 and their detailed discussion is given in the succeeding section.

TABLE 4.2
Soil OM Function in the Soil for the Sustainability of Cropping Systems

Soil Property	Changes in Soil Property in Favor of Sustainability of Cropping Systems
Physical	i. Texture
	ii. Structure
	iii. Bulk density
	iv. Water-holding capacity
Chemical	i. Availability of macronutrients
	ii. Availability of micronutrients
	iii. Cation exchange capacity
	iv. Aluminum toxicity
	v. Allelopathy
	vi. Heavy metal toxicity
Biological	i. Nitrogen mineralization bacteria
	ii. Dinitrogen-fixing bacteria
	iii. Mycorrhizae fungi
	iv. Microbial biomass

4.5.1 TEXTURE

Soil texture is the relative proportion of the various soil separates in a soil. The three soil separates that make soil texture are sand (2–0.02 mm in diameter), silt (0.02–0.002 mm), and clay (0.002 mm or less in diameter). Soil texture is a fundamental parameter in soil science and a major component of the soil natural capital (Robinson et al., 2009). Texture is widely used in agriculture and engineering as well as in basic research to estimate, for example, the water release curves in flow and transport modeling (Schaap et al., 2001; Wuddivira et al., 2012). Soil texture, especially the clay content, controls the magnitude and rates of many physical, chemical, and hydrological processes in soils. The important soil phenomena such as nutrient storage, nutrient availability, water retention, and stability of aggregates may vary across the field in response to the spatial variability of clay percentage. Soil moisture, which is the major control for the rainfall–runoff response in a watershed, has been directly linked to clay variability (Crave and Gascuel-Odoux, 1997; Wuddivira et al., 2012).

Particle size fractionation is a useful indicator of SOM dynamics (Borchers and Perry, 1992; Parker et al., 2002; Gartzia-Bengoetxea et al., 2009; Norris et al., 2011; Grand and Lavkulich, 2012). SOM associated with the clay fraction is considered to be the most stable fraction, with physical occlusion and the formation of complexes with mineral elements contributing to its stabilization (Paul, 1984; Sollins et al., 1996; Eusterhues et al., 2003). In contrast, silt- and sand-sized SOM fractions are considered to be more reactive due to weaker interactions with minerals (Tiessen and Stewart, 1983; Six et al., 2002).

Soil texture is unchanged by cultural and management practices. The OM content of the soil is highly related to its clay content. Kadebra (1978) found a correlation coefficient of 0.84 for organic C and percent of clay in soils developed from basaltic rocks. No such relationship was found for soils from acidic rocks. Broersma and Lavkulich (1980) reported that 24–48% of OM in selected soils of Canada was associated with the clay fraction and 40–60% was associated with the fine silt fraction. The close correlation between clay mineral concentration and OM in soils may be accounted for by the binding of organic C to clay minerals (Loll and Bollag, 1983; Mengel et al., 2001).

Nichols (1984) studied the relationship of organic carbon to soil properties of Mollisols of the southern Great Plains of the United States. Stepwise multiple regression analyses were made on the percent of organic carbon versus the percent of clay, percent of silt, base saturation, mean annual precipitation, and mean annual temperature. Results indicate a significant relationship between organic carbon and clay with a lesser relationship with precipitation. The other variable did not improve the predictive equation. Six et al. (2000a) confirmed the conclusions of Hassink (1997) showing that the quantity of clay- and silt-size particles is positively correlated with SOM in the same size fractions across a broad range of climates and soil types. Texture has an effect on aggregation (Six et al., 2000b), and there is evidence that the extent of physical protection of SOM increases with increasing clay content (Six et al. 2000a; Vandenbygaart and Kay, 2004). Similarly, many studies have reported that a major part of OM in predominantly inorganic soils is usually found in the clay- and silt-size fractions (Anderson et al., 1981; Tiessen and Stewart, 1983; Catroux and Schnitzer, 1987).

Soil compaction affects the physical, chemical, and biological properties of soils, and has been considered one of the main causes of agricultural soil degradation (Imhoff et al., 2004). The resistance to compaction depends on the intrinsic soil attributes, with texture being one of the most relevant (Larson and Gupta, 1980; McBride, 1989). OM plays an important role in soil compaction (Larson and Gupta, 1980; McBride, 1989; McBridge and Watson, 1990). The susceptibility to compaction decreases as the SOM content increases (Zhang et al., 1997).

4.5.2 STRUCTURE

The soil is a porous mixture of inorganic particles, OM, air, and water. This mixture also contains a large variety of living microorganisms. The inorganic particles and OM make up the soil solids, while the soil pore space is occupied by air and water. Soil structure is the combination or arrangement of primary soil particles into secondary units or peds. The secondary units are characterized on the basis of size, shape, and grade. Each ped is in turn made up of small clusters or aggregates of soil particles. Exceptions are sandy soils that exhibit single-grain characteristic.

Soil structure is an important property that mediates many physical and biological processes and controls SOM decomposition (Van Veen and Kuikman, 1990; Mikha and Rice, 2004). From the agronomic standpoint, soil structure affects plant growth through its influence on infiltration, percolation and retention of water, soil aeration, and mechanical impedance to root growth. The major binding agents responsible for aggregate formation are the silicate clays, oxides of iron and aluminum, and OM and its biological decomposition products (Duiker et al., 2003; Denef et al., 2004). The stability of aggregates is fundamental for the favorable physical conditions in the soil (Tisdall and Oades, 1982; Six et al., 1999; Carter, 2002) and for sustaining crop productivity (Eynard et al., 2004). Iron and aluminum oxides and OM increase the stability of soil aggregates. SOM compounds bind the primary particles in the aggregate, physically and chemically, and this, in turn, increases the stability of the aggregates and limits their breakdown during the wetting process (Emerson, 1977; Lado et al., 2004a,b). Chaney and Swift (1984) used wet sieving to measure the aggregate stability of 26 agricultural soils with differing properties, and they found a high positive correlation between aggregate stability and OM content, suggesting that OM is an important controlling factor. Benito and Diaz-Fierros (1992) studied the effects of various cropping systems on the structural stability of soils containing various OM contents. They found that a decrease of OM content in the soil led to a decrease in soil structural stability. Emerson (1977) suggested that OM stabilized the aggregates mainly by forming and strengthening bonds between the particles within them.

In addition, activities of plant roots and soil fauna and microorganisms also influence soil aggregate stability (Tisdall and Oades, 1982; Marquez et al., 2004). Several studies have elucidated the relationship between aggregates and associated SOM dynamics (Elliott, 1986; Jastrow, 1996; Six et al., 1998, 2000a,b; Denef et al., 2004). Surface soil with high OM content has a well granulated and stable structure. Carter (2002) reported that soil mineralogy and particle size distribution regulate the capacity of a soil to preserve OM and control soil aggregation (Carter, 2002). A marked reduction in the quantity and/or quality of OM often leads to a deterioration of the soil structure (Nemati et al., 2000; Zaher et al., 2005).

4.5.3 BULK DENSITY

The bulk density of soil is defined as the mass of dry soil per unit bulk volume. The unit of soil bulk density is $Mg\ m^{-3}$. Bulk density is a physical property of the soil that can be used as a simple index to the general structural condition of the soil. Although it cannot be interpreted in a specific manner as with the degree of aggregation, aggregate stability, or pore size distribution, bulk density does provide a general index to air–water relations and impedance to root growth. The bulk density of most surface soils usually ranges from 1.0 to 1.6 $Mg\ m^{-3}$ (Fageria and Gheyi, 1999). SOM significantly influences soil bulk density. The higher the OM, the lower is the bulk density and vice versa. Bockheim et al. (2003) reported that bulk density significantly decreased in a quadratic manner with increasing organic soil carbon in the tundra soils of arctic Alaska. Prevost (2004) reported that the SOM content was found to be closely related to bulk density and porosity after clear cutting and mechanical site preparation.

Soil bulk density significantly influences the physical, chemical, and biological properties of soil–plant systems and consequently the nutrient uptake. Bulk density is often used as an index of assessing soil compaction and productivity. Heuscher et al. (2005) reported that organic carbon content was the strongest contributor to bulk density prediction. These authors reported that organic carbon shows a negative relationship with bulk density, indicating that bulk density decreases as organic carbon increases.

4.5.4 WATER-HOLDING CAPACITY

The knowledge of water dynamics in soil is essential for a better management of irrigation, fertilization, and leaching of nutrients and heavy metals from a soil profile (Gerard et al., 2004). One of the most important effects of OM addition to the soil is that it changes the soil's water retention characteristics, which are generally positively related to crop production. A reduction in the available water capacity is considered the foremost contributing factor in the loss of soil productivity caused by erosion. This reduction in the available water capacity is attributed to changes induced in the soil's water-holding characteristics of the root zone or by reduction in the depth (thickness) of the rooting zone (Bauer and Black, 1992). Most of the OM is generally concentrated in the plow layer of the soils (Fageria et al., 1991). Frye et al. (1982), in Kentucky, reported that the available water-holding capacity of an eroded Maury soil (fine–silty, mixed, misic typic paleudalf) was 4×6.1 lower in the upper 15 cm on a volume basis than its noneroded counterpart. Biswas and Khosla (1971) also observed a similar increase in the soil–water retention characteristics and hydraulic conductivity from applying farmyard manure to soils over a 20-year period. Mays et al. (1973) found an increase from 11.1% to 15.3% in the water content corresponding to the 0.33 bar (in modern terminology, the pascal (Pa) is the unit of pressure, 1 bar = 10^5 Pa = 10^2 kPa = 0.1 MPa) suction of a silt loam soil after an application of 327 metric tons ha^{-1} of municipal compost for 2 years.

Gupta et al. (1977) also reported that the amount of water retained at 15 bars increased linearly with the increase in sludge addition of OM in a coarse sandy soil. Scoot and Wood (1989) reported a linear increase in the water retention of Crowley silt loam soil with increasing OM content. OM content alone could account for 84.4% of the variability of water retained at 10 kPa. The slope of the line indicates that each 10 g kg^{-1} of OM could account for an increase of 5.6% by volume of water retained (Scoot and Wood, 1989). Martens and Frankenberger Jr. (1992) studied the effects of different organic amendments on the soil physical parameters and water infiltration rates on irrigated soil. The incorporation of three loadings (75 Mg ha^{-1} each) of poultry manure, sewage sludge, barley straw, and alfalfa into an Arlington soil for 2 years increased the soil respiration rates (139–290%), soil aggregate stability (22–59%), organic C content (13–84%), soil saccharide content (25–41%), and soil moisture content (3–25%), and decreased soil bulk density (7–11%). The changes in soil physical properties resulted in significantly increased cumulative water infiltration rates (18–25%) in the organic-amended plots as compared with unamended plots. The increase in water retention of soil due to the addition of OM may be related to the following factors: (i) decreased bulk density and increased total porosity, (ii) change in the aggregate size distribution (which may change the pore size distribution), and (iii) increased absorptive capacity of the soil (increase in the total surface area) (Fageria and Gheyi, 1999).

Lado et al. (2004a) reported that the saturated hydrolic conductivity of soil with high OM content (3.5%) was higher than that with low OM content (2.3%). Similarly, Lado et al. (2004b) reported that in sandy loam soil, an increase of OM content from 2.3% to 3.5% reduced the aggregate breakdown, soil dispersivity, and the seal formation at the soil surface under raindrop impact conditions. These authors suggested that the final infiltration values were lower in the low than in the high OM soil because (i) there was a more extensive breakdown and dispersion of the aggregate at the surface of the low OM soil than at that of the high OM soil, so that a more continuous crust was formed on the former soil, and (ii) the rearrangement of the detached and dispersed particles in the crust differed between the two soils, so that a thicker, higher-density crust was formed on the low than on the high OM soil.

4.6 OM VERSUS SOIL CHEMICAL PROPERTIES

OM brings many significant changes in soil chemical properties such as reducing Al toxicity and decreasing allelopathy in crop plants. It improves the availability of macro- and micronutrients to crop plants. OM in soils also controls fluctuations in the pH buffering capacity. The main soil chemical properties, which are influenced by SOM, are summarized in Table 4.2 and their detailed discussion is given in the succeeding section.

4.6.1 Availability of Macronutrients

Soil organic C and N cycling are strongly linked to agroecosystems (Spargo et al., 2012). Practices that build SOM tend to increase a soil's capacity to meet crop N needs through mineralization of soil organic N. Furthermore, OM is a major indigenous source of available N that contains as much as 65% of the total soil P, and provides significant amounts of S and other nutrients essential for plant growth (Bauer and Black, 1994). Also universally accepted is that the C fraction is used by microorganisms as a major energy source for metabolic activity, in the process altering nutrient availability (Bauer and Black, 1994). OM has many of the characteristics of an ideal N fertilizer. Organic N is not readily leached or denitrified and its mineralization rate is dependent on many of the same factors that affect plant growth, such as temperature and water availability. The N-supplying power of both OM and legumes is particularly important in today's economy, as the cost of N fertilizer has increased dramatically in recent years. The National Research Council has estimated its availability in the United States at about 3.4 million tons per year, approximately 15% of the total annual N input to soil (Darst and Murphy, 1990). The level of OM is an excellent predictor of the amount of total N in the soil. According to Kapland and Estes (1985), there was a linear relationship between the total N and OM contents in 24 soils of New Hampshire. The linear regression model was total g kg^{-1} N = $-0.25 + 0.056$ (g kg^{-1} OM). According to this equation, approximately 5.6% of OM was made up of total N. This value is somewhat higher than the empirically derived value of 5% by Read and Ridgell (1921). Essentially, the entire N in the plow layer of the soil, of the order of 93–97%, occurs in organic combinations. Most of the remainder can be accounted for as nonexchangeable (fixed) NH_4^+ at any one time (Stevenson, 1982).

The supply of N to a crop by OM depends on the rate of mineralization. Schepers and Mosier (1991) reported that, for a given climatic region, a general estimate of N mineralization could be made based on the SOM content. They estimated that, assuming 2% of the total organic N on the surface, 30 cm is mineralized annually; a soil with 1% SOM content could be expected to mineralize approximately 45 kg N ha^{-1} year^{-1}. It is important to remember that these are general estimates because the amount of organic N made available through mineralization processes will vary greatly over time due to factors such as temperature, precipitation, and tillage (Doran, 1980; Franzluebbers et al., 1995; Wienhold and Halvorson, 1999).

Soil management practices such as the addition of OM to the soil may modify the amounts of available P found in soils. Organic materials that are generally returned to the soil contain P, ranging from 0.1% to 0.5% (Dalton et al., 1952; Singh and Jones, 1976). Broadbent (1953) presented a comprehensive review on the role of OM in the release and tie-up of P from the soil. A number of researchers report a decrease in P sorption by soils in the presence of OM (Frossard et al., 1986; Kuo, 1983; Reddy et al., 1980). Sah and Mikkelsen (1989) reported that flooded rice soils without added OM increased P sorption by 10–70% in half of the 10 soils of California. The common belief of these workers is that OM decomposition produces organic acids, which form stable complexes with Fe and Al and, consequently, block P retention by them. Other workers report that OM increases P retention by the soil (Harter, 1969; Sen Gupta, 1969). Some have suggested that this results from microbial assimilation. P becomes available after the decomposition of organic P compounds following the death of microbial cells (Fageria and Gheyi, 1999). Diaz et al. (1993) reported that P release from OM oxidation has been estimated at 72 kg ha^{-1} year^{-1} in Histosols of the Everglades (USA).

Kapland and Estes (1985) reported that increases in OM levels resulted in corresponding increases in exchangeable soil calcium, magnesium, potassium, extractable phosphorous (Bray 1), and total nitrogen in agricultural soils from New Hampshire. Within terrestrial ecosystems, sulfur (S) is a macronutrient that is required for the metabolism of all organisms, and it may be deficient in certain soils and thus limits crop production. The amount and form of S present within some soils, especially in humid regions, is related to the quantity and type of OM since much of the S is in organic forms (David et al., 1982). In general, a linear relationship has been demonstrated between the content of SOM and the content of total S in surface soils (Freney, 1986). Total soil S decreased in soils where no organic material was added (Kirchmann et al., 1996).

4.6.2 Availability of Micronutrients

OM plays a key role in the soil micronutrient cycle. Knowledge of the nature of the organic ligands that form complexes with metal ions and of the properties of the complexes thus formed will lead to a better understanding of the factors that affect trace element availability to plants (Stevenson, 1991). Organic chemicals with two or more functional groups that can bind with metals to form a ring structure are known as chelating agents (Soil Science Society of America, 2008). OM fractions such as FAs can form chelate structures with some metals. These chelates can bind micronutrients such as copper, iron, zinc, and manganese and improve their availability to plants.

Stevenson (1991) summarized the formation of metal–organic complexes that have the following effects on the soil micronutrient cycle: (i) micronutrient cations that would ordinarily precipitate at the pH values found in most soils are maintained in solution through complexation with soluble OM. Many biochemicals synthesized by microorganisms form water-soluble complexes with trace elements. Complexes of the trace element with FA are also water soluble, (ii) under certain conditions, metal ion concentrations may be reduced to a nontoxic level through complexation with soil OM. This is particularly true when the metal–organic complex has low solubility, such as in the case of complexes with HA and other high-molecular-weight components of OM, (iii) various complexing agents mediate the transport of trace elements to plant roots and, in some cases, to other ecosystems, such as lakes and streams, (iv) organic substances can enhance the availability of insoluble phosphates through complexation of Fe and Al in acid soils, and Ca in calcareous soils, and (v) chelation plays a major role in the weathering of rocks and minerals. Lichens, for example, enhance the disintegration of rock surfaces to which they are attached through the production of chelating agents.

Loeppert and Hallmark (1985) reported that correlation coefficients for visual evaluation of iron deficiency in sorghum plants and SOM content were highly significant. In all cases, regression models indicate that the tendency toward chlorosis decreased with increasing OM content, suggesting that OM may stabilize soil Fe in a form that is more readily available to the sorghum plant. Also, a higher OM content would be conducive to a more active microbial population, which may cycle Fe into the biosphere in a form that is more readily available to growing plants. It is well documented that SOM can readily complex Fe^{3+} (Schnitzer and Khan, 1972; Bloom, 1981) and is a good reducing agent (Szilagyi, 1971) at pH <4.0. There are only a few studies dealing with the Fe^{3+}–OM complex at pH >7.0. Mossbaur spectroscopy (Goodman and Cheshire, 1979) has provided evidence that Fe^{3+} on soil HA is readily hydrolyzed and precipitated as ferric hydroxide as the pH is increased above pH 4.0. On the other hand, there is evidence that the marine $Fe(OH)_x$–HA complex may be soluble (Picard and Felbeck, 1976). Therefore, even though Fe^{3+} may be hydrolyzed at pH >7.0, it may be stabilized as absorbed $Fe(OH)_x$ species, which are more readily available to growing plants (Loeppert and Hallmark, 1985). Cellulosic OM added before flooding increased amorphous Fe concentration in the flood soils for lowland rice production (Sah and Mikkelsen, 1986). Iron chlorosis in grain sorghum, soybeans, and cowpeas occurs on calcareous soils in spots throughout the Great Plains. One reason for this deficiency is that the OM content and cation exchange capacity (CEC) of these soils is low compared with that of other soils in the region (Mathers et al., 1980). Manure supplied Fe as an organic complex available to plants. Tan et al. (1971) found that extracts from poultry litter complexed Fe and other ions. Miller et al. (1969) demonstrated that poultry manure was beneficial for correcting Fe deficiency in plants. An application of 11 tons ha^{-1} of farmyard manure produced larger-grain sorghum than a similar application of Fe from $FeSO_4$ (Mathers et al., 1980).

The role of SOM in B absorption and desorption processes in soil is not yet clearly understood. Elrashidi and O'Connor (1982) showed a positive relationship between SOM content and B absorption, but they were not able to find any precise correlation between the content of OM and the occurrence of significant hysteresis during B desorption from soil. Gupta (1968) found a positive correlation between OM and hot-water-soluble B contents in soils. Since solubility in hot water is considered an availability index, this result supports the hypothesis that OM is the main reserve of B easily available to plants. Marzadori et al. (1991) reported that OM present in the soil has a characteristic influence on the behavior of B in the soil. SOM appears to be responsible for occluding important adsorption sites and hinders the possible hysteric behavior, that is, it plays a positive role in B release from soil surfaces by conferring reversibility characteristics on adsorption processes.

Usually, lower Cu application rates are required with the organic than the inorganic carrier to correct Cu deficiency by either a hard or broadcast application (Fageria and Gheyi, 1999). Similarly, lower rates of broadcast Zn as an organic and as an inorganic carrier can generally be used to correct Zn deficiency (Moraghan, 1983). Humic substances are ubiquitous in soils, sediments, and natural waters. They play a vital role in enhancing the availability of Cu (II) and other micronutrient cations to plants and other living organisms and in reducing toxicity effects due to free Cu (II) (Stevenson and Chen, 1991). Manganese plays a very important role in enzyme synthesis and in the growth and development of crop plants. OM is the storehouse for Mn availability. Shuman (1985a,b) studied the effect of texture and other soil properties on fractionation of Mn in soils and found that Mn was primarily in the organic and manganese oxide fractions in fine-textured soils. Xiang and Banin (1996) reported that Mn may be present in many forms in the soils such as OM, Fe oxides, and primary and secondary minerals, and that these may also be partly available to plants.

4.6.3 Cation Exchange Capacity

OM, depending on its level in the soil, can make a significant contribution to the soil's CEC. Increasing the OM level in the soil increased the soil CEC (Kapland and Estes, 1985; Fageria and Gheyi, 1999). The marked effect of OM on soil CEC can be explained by the high CEC of OM (Helling et al., 1964). Kapland and Estes (1985) reported that an incremental 1% increase in SOM on a dry weight basis (starting near zero) resulted in a corresponding increase of 1.7 cmol CEC kg^{-1} of soil.

Martel et al. (1978) also reported that CEC was highly correlated with OM in the surface horizon of clay-rich soils in lowland Quebec, Canada. Results of the above-cited studies have shown that OM makes a significant contribution to the CEC of the soil, but the actual contribution depends on the soil pH. The results of Helling et al. (1964) show that, for each unit change in pH, the change in CEC of OM was several times greater than for clay. At pH 2.5, only 19% of the CEC of several grassland and forest soils was caused by OM, but this value rose to 45% at pH 8.0.

4.6.4 Aluminum Toxicity

OM plays an important role in controlling the level of aluminum in the soil solution (Bloom et al. 1979a,b). When grown at the same pH, plants from soils high in OM do not exhibit the symptoms of Al toxicity common to plants grown in soils low in OM (Thomas, 1975; Coleman and Thomas, 1964). Hargrove and Thomas (1981) established that OM can, in part, alleviate Al toxicity. In the 24 soils of this study that contained no amendments, the variation in top dry weight and root dry weight was attributed to the interaction of OM and Al (significant at $P < 0.01$). Foy (1964) suggested that the reason alfalfa could grow in a Bayboro soil with high Al was because the Al was chelated by OM, thus reducing the amount of Al in solution. Kapland and Estes (l985) reported that the critical Al level of alfalfa was correlated with SOM levels ($r = 0.88$). An increase of 1% in SOM on a dry weight basis (starting from about zero) increased the critical Al level by 0.3 cmol kg^{-1}. Hargrove and Thomas (1981) found in a study using artificial media that, by increasing SOM with peat, the critical soil Al level for barley increased. OM such as sewage sludge or animal manures added to acid soils improves fertility as well as reduces Al toxicity problems (Kapland and Estes, 1985).

Surface application or surface incorporation of OM also decreased phytotoxic subsoil Al^{3+} activities because dissolved organic matter (DOM) that leached into the subsoil formed nontoxic Al-DOM complexes (Hue, 1992; Liu and Hue, 1996; Hue and Licudine, 1999; Willert and Stehouwer, 2003). The combined application of $CaCO_3$ and OM in lime-stabilized biosolids decreased subsoil acidity and increased subsoil Ca saturation, compared with $CaCO_3$ alone (Tan et al., 1985; Brown et al., 1997; Tester, 1990; Willert and Stehouwer, 2003). This effect was attributed to increases in Ca mobility caused by Ca–DOM complexes (Willert and Stehouwer, 2003).

Aluminum–OM has been shown to be an important factor in determining pH buffering and the relationship between pH and Al^{3+} activity in soil solution (Bloom et al., 1979a). Bloom et al. (1979a) also reported that the fraction of Al bound by OM is important in determining the quantity of Al extracted from surface soils by neutral salts and that the exchangeable aluminum and effective exchange capacity are defined by the cation and concentration of the extracting salt. Several investigations have shown that OM can influence the relationship between pH and the quantity of Al in soil solution (Clark and Nichol, 1966; Evans and Kamprath, 1970). The addition of OM to an acid soil decreases the concentration of Al in soil solution and also the effects of Al toxicity compared to the original soil adjusted to the same pH. Field observations also indicate beneficial effects for the addition of OM to highly acidic soils (Thomas, 1975).

Adams and Moore (1983) found that Al was toxic at a solution concentration of >0.4 µM in the Bt horizon low in organic C (0.14–0.30%) but not in horizons higher in organic C (0.28–1.22%) with the solution of Al between 9 and 134 µM. Ahmad and Tan (1986) found that adding OM (up to 10% wheat straw) can be as effective as adding lime in reducing Al toxicity to wheat and soybean in acid soils. Bloom et al. (1979b) reported that the management of OM may be effective in lowering Al^{3+} activity in acid soil solutions. Additions of OM may, however, lower the apparent solubility of $Al(OH)_3$ without lowering the Al^{3+} activity. If the degree of neutralization of the added organic acids is low, the release of H^+ ions from OM may result in a decrease in pH without a decrease of Al^{3+}. This effect is more important at higher pH values. Thus, the effectiveness of OM in lowering Al^{3+} depends on the soil pH and the base saturation of OM. Since equilibration with soil minerals is very low, yearly additions of OM may be effective in reducing Al toxicity.

Thomas (1975) reported that OM, which usually accumulates near the surface of no-tillage (NT) soils, tends to ameliorate the effects of soil acidity. Phytotoxic Al species have a strong affinity with HA and form insoluble Al–HA complexes (Vance et al., 1996; Hiradate and Yamaguchi, 2003; Yamaguchi et al., 2004). Since the formation of insoluble Al–HA complexes reduces the concentration of Al in soil solution, HA has a role in preventing Al toxicity in corn (Tan and Binger, 1986).

4.6.5 ALLELOPATHY

Allelopathy is defined as any direct or indirect harmful or beneficial effect by one plant on another through the production of chemical compounds that escape into the environment (Rice, 1974). IAS—The International Allelopathy Society (1996)—defined allelopathy as any process involving secondary metabolites produced by plants, algae, bacteria, and fungi that influence the growth and development of agricultural and biological systems. This definition considers all biochemical interactions between living systems, including plants, algae, bacteria, and fungi and their environment (Macias et al., 1988). Willis (1985) reported that the basic conditions necessary to demonstrate allelopathy in natural systems are (i) a pattern of inhibition of one species or plant by another must be shown, (ii) the putatively aggressive plant must produce a toxin, (iii) there must be a mode of toxin release from the plant into the environment, (iv) there must be toxin transport and/or accumulation in the environment, (v) the afflicted plant must have some means of toxin uptake, and (vi) the observed pattern of inhibition cannot be explained solely by physical or other biotic factors, especially competition and herbivory. However, Blum et al. (1999) reported that no study has ever demonstrated all of these criteria. In Brazilian Oxisols, upland rice yield is significantly reduced after 2 or 3 years of consecutive planting on the same area in monoculture due to autoallelopathy (Fageria and Baligar, 2003a). Fageria and Souza (1995) also reported a yield reduction of upland rice in the third year when it was grown in rotation with dry bean on a Brazilian Oxisol.

The organic compounds involved in allelopathy are collectively called allelochemicals (Olofsdotter et al., 1995; Olofsdotter, 2001). Harborne (1977) and Rice (1979) have reviewed the chemistry of various allelochemic compounds in terrestrial ecosystems. These include simple phenolic acids, aliphatic acids, coumarins, terpenoids, lactones, tannins, flavonoids, alkaloids, cyanogenic glycosides, and glucosinolates. Phenolic acids have been identified in an allelopathic rice germplasm (Rimando et al., 2001) and have previously been described as allelochemicals (Blum et al., 1999). Most of them are secondary metabolites released into the environment by leaching, volatilization, or exudation from shoots and roots. Many compounds are degradation products released during the decomposition of dead tissues. Once these chemicals are released into the immediate environment, they must accumulate in sufficient quantity to affect other plants, persist for some period of time, or be constantly released to have lasting effects (Putnam and Duke, 1978). Abiotic (physical and chemical) and biotic (microbial) factors can influence the phytotoxicity of chemicals in terms of the quality and quantity required to cause injury (Fageria and Baligar, 2003a). After entering the soil, allelochemicals encounter millions of soil microbes. The accumulation of chemicals at phytotoxic levels and their fate and persistence in soil are important determining factors for allelochemical interference. After entry into the soil, all chemicals undergo processes such as retention, transport, and transformation, which influence their phytotoxic levels (Fageria and Baligar, 2003a).

Maintaining an adequate level of OM in the soil is a useful method for neutralizing toxic chemicals produced in the process of allelopathy by plants. SOM content can be improved through the application of animal manures and green manuring, use of crop rotation, and conservation tillage (Fageria and Baligar, 2003a). The beneficial effects of OM in detoxifying chemical substances depend on the concentration and the type of chemical compounds and also on other soil chemical properties. SOM may coat mineral surfaces (e.g., Mn^{2+} and Fe^{2+}), which prevents phenolic acids from directly contacting mineral ions and, thus, oxidation of phenolic acids.

Guimares and Yokoyama (1998) studied the effects of upland rice–soybean rotation and upland rice planted in monoculture on a Brazilian Oxisol (Table 4.3). The data in Table 4.3 show that the upland rice yield was significantly reduced in monoculture as compared to rice grown in rotation with soybean. This illustrates that the negative effects of rice allelopathy on the rice crop can be reduced by crop rotation. Similarly, Fageria (2001) reported that a substantially good yield of upland rice can be obtained when grown in upland rice-common bean–corn–soybean rotation on an Oxisol.

TABLE 4.3
Upland Rice Grain Yield in Rotation with Soybean and Monoculture Grown on a Brazilian Oxisol

Crop Rotation	Grain Yield (kg ha^{-1})
Rice after 3 years of soybean	4325
Rice after 1 year of soybean	2577
Rice in monoculture for 5 years	1160

Source: Adapted from Guimaraes, C. B. and L. P. Yokoyama. 1998. *Technology for Upland Rice*, pp. 19–24. Santo Antonio de Goias, Brazil: Embrapa Arroz e Feijão. With permission.

4.7 OM VERSUS SOIL BIOLOGICAL PROPERTIES

SOM contents significantly influence the soil biological properties such as nitrogen mineralization bacteria, dinitrogen-fixing bacteria, mycorrhizae fungi, and total microbial biomass. These properties are summarized in Table 4.2 and a detailed discussion is given in the succeeding sections.

4.7.1 NITROGEN-MINERALIZING BACTERIA

Nitrogen mineralization is the conversion of organic N into inorganic N by microbial activity. Urea and ammonium sulfate are dominant nitrogen carriers used for crop production around the world. The oxidation of the ammonium form of nitrogen fertilizers, which form NO_3^-, can be explained by the following equation:

$$NH_4^+ + 2O_2 \Leftrightarrow NO_3^- + H_2O + 2H^+$$

The oxidation of NH_4^+ in the above equation is known as nitrification and heterotrophic and autotrophic bacteria can carry it out. The most important autotrophic genera of bacteria are *Nitrosomonas* and *Nitrobacter*. An adequate quantity of OM in the soil reduces the soil acidity and improves activities of these nitrogen mineralization bacteria. With the reduction of soil acidity, there is improvement in the nodule formation of clover by an indigenous rhizobial strain (Almendras and Bottomley, 1987; Howieson et al., 1993). OM also influences mineralization of N through its higher water-holding capacity. Since nitrifying bacteria are generally more sensitive to water deficits than fungi, the bacteria-dependent nitrification process ($NH_4^+ \Rightarrow NH_2^- \Rightarrow NH_3^-$) may essentially cease to operate in a dry soil, whereas the ammonification step (inorganic $N \Rightarrow NH_4^+$), accomplished predominantly by more drought-tolerant fungi, may still proceed (Power, 1990).

4.7.2 DENITRIFICATION

Denitrification is the reduction of nitrogen oxides (usually nitrate and nitrite) to molecular nitrogen or nitrogen oxides with a lower oxidation state of nitrogen by bacterial activity (denitrification) or by chemical reactions involving nitrite (chemodenitrification) (Soil Science Society of America, 2008). Denitrification is one of the major mechanisms for N loss from the soil. Hauck (1981) reported that denitrification can lose as much as 30% of the applied N under field conditions. The process of denitrification can be expressed in the form of the following equation:

$$NO_3^- \text{ (nitrate)} \Rightarrow NO_2 \text{ (nitrite)} \Rightarrow NO \text{ (nitric oxide)} \Rightarrow \text{ (nitric oxide)}$$
$$\Rightarrow N_2O \text{ (nitrous oxide)} \Rightarrow N_2 \text{ (dinitrogen)}$$

The majority of denitrification is biologically catalyzed and closely linked to the bacterial respiratory metabolism (Aulakh et al., 1992). In chemodenitrification, generation of N gas is catalyzed by abiotic agents, but this process may only be of importance in acidic or frozen soils (Christianson and Cho, 1983). SOM has a significant influence on the denitrification process in the soil–plant system. In addition, soil pH, temperature, nitrate concentration, aeration, and water status control the denitrification rate in the soils. Denitrifying organisms use organic C as electron donors for energy and for synthesis of cellular constituents (Aulakh et al., 1992). Hence, denitrification strongly depends on the availability of organic compounds such as native SOM, crop residues, or root biomass. The higher the organic carbon in the soil, the higher is the denitrification rate (Aulakh et al., 1983, 1984, 1992). The denitrification rate in the soil also depends on the composition of organic residues added to the soil. The C:N ratio of organic residues is one of the best criteria of its decomposition rate and supplying energy to the denitrifying bacteria (Aulakh et al., 1992). The lower the C:N ratio, the higher is the denitrification rate (Aulakh et al., 1991). The relationship between SOC and denitrification is discussed by Bremner and Shaw (1958), McGarity (1961), and Aulakh et al. (1992).

4.7.3 Dinitrogen Fixation

Dinitrogen fixation is the conversion of molecular nitrogen (N_2) into ammonia and subsequently into organic nitrogen utilizable in biological processes (Soil Science Society of America, 2008). Although mixed cropping and crop rotation with legumes were practiced for centuries, the basis of their benefit was not recognized until Boussingault (1838), a French scientist, presented evidence that the legumes fixed nitrogen from the air (Burris, 1998). Rhizobia encompass a range of bacterial genera, including *Rhizobium, Bradyrhizobium, Sinorhizobium, Mesorhizobium, Allorhizobium,* and *Azorhizobium,* which are able to establish a symbiosis with leguminous plants (Sessitsch et al., 2002). Rhizobia form nodules on roots or stems of their hosts, in which they reduce atmospheric nitrogen and make available to the plants. Biological nitrogen fixation is an important component of sustainable agriculture, and rhizobium inoculants have been applied frequently as biofertilizers (Sessitsch et al., 2002). In addition, rhizobium is a frequent rhizosphere colonizer of a wide range of plants and may also inhabit nonleguminous plants endophytically. In these rhizospheric and endophytic habitats, they may exhibit several plant growth-promoting effects, such as hormone production, phosphate solubilization, and the suppressions of pathogens (Sessitsch et al., 2002).

With the advancement of research on dinitrogen fixation, many aspects of the biochemistry of the process were elucidated between 1930 and 1980, and it was shown that two catalytic proteins plus a reductant and adenosine triphosphate were involved in the reduction of N_2 to ammonia in all the systems studied (Burris, 1998). Biological nitrogen fixation had a significant economic and environmental impact on crop production worldwide. Biological N fixation by legumes offers the potential to reduce and sometimes eliminate the need for N fertilizers for the following crop (Singh et al., 2004a). The quantity of nitrogen fixed by legumes varied from species to species and was also influenced by environmental factors. However, Peoples et al. (1995) reported that various legume crops and pasture species often fix as much as 200–300 kg N ha^{-1}. Globally, symbiotic nitrogen fixation has been estimated to amount to at least 70 million metric tons nitrogen per year (Brockwell and Bottomley, 1995). SOM improved nitrogen fixation by improving the soil physical and chemical properties. SOM improves the soil structure and pH, and it has been reported by Moawad et al. (1984), Palanipappan et al. (1997), and Howieson et al. (1992) that improved soil structure and pH improves activities of dinitrogen-fixing bacteria in the soil.

4.7.4 Mycorrhizae Fungi

The mycorrhizal association is an important factor in the retention or loss of C in terrestrial ecosystems as well as in plant nutrition (Calderon et al., 2012). One of the most important groups of soil microorganisms is mycorrhizal fungi. Vesicular–arbuscular mycorrhizal (VAM) fungi are present in nearly all-natural soils, and these fungi infect the greater majority of plants including the major food crops (Fageria et al., 2011). Mycorrhizae fungi have been shown to improve the nutrition of the host plants for nutrients that are diffusion limited, such as P, Zn, Cu, and Fe (Tinker, 1982; Marschner and Dell, 1994; Smith and Read, 1997). Mycorrhizae fungi receive carbohydrates from the host plant in return for the development of an extensive hyphal network that effectively provides the plant with a substantial increase in the root surface area (Smith and Read, 1997; Richardson, 2001). The symbiosis may also enhance the plant's resistance to biotic and abiotic stresses. Additionally, VAM fungi develop an extensive external hyphal network, which makes a significant contribution to the improvement of soil structure. Therefore, these fungi constitute an integral and important component of agricultural systems (Harrier and Watson, 2003). An adequate amount of OM improves the VAM association of plants due to the improved physical and chemical soil properties.

4.7.5 MICROBIAL BIOMASS

OM is one of the essential components of soil quality which support soil microbial life. The microbial biomass mediates many important functions in soils that include nutrient mineralization and cycling, and decomposition and formation of SOM as they are the main sources of enzymes in soils (Tabatabai, 1994; Acosta-Martinez et al., 2004). Transformation and storage of soil nutrients is regulated by the microbial biomass present and the flow of nutrients through the soil microbial fraction can be substantial (Martens, 1995; Prenger and Reddy, 2004). Enzymes are present in the soil within various biotic and abiotic components (Burns, 1982). Microbial biomass C and N comprise only 1–3% of total soil C and up to 5% of total N in soils, respectively, but are biologically the most active fraction of SOM (Smith and Paul, 1990; Franzluebbers et al., 2001; Acosta-Martinez et al., 2004). Several studies highlighted the role of microbial biomass in decomposition of substances such as carbohydrates and lipids originating from plant and microbial activity in the improvement of soil quality (Tisdall, 1994; Wright et al., 1999; Morse et al., 2000; Zaher et al., 2005).

Microorganisms play an important role in the acquisition and transfer of nutrients in soil. For phosphorus, soil microorganisms are involved in a range of processes that affect P transformation and thus influence subsequent availability to plant roots. In particular, microorganisms can solubilize and mineralize P from inorganic and organic pools of total P (Richardson, 2001). In addition, microorganisms may effectively increase the surface area of roots. The increase in surface area may increase the availability of water and nutrients. Extensive ranges of soil bacteria and fungi that are able to solubilize various forms of precipitated P have been reported (Rodriguez and Fraga, 1999). Predominant among these organisms are *Bacillus, Pseudomonas, Penicillium,* and *Aspergillus* spp. (Richardson, 2001). There is also a need for better management of P fertilizer in agricultural systems so as to minimize the adverse environmental effects of P loss (Tunney et al., 1997). Improving the efficiency of P uptake through a microbial biomass is an important strategy both economically and environmentally (Richardson, 2001).

Humic substances extracted from manures increase the efficiency of N-fixing organisms such as *Rhizobium* and *Azotobacter.* OM serves as a source of energy for both macro- and microfaunal organisms (Fageria and Gheyi, 1999). The numbers of bacteria, actinomycetes, and fungi in the soil are related in a general way to the humus content. Earthworms and other faunal organisms are strongly affected by the quantity of plant residue material returned to the soil (Stevenson, 1982). The OM content of the soils also influences the pathogenic microorganisms. An adequate supply of OM favors the growth of saprophytic organisms relative to parasitic organisms and thereby reduces the population of the latter. Biologically active compounds in soils, such as antibiotic and certain phenotic acids, may enhance the ability of certain plants to resist attack by pathogens (Stevenson, 1982).

4.8 MANAGEMENT PRACTICES TO IMPROVE/STABILIZE OM IN SOIL

Pools of SOC account for about 68% of C in forest biomes (Kimble et al., 2003). Carbon may be sourced from or stored in the soils of terrestrial ecosystems (Kirschbaum, 2000; Post and Kwon, 2000; Wets and Post, 2002; Clark and Johnson, 2011), and outcomes are determined by land management decisions, anthropogenic or natural disturbances, and environmental conditions (e.g., climate) (West and Post, 2002). Changes in SOC pools represent the balance between inputs from litter, root turnover and exudates, faunal necrotic mass production, and losses via respiration by soil organisms, erosion, and leaching (Clark and Johnson, 2011).

Adopting appropriate soil and crop management practices can improve and/or stabilize the SOM content of soils. These practices include liming acid soils, use of organic manures including cover crops, use of adequate fertilizer rates for annual crops, and use of appropriate crop rotation. In addition, the use of conservation tillage can also improve the OM content of the soil. The major management practices to improve or stabilize OM in the soil are summarized in Table 4.4 and their detailed discussions are given in the succeeding sections.

TABLE 4.4

Management Practices to Improve or Stabilize OM Content of Soils

   i. Conservation tillage

  ii. Crop rotation

 iii. Use of an adequate rate of fertilizers

 iv. Liming acid soils

  v. Use of organic manures

    a. Cover crops/green manuring

    b. Farmyard manures

    c. Municipality compost

    d. Recycling crop residues

 vi. Keeping land under pasture

4.8.1 ADOPTING CONSERVATION TILLAGE SYSTEM

Conservation tillage is defined as any tillage sequence, the objective of which is to minimize or reduce the loss of soil and water operationally, a tillage or tillage and planting combination that leaves a 30% or greater cover of crop residues on the surface (Soil Science Society of America, 2008). Minimum tillage, NT, or zero-tillage terms are also used in the literature. According to the Soil Science Society of America (2008), minimum tillage is defined as the minimum use of primary and/or secondary tillage necessary for meeting crop production requirements under the existing soil and climatic conditions, usually resulting in fewer tillage operations for conventional tillage (CT). Similarly, minimum tillage or zero tillage is defined as a procedure whereby a crop is planted directly into the soil with no primary or secondary tillage since harvest of the previous crop. In this process, usually a special planter is necessary to prepare a narrow shallow seedbed immediately surrounding the seed being planted. NT is sometimes practiced in combination with subsoiling to facilitate seeding and early root growth, whereby the surface residue is left virtually undisturbed except for a small slot in the path of the subsoil shank.

Conservation, minimum tillage, or NT is widely adopted in developed as well as developing countries in recent years for crop production. Horowitz et al. (2010) reported that about 35% of row crop hectares in the United States was planted using no till in 2009 and that the median rate of adoption from 2000 to 2007 was roughly 1.5% year^{-1}, a trend that shows no sign of declining. It is projected that conservation tillage will be practiced on 75% of cropland in the United States by 2020 (Lal, 1997). Kern and Johnson (1993) reported that increasing conservation tillage to 76% of planted cropland would change agricultural systems from C sources to C sink. No-till management is generally recognized as an effective strategy to sequester C (West and Post, 2002; Franzluebbers, 2005; Spargo et al., 2012), and it has also been shown to conserve soil organic N (Franzluebbers, 2004; Spargo et al., 2008).

Intensive tillage disrupts the soil structure and also increases the oxidation of SOM because of the increased aeration and microbial activity (Vance, 2000). Hence, it depletes the SOC stocks through erosion (Blanco-Canqui and Lal, 2008). It has been reported that conversion of natural systems to agroecosystems can deplete the SOC stock by as much as 40% for soils under the forest to about 60% for those under grassland (West and Post, 2002; Blanco-Canqui and Lal, 2008). The NT system has been used in the Corn Belt soils of the U.S. Midwest for more than a century (Wander et al., 1998) to enhance corn and soybean production (Kumar et al., 2012). Therefore, this system is being promoted as an alternative to more intensive tillage systems to restore SOC stocks and to improve soil properties (Blanco-Canqui et al., 2009).

There is a general concept that tillage decreases aggregate stability by increasing mineralization of OM and exposing aggregates to additional raindrop impact energies (Tisdall and Oades, 1982; Elliott, 1986; Angers et al., 1992; Amezketa, 1999; Balesdent et al., 2000; Park and Smucker,

2005). Tillage promotes SOM loss through crop residue incorporation into soil, physical break-down of residues, and disruption of macroaggregates (Beare et al., 1994; Paustian et al., 2000; Six et al., 2000a,b; Wright and Hons, 2004). In contrast, conservation or NT reduces soil mixing and soil disturbance, which allows SOM accumulation (Blevins and Frye, 1993). Many studies have shown that conservation tillage improves soil aggregation and aggregate stability (Beare et al., 1994; Six et al., 1999). Conservation or minimum tillage promotes soil aggregation through enhanced binding of soil particles as a result of greater SOM content (Jastrow, 1996; Paustian et al., 2000; Six et al., 2002). Microaggregates often form around particles of undecomposed SOM, providing protection from decomposition (Gupta and Germida, 1988; Gregorich et al., 1989; Six et al., 2002; Wright and Hons, 2004). Microaggregates are more stable than macroaggregates, and thus tillage is more disruptive of large aggregates than smaller aggregates, making SOM from large aggregates more susceptible to mineralization (Cambardella and Elliott, 1993; Six et al., 2002; Wright and Hons, 2004). Since tillage often increases the proportion of microaggregates to macroaggregates, there may be less crop-derived SOM in CT than conservation or NT (Six et al., 2000a; Wright and Hons, 2004). Fungal growth and mycorrhizal fungi, which are promoted by NT, contribute to the formation and stabilization of macroaggregates (Tisdall and Oades, 1982; Beare and Bruce, 1993).

The larger SOM accumulation in conservation tillage had been observed in intensive cropping systems, where multiple crops are grown yearly (Ortega et al., 2002; Wright and Hons, 2004). The use of conservation tillage, including no till, is being considered as part of a strategy to reduce C loss from agricultural soils (Kern and Johnson, 1993; Paustian et al., 1997; Denef et al., 2004). Crop species also influence SOM accumulation in the soil. Residue quality often plays an important role in regulating long-term SOM storage (Lynch and Bragg, 1985). Crop residues having a low N concentration, such as wheat, generally decompose at slower rates than residues with higher N, such as sorghum and soybean (Franzluebbers et al., 1995; Wright and Hons, 2004), since wheat residues often persist longer and increase SOM more than does sorghum or soybean (Wright and Hons, 2005).

4.8.2 Adopting Appropriate Crop Rotation

Conventional monoculture agriculture systems can reduce the quality of soils by loss of OM and structure because of the low level of organic inputs and regular disturbance from tillage practices (Acosta-Martinez et al., 2004). Crop rotation may have many positive effects on soil quality and consequently on crop production. Crop rotation is defined as a planned sequence of crops growing in a regularly recurring succession on the same area of land, as contrasted to the continuous culturing of one crop or growing of a variable sequence of crops (Soil Science Society of America, 2008). Bullock (1992) defined crop rotation as a system of growing different types of crops in a recurrent succession and in an advantageous sequence on the same land. Crop rotations are a key component of successful organic arable systems (Robson et al., 2002). Rotations can be optimized to conserve and recycle nutrients and minimize pest, disease, and weed problems (Lampkin, 1990; Robson et al., 2002). Appropriate crop rotation has a significant influence on the SOM content of soils. The results of long-term field trials in Illinois (USA) showed that crop rotation influenced the content of SOM (Odell et al., 1984). The level of soil C and N was highest in the rotation of maize–oats–clover, and lowest in the permanent corn rotation (Mengel et al., 2001). The fundamental prerequisite of an appropriate crop rotation is listed in Table 4.5.

Crop rotations under CT that provide residues with low C/N ratios stimulate the decomposition of native SOM to a greater extent than do rotations providing residues with high C/N ratios (Ghidey and Alberts, 1993; Sisti et al., 2004; Wright and Hons, 2004). Under NT, crop rotations have been shown to have a minimal effect on native SOM decomposition (Sisti et al., 2004). Wright and Hons (2004) also reported that greater differences in SOM between crop species occurred under CT rather than NT, especially in subsurface soil. Wani et al. (1994) reported that green manures and

TABLE 4.5

Fundamental Prerequisite of an Appropriate Crop Rotation

Prerequisite	Advantage
Including legume crops in rotation	Supply N to succeeding crops.
Including crops of different root architecture	Improve soil physical and chemical properties, such as nutrient uptake from deeper soil layers, soil porosity, aeration and drainage, and improve SOM content.
Including appropriate cover/green manure crops in the rotation	Protect soil from erosion, control weeds, supply OM, conserve moisture, improve soil hydraulic conductivity, and control diseases and insects.
Include crops with different resistance to diseases and insects	Break the cycle of diseases and insects, and reduce the host plant presence in rotation.
Include crops suited to a given agroecological region	Better economic return and ecologically viable.

organic amendments in crop rotations provided a measurable increase in SOM quality and other soil quality attributes compared with continuous cereal systems.

Crop rotations have positive effects on soil properties related to the higher C inputs and diversity of plant residues to soils in comparison with continuous systems (Miller and Dick, 1995; Entry et al., 1996; Moore et al., 2000; Acosta-Martinez et al., 2004). Conservation tillage increases soil organic C (Franzluebbers et al., 1995; Deng and Tabatabai, 1997; Acosta-Martinez et al., 2003) and microbial biomass (Angers et al., 1993; Franzluebbers et al., 1994, 1995) to modify the soil microbial community (Acosta-Martinez et al., 2004).

4.8.3 Use of Adequate Rate of Fertilizers

The use of an adequate rate of fertilizers for annual crops is an important factor in increasing crop productivity and SOM content. Bremer et al. (2011) reported that the application of synthetic fertilizers consistently increased the crop yield and SOC. Hsieh (1992, 1996) reported that yields of plots with no fertilizer and continuous corn were consistently the lowest and corn yields of plots with fertilizer and rotation were consistently the highest. A higher grain yield generally produced higher straw yields as well as higher roots biomass (Baligar et al., 1998; Fageria and Baligar, 2005). A higher straw yield and root biomass may improve the OM content of the soil.

Modern agriculture is characterized by an exponential increase in the use of N fertilizers (Vitousek et al., 1997). Increased rates of N_2O evolution by N-fertilized soils are well documented in field and laboratory studies (Sarawat and Keeney, 1986). Thus, the accelerated application of N fertilizers in crop production is regarded as the major reason for enhanced N_2O release from soils, and agriculture is currently estimated to contribute 90% of the total anthropogenic N_2O emissions (Duxbury, 1994). Under these situations, the use of an adequate rate of N is an important component of precise agriculture and one of the most critical environmental challenges related to soil quality. The use of an adequate N rate in cropping systems can improve soil carbon (Potter et al., 1997; Salinas-Garcia et al., 1997; Halvorson et al., 2002). Rasmussen and Rohde (1988) and Robinson et al. (1996) reported linear increases in SOC with applied N fertilizer, and they also noted that crop residue has a positive impact on SOC.

Ladha et al. (2012) evaluated the impact of commercial fertilizer N on SOM from long-term experiments and concluded that the application of N fertilizer leads to a slower decrease in SOM contents, or may cause a small increase, after a new equilibrium is reached following N application. Like N, the use of an adequate rate of P is also important for improving the OM content of soils and consequently crop yields and soil quality. The lack of an adequate level of P in the soil may contribute to land degradation and subsequent water pollution in vast areas, mostly in the lesser-developed countries of tropical and subtropical regions. Phosphorus deficiency often limits the growth of crops, and may even cause a crop failure, which forces farmers to clear more land in order to survive. Without adequate phosphorus, regrowth of natural vegetation on disturbed forest and savanna sites is often too slow to prevent soil erosion and depletion of SOM (Fageria, 2002).

4.8.4 LIMING ACID SOILS

Acidic soils, which occupy 30–40% of agricultural land in the world, compromise plant growth and have a negative impact on agriculture (Yamaguchi et al., 2004). Liming is one of the most effective and dominant practices to reduce soil acidity (Fageria and Baligar, 2003b). Liming had many beneficial effects on acid soils when it was applied in adequate amounts. It improves soil pH and base saturation, neutralizes H^+ and Al^{3+} ions toxicity, and improves the biological activity of beneficial microorganisms. Liming also improves P uptake by crops grown on limed acid soils (Fageria, 1989; Fageria and Baligar, 2003b). All these chemical changes create a favorable environment for plant growth and improve the OM content of soils and bring sustainability of cropping systems. Data in Table 4.6 show changes in soil chemical properties at two soil depths of a Brazilian Oxisol with liming under NT cropping system. The pH, base saturation, and Ca and Mg saturation increased significantly with increasing lime rates at two soil depths. The H + Al and acidity saturation decreased significantly with increasing lime rates at two soil depths.

TABLE 4.6
Selected Soil Chemical Properties after Harvest of Dry Bean Crops at Two Soil Depths as Influenced by Liming Treatments

Soil Property	Lime Rate (Mg ha⁻¹)			F-Test	CV(%)
	0	12	24		
(0–10 cm Depth)					
pH	5.4c	6.7b	7.1a	[a]	2
Base saturation (%)	27.9c	70.5b	84.3a	[a]	21
H + Al (cmol$_c$ kg⁻¹)	7.0a	2.3b	1.2c	[a]	19
Acidity saturation (%)	7.2a	2.1b	1.6c	[a]	25
Ca (cmol$_c$ kg⁻¹)	1.9c	3.9b	4.7a	[a]	10
Mg (cmol$_c$ kg⁻¹)	0.5b	1.4a	1.4a	[a]	13
(10–20 cm Depth)					
pH	5.3c	6.1b	6.5a	[a]	2
Base saturation (%)	24.6c	49.5b	62.8a	[a]	21
H + Al (cmol$_c$ kg⁻¹)	7.1a	4.4b	3.1c	[a]	14
Acidity saturation (%)	7.6a	5.2b	3.7c	[a]	7
Ca (cmol$_c$ kg⁻¹)	1.7c	2.9b	3.8a	[a]	15
Mg (cmol$_c$ kg⁻¹)	0.4c	1.1b	1.3a	[a]	15

Source: Adapted from Fageria, N. K., L. F. Stone, and A. Moreira. 2008. *J. Plant Nutr.* 31:1723–1735. With permission.

Note: Values are averages of three bean crops at harvest.

[a] Significant at the 1% probability level. Means followed by the same letter in the same line for the same parameter under different lime treatments are statistically not significant at the 5% probability level by Tukey's test.

Soil pH is one of the most important chemical properties in determining the availability of macro- and micronutrients. Soil pH buffering caused by the protonation and deprotonation of minerals and organic materials reduces the change in soil pH when acids or bases are added to the soils (Weaver et al., 2004). Aitken et al. (1990) reported that the use of multiple linear regression analysis showed that organic carbon was most significantly variable, accounting for 78% of the variance in the pH buffering capacity, although clay accounted for 32% of the variance. The high organic C content will also determine CEC and have a large impact on the pH buffering capacity (Giesler et al., 2005).

Liming acid soils is an effective and dominant practice to improve the mineralization rate of SOM. When soil pH is lower than 5.5, the microbial activities of nitrifying bacteria are lower and the SOM mineralization rate is reduced. In dry bean grown in an NT system in Brazilian Oxisol, the author detected N deficiency 2 weeks after germination in plots that did not receive lime (pH 5.2), but in plots that received lime (pH 6), N deficiency was not observed. In this case, it is hypothesized that OM mineralization liberated sufficient N in the plots that received lime. According to Andrew (1978), the H^+ ion is particularly important in legumes grown without fertilizer N. It affects rhizobial survival and multiplication in soils, root infection and nodule initiation, legume rhizobial efficiency, and nutrition of the host plant (Foy, 1984). Franco and Munns (1982) reported that decreasing the pH of a nutrient solution from 5.5 to 5.0 decreased the number of nodules formed by dry bean. Ve et al. (2004) also reported that the degree of humification was positively associated with exchangeable Ca^{2+} and Mg^{2+} (main component of dolomitic lime), suggesting a positive influence of these basic cations on the turnover of young OM.

Ammonification and nitrification are two important reactions in soils affecting N availability. Ammonification is the biological process leading to ammoniacal nitrogen formation from nitrogen-containing organic compounds, whereas nitrification is a biological oxidation of ammonium to nitrite and nitrate. Ammonification can occur over a wide range of soil pH, but nitrification is markedly reduced at pH values <6.0 and >8.0 (Alexander, 1980). Alexander (1980) reported that nitrification of organic materials stops at pH below 4.5. Morrill and Dawson (1967) reported that soil pH is the best indicator of nitrification in various types of soils. Haynes and Sherlock (1986) and Clough et al. (2004) reported that liming enhances nitrification and that cumulative N_2O emissions under field capacity conditions are reduced with liming. Clough et al. (2003, 2004), however, reported that under saturated soil conditions, the cumulative fluxes of N_2O and N_2 are much greater than under field capacity conditions and are enhanced following liming. Since pH has a potential effect on N_2O production pathways and the reduction of N_2O to N_2, it has been suggested that liming may provide an option for the mitigation of N_2O from soils (Stevens et al., 1998).

4.8.5 Use of Organic Manures

The use of organic manures and/or crop residues improves the OM content of soils (Rochette and Gregorich, 1998; Singh et al., 2004a). These management practices bring several favorable physical, chemical, and biological changes in the soil–plant systems and consequently improve the sustainability of cropping systems. The beneficial effects of these management practices are discussed in the succeeding sections.

4.8.6 Use of Cover Crops/Green Manuring

Cover crops are defined as the close-growing crops that provide soil protection, and soil improvement between periods of normal crop production, or between trees in orchards and vines in vineyards. When plowed under and incorporated into the soil, cover crops may be referred to as green manure crops (Soil Science Society of America, 2008). The positive role of cover crop/green manuring in crop production has been known since ancient times. The importance of this soil-ameliorating practice is increasing in recent years because of the high cost of chemical fertilizers, increased risk of environmental pollution, and need of sustainable cropping systems (Fageria, 2007). Cover crop/green manuring can improve the soil physical, chemical, and biological properties and, consequently, the crop yields. Cover crops or green manuring can increase cropping system sustainability by reducing soil erosion (MacRae and Mehuys, 1985; Smith et al., 1987), by increasing SOM and fertility levels (Cavigelli and Thien, 2003), and by reducing the global warming potential (Robertson et al., 2000). Cavigelli and Thien (2003) reported that incorporating green manure crops into the soil may increase P bioavailability for succeeding crops. Furthermore, the potential benefits of cover crop/green manuring are reduced NO_3^- leaching risk and lower fertilizer N requirements for succeeding crops (Fageria et al., 2005). However, its influence may vary from soil to soil, crop to crop, environmental variables, and the type of green manure crop used and its management. The beneficial effects of cover crop/green manuring in crop production should not be evaluated in isolation, however, in integration with chemical fertilizers (Fageria et al., 2005; Baligar and Fageria, 2007; Fageria, 2007) (Table 4.7).

Soil compaction is a major problem for crop production around the world. Deep-rooted cover crops are one possible solution to compaction problems, especially in no-till farming systems (Unger and Kaspar, 1994; Williams and Weil, 2004). The deep-growing tap roots of the perennial alfalfa (*Medicago sativa* L.) can increase the infiltration rate on compacted no-till soils (Meek et al., 1990), and the recolonization of root channels left by alfalfa has been shown to benefit the corn (*Zea mays* L.) root systems that follow (Rasse and Smucker, 1998). Similarly, Williams and Weil (2004) reported that soybean (*Glycine max.* L. Merr.) roots were observed to take advantage of the root channels left by the decomposition of cover crop roots of cereal rye (*Secale cereale* L.) and forage radish (*Raphanus sativus* L. Diachon).

TABLE 4.7

Response of Upland Rice and Common Bean to Chemical Fertilization and Green Manuring Grown in Rotation on a Brazilian Oxisol

Fertility Level[a]	First Upland Rice Crop[b]	First Bean Crop	Second Upland Rice Crop[b]	Second Bean Crop	Third Upland Rice Crop	Third Bean Crop
			Grain Yield (kg ha^{-1})			
Low	2188a	1935b	2383a	866c	480c	890c
Medium	2428a	2382a	2795a	1831ab	1127b	1242ab
High	2330a	2568a	2657a	2432a	1324b	1486a
Medium + green manure	—	2344a	—	1202bc	2403a	1065bc

Source: Adapted from Fageria, N. K. and N. P. Souza. 1995. *Pesq. Agropec. Bras.* 30:359–368. With permission.

Note: Values followed by the same letter in the same column are statistically not different by Tukey's test at the 5% probability level.

[a] Soil fertility levels for rice were low (without addition of fertilizers); medium (50 kg N ha^{-1}, 26 kg P ha^{-1}, 33 kg K ha^{-1}, and 30 kg ha^{-1} fritted glass material as a source of micronutrients); and high (all the nutrients were applied at double the medium level). *Cajanus cajan* L. was used as a green manure at the rate of 25.6 t ha^{-1} green matter. For the common bean, the fertility levels were low (without addition of fertilizers); medium (35 kg N ha^{-1}, 44 kg P ha^{-1}, 42 kg K ha^{-1}, and 30 kg ha^{-1} fritted glass material as a source of micronutrients); and high (all the nutrients were applied at double the medium level).

[b] The first and third rice crop plots with a medium + green manure fertility level were planted with green manure and incorporated about 90 days after sowing (at flowering) and hence the grain yield was not presented.

4.8.7   Use of Farmyard Manures

When applied to the soil, manure is a valuable resource, as a fertilizer and soil amendment, in crop production (Campbell et al., 1986). When applied in excessive amounts, however, manure has a pollution potential to soil and water (McCalla and Norstadt, 1974). Of particular concern is N in its several forms. Nitrogen found in manure can be converted into nitrate, a form that is mobile in the soil and can be leached into groundwater. Nitrate–nitrogen, when present in excess of 10 mg kg^{-1} in drinking water, is regarded as a pollutant and a health hazard (Hendry et al., 1984). However, the appropriate amount and method of application brings more beneficial effects of farmyard manures in crop production without having detrimental effects on environmental pollution. Animal manures contain a considerable amount of N, P, and K and their disposal on land for crop production may be beneficial from the standpoint of nutrient recycling and reduced use of commercial fertilizers (Kuo and Baker, 1982). Aoyama et al. (1999) reported that manure application contributes to the accumulation of macroaggregate-protected C and N. Jenkinson et al. (1994) reported that the long-term application of farmyard manure increases SOC considerably as was found in the classical experiments over 150 years at Rothamsted in the United Kingdom. A microbial biomass is the biomass of living and dead biomass of soil bacteria and fungi were also doubled by farmyard manure application in the Rothamsted field trials (Jenkinson and Rayner, 1977; Mengel et al., 2001). Yang and Janssen (1997) reported that straw and farmyard manure are important organic inputs in northern China to maintain or even increase the SOM content.

Manure from cattle and other livestock supplies an estimated 30% of the N needs for crop production in the northeastern United States and is an important source of N in other livestock-intensive regions (Bandel and Fox, 1984). In Asian countries such as India, Pakistan, Bangladesh, and Sri Lanka, the Philippines and China farmyard manures are the major source of nutrient supply on small farm holdings (Fageria and Gheyi, 1999). Proper and efficient management of manure on cropland is important for improving the economics of crop production and for minimizing adverse impacts on water quality (Jokela, 1992).

Manure has long been considered a desirable soil amendment and reports of its effects on soil properties are numerous (Sommerfeldt and Chang, 1985; Campbell et al., 1986; Tester, 1990). Manure increases the availability, persistence, and movement of phosphorus (Abbott and Tucker, 1973). Abbott and Tucker (1973) found that P concentration was higher under manured cotton than under either the control or inorganic nitrogen or P treatments. Manure applied in 1965 and 1967 increased by 35% the P uptake of alfalfa over that of control in 1970. Hannapel et al. (1964) found that OM treatments increased the movement of P in calcareous soils.

Exchangeable K builds up in soil when large amounts of manure are applied (Mathers and Stewart, 1974). Pratt and Laag (1977) reported that manure containing 1.6–2.2% N appeared to mineralize at the rate of 40–50% during the first year after application, 10–20% during the second year, and 5% during the third year in the irrigated soils of southern California. Meek et al. (1982) reported that the application of manure resulted in an increase of OM, increase of K and P availability in a silty clay soil, and an N mineralization rate of about 5% after the first year. The prerequisite of the use of farmyard manure is that it should be well decomposed to bring beneficial effects for plant growth. An undecomposed manure may have a high C/N ratio, may liberate some harmful products during decomposition, and may consequently bring harmful effects for plant growth.

4.8.8   Use of Municipality Compost

Municipality compost is a good source of OM, if it is adequately transformed into compost material. Composting manure produces a stabilized product that can be stored or spread with little odor or fly-breeding potential (Fageria, 2002). Brady and Weil (2002) defined composting as the practice of creating humus-like organic materials outside the soil by mixing, piling, or otherwise storing organic materials under conditions conducive to aerobic decomposition and nutrient conservation.

The other advantages of composting include killing pathogens and most weed seeds, and improving the handling characteristics of manure by reducing the volume and weight. The disadvantages of composting include nutrient loss, specifically N, and requirements for time, money, equipment, and labor. Eghball et al. (1997) found that as much as 40% of the total beef feedlot manure N can be lost during composting, and significant losses of K and Na (>6.5% of total K and Na) occur in a runoff from composting windrows not protected from rainfall. Overall, the benefits of composting, however, have many disadvantages. Besides, intensively cropped systems may also slowly increase the SOM content, thereby improving the long-term plant environment (Bowman et al., 1999).

4.8.9 RECYCLING CROP RESIDUES

Crop residues are portion of plants remaining after seed harvest, refers mainly to grain crop residue, such as corn stover, or of small grain straw and stubble (Crop Science Society of America, 1992). Crop residue management affects the biological and chemical processes that govern the conversion of C and N into SOM and the residual availability of N into succeeding crops (Bird et al., 2003). In general, the incorporation of plant residues in soil can affect the soil microclimate and increase the plant residue contact with soil. This will increase the residue decomposition and OM transformation (Beare et al., 1992; Cambardella and Elloitt, 1993). Bird et al. (2001, 2002) reported that sustained increase in soil microbial biomass, C, N, and SOM pools after four seasons of straw incorporation compared with straw burned. Crop residues are important in the formation of SOM (Wilhelm et al., 2004).

Recycling or incorporating crop residues into the soil after harvest of grain is an important management practice to improve organic carbon. Crop residues not only add organic carbon but also supply essential nutrients depending on crop species or genotypes within species. Carter et al. (1998) and Kay (1998) reported that the addition of OM to the soil in the form of crop residues increases the level of low-density macroorganic matter, which can represent up to 45% of total SOM. This form of SOM functions in improving the mechanical properties of soil (Carter, 2002). Kern and Johnson (1993) and Paustian et al. (2000) reported that agricultural management practices, such as crop residue incorporation, are being promoted to increase biomass incorporation into SOC pools, enhance soil quality, and sequester atmospheric CO_2. The frequent incorporation of straw into soils, particularly when combined with mineral nitrogen fertilizer, results in an increase of SOM (Amberger and Schweiger, 1991). Johnson et al. (2004) reported that crop residue is an important source of new C for building and maintaining SOM.

Singh et al. (2004b) reported that burning of rice straw is cost-effective and the predominant method of disposal is in areas under combined harvesting in the Indo-Gangetic Plains (IGP) of India. Burning of straw can accelerate losses of SOM and nutrients, increasing C emissions, causing intense air pollution, and reducing soil microbial activity (Biederbeck et al., 1980; Rasmussen et al., 1980; Kumar and Goh, 2000). Singh et al. (2004b) reported that 113.6 million Mg of rice and wheat residues, containing about 1.90 million Mg of nutrients, are available in the IGP of India. In the Punjab state of India alone, about 12 million Mg of rice straw are burned annually, which causes about 0.7 million Mg of N loss. The gas emissions from burning of rice straw are 70% CO_2, 7% CO, 0.66% CH_4, and 2.09% N_2O (Singh et al., 2004b). These figures show that burning crop straw not only reduces the soil quality (SOM) but is also responsible for environmental pollution. Hence, it should be incorporated into the soil rather than burned.

Mann et al. (2002), Wilhelm et al. (2004), and Hooker et al. (2005) reported that the use of corn stover as a source of ethanol for fuel in the United States may negatively impact agricultural sustainability if SOC levels and soil quality decline as a result. Wilhelm et al. (2004) reported that crop residue as a feedstock for biomass ethanol production is an appropriate solution for fuel production and reducing greenhouse gases in the atmosphere. However, residues are necessary to protect the soil from erosion and contribute to SOC levels, a key factor in most desirable characteristics of soil quality, and are positively related to soil and crop productivity. Supply of

SOC by crop residues depends on crop species. For example, corn (*Zea mays* L.) residues can provide about 1.7 times more C than barley (*Hordeum vulgare* L.), oat (*Avena sativa* L.), sorghum (*Sorghum bicolor* L. Moench), soybean (*Glycine max.* L. Merr.), sunflower (*Helianthus annuus* L.), and wheat (*Triticum aestivum* L.) residues based on production levels (Allmaras et al., 2000; Wilhelm et al., 2004). Crop residues on the soil surface protect OM-rich topsoil (Gilley et al., 1986; Gregorich et al., 1998; Nelson, 2002).

4.8.10 Keeping Land under Pasture

The lower total soil C and N observed in cultivated cropland soils compared with native rangeland soils have been well documented (Bauer and Black, 1981; Woods and Schuman, 1988; Bronson et al., 2004). Grasses in pasture are generally considered as maintaining or increasing SOM contents (Dalal et al., 1995). Gebhart et al. (1994) reported that the conversion of cropland into perennial grass cover through the Conservation Reserve Program in the United States increased SOC in the Great Plains and potentially sequestrated Cerri et al. (1991) and Feigl et al. (1995) observed increased C in surface soil after 1 and 2 years of pasture use in central Amazon of Brazil. Similarly, increase in C and N has been reported in surface soils of eastern Amazon of Brazil after several years of pasture (Feigl et al., 1995). Webb et al. (2003) also reported that SOC content increases in a quadratic manner in many soils under grasslands. Wilcke and Lilienfein (2004) reported that the pasture replaced the original C faster and to a greater depth in the Cerredo soils of Brazil than the *Pinus* plantation. These authors explained that this could be attributed to the differences in root architecture and the amount of root litter. Furthermore, lime and fertilizer applications to the pastures increase the biomass production and thus the root litter input into the soil.

When animals are fed on pastureland, a large proportion of the nutrients ingested by animals is returned to the soil in the form of urine and feces (West et al., 1989). Animals retain only a small proportion, about 20%, of the nutrients they ingest, and the rest is returned to the soil through excreta (Rao et al., 1992). The expected buildup in soil fertility in a grass–legume pasture under grazing could result from a more rapid cycling and greater proportion of nutrients in a plant-available form. The appropriate management of pastureland improves the soil biological activity and reduces soil erosion. The increased biological activity is beneficial to the soil properties such as mineralization, humification, texture, porosity, water infiltration, and retention. Rao et al. (1992) reported that the contribution of legume residues to SOM quality and turnover, together with improved soil fertility, soil structure, and biological activity, was associated with a 1 t ha^{-1} yield increase in a rice crop following 10-year-old grass + legume plots that did not require any N fertilizer when compared with rice following a grass-alone pasture of the same age.

4.9 OM VERSUS ENVIRONMENT

Maintaining an adequate level of OM in the arable lands is crucial to keep the soil quality high, which is responsible for improved environmental quality. OM reduces soil erodibility, which is one of the main factors in soil degradation and contamination of water bodies by pesticides and other chemicals. OM also functions as a filter to retain heavy metals and nutrients, which reduces environmental contamination.

4.9.1 Reduction in Nutrient/Heavy Metals Leaching

The leaching of nutrients or heavy metals in groundwater is a serious environmental problem in modern high-input agriculture. Sediment, N, and P in a runoff are the major sources of nonpoint-source pollution (Blanco-Canqui et al., 2004). Several studies have shown that fine soil fractions are often preferentially transported to surface water through a runoff, and nutrients and toxic heavy metals or pesticides attached in the fine fractions are discharged along with the runoff (Ghadiri and Rose, 1991; Uusitalo et al., 2001; Zhang et al., 2003a). Despite the current use of soil and water conservation practices, losses of sediment, N, and P from rural lands remain high (USEPA, 1996). Annual sediment loss in the United States exceeds 10 billion metric tons and costs the society 44 billion US\$ of degraded water resources (USEPA, 1996; Blanco-Canqui et al., 2004). Similarly, heavy metals also contaminate vast land areas. Heavy metal inputs included those from commercial fertilizers, liming materials and agrochemicals, sewage sludges, and other wastes used as soil amendments, irrigation waters, and atmospheric deposition (Senesi et al., 1999; He et al., 2004). Soils receiving repeated applications of organic manures, fungicides, and pesticides have exhibited high concentrations of extractable heavy metals (Sims and Wolf, 1994; Han et al., 2000) and subsequently resulted in increased heavy metal concentrations in the runoff (Moore et al., 1998). The mobility of heavy metals depends not only on the total concentration in the soil but also on soil properties, metal properties, and environmental factors (He et al., 2004). Dowdy and Volk (1983) reported that the movement of heavy metals in soils could occur in sandy, acidic, and low OM soil subject to heavy rainfall or irrigation.

Yamaguchi et al. (2004) reported that HA is a representative ligand that controls the mobility and fate of metal species in soils and aquatic systems. Li et al. (2003) reported that the amphipathic nature of HAs enables them to interact with a wide variety of inorganic and organic pollutants, including heavy metals and charged organic pollutants via chemical bonding and less polar organic pollutants through nonspecific physical interactions. Owing to their ubiquity in surface aquatic and groundwater systems, HAs often play important roles in environmental processes governing the fate and transport of organic and inorganic pollutants in soil–plant systems (Weber, 1988; Bartschat et al., 1992; Stevenson, 1994; Li et al., 2003). OM increases the CEC of soils, which may be responsible for reducing the leaching of essential plant nutrients and heavy metals in the soil profile. In addition, OM constituents carry a negative charge and are hence able to adsorb heavy metal cations (Mengel et al., 2001).

4.9.2 REDUCTION IN SOIL EROSION

Soil erosion has been defined as the process of detachment and transportation of soil material by erosive agents (Ellison, 1947). Soil detachment is the subprocess of dislodgment of soil particles from the soil mass at a particular location on the soil surface (Zhang et al., 2003b). Soil erosion by wind and water is a worldwide concern in reducing soil quality, maintaining farming systems sustainability, and increasing environmental pollution (Fageria, 2002). Soil aggregates low in OM and clay contents are generally susceptible to disintegration at low rainfall energies and subject to erosion (Rhoton et al., 2003). Several million hectares of soils in the United States are subject to excessive runoff and erosion losses because of the low levels of OM and clay content (Rhoton and Tayler, 1990; Rhoton et al., 2003). Soil erosion from irrigation, especially furrow irrigation, contributes to nonpoint-source pollution (Lentz et al., 1996) and is a serious threat to crop productivity in many regions (Carter, 1993).

Accelerated soil erosion also removes the layer of topsoil that is richest in OM (Hillel and Rosenzweig, 2002). Bauer and Black (1994) reported that soil erosion in the northern Great Plains of the United States is deemed to diminish soil productivity through the concomitant diminution of SOM content. Bauer and Black (1994) reported that loss of soil productivity resulting from a decline in SOM content associated with soil loss by erosion in the northern Great Plains of the United States is a consequence of the concomitant loss of fertility. Carpenter et al. (1998) reported that P is primarily transported through surface runoff from agricultural lands and was a major cause of eutrophication in surface waters. Pennock and Van Kessel (1997) assessed soil redistribution in landscapes of southern Saskatchewan with different cultivation histories and showed that SOM was the major soil quality indicator influenced by erosion. Gregorich et al. (1998) reviewed the relation between soil erosion and deposition processes and distribution and loss of SOM using the century model.

4.9.3 Mitigation of CO_2 Release in the Atmosphere

The atmospheric CO_2 concentration has gained much attention in recent years due to its potential contribution to global warming (Robinson et al., 1996). Soil is an important sink of CO_2 released to the atmosphere. Many studies have shown that soil CO_2 originates primarily from microbial oxidation of OM and root respiration (Hanson et al., 2000; Pumpanen et al., 2003). Approximately twice as much C is stored in the soil as in the atmosphere (Lal et al., 1995; Qualls et al., 2003). Brady and Weil (2002) reported that OM in the world's soils contains about 3 times as much carbon as is found in all worlds' vegetation. SOM plays a critical role in the global carbon balance that is thought to be the major factor affecting global warming (Brady and Weil, 2002). Similarly, Qualls et al. (2003) reported that the factors that control the storage of C in the soil are among those that play an important role in regulating the concentration of CO_2 in the atmosphere. Qualls (2004) reported that humic substances are inherently difficult for microbes to mineralize, and this property can contribute to the sequestration of C in soil. Schlesinger (1990) and Post and Kwon (2000) reported that currently there has also been an added interest in the role of SOM as a potential sink for atmospheric CO_2. Hence, the OM content of the soil can be considered as one of the vital components of CO_2 mitigation in the atmosphere.

The types of crop residues play important roles in C sequestration and soil aggregation because of the C:N ratios or quality of the residues (Potter et al., 1998; Wright and Hons, 2004). The degradation of fresh crop residues is often governed by C:N ratios (Oades, 1988; Chesire and Chapman, 1996), but as the residue undergoes decomposition, it becomes more recalcitrant and degradation 10 is controlled by the lignin content or lignin/N ratios (Tian et al. 1992). Hence, the ability of soils to sequestration is closely related to N (Wright and Hons, 2004).

4.9.4 Adsorption of Herbicides

Herbicides are commonly used in modern agriculture to control weeds. If agricultural lands are having low OM content, applied herbicides may leach easily and contaminate the surface or groundwater. For example, atrazine (2-chloro-4-ethylamino-6-isopropylaminos-triazine) is a common herbicide applied to agricultural lands worldwide to control weeds (Ben-Hur et al., 2003). Atrazine is one of the most widely detected herbicides in surface and groundwater in the United States (Poinke et al., 1988; Ben-Hur et al., 2003). SOM can be divided into solid- and water-dissolved fractions, both of which can associate with herbicides (Ben-Hur et al., 2003). Adsorption of herbicide on SOM should decrease its transport in the soil profile (Moorman et al., 2001). Ben-Hur et al. (2003) reported that the atrazine-dissolved OM complex decreases the mobility of atrazine if the lower horizons are lower in OM than the upper horizons (the more typical field case) because of the dissolved OM adsorption on the solids.

4.10 SOM VERSUS CROP YIELDS

The importance of soil physical, chemical, and biological properties in crop production should be evaluated by their effect or relationship with crop yield. If the association is positive, that means it is highly important in improving or maintaining crop productivity. Some studies have related SOM and SOM-associated parameters with crop productivity (Lucas and Weil, 2012). Positive correlations between SOM content and crop yields have been observed (Kravchenko and Bullock, 2000; Majchrzak et al., 2001; Alvarez et al., 2002). Stine and Weil (2002) found SOM, macroaggregate stability, and soil porosity to be related to crop productivity under different tillage regimes.

Alvarez et al. (2002) reported light-fraction OM and mineralizable N to be related to wheat yield variability. Other studies found that gains in SOM coincide with crop yield gains (Bauer and Black, 1994) and that losses in SOM coincided with crop yield losses (Diaz-Zorita et al., 1999). When crop residues were returned to the surface of degraded soils, Bruce et al. (1995) found the restoration of soil productivity to coincide with an increase in SOM. Lal (2006) found that increasing total SOC by 1 Mg ha[-1] translated to yield increases of 20–70 kg ha^{-1} for wheat, 10 –50 kg ha^{-1} for rice, and 30–300 kg ha^{-1} for corn. Weil and Magdoff (2004) successfully isolated SOM effects on crop yields; hence crop rotation was imposed to alter SOM for 20 years, and then all plots were treated alike for 2 years. They found SOM levels to account for 82–84% of the variation in corn yields during these 2 years.

Strickling (1975) suggested that the influence of SOM on soil aggregation enhanced water infiltration and, consequently, crop yields. Weil and Magdoff (2004) reported that SOM is a key determinant of soil quality because it influences nutrient holding and exchange, soil structure, erosion, resistance, and biological processes such as N mineralization. All these soil properties are associated with crop yield improvement (Fageria and Gheyi, 1999). Brady and Weil (2002) reported that small quantities of both fulvic and HAs in soil solution are known to enhance certain aspects of plant growth. Components of these humic substances probably act as regulators of specific plant growth functions, such as cell elongation and lateral root initiation (Brady and Weil, 2002).

4.11 CONCLUSIONS

A major part of nitrogen is taken up from the organic substances of the soil during crop growth. The SOM content of the soils is also related to crop productivity. Hence, a discussion on OM management in crop production is very pertinent here. SOM is a heterogeneous and dynamic soil component that varies in molecular structure, decomposition rate, and turnover time and exerts a major influence on soil quality and the global C cycle. Its role in improving the crop productivity and sustainability of agricultural systems is enormously high. OM modifies the soil physical, chemical, and biological properties in favor of better soil quality and, consequently, higher crop yields. A substantial amount of N requirements of the plants is satisfied from the mineralization of SOM. The pool sizes of SOM are soil specific, while their mineralization rate constants vary with environmental conditions. Plant and animal residues are the major sources of OM formation in the soil. SOM formation and accumulation is, however, highly dependent on management practices and the amount and placement of organic materials. For example, improving SOM with CT in crop production is difficult due to the rapid decomposition of organic materials by microbial action and loss of soil carbon, especially under tropical conditions. However, soil characteristics are improved with the addition of organic materials to the soil, which are in favor of higher crop yields. Adopting appropriate soil and crop management practices can be helpful in improving and/or stabilizing SOM of soils. These practices are the use of appropriate crop rotation, adoption of conservation or minimum tillage, and application of farmyard and compost manures and liming acid soils. In addition, the use of an adequate rate of fertilizers can also improve the OM content of soils by increasing the shoot and root biomass. SOM is a storehouse for slow release of essential plant nutrients that also reduces their leaching to groundwater.

Heavy metals are important environmental pollutants threatening the health of man, animal, and agroecosystems. The fate of heavy metals in the soil–plant systems is largely controlled by sorption reactions with soil colloids. The SOM owing primarily to higher CEC and to form inner-sphere complexes through surface reactions groups, important sorbent of heavy metals. Thus, decreasing their toxicity to crop plants in heavy metal contaminated soils as well as inhibited their leaching to groundwater. SOM also adsorbs herbicides and prevents their leaching to groundwater. In addition, SOM stores a large amount of C in the soil and avoids CO_2 escape into the atmosphere. Hence, OM plays an important role in reducing global warming or in greenhouse effects and environmental pollution.

REFERENCES

Abbott, J. L. and T. C. Tucker. 1973. Persistence of manure phosphorus in calcareous soil. *Soil Sci. Soc. Am. Proc.* 37:60–63.

Acosta-Martinez, V., T. M. Zobeck, and V. Allen. 2004. Soil microbial, chemical and physical properties in continuous cotton and integrated crop–livestock systems. *Soil Sci. Soc. Am. J.* 68:1875–1884.

Acosta-Martinez, V., T. M. Zobeck, T. E. Gill, and A. C. Kennedy. 2003. Enzyme activities and microbial community structure in semiarid agricultural soils. *Biol. Fert. Soils* 38:216–227.

Adams, F. and B. L. Moore. 1983. Chemical factors affecting root growth in subsoil horizons of coastal plain soils. *Soil Sci. Soc. Am. J.* 47:99–102.

Ahmad, F. and K. H. Tan. 1986. Effect of lime and organic matter on soybean seedlings grown in aluminum-toxic soils. *Soil Sci. Soc. Am. J.* 50:656–661.

Aitken, R. L., P. W. Moody, and P. G. McKinley. 1990. Lime requirement of acidic Queensland soils. I. Relationships between soil properties and pH buffer capacity. *Aust. J. Soil Res.* 28:695–701.

Alexander, M. 1980. Effects of acidity on microorganisms and microbial processes in soils. In: *Effects of Acid Precipitation of Terrestrial Ecosystems*, eds., T. Hutchinson, and M. Havas, pp. 363–364. New York: Plenum Publishing Corp.

Allison, L. E. 1965. Organic carbon. In: *Methods of Soil Analysis*, Part 2, ed., C. A. Black, pp. 1367–1378. Madison, Wisconsin: ASA.

Allison, F. E. 1973. *Soil Organic Matter and Its Role in Crop Production*. Amsterdam: Elsevier.

Almendras, A. S. and P. J. Bottomley. 1987. Influence of lime and phosphate on nodulation of soil-grown *Trifolium subterraneum* L. by indigenous *Rhizobium trifolii*. *Appl. Environ. Microbiol.* 53:2090–2097.

Allison, L. E. 1965. Organic carbon. In: *Methods of Soil Analysis*, Part 2, ed., C. A. Black, pp. 1367–1378. Madison, Wisconsin: ASA.

Allmaras, R. R., H. H Schomberg, C. L. Jr. Douglas, and T. H. Dao. 2000. Soil organic carbon sequestration potential of adopting conservation tillage in U.S. croplands. *J. Soil Water Cons.* 55:365–373.

Alvarez, R., C. R. Alvarez, and H. S. Steinbach. 2002. Association between soil organic matter and wheat yield in humid Pampas of Argentina. *Commun. Soil Sci. Plant Anal.* 33:749–757.

Amberger, A. and P. Schweiger. 1991. Effect of straw incorporation combined with an application of Ca cyanamide in long-term field trials. *Zeitschrift Fur Acker und Pflanzenbau* 134:323–334.

Amezketa, E. 1999. Soil aggregate stability: A review. *J Sustain. Agric.* 14:83–151.

Anderson, D. W., S. Saggar, J. R. Bettany, and J. W. Stewart. 1981. Particle size fractions and their use in studies of soil organic matter: I. The nature and distribution of forms of carbon, nitrogen, and sulfur. *Soil Sci. Soc. Am. J.* 45:767–772.

Andrew, C. S. 1978. Mineral characterization of tropical forage legumes. In: *Mineral Nutrition of Legumes in Tropical and Subtropical Soils*, eds., C. S. Andrew and E. J. Kamprath, pp. 93–112. East Melbourne, Australia: CSIOR.

Angers, D. A., A. N'Dayegamiye, and D. Cote. 1993. Tillage-induced differences in organic matter of particle size fractions and microbial biomass. *Soil Sci. Soc. Am. J.* 57:512–516.

Angers, D. A., A. Pesant, and J. Vigneux. 1992. Early cropping induced changes in soil aggregation, organic matter, and microbial biomass. *Soil Sci. Soc. Am. J.* 56:115–119.

Aoyama, M., D. A. Angers, and A. N'Dayegamiye. 1999. Particulate and mineral associated organic matter in water-stable aggregates as affected by mineral fertilizer and manure applications. *Can. J. Plant Sci.* 79:295–302.

Aulakh, M. S., J. W. Doran, and A. R. Mosier. 1992. Soil denitrification—Significance, measurement, and effects of management. *Adv. Soil Sci.* 18:1–57.

Aulakh, M. S., J. W. Doran, D. T. Walters, A. R. Mosier, and D. D. Francis. 1991. Crop residue type and placement effects on denitrification and mineralization. *Soil Sci. Soc. Am. J.* 55:1020–1025.

Aulakh, M. S., D. A. Rennie, and E. A. Paul. 1983. Field studies on gaseous nitrogen losses from soils under continuous wheat versus a wheat-fallow rotation. *Plant Soil* 75:15–27.

Aulakh, M. S., D. A. Rennie, and E. A. Paul. 1984. Gaseous nitrogen losses from soils under zero-till as compared with conventional-till management systems. *J. Environ. Qual.* 13:130–136.

Aust, W. M. and R. Lea. 1991. Soil temperature and organic matter in a disturbed–forested wetland. *Soil Sci. Soc. Am. J.* 55:1741–1746.

Bailey, J. S., R. J. Stevens, and D. J. Kilpatrick. 1991. A rapid method for predicting the time requirement of acidic temperate soils with widely varying organic matter contents. In: *Plant–Soil Interactions at Low pH*, eds., R. J. Wright, V. C. Baligar, and R. P. Murrmann, pp. 253–282. Dordrecht: Kluwer Academic Publishers.

Baldock, J. A. and K. Broos. 2012. Soil organic matter. In: *Handbook of Soil Sciences: Properties and Processes*, 2nd edition, eds., P. M. Huang, Y. Li, and M. E. Sumner, pp. 11-1–11-51. Boca Raton, Florida: CRC Press.

Balesdent, J., C. Chenu, and M. Balabane. 2000. Relationship of soil organic matter dynamics to physical protection and tillage. *Soil Tillage Res.* 53:215–230.

Baldock, J. A. and P. N. Nelson. 2000. Soil organic matter. In: *Handbook of Soil Science*, ed., M. E. Sumner, pp. B25–B84. Boca Raton, Florida: CRC Press.

Baligar, V. C. and N. K. Fageria. 2007. Agronomy and physiology of tropical cover crops. *J. Plant Nutr.* 30:1287–1339.

Baligar, V. C., N. K. Fageria, and M. A. Elrashidi. 1998. Toxicity and nutrient constraints on root growth. *HortScience* 33:960–965.

Bandel, V. A. and R. H. Fox. 1984. Management of nitrogen in New England and Middle Atlantic States. In: *Nitrogen in Crop Production*, ed., R. D. Hauck, pp. 677–689. Madison, Wisconsin: ASA, CSSA and SSSA.

Bartschat, B. M., S. E. Cabaniss, and F. M. M. Morel. 1992. Oligoelectrolyte model for cation binding by humic substances. *Environ. Sci. Technol.* 26:284–294.

Bauer, A. and A. L. Black. 1981. Soil carbon, nitrogen, and bulk density comparisons in two cropland tillage systems after 25 years and in virgin grassland. *Soil Sci. Soc. Am. J.* 45:1166–1170.

Bauer, A. and A. L. Black. 1992. Organic carbon effects on available water capacity of three soil textural groups. *Soil Sci. Soc. Am. J.* 56:248–254.

Bauer, A. and A. L. Black. 1994. Quantification of the effect of soil organic matter content on soil productivity. *Soil Sci. Soc. Am. J.* 58:15–193.

Beare, M. H. and R. R. Bruce. 1993. A comparison of methods for measuring water-stable aggregates: Implications for determining environmental effects on soil structure. *Geoderma* 56:87–104.

Beare, M. H., M. L. Cabrera, P. F. Hendrix, and D. C. Coleman. 1994. Aggregate protected and unprotected organic matter pools in conventional and no-tillage soils. *Soil Sci. Soc. Am. J.* 58:787–795.

Beare, M. H., R. W. Parmelee, P. F. Hendrix, W. Cheng, D. C. Coleman, and D. A. Crossley. 1992. Microbial and faunal interactions and effects on litter nitrogen and decomposition in agroecosystms. *Ecol. Monogr.* 62:569–591.

Benito, E. and F. Diaz-Fierros. 1992. Effects of cropping on the structural stability of soils rich in organic matter. *Soil Tillage Res.* 23:153–161.

Ben-Hur, M., J. Letey, W. J. Farmer, C. F. Williams, and S. D. Nelson. 2003. Soluble and solid organic matter effects on atrazine adsorption in cultivated soils. *Soil Sci. Soc. Am. J.* 67:1140–1146.

Bhogal, A., S. D. Young, R. Sylvester-Bradley, F. M. O'Donnell, and R. B. Ralph. 1997. Cumulative effects of nitrogen application to winter wheat at Ropsley, UK from 1978 to 1990. *J. Agric. Sci.* 129:1–12.

Biederbeck, V., C. A. Campbell, K. E. Bowren, M. Schnitzer, and R. N. McIver. 1980. Effect of burning cereal straw on soil properties and grain yields in Saskatchewan. *Soil Sci. Soc. Am. J.* 44:103–111.

Bird, J. A., W. R. Horwath, A. J. Eagle, and C. V. Kessel. 2001. Immobilization of fertilizer N in rice: Effects of straw management practices. *Soil Sci. Soc. Am. J.* 65:1143–1152.

Bird, J. A., C. V. Kessel, and W. R. Horwath. 2002. Nitrogen dynamics in humic fractions in temperate rice. *Soil Sci. Soc. Am. J.* 66:478–488.

Bird, J. A., C. V. Kessel, and W. R. Horwath. 2003. Stabilization of ^{13}C-carbon and immobilization of ^{15}N-nitrogen from rice straw in humic fractions. *Soil Sci. Soc. Am. J.* 67:806–816.

Biswas, T. D. and B. K. Khosla. 1971. Building up of organic matter status of soil and its relation to the soil physical properties. *Int. Symp. Soil Fertil. Eval. Proc.* 1:831–842.

Blanco-Canqui, H. B., C. J. Gantzer, S. H. Anderson, E. E. Alberts, and A. L. Thompson. 2004. Grass barrier and vegetative filter strip effectiveness in reducing runoff, sediment, nitrogen and phosphorus loss. *Soil Sci. Soc. Am. J.* 68:1670–1678.

Blanco-Canqui, H. and R. Lal. 2008. No-tillage and soil profile carbon sequestration: Na on farm assessment. *Soil Sci. Soc. Am. J.* 72:693–701.

Blanco-Canqui, H., M. M. Mikha, J. G. Benjamin, L. R. Stone, A. J. Schlegel, and D. J. Lyon. 2009. Regional study of no-till impacts on near surface aggregate properties that influence soil erodibility. *Soil Sci. Soc. Am. J.* 73:1361–1368.

Blevins, R. L. and W. W. Frye. 1993. Conservation tillage: An ecological approach to soil management. *Adv. Agron.* 51:33–78.

Bloom, P. R. 1981. Metal organic matter interactions in soil. In: *Chemistry in the Soil Environment*, ed., ASA, pp. 129–150. Madison, Wisconsin: ASA.

Bloom, P. R., M. B. McBridge, and R. M. Weaver. 1979a. Aluminum and organic matter in acid soils: Salt extractable aluminum. *Soil Sci. Soc. Am. J.* 43:813–815.

Bloom, P. R., M. B. McBridge, and R. M. Weaver. 1979b. Aluminum organic matter in acid soils: Buffering and solution aluminum activity. *Soil Sci. Soc. Am. J.* 43:488–493.

Blum, U., S. R. Shafer, and M. E. Lehman. 1999. Evidence for inhibitory allelopathic interactions involving phenolic acids in field soils: Concepts vs. an experimental model. *Crit. Rev. Plant Sci.* 18:673–693.

Bockheim, J. G., K. M. Hinkel, and F. E. Nelson. 2003. Predicting carbon storage in tundra soils of arctic Alaska. *Soil Sci. Soc. Am. J.* 67:948–950.

Borchers, J. G. and D. A. Perry. 1992. The influence of soil texture and aggregation on carbon and nitrogen dynamics in Southwest Oregon forests and clearcuts. *Can. J. For. Res.* 22:298–305.

Bornemisza, E. and R. Pineda. 1969. The amorphous minerals and the mineralization of nitrogen in volcanic ash soils. In: *Panel on Soil Derived from Volcanic Ash in Latin America*, pp. 7.1–B7.7. Turrialba, Costa Rica: IICA.

Boussingault, J. 1838. Recherches chimiques sur la végétation enterprises dans le but déxaminer si les plantes prennent de lazote de latmosphere. *Annales de Chimie et de Physique* 67:1–54.

Bowman, R. A., M. F. Vigil, D. C. Nielsen, and R. L. Anderson. 1999. Soil organic matter changes in intensively cropped dryland systems. *Soil Sci. Soc. Am. J.* 63:186–191.

Brady, N. C. and R. R. Weil. 2002. *The Nature and Properties of Soils*, 13th edition. Upper Saddle River, New Jersey: Prentice-Hall.

Bremer, E., H. H. Janzen, B. H. Ellert, and R. H. McKenzie. 2011. Carbon, nitrogen, and greenhouse gas balances in an 18-year cropping system study on the northern Great Plains. *Soil Sci. Soc. Am. J.* 75:1493–1502.

Bremner, J. M. and K. Shaw. 1958. Denitrification in soil. II. Factors affecting denitrification. *J. Agric. Sci.* 51:40–52.

Broadbent, F. E. 1953. The soil organic fraction. *Adv. Agron.* 5:153–183.

Brockwell, J. and P. J. Bottomley. 1995. Recent advances in inoculant technology and prospects for the future. *Soil Biol. Biochem.* 27:683–697.

Broersma, K. and L. M. Lavkulich. 1980. Organic matter distribution with particle-size in surface horizons of some sombric soils in Vancouver Island. *Can. J. Soil Sci.* 60:583–586.

Bronick, C. J. and R. Lai. 2005. Soil structure and management: A review. *Geoderma* 124:3–22.

Bronson, K. F., T. M. Zobeck, T. T. Chua, V. Acosta-Martinez, R. S. V. Pelt, and J. D. Booker. 2004. Carbon and nitrogen pools of southern High Plains cropland and grassland soils. *Soil Sci. Soc. Am. J.* 68:1695–1704.

Brown, S. R., R. Chaney, and J. S. Angle. 1997. Subsurface liming and metal movement in soils amended with lime-stabilized biosolids. *J. Environ. Qual.* 26:724–732.

Bruce, R. R., G. W. Langdale, L. T. West, and W. P. Miller. 1995. Surface soil degradation and soil productivity restoration and maintenance. *Soil Sci. Soc. Am. J.* 59:654–660.

Bullock, D. G. 1992. Crop rotation. *Crit. Rev. Plant Sci.* 11:309–326.

Burris, R. H. 1998. Discoveries in biological nitrogen fixation. In: *Discoveries in Plant Biology, Vol. 1*, eds., S. D. Kung, and S. F. Yang, pp. 257–278. London: World Scientific.

Burke, I. C., C. M. Yonker, W. J. Parton, C. V. Cole, K. Flach, and D. S. Schimel. 1989. Texture, climate and cultivated effects on soil organic matter contents in U. S. grassland soils. *Soil Sci. Soc. Am. J.* 53:800–805.

Burns, R. G. 1982. Enzymes activities in soil: Location and a possible role in microbial ecology. *Soil Biol. Biochem.* 14:423–427.

Cadisch, G. and K. E. Giller. 1997. *Driven by Nature: Plant Litter Quality and Decomposition*. Wallingford, UK: CAB.

Calderon, F. J., D. J. Schultz, and E. A. Paul. 2012. Carbon allocation, belowground transfers, and lipid turnover in plant–microbial association. *Soil Sci. Soc. Am. J.* 76:1614–1623.

Cambardella, C. A. and E. T. Elloitt. 1993. Carbon and nitrogen distribution in aggregates from cultivated and native grassland soils. *Soil Sci. Soc. Am. J.* 57:1071–1076.

Campbell, C. A. 1978. Soil organic carbon nitrogen and fertility. In: *Soil Organic Matter*, eds., M. Schnitzer, and S. U. Khan, pp. 173–271. New York: Elsevier.

Campbell, C. A., M. Schnitzer, J. W. B. Stewart, V. O. Biederbeck, and F. Selles. 1986. Effect of manure and P fertilizer on properties of a black Chernozem in southern Saskatchewan. *Can. J. Soil Sci.* 66:601–613.

Carpenter, S. R., N. F. Caraco, D. L. Correll, R. W. Howarth, A. N. Sharpley, and V. H. Smith. 1998. Nonpoint pollution of surface waters with phosphorus and nitrogen. *Appl. Soil Ecol.* 8:559–568.

Carter, D. L. 1993. Furrow irrigation erosion lowers soil productivity. *J. Irrigation Drain Eng. Am. Soc. Civil Eng.* 119:964–974.

Carter, M. R. 2002. Soil quality for sustainable land management: Organic matter and aggregation interactions that maintain soil functions. *Agron. J.* 94:38–47.

Carter, M. R., E. G. Gregorich, D. A. Angers, R. G. Donald, and M. A. Bolinder. 1998. Organic C and N storage, and organic C fractions, in adjacent cultivated and forested soils of eastern Canada. *Soil Tillage Res.* 47:253–261.

Carter, M. R. and B. A. Stewart. 1996. *Structure and Organic Matter Storage in Agriculture Soils*. Boca Raton, Florida: CRC Press.

Cartroux, G. and M. Schnitzer. 1987. Chemical, spectroscopic, and biological characteristics of the organic matter in particle size fractions separated from an Aquoll. *Soil Sci. Soc. Am. J.* 51:1200–1207.

Cavigelli, M. A. and S. T. Thien. 2003. Phosphorus bioavailability following incorporation of green manure crops. *Soil Sci. Soc. Am. J.* 67:1186–1194.

Cerri, C. C., B. Volkoff, and F. Andreaux. 1991. Nature and behaviour of organic matter in soils under natural forest, and after deforestation, burning and cultivation, near Manus. *Forest Ecol. Manag.* 38:247–257.

Chaney, K. and R. S. Swift. 1984. The influence of organic matter on aggregate stability in some British soils. *J. Soil Sci.* 35:223–230.

Cheng, H. H. and J. M. Kimble. 2001. Characterization of soil organic carbon pools. In: *Assessment Methods for Soil Carbon*, ed., R. Lal et al., pp. 117–129. Boca Raton, Florida: CRC Press.

Chesire, M. V. and S. J. Chapman. 1996. Influence of n and P status of plant material and of added N and P on the mineralization of C from ^{14}C-labelled ryegrass in soil. *Biol. Fert. Soils* 21:166–170.

Christianson, C. B. and C. M. Cho. 1983. Chemical denitrification of nitrite in frozen soils. *Soil Sci. Soc. Am. J.* 47:38–42.

Clark, J. D. and A. H. Johnson. 2011. Carbon and nitrogen accumulation in post-agricultural forest soils of western New England. *Soil Sci. Soc. Am. J.* 75:1530–1542.

Clark, J. S. and W. E. Nichol. 1966. The lime potential base saturation relations of acid surface horizons of mineral and organic soil. *Can. J. Soil Sci.* 46:281–285.

Clough, T. J., F. M. Kelliher, R. R. Sherlock, and C. D. Ford. 2004. Lime and soil moisture effects on nitrous oxide emissions from a urine patch. *Soil Sci. Soc. Am. J.* 68:1600–1609.

Clough, T. J., R. R. Sherlock, and F. M. Kelliher. 2003. Can liming mitigate N$_2$O fluxes from urine-amended soil? *Aust. J. Soil Res.* 41:439–457.

Cochrane, T. T., L. G. Sanchez, J. A. Porras, L. G. Azevedo, and C. L. Garver. 1985. *Land in Tropical America. Planaltina*. Brasilia: CIAT-EMBRAPA/CPAC.

Coleman, N. T. and G. W. Thomas. 1964. Buffer curves of acid clays as affected by the presence of ferric iron and aluminum. *Soil Sci. Soc. Am. Proc.* 28:187–190.

Crave, A. and C. Gascuel-Odoux. 1997. The influence of topography on time and space distribution of soil surface water content. *Hydrol. Process.* 11:203–210.

Crop Science Society of America. 1992. *Glossary of Crop Science Terms*. Madison, Wisconsin: Crop Science Society of America.

Dalal, R. C., R. C. Strong, E. J. Weston, J. E. Cooper, K. J. Lehane, A. J. King, and C. J. Chicken. 1995. Sustaining productivity of a Vertisol at Warra, Queensland, with fertilizers, no tillage, or legumes. I. Organic matter status. *Aust. J. Exp. Agric.* 35:903–913.

Dalton, J. D., G. C. Russel, and D. H. Sieling. 1952. Effect of organic matter on phosphate availability. *Soil Sci.* 73:173–181.

Darst, B. C. and L. S. Murphy. 1990. Organic matter: An integral ingredient in crop production. *Better Crops* 74:4–5.

David, M. B., M. J. Mitchell, and J. P. Nakas. 1982. Organic and inorganic sulfur constituents of a forest soil and their relationship to microbial activity. *Soil Sci. Soc. Am. J.* 46:847–852.

Denef, K., J. Six, R. Merckx, and K. Paustian. 2004. Carbon sequestration in microaggregates of no-tillage soils with different clay mineralogy. *Soil Sci. Soc. Am. J.* 68:1935–1944.

Deng, S. P. and M. A. Tabatabai. 1997. Effect of tillage and residue management on enzyme activities in soils. II. Glycosides. *Biol. Fert. Soils* 22:208–213.

Dent, F. J. 1980. Major production systems and soil-related constraints in southeast Asia. In: *Properties for Alleviating Soil Related Constraints to Food Production in the Tropics*, ed., IRRI, pp. 79–106. Los Baños, the Philippines: IRRI.

Diaz, O. A., D. L. Anderson, and E. A. Hanlon. 1993. Phosphorus mineralization from Histosols of the Everglades agricultural area. *Soil Sci.* 156:178–185.

Diaz-Zorita, M., D. F. Buschiazzo, and N. Peinemann. 1999. Soil organic matter and wheat productivity in the semiarid Argentina Pampas. *Agron. J.* 91:276–279.

Dodla, S. K., J. J. Wang, and R. L. Cook. 2012. Molecular composition of humic acids from coastal wetland soils along a salinity gradient. *Soil Sci. Soc. Am. J.* 76:1592–1605.

Doran, J. W. 1980. Soil microbial and biochemical changes associated with reduced tillage. *Soil Sci. Soc. Am. J.* 44:765–771.

Doran, J. W. 2002. Soil health and global sustainability: Translating science into practice. *Agric. Ecosyst. Environ.* 88:119–127.

Doran, J. W., E. T. Elliot, and K. Paustain. 1998. Soil microbial activity, nitrogen cycling, and long term changes in organic carbon pools as related to fallow tillage management. *Soil Tillage Res.* 49:3–18.

Doran, J. W., D. G. Fraser, M. N. Culick, and W. C. Liebhardt. 1987. Influence of alternative and conventional agricultural management on soil microbial processes and nitrogen availability. *Am. J. Alter. Agric.* 2:99–109.

Doran, J. W. and T. B. Parkin. 1994. Defining and assessing soil quality. In: *Defining Soil Quality for a Sustainable Environment*, eds., J. W. Doran, D. C. Coleman, D. F. Bezdicek, and B. A. Stewart, pp. 3–22. Madison, Wisconsin: SSSA.

Dowdy, R. H. and V. V. Volk. 1983. Movement of heavy metals in soils. In: *Chemical Mobility and Reactivity in Soil Systems*, eds. D. W. Nelsen, D. E. Elrick, and K. K. Tanji, pp. 229–240. Madison, Wisconsin: SSSA.

Driessen, P. M. 1978. Peat soils. In: *Soils and Rice*, ed., IRRI, pp. 763–779. Los Baños, the Philippines: IRRI.

Duiker, S. W., F. E. Rhoton, J. Torrent, N. E. Smeck, and R. Lal. 2003. Iron hydroxide crystallinity effects on soil aggregation. *Soil Sci. Soc. Am. J.* 67:606–611.

Duxbury, J. M. 1994. The significance of agricultural sources of greenhouse gases. *Nutr. Cycl. Agroecosyst.* 38:151–163.

Echeverria, M. E., D. Markewitz, L. A. Morris, and R. L. Hendrick. 2004. Soil organic matter fractions under managed pine plantations of the southeastern USA. *Soil Sci. Soc. Am. J.* 68:950–958.

Eghball, B., J. F. Power, J. E. Gilley, and J. W. Doran. 1997. Nutrient, carbon, and mass loss of beef cattle feedlot manure during composting. *J. Environ. Qual.* 26:189–193.

Ellerbrock, R. H., H. H. Gerek, J. Bachmann, and M. O. Goebel. 2005. Composition of organic matter fractions for explaining wettability of three forest soils. *Soil Sci. Soc. Am. J.* 69:57–66.

Elliott, E. T. 1986. Aggregate structure and carbon, nitrogen, and phosphorus in native and cultivated soils. *Soil Sci. Soc. Am. J.* 50:627–633.

Ellison, W. D. 1947. Soil erosion studies. Part I. *Agric. Eng.* 28:145–146.

Elrashidi, M. A. and G. A. O'Connor. 1982. Boron desorption and adsorption in soils. *Soil Sci. Soc. Am. J.* 46:27–31.

Emerson, W. W. 1977. Physical properties and structure. In: *Soil Factors in Crop Production in a Semi-Arid Environment*, eds., J. S. Russell, and E. L. Greacen, pp. 78–104. St. Lucia, Australia: University of Queensland Press.

Entry, J. A., C. C. Mitchell, and C. B. Backman. 1996. Influence of management practices on soil organic matter, microbial biomass and cotton yield in Alabamas old rotation. *Biol. Fert. Soil* 23:353–358.

Eusterhues, K., C. Rumpel, M. Kleber, and I. Kogel-Knabner. 2003. Stabilisation of soil organic matter by interactions with minerals as revealed by mineral dissolution and oxidative degradation. *Org. Geochem.* 34:1591–1600.

Evans, C. E. and E. J. Kamprath. 1970. Lime response as related to percent Al saturation, solution Al and organic matter content. *Soil Sci. Soc. Am. Proc.* 34:893–896.

Eynard, A., T. E. Schumacher, M. J. Lindstrom, and D. D. Malo. 2004. Porosity and pore-size distribution in cultivated ustolls and usterts. *Soil Sci. Soc. Am. J.* 68:1927–1934.

Fageria, N. K. 1989. *Tropical Soils and Physiological Aspects of Cro*ps. Brasilia: EMBRAPA.

Fageria, N. K. 2001. Response of upland rice, dry bean, corn, and soybean to base saturation in Cerredo soil. *Rev. Bras. Eng. Agríc. Amb.* 5:416–424.

Fageria, N. K. 2002. Soil quality versus environmentally based agricultural management practices. *Commun. Soil Sci. Plant Anal.* 33:2301–2329.

Fageria, N. K. 2007. Green manuring in crop production. *J. Plant Nutr.* 30:691–719.

Fageria, N. K. 2012. Role of soil organic matter in maintaining sustainability of cropping systems. *Commun. Soil Sci. Plant Anal.* 43:2063–2113.

Fageria, N. K. and V. C. Baligar. 2003a. Upland rice and allelopathy. *Commun. Soil Sci. Plant Anal.* 34:1311–1329.

Fageria, N. K. and V. C. Baligar. 2003b. Fertility management of tropical acid soils for sustainable crop production. In: *Handbook of Soil Acidity*, ed., Z. Rengel, pp. 359–385. New York: Marcel Dekker.

Fageria, N. K. and V. C. Baligar. 2005. Enhancing nitrogen use efficiency in crop plants. *Adv. Agron.* 88: 97–185.

Fageria, N. K., V. C. Baligar, and C. A. Jones. 2011. *Growth and Mineral Nutrition of Field Crops*, 3rd edition. Boca Raton, Florida: CRC Press.

Fageria, N. K., V. C. Baligar, and B. A. Bailey. 2005. Role of cover crops in improving soil and row crop productivity. *Commun. Soil Sci. Plant Anal.* 36:2733–2757.

Fageria, N. K. and F. Breseghello. 2004. Nutritional diagnostic in upland rice production in some municipalities of State of Mato Grosso, Brazil. *J. Plant Nutr.* 27:15–28.

Fageria, N. K. and H. R. Gheyi. 1999. *Efficient Crop Production*. Campina Grande, Brazil: Federal University of Paraiba.

Fageria, N. K., A. B. Santos, D. G. Lins, and S. L. Camargo. 1997. Characterization of fertility and particle size of varzea soils of Mato Grosso and Mato Grosso do Sul States of Brazil. *Commun. Soil Sci. Plant Anal.* 28:37–47.

Fageria, N. K. and N. P. Souza. 1995. Response of rice and common bean crops in succession to fertilization in Cerredo soil. *Pesq. Agropec. Bras.* 30:359–368.

Fageria, N. K., R. J. Wright, V. C. Baligar, and C. M. R. Sousa. 1991. Characterization of physical and chemical properties of varzea soils of Goias State of Brazil. *Commun. Soil Sci. Plant Anal.* 22:1631–1646.

Feigl, B. J., J. Melillo, and C. C. Cerri. 1995. Changes in the origin and quality of soil organic matter after pasture introduction in Rondonia, Brazil. *Plant Soil* 175:21–29.

Follett, R. F. 2001. Soil management concepts and carbon sequestration in cropland soils. *Soil Tillage Res.* 61:77–92.

Foy, C. D. 1964. Toxic factors in acid soils of the southern United States as related to the response of alfalfa to lime, USDA Research Report 80. U. S. Government Printing Office, Washington DC.

Foy, C. D. 1984. Physiological effects of hydrogen, aluminum, and manganese toxicities in acid soils. In: *Soil Acidity and Liming*, 2nd edition, ed., F. Adams, pp. 57–97. Madison, Wisconsin: ASA, CSSA, and SSSA.

Franco, A. A. and D. A. Munns. 1982. Acidity and aluminum restraints on nodulation, nitrogen fixation and growth of *Phaseolus vulgaris* in nutrient solution. *Soil Sci. Soc. Am. J.* 46:296–301.

Franzluebbers, A. J., F. M. Hons, and D. A. Zuberer. 1994. Long-term changes in soil carbon and nitrogen pools in wheat management systems. *Soil Sci. Soc. Am. J.* 58:1639–1645.

Franzluebbers, A. J., F. M. Hons, and D. A. Zuberer. 1995. Soil organic carbon, microbial biomass, and mineralizable carbon and nitrogen in sorghum. *Soil Sci. Soc. Am. J.* 59:460–466.

Franzluebbers, A. J., R. L. Haney, C. W. Honeycutt, M. A. Arshad, H. H. Schomberg, and F. M. Hons. 2001. Climatic influences on active fractions of soil organic matter. *Soil Biol. Biochem.* 33:1103–1111.

Franzluebbers, A. J. 2004. Tillage and residue management effects on soil organic matter. In: *Soil Organic Matter in Sustainable Agriculture*, eds., F. R. Magdoff, and R. R. Weil, pp. 227–268. Boca Raton, Florida: CRC Press.

Franzluebbers, A. J. 2005. Soil organic carbon sequestration and agricultural greenhouse gas emissions in the southern USA. *Soil Tillage Res.* 83:120–147.

Freney, J. R. 1986. Forms and reactions of organic sulfur compounds in soils. In: *Sulfur in Agriculture*, ed., M. A. Tabatabai, pp. 207–232. Madison, Wisconsin: ASA, CSSA and SSSA.

Frossard, E., B. Truong, and F. Jaquin. 1986. Effect of organic matter and adsorption and desorption of phosphorous in Oxisol. *Agronomie* 6:503–508.

Frye, W. W., S. A. Ebelhar, L. W. Murdock, and R. L. Blevins. 1982. Soil erosion effects on properties and productivity of two Kentucky soils. *Soil Sci. Soc. Am. J.* 46:1051–1055.

Gartzia-Bengoetxea, N., A. Gonzalez-Arias, A. Merino, and I. M. Arano. 2009. Soil organic matter in soil physical fractions in adjacent semi-natural and cultivated stands in temperate Atlantic forests. *Soil Biol. Biochem.* 41:1674–1683.

Gebhart, D. L., H. B. Johnson, H. S. Mayeux, and H. W. Polleuy. 1994. The CRP increases soil organic carbon. *J. Soil Water Cons.* 49:488–492.

Gerard, F., M. Tinsley, and K. U. Mayer. 2004. Preferential flow revealed by hydrologic modeling based on predicted hydraulic properties. *Soil Sci. Soc. Am. J.* 68:1526–1538.

Ghadiri, H. and C. W. Rose. 1991. Sorbed chemical transport in overland flow. I. A nutrient and pesticide enrichment mechanism. *J. Environ. Qual.* 20:628–633.

Ghidey, F. and E. E. Alberts. 1993. Residue type and placement effects on decomposition: Field study and model evaluation. *Trans. ASAE* 36:1611–1617.

Giesler, R., T. Anderson, L. Lovgren, and P. Persson. 2005. Phosphate sorption in aluminum and iron rich humus soils. *Soil Sci. Soc. Am. J.* 69:77–86.

Gilley, J. E., S. C. Finkner, R. G. Spomer, and L. N. Mielke. 1986. Runoff and erosion as affected by corn residues. Part I. Total losses. *Trans. ASAE* 29:157–160.

Goodman, B. A. and M. V. Cheshire. 1979. A Mossbauer spectroscopic study of the effect of pH on the reaction between iron and humic acid in aqueous media. *J. Soil Sci.* 30:85–91.

Grace, P. R., J. M. Oades, H. Keith, and T. W. Hancock. 1995. Trends in wheat yields and soil organic carbon in the permanent rotation trial in the Waite Agriculture Research Institute, South Australia. *Aust. J. Exp. Agric.* 35:857–864.

Grand, S. and L. M. Lavkulich. 2012. Effects of forest harvest on soil carbon and related variables in Canadian Spodosols. *Soil Sci. Soc. Am. J.* 76:1816–1827.

Gregorich, E. G., K. R. Greer, D. W. Anderson, and B. C. Liang. 1998. Carbon distribution and losses: Erosion and deposition effects. *Soil Tillage Res.* 47:291–302.

Gregorich, E. G. and H. H. Janzen. 1996. Storage of soil carbon in the light fraction and macroorganic matter. In: *Structure and Organic Matter Storage in Agricultural Soils*, eds., M. R. Carter, and D. A. Stewart, pp. 167–190. Boca Raton, Florida: CRC Press.

Gregorich, E. G., R. G. Kachanoski, and R. P. Voroney. 1989. Carbon mineralization in soil size fractions after various amounts of aggregate disruption. *J. Soil Sci.* 40:649–659.

Guimaraes, C. B. and L. P. Yokoyama. 1998. Upland rice in rotation with soybean. In: *Technology for Upland Rice*, eds., F. Breseghello, and L. F. Stone, pp. 19–24. Santo Antonio de Goias, Brazil: Embrapa Arroz e Feijão.

Gupta, U. C. 1968. Relationship of total and hot water soluble boron and fixation of added B to properties of podzol soils. *Soil Sci. Soc. Am. Proc.* 32:45–48.

Gupta, S. C., R. H. Dowdy, and W. E. Larson. 1977. Hydraulic and thermal properties of a sandy soil as influenced by incorporation of sewage sludge. *Soil Sci. Soc. Am. J.* 41:601–605.

Gupta, V. V. S. R. and J. J. Germida. 1988. Distribution of microbial biomass and its activity in different soil aggregate size classes as affected by cultivation. *Soil Biol. Biochem.* 20:777–786.

Halvorson, A. D., B. J. Wienhold, and A. L. Black. 2002. Tillage, nitrogen, and cropping system effects on soil carbon sequestration. *Soil Sci. Soc. Am. J.* 66:906–912.

Han, F. X., W. L. Kingery, H. M. Selim, and P. D. Derard. 2000. Accumulation of heavy metals in a long term poultry waste-amended soil. *Soil Sci.* 165:260–268.

Hannapel, R. J., W. H. Fuller, S. Basma, and J. S. Bullock. 1964. Phosphorus movement in a calcarious soil: I. Predominance of organic forms of phosphorus in phosphorus movement. *Soil Sci.* 9:350–357.

Hanson, P. J., N. T. Edwards, C. T. Garten, and J. A. Andrews. 2000. Separating root and soil microbial contributions to soil respiration: A review of methods and observations. *Biogeochemistry* 48:115–146.

Harborne, J. B. 1977. *Introduction to Ecological Biochemistry*. London: Academic Press.

Harrier, L. A. and Watson, C. A. 2003. The role of arbuscular mycorrhizal fungi in sustainable cropping systems. *Adv. Agron.* 20:185–225.

Hargrove, W. L. and G. W. Thomas. 1981. Effects of organic matter on exchangeable aluminum and plant growth in acid soils. In: *Chemistry in the Soil Environment*, ed., D. E. Baker, pp. 151–166. Madison, Wisconsin: ASA and SSSA.

Harter, R. D. 1969. Phosphorus adsorption sites in soils. *Soil Sci. Soc. Am. Proc.* 33:630–632.

Hassink, J. 1997. The capacity of soils to physically protect organic C and N. *Plant Soil* 191:77–87.

Hauck, R. D. 1981. Nitrogen fertilizer effects on nitrogen cycle processes. In: *Terrestrial N Cycle*, eds., F. E. Clark, and T. Rosswall, pp. 551–562. Stockholm: Ecolgical Bulletin, 33.

Hayes, M. H. B. and C. C. Edward. 2001. Humic substances: Considerations of compositions, aspects of structure and environmental influences. *Soil Sci.* 166:723–737.

Haynes, R. J. 2005. Labile organic matter fractions as central components of the quality of agricultural soils: An overview. *Adv. Agron.* 85:221–268.

Haynes, R. J. and P. H. Sherlock. 1986. Gaseous losses of nitrogen. In: *Nitrogen in Plant–Soil Systems*, ed., R. J. Haynes, pp. 242–302. New York: Academic Press.

Hayes, M. H. B. and R. S. Swift. 1983. The chemistry of soil organic colloids. In: *The Chemistry of Soil Constituents*, eds., D. J. Greenland, and M. H. B. Hayes, pp. 179–320. New York: John Wiley & Sons.

He, Z. L., M. K. Zhang, D. V. Calvert, P. J. Stoffella, X. E. Yang, and S. Yu. 2004. Transport of heavy metals in surface runoff from vegetable and citrus fields. *Soil Sci. Soc. Am. J.* 68:1662–1669.

Heenan, D. P., W. J. McGhie, F. M. Thompson, and K. Y. Chan. 1995. Decline in soil organic carbon and total nitrogen in relation to tillage, stubble management, and rotations. *Aust. J. Exp. Agric.* 35:877–884.

Helling, C. S., G. Chester, and R. B. Corey. 1964. Contribution of organic matter and clay soil to cation exchange capacity as affected by the pH of the saturation solution. *Soil Sci. Soc. Am. Proc.* 28: 517–520.

Hendry, M. J., R. G. L. McCready, and W. D. Gould. 1984. Distribution, source and evolution of nitrate in a glacial till of Southern Alberta. *Can. J. Hydrol.* 70:177–198.

Heuscher, S. A., C. C. Brandt, and P. M. Jardine. 2005. Using soil physical and chemical properties to estimate bulk density. *Soil Sci. Soc. Am. J.* 69:51–56.

Hillel, D. and C. Rosenzweig. 2002. Desertification in relation to climate variability in change. *Adv. Agron.* 77:1–38.

Hiradate, S. and N. U. Yamaguchi. 2003. Chemical species of Al reacting with soil humic acids. *J. Inorg. Biochem.* 97:26–31.

Hooker, B. A., T. F. Morris, R. Peters, and Z. G. Cardon. 2005. Long-term effects of tillage and corn stalk return on soil carbon dynamics. *Soil Sci. Soc. Am. J.* 69:188–196.

Horowitz, J., R. Ebel, and K. Ueda. 2010. *No-Tillage Farming Is a Growing Practice. Econ. Info. Bull.* 70. Washington, DC: USDA-ERA.

Howieson, J. G., L. K. Abbott, and A. D. Robson. 1992. Calcium modifies pH effects on acid tolerant and acid sensitive strains of *Rhizobium meliloti*. *Aust. J. Agric. Res.* 43:565–572.

Howieson, J. G., A. D. Robson, and M. A. Ewing. 1993. External phosphate and calcium concentrations, but not the products of rhizobial nodulation genes, affect the attachment of *Rhizobium meliloti* to roots of annual medics. *Soil Biol. Biochem.* 25:567–573.

Hsieh, Y. P. 1992. Size and mean ages of stable soil organic carbon in cropland. *Soil Sci. Soc. Am. J.* 56:460–464.

Hsieh, Y. P. 1996. Soil organic carbon pools of two tropical soils inferred by carbon signatures. *Soil Sci. Soc. Am. J.* 60:1117–1121.

Hue, N. V. 1992. Correcting soil acidity of a highly weathered Ultisol with chicken manure and sewage sludge. *Commun. Soil Sci. Plant Anal.* 23:241–264.

Hue, N. V. and D. L. Licudine. 1999. Amelioration of subsoil acidity through surface application of organic manures. *J. Environ. Quality* 28:623–632.

IAS (International Allelopathy Society). 1996. *First World Congress on Allelopathy: A Science for the Future.* Cadiz, Spain: International Allelopathy Society.

Imhoff, S., A. V. Da Silva, and D. Fallow. 2004. Susceptibility to compaction, load support capacity, and soil compressibility of hapludox. *Soil Sci. Soc. Am. J.* 68:17–24.

Jastrow, J. D. 1996. Soil aggregate formation and the accrual of particulate and mineral-associated organic matter. *Soil Biol. Biochem.* 28:656–676.

Jenkinson, D. S. 1988. Soil organic matter and its dynamics. In: *Soil Conditions and Plant Growth,* 11th edition, ed., A. W. Russel, pp. 564–607. London: Longman.

Jenkinson, D. S. and A. Ayanaba. 1977. Decomposition of carbon-14 labelled plant material under tropical conditions. *Soil Sci. Soc. Am. J.* 41:912–915.

Jenkinson, D. S., D. E. Adams, and A. Wild. 1991. Model estimates of CO_2 emissions from soil in response to global warming. *Nature* 351:304–306.

Jenkinson, D. S., N. J. Bradbury, and K. Coleman. 1994. How the Rothamsted classical experiments have been used to develop and to test models for the turnover of carbon and nitrogen in soil. In: *Long-term Experiments in Agricultural and Ecological Sciences*, eds., R. A. Leigh, and A. E. Johnston, pp. 117–138. Oxford, UK: CAB International.

Jenkinson, D. S. and J. H. Rayner. 1977. The turnover of soil organic matter in some of the Rothamsted classical experiments. *Soil Sci.* 123:298–305.

Jokela, W. E. 1992. Nitrogen fertilizer and dairy manure effects on corn yield and soil nitrate. *Soil Sci. Soc. Am. J.* 56:148–154.

Johnson, J. M. F., D. Reicosky, B. Sharratt, M. Lindstrom, W. Voorhees, and L. Carpenter-Boggs. 2004. Characterization of soil amended with the by-product of corn stover fermentation. *Soil Sci. Soc. Am. J.* 68:139–147.

Kadebra, O. 1978. Organic matter status of some savanna soils of northern Nigeria. *Soil Sci.* 125:122–127.

Kapland, D. I. and G. O. Estes. 1985. Organic matter relationship to soil nutrient status and aluminum toxicity in alfalfa. *Agron. J.* 77:735–738.

Kay, B. D. 1998. Soil structure and organic carbon: A review. In: *Soil Processes and the Carbon*, eds., R. Lal, J. M. Kimble, R. F. Follett, and B. A. Stewart, pp. 169–197. Boca Raton, Florida: CRC Press.

Kern, J. S. and M. G. Johnson. 1993. Conservation tillage impacts on national soil and atmospheric carbon levels. *Soil Sci. Soc. Am. J.* 57:200–210.

Kimble, J. M., L. S. Heath, R. A. Birdsey, and R. Lal. 2003. *The Potential of U.S. Forest Soils to Sequester Carbon and Mitigate the Greenhouse Effect.* Boca Raton, Florida: Lewis Publication.

Kirschbaum, M. U. F. 2000. Will changes in soil organic carbon act as a positive or negative feedback on global warming? *Biogeochemistry* 48:21–51.

Kirchmann, H., F. Pichlmayer, and M. H. Gerzabek. 1996. Sulfur balances and sulfur-34 abundance in a long-term fertilizer experiment. *Soil Sci. Soc. Am. J.* 59:174–178.

Kravchenko, A. N. and D. G. Bullock. 2000. Correlation of corn and soybean grain yield with topography and soil properties. *Agron. J.* 92:75–83.

Kumar, K. and K. M. Goh. 2000. Crop residue management: Effects on soil quality, soil nitrogen dynamics, crop yield, and nitrogen recovery. *Adv. Agron.* 68:197–319.

Kumar, S., A. Kadono, R. Lal, and W. Dick. 2012. Long-term no-till impacts on organic carbon and properties of two contrasting soils and corn yields in Ohio. *Soil Sci. Soc. Am. J.* 76:1798–1809.

Kuo, S. 1983. *Effects of Organic Residues and Nitrogen Transformations on Phosphorus Sorption and Desorption by Soil.* Madison, Wisconsin: *Agronomy Abstract.* ASA.

Kuo, S. and A. S. Baker. 1982. The effect of soil drainage on phosphorus status and availability to corn in long-term manure amended soils. *Soil Sci. Soc. Am. J.* 46:744–747.

Ladha, J. K., C. K. Reddy, A. T. Padre, and C. V. Kessel. 2012. Role of nitrogen fertilization in sustaining organic matter in cultivated soils. *Better Crops Plant Food* 96:24–25.

Lado, M., A. Paz, and M. Ben-Hur. 2004a. Organic matter and aggregate size interactions in saturated hydraulic conductivity. *Soil Sci. Soc. Am. J.* 68:234–242.

Lado, M., A. Paz, and M. Ben-Hur. 2004b. Organic matter and aggregate size interactions in infiltration, seal formation and soil loss. *Soil Sci. Soc. Am. J.* 68:935–942.

Lal, R. 1997. Residue management, conservation tillage and soil restoration for mitigating greenhouse effects by CO_2-enrichment. *Soil Tillage Res.* 43:81–107.

Lal, R. 1998. Soil erosion impact on agronomic productivity and environmental quality. *Crit. Rev. Plant Sci.* 17:319–464.

Lal, R. 2006. Enhancing crop yields in the developing countries through restoration of the soil organic carbon pool in agricultural lands. *Land Degrade. Dev.* 17:197–209.

Lal, R., J. Kimble, E. Levine, and C. Whitman. 1995. World soils and greenhouse effect: An overview. In: *Soils and Global Change*, eds., R. Lal, J. Kimble, E. Levine, and B. A. Stewart, pp. 131–142. Boca Raton, Florida: CRC Press.

Lampkin, N. 1990. *Organic Farming.* Ipswich, UK: Farming Press Books.

Larson, W. F. and S. C. Gupta. 1980. Estimating critical stress in unsaturated soils from changes in pore water pressure during confined compression. *Soil Sci. Soc. Am. J.* 44:1127–1132.

Lentz, R. D., R. E. Sojka, and D. L. Carter. 1996. Furrow irrigation water-quality effects on soil loss and infiltration. *Soil Sci. Soc. Am. J.* 60:238–245.

Li, L., W. Huang, P. Peng, G. Sheng, and J. Fu. 2003. Chemical and molecular heterogeneity of humic acids repetitively extracted from a peat. *Soil Sci. Soc. Am. J.* 67:740–746.

Liebhardt, W. C., R. W. Andrews, M. N. Culik, R. R. Harwood, R. R. Janeke, J. K. Radke, and S. L. Rieger-Schwartz. 1989. Crop production during conversion from conventional to low-input methods. *Agron. J.* 81:150159.

Liu, J. and N. V. Hue. 1996. Ameliorating subsoil acidity by surface application of calcium fulvates derived from common organic materials. *Biol. Fert. Soils* 21:264–270.

Loeppert, R. H. and C. T. Hallmark. 1985. Indigenous soil properties influencing the availability of iron in calcareous soils. *Soil Sci. Soc. Am. J.* 49:597–603.

Loll, M. J. and J. M. Bollag. 1983. Protein transformation in soil. *Adv. Agron.* 36:351–382.

Lopes, A. S. 1983. *Soils under Cerredo: Characteristics, Properties, and Management.* Piracicaba, Brazil: Potash and Phosphate Institute.

Lucas, S. T. and R. R. Weil. 2012. Can a labile carbon test be used to predict crop responses to improve soil organic matter management? *Agron. J.* 104:1160–1170.

Lynch, J. M. and E. Bragg. 1985. Microorganisms and soil aggregate stability. *Adv. Soil Sci.* 2:133–171.

Macias, F. A., R. M. Olive, A. M. Simone, and J. C. G. Galindo. 1988. What are allelochemicals? In: *Allelopathy in Rice,* ed., IRRI, pp. 69–79. Los Baños, the Philippines: IRRI.

MacRae, R. J. and G. R. Mehuys. 1985. The effect of green manuring on the physical properties of temperate-area soils. *Adv. Soil Sci.* 3:71–94.

Majchrzak, R. N., K. N. Olson, G. Bollero, and E. D. Nafziger. 2001. Using soil properties to predict wheat yields on Illinois soils. *Soil Sci.* 166:267–280.

Mann, L. V., V. Tolbert, and J. Cushman. 2002. Potential environmental effects of corn (*Zea mays* L.) stover removal with emphasis on soil organic matter and erosion. A review. *Agric. Ecosyst. Environ.* 89:149–166.

Marschner, H. and B. Dell. 1994. Nutrient uptake in mycorrhizal symbiosis. *Plant Soil* 159:89–102.

Marquez, C. O., V. J. Garcia, C. A. Cambardella, R. C. Schultz, and T. M. Isenhart. 2004. Aggregate size stability distribution and soil stability. *Soil Sci. Soc. Am. J.* 68:725–735.

Martel, Y. A., C. R. De Kimpe, and M. R. Laverdiere. 1978. Carbon-exchange capacity of clay-rich soils in relation to organic matter, mineral composition and surface area. *Soil Sci. Soc. Am. J.* 42:764–767.

Martens, R. 1995. Current methods for measuring microbial biomass C in soil: Potentials and limitations. *Biol. Fert. Soils* 19:87–99.

Martens, D. A. and W. T. Jr. Frankenberger. 1992. Modification of infiltration rates in an organic-amended irrigated soil. *Agron. J.* 84:707–717.

Marzadori, C., L. V. Anlisari, C. Ciavatta, and P. Sequi. 1991. Soil organic matter influence on adsorption and desorption of boron. *Soil Sci. Soc. Am. J.* 55:1582–1585.

Mathers, A. C. and B. A. Stewart. 1974. Corn silage yield and soil chemical properties as affected by cattle feedlot manure. *J. Environ. Qual.* 3:143–147.

Mathers, A. C., J. D. Thomas, B. A. Stewart, and J. E. Herring. 1980. Manure and inorganic fertilizer effects on sorghum and sunflower growth on iron-deficient soil. *Agron. J.* 72:1025–1029.

Mays, D. A., G. L. Terman, and J. C. Duggan. 1973. Municipal compost: Effect on crop yield and soil properties. *J. Environ. Qual.* 2:89–91.

Mcbride, R. A. 1989. Estimation of density moisture stress functions from uniaxial compression of unsaturated, structured soils. *Soil Tillage Res.* 1:383–397.

Mcbridge, R. A. and G. C. Watson. 1990. An investigation of re-expansion of unsaturated, structured soils during cyclic static loading. *Soil Tillage Res.* 17:241–253.

McCalla, T. M. and F. A. Norstadt. 1974. Toxicity problems in mulch tillage. *Agric. Environ.* 1:153–174.

McGarity, J. W. 1961. Denitrification studies on some South Australian soils. *Plant Soil* 14:1–21.

McLauchlan, K. K. and S. E. Hobbie. 2004. Comparison of labile soil organic matter fractionation techniques. *Soil Sci. Soc. Am. J.* 68:1616–1625.

Meek, B., L. Graham, and T. Donovan. 1982. Long-term effects of manure on soil nitrogen, phosphorus, potassium, sodium, organic matter and water infiltration rate. *Soil Sci. Soc. Am. J.* 46:1014–1019.

Meek, B. D., W. R. Detarr, D. Rolph, E. R. Rechel, and L. M. Carter. 1990. Infiltration rate as affected by an alfalfa and no-till cotton cropping system. *Soil Sci. Soc. Am. J.* 54:505–508.

Mengel, K., E. A. Kirkby, H. Kosegarten, and T. Appel. 2001. *Principles of Plant Nutrition*, 5th edition. Dordrecht: Kluwer Academic Publishers.

Mikha, M. M. and C. W. Rice. 2004. Tillage and manure effects on soil and aggregate associated carbon and nitrogen. *Soil Sci. Soc. Am. J.* 68:809–816.

Miller, M. and R. P. Dick. 1995. Thermal stability and activities of soil enzymes influenced by crop rotations. *Soil Biol. Biochem.* 27:1161–1166.

Miller, B. F., W. L. Lindsay, and A. A. Parsa. 1969. Use of poultry manure for correction of Zn and Fe deficiencies in plants. In: *Proceedings of Animal Waste Management Conference, January 1969*, ed., Cornell University, pp. 120–123. Ithaca, New York: Cornell University.

Moawad, H. A., W. R. Ellis, and E. L. Schmidt. 1984. Rhizosphere response as a factor in competition among three serogroups of indigenous *Rhizobium japonicum* for nodulation of field grown soybean. *Appl. Environ. Microbiol.* 47:607–612.

Moraghan, J. T. 1983. Zinc deficiency of flax in North Dakota. *North Dakota Farm Res.* 40:23–26.

Moore, P. A., T. C. Daniel, J. T. Gilmour, B. R. Shreve, D. R. Edwards, and B. H. Wood. 1998. Decreasing metal runoff from poultry litter with aluminum sulfate. *J. Environ. Qual.* 27:92–99.

Moore, J. M., S. Klose, and M. A. Tabatabai. 2000. Soil microbial biomass carbon and nitrogen as affected by cropping systems. *Biol. Fert. Soils* 31:200–210.

Moorman, T. B., K. Jayachandran, and A. Reungsang. 2001. Adsorption and desorption of atrazine in soils and subsurface sediments. *Soil Sci.* 166:921–929.

Morrill, L. G. and J. E. Dawson. 1967. Patterns observed for the oxidation of ammonium to nitrate by soil organisms. *Soil Sci. Soc. Am. Proc.* 31:757–760.

Morse, C. C., I. V. Yevdokimov, and J. C. DeLuca. 2000. *In Situ* extraction of rhizosphere organic compounds from contrasting plant communities. *Commun. Soil Sci. Plant Anal.* 31:725–742.

Nelson, R. G. 2002. Resource assessment and removal analysis for corn stover and wheat straw in the eastern and Midwestern United States—Rainfall and wind induced soil erosion methodology. *Biomass Bioenergy* 22:349–363.

Nemati, M. R., J. Caron, and J. Gallichand. 2000. Using paper deinking sludge to maintain soil structure form: Field measurements. *Soil Sci. Soc. Am. J.* 64:275–285.

Nichols, J. D. 1984. Relation of organic carbon to soil properties and climate in the southern Great Plains. *Soil Sci. Soc. Am. J.* 48:1382–1384.

Norris, C. E., S. A. Quideau, J. S. Bhatti, and R. E. Wasylishen. 2011. Soil carbon stabilization in jack pine stands along the boreal forest transect case study. *Glob. Change Biol.* 17:480–494.

Oades, J. M. 1988. The retention of organic matter in soils. *Biogeochemistry* 5:35–70.

Odell, R. T., S. W. Melsted, and W. M. Walker. 1984. Changes in organic carbon and nitrogen of Morrow plot soils under different treatments, 1904–1973. *Soil Sci.* 137:160–171.

Olofsdotter, M. 2001. Rice—A step toward use of allelopathy. *Agron. J.* 93:3–8.

Olofsdotter, M., D. Navarez, and K. Moody. 1995. Allelopathic potential in rice (*Oryza sativa* L.) germplasm. *Ann. Appl. Biol.* 127:543–560.

Ortega, R. A., G. A. Peterson, and D. G. Westfall. 2002. Residue accumulation and changes in soil organic matter as affected by cropping intensity in no-till dryland agroecosystems. *Agron. J.* 94:944–954.

Palanipappan, S. P., P. S. Sreedhar, P. Loganathan, and J. Thomas. 1997. Competitiveness of native *Bradyrhizobium japonicum* strains in two different soil types. *Biol. Fert. Soils* 25:279–284.

Park, E. J. and A. J. M. Smucker. 2005. Saturated hydraulic conductivity and porosity within macroaggregates modified by tillage. *Soil Sci. Soc. Am. J.* 69:38–45.

Parker, J. L., I. J. Fernandez, L. E. Rustad, and S. A. Norton. 2002. Soil organic matter fractions in experimental forested watershed. *Water Air Soil Pollut.* 138:101–121.

Pauel, E. 1984. Dynamics of organic matter in soils. *Plant Soil* 76:275–285.

Paul, E. A. and F. E. Clark. 1996. *Soil Microbiology and Biochemistry*. San Diego, California: Academic Press.

Paustian, K., O. Andren, H. Janzen, R. Lal, P. Smith, G. Tian, H. Tiessen, M. V. Noordwijk, and P. Woomer. 1997. Agricultural soil as a C sink to offset CO_2 emissions. *Soil Use Manage.* 13:230–244.

Paustian, K., H. P. Collins, and E. A. Paul. 1997. Management controls on soil carbon. In: *Soil Organic Matter in Temperate Agroecosystems*, eds. E. A. Paul, K. Paustian, E. T. Elliot, and C. V. Cole, pp. 15–49. Boca Raton, Florida: CRC Press.

Paustian, K., W. J. Parton, and J. Persson. 1992. Modeling soil organic matter in organic-amended and nitrogen fertilized long term plots. *Soil Sci. Soc. Am. J.* 56:476–488.

Paustian, K., J. Six, E. T. Elliott, and H. W. Hunt. 2000. Management options for reducing CO_2 emissions from agricultural soils. *Biochemistry* 48:147–163.

Pavan, M. A. and M. Miyazawa. 1984. Manganese disponibility in soil: Problems and difficulties in interpretation of soil analysis for soil fertility. *Rev. Bras. Ciênc. Solo* 8:285–290.

Pennock, D. J. and C. Van Kessel. 1997. Clear-cut forest harvest impacts on soil quality indicators in the mixed wood forest of Saskatchewan, Canada. *Geoderma* 75:13–32.

Peoples, M. B., D. F. Herridge, and J. K. Lahda. 1995. Biological nitrogen fixation: An efficient source of nitrogen for sustainable agricultural production. *Plant Soil* 174:3–28.

Picard, G. L. and G. T. Felbeck. 1976. The complexation of iron by marine humic acid. *Geochim. et Cosmochim. Acta* 40:1347–1350.

Poinke, H. B., D. E. Glotfelty, A. D. Lucas, and B. Urban. 1988. Pesticide contamination of ground waters in the Mahantango Creek watershed. *J. Environ. Qual.* 17:76–84.

Post, W. M. and K. C. Kwon. 2000. Soil carbon sequestration and land use change: Processes and potential. *Glob. Change Biol.* 6:317–327.

Potter, K. N., O. R. Jones, H. A. Torbert, and P. W. Unger. 1997. Crop rotation and tillage effects on organic carbon sequestration in the semiarid southern Great Plains. *Soil Sci.* 162:140–147.

Potter, K. N., H. A. Tolbert, O. R. Jones, J. E. Matocha, J. E. Morrison, and P. W. Unger. 1998. Distribution and amount of soil organic C in long-term management systems in Texas. *Soil Tillage Res.* 47:309–321.

Power, J. F. 1990. Fertility management and nutrient cycling. *Adv. Soil Sci.* 13:131–149.

Pratt, P. F. and A. E. Laag. 1977. Potassium accumulation and movement in an irrigated soil treated with animal manures. *Soil Sci. Soc. Am. J.* 41:1130–1133.

Prenger, J. P. and K. R. Reddy. 2004. Microbial enzyme activities in a freshwater marsh after cessation of nutrient loading. *Soil Sci. Soc. Am. J.* 68:1796–1804.

Prevost, M. 2004. Predicting soil properties from organic matter content following mechanical site preparation of forest soils. *Soil Sci. Soc. Am. J.* 68:943–949.

Pumpanen, J., H. Ilvesniemi, and P. Hari. 2003. A process-based model for predicting soil carbon dioxide efflux and concentration. *Soil Sci. Soc. Am. J.* 67:402–413.

Putnam, A. R. and W. B. Duke. 1978. Allelopathy in agroecosystems. *Annu. Rev. Phytopathol.* 16:431–451.

Qualls, R. G. 2004. Biodegradability of humic substances and other fractions of decomposing leaf litter. *Soil Sci. Soc. Am. J.* 68:1705–1712.

Qualls, R. G., A. Takiyama, and R. L. Wershaw. 2003. Formation and losses of humic substances during decomposition in a pine forest floor. *Soil Sci. Soc. Am. J.* 67:899–909.

Rao, I. M., M. A. Ayarza, R. J. Thomas, M. J. Fisher, J. I. Sanz, J. M. Spain, and C. E. Lascano. 1992. Soil–plant factors and processes affecting productivity in ley farming. In: *Pasture for the Tropical Lowlands: CIAT's Contribution*, ed., CIAT, pp. 145–175. Cali, Colombia: CIAT.

Rasse, D. P. and A. J. M. Smucker. 1998. Root recolonization of previous root channels in corn and alfalfa rotation. *Plant Soil* 204:203–212.

Rasmussen, P. E., R. R. Allmaras, C. R. Rohde, and N. C. Jr. Roager. 1980. Crop residue influences on soil carbon and nitrogen in a wheat-fallow system. *Soil Sci. Soc. Am. J.* 44:596–600.

Rasmussen, P. E. and C. R. Rohde. 1988. Long-term tillage and nitrogen fertilization effects on organic nitrogen and carbon in a semiarid soil. *Soil Sci. Soc. Am. J.* 52:1114–1117.

Rimando, A. M., M. Olofsdotter, F. E. Dayan, and S. O. Duke. 2001. Searching for rice allelochemicals: An example of bioassay-guided isolation. *Agron. J.* 93:16–20.

Read, J. W. and R. H. Ridgell. 1921. On the use of the conventional carbon factor in estimating soil organic matter. *Soil Sci.* 13:1–6.

Reddy, K. R., M. R. Overcash, R. Khaleal, and P. W. Westerman. 1980. Phosphorus sorption–desorption characteristics of two soil utilized for disposal of animal manure. *J. Environ. Qual.* 9:86–92.

Reicosky, D. C. and F. Forcella. 1988. Cover crop and soil quality interaction in agroecosystems. *J. Soil Water Conser.* 53:224–229.

Rhoton, F. E. M. J. M. Romkens, J. M. Bigham, T. M. Zobeck, and D. R. Upchurch. 2003. Ferrihydrite influence on infiltration, runoff, and soil loss. *Soil Sci. Soc. Am. J.* 67:1220–1226.

Rhoton, F. E. and D. D. Tyler. 1990. Erosion-induced changes in the properties of a fragipan soil. *Soil Sci. Soc. Am. J.* 54:223–228.

Rice, E. L. 1974. *Allelopathy.* New York: Academic Press.

Rice, E. L. 1979. Allelopathy: An update. *Botan. Rev.* 45:15–109.

Richardson, A. E. 2001. Prospects for using soil microorganisms to improve the acquisition of phosphorus by plants. *Aust. J. Plant Physiol.* 28:897–906.

Robertson, G. P., E. A. Paul, and R. R. Harwood. 2000. Greenhouse gases in intensive agriculture: Contributions of individual gases to the radiative forcing of the atmosphere. *Science* 289:1922–1925.

Robinson, C. A., R. M. Cruse, and M. Ghaffarzadeh. 1996. Cropping system and nitrogen effects on Mollisol organic carbon. *Soil Sci. Soc. Am. J.* 60:264–269.

Robinson, D. A., L. Lebron, and H. Vereecken. 2009. On the definition of the natural capital of soils: A framework for description, evaluation, and monitoring. *Soil Sci. Soc. Am. J.* 73:1904–1911.

Robson, M. C., S. M. Fowler, N. H. Lampkin, C. Leifert, M. Leitch, D. Robinson, C. A. Watson, and A. M. Litterick. 2002. The agronomic and economic potential of break crops for ley/arable rotations in temperate organic agriculture. *Adv. Agron.* 77:369–427.

Rochette, P. and E. G. Gregorich. 1998. Dynamics of soil microbial biomass C, soluble organic C, and CO_2 evolution after three years of manure application. *Can. J. Soil Sci.* 78:283–290.

Rodriguez, H. and R. Fraga. 1999. Phosphate solubilizing bacteria and their role in plant growth promotion. *Biotech. Adv.* 17:319–339.

Sah, R. N. and D. S. Mikkelsen. 1986. Effects of anaerobic decomposition of organic matter on sorption and transformation of P. I. Effects on P sorption. *Soil Sci.* 142:267–274.

Sah, R. N. and D. S. Mikkelsen. 1989. Phosphorus behavior in flooded–drained soils. I. Effects on phosphorus sorption. *Soil Sci. Soc. Am. J.* 53:1718–1722.

Sahrawat, K. L. and D. R. Keeney. 1986. Nitrous oxide emission from soils. *Adv. Soil Sci.* 4:103–148.

Salinas-Garcia, J. R., F. M. Hons, J. E. Matocha, and D. A. Zuberer. 1997. Soil carbon dynamics as affected by long term tillage and nitrogen fertilization. *Biol. Fert. Soils* 25:182–188.

Sanchez, P. A. and R. H. Miller. 1986. Organic matter and soil fertility management. *Transactions of the 13th Congress of the International Society of Soil Science*, Hamburg. 6:609–625.

Schaap, M. G., F. J. Leij, and M. T. V. Genuchten. 2001. ROSETTA: A computer program for estimating soil hydraulic parameters with hierarchical pedotransfer functions. *J. Hydrol.* 251:163–176.

Schepers, J. S. and A. R. Mosier. 1991. Accounting for nitrogen in nonequilibrium soil–crop systems. In: *Managing Nitrogen for Groundwater Quality and Farm Profitability*, eds., R. F. Follett, D. R. Keeney, and R. M. Cruse, pp. 125–138. Madison, Wisconsin: SSSA.

Schimel, D. S., D. C. Coleman, and K. A. Horton. 1985. Soil organic matter dynamics in paired rangeland and crop toposequences in North Dakota. *Geoderma* 36:201–214.

Schimel, D. S., B. H. Braswell, and E. A. Holland. 1994. Climatic edaphic and biotic controls over storage and turnover of carbon in soils. *Glob. Biogeochem. Cycles* 8:279–293.

Schlesinger, W. H. 1990. Evidence from chronosequence studies for a low carbon-storage potential of soils. *Nature* 348:232–234.

Schnitzer, M. and S. V. Khan. 1972. *Humic Substances in the Environment.* New York: Marcel Dekker.

Scholes, R. J. and M. C. Scholes. 1995. The effect of land use on nonliving organic matter in the soil. In: *The Role of Nonliving Organic Matter in the Earths C Cycle*, eds., R. G. Zepp, and C. Sonntag, pp. 210–225. Chichester, UK: John Wiley & Sons.

Scoot, H. D. and L. S. Wood. 1989. Impact of crop production on the physical status of a typic Albaqualf. *Soil Sci. Soc. Am. J.* 53:1819–1825.

Sen Gupta, M. B. 1969. Phosphorus mobility in alkali soil. 1. Effect of SAR and organic matter addition. *J. Ind. Soc. Soil Sci.* 17:115–118.

Senesi, G. S., G. Baldassarre, N. Senesi, and B. Radina. 1999. Trace element inputs into soils by anthropogenic activities and implications for human health. *Chemosphere* 39:343–377.

Sessitsch, A., J. G. Howieson, X. Perret, H. Antoun, and E. Martinez. 2002. Advances in rhizobium research. *Crit. Rev. Plant Sci.* 21:323–378.

Shuman, L. M. 1985a. Effects of tillage on the distribution of manganese, copper, iron, and zinc in soil fractions. *Soil Sci. Soc. Am. J.* 49:1117–1122.

Shuman, L. M. 1985b. Fractionation method for soil micronutrients. *Soil Sci.* 140:11–22.

Sims, J. T. and D. C. Wolf. 1994. Poultry manure management: Agricultural and environmental issues. *Adv. Agron.* 52:1–83.

Singh, B. B. and J. P. Jones. 1976. Phosphorus sorption and desorption characteristics of soil as affected by organic residues. *Soil Sci. Soc. Am. J.* 40:389–394.

Singh, Y., B. Singh, J. K. Ladha, C. S. Khind, R. K. Gupta, O. P. Meelu, and E. Pasuquin. 2004a. Long-term effects of organic inputs on yield and soil fertility in the rice–wheat rotation. *Soil Sci. Soc. Am. J.* 68:845–853.

Singh, Y., B. Singh, J. K. Ladha, C. S. Khind, T. S. Khera, and C. S. Bueno. 2004b. Effects of residue decomposition on productivity and soil fertility in rice–wheat rotation. *Soil Sci. Soc. Am. J.* 68:854–864.

Sisti, C. P. J., H. P. Santos, R. Kohhann, B. J. R. Alves, S. Urquiaga, and R. M. Boddey. 2004. Change in carbon and nitrogen stocks in soil under 13 years of conventional and zero tillage in southern Brazil. *Soil Tillage Res.* 76:39–58.

Six, J., E. T. Elliott, K. Paustian, and J. W. Doran. 1998. Aggregation and soil organic matter accumulation in cultivated and native grassland soils. *Soil Sci. Soc. Am. J.* 62:1367–1377.

Six, J., E. T. Elliott, and K. Paustain. 1999. Aggregate and soil organic matter dynamics under conventional and no-tillage systems. *Soil Sci. Soc. Am. J.* 63:1350–1358.

Six, J., E. T. Elliott, and K. Paustain. 2000a. Soil microaggregate turnover and microaggregate formation: A mechanism for C sequestration under no-tillage agriculture. *Soil Biol. Biochem.* 32:2099–2013.

Six, J., K. Paustian, E. T. Elliott, and C. Combrink. 2000b. Soil structure and organic matter: I. Distribution of aggregate size classes and aggregate-associated carbon. *Soil Sci. Soc. Am. J.* 64:681–689.

Six, J., C. Feller, K. Denef, S. M. Ogle, J. C. Moraes, and A. Albrecht. 2002. Soil organic matter, biota and aggregation in temperate and tropical soils—Effects of no-tillage. *Agronomie* 22:755–775.

Smith, M. S., W. W. Frye, and J. J. Varco. 1987. Legume winter cover crops. *Adv. Soil Sci.* 7:95–139.

Smith, J. L. and E. A. Paul. 1990. The significance of soil microbial biomass estimations. In: *Soil Biochemistry*, Vol. 6 eds., J. M. Bolag, and G. Strotzky, pp. 357–396. New York: Marcel Dekker.

Smith, S. E. and R. J. Read. 1997. *Mycorrhizal Symbiosis*. San Diego, California: Academic Press.

Soil Science Society of America. 2008. *Glossary of Soil Science Terms*. American Society of Soil Science, Madison, Wisconsin.

Sollins, P., P. Homann, and B. A. Caldwell. 1996. Stabilization and destabilization of organic matter: Mechanisms and controls. *Geoderma* 74:65–105.

Sommerfeldt, T. G. and C. Chang. 1985. Changes in soil properties under annual applications of feedlot manure and different tillage practices. *Soil Sci. Soc. Am. J.* 49:983–987.

Spaccini, R., A. Piccolo, P. Conte, G. Haberhauer, and M. H. Gerzabek. 2002. Increasing soil organic carbon sequestration through hydrophobic protection by humic substances. *Soil Biol. Biochem.* 34:1839–1851.

Spargo, J. T., M. M. Alley, R. F. Follett, and J. V. Wallace. 2008. Soil nitrogen conservation with continuous no-till management. *Nutr. Cycl. Agroecosyst.* 82:283–297.

Spargo, J. T., M. A. Cavigelli, M. M. Alley, J. E. Maul, J. S. Buyer, C. H. Sequeira, and R. F. Follett. 2012. Changes in soil organic carbon and nitrogen fractions with duration of no-tillage management. *Soil Sci. Soc. Am. J.* 76:1624–1633.

Stanford, G. and S. J. Smith. 1972. Nitrogen mineralization potentials of soils. *Soil Sci. Soc. Am. Proc.* 36:465–472.

Stevens, R. J., R. J. Laughlin, and J. P. Malone. 1998. Soil pH affects the process reducing nitrate to nitrous oxide and di-nitrogen. *Soil Biol. Biochem.* 30:1119–1126.

Stevenson, F. J. 1982. *Humus Chemistry: Genesis, Composition, Reactions*. New York: John Wiley & Sons.

Stevenson, F. J. 1991. Organic matter–micronutrient reactions in soil. In: *Micronutrients in Agriculture*, 2nd edition, ed., R. R. Mortvedt, pp. 145–186. Madison, Wisconsin: SSSA.

Stevenson, F. J. 1994. *Humus Chemistry: Genesis, Composition, Reactions*, 2nd edition. New York: John Wiley & Sons.

Stevenson, F. J. and Y. Chen. 1991. Stability constants of copper (II)–humate complexes determined by modified potentiometric titration. *Soil Sci. Soc. Am. J.* 55:1586–1591.

Stine, M. A. and R. R. Weil. 2002. The relationship between soil quality and crop productivity across three tillage systems in south central Honduras. *Am. J. Alter. Agric.* 17:2–8.

Strickling, E. 1975. Crop sequences and tillage in efficient crop production. In: *Agronomy Abstracts*, ed., ASA, pp. 20–29. Madison, Wisconsin: ASA.

Szilagyi, M. 1971. Reduction of Fe^{3+} ion by humic acid preparations. *Soil Sci.* 111:233–235.

Tabatabai, M. A. 1994. Soil enzymes. In: *Methods of Soil Analysis*, Part 2, ed., R. W. Weaver et al., pp. 775–833. Madison, Wisconsin: SSSA.

Tan, K. H. and A. Binger. 1986. Effect of humic acid on aluminum toxicity in corn plants. *Soil Sci.* 141:20–25.

Tan, K. H., J. H. Edwards, and O. L. Bennett. 1985. Effect of sewage sludge on mobilization of surface applied calcium in a Greenville soil. *Soil Sci.* 139:262–268.

Tan, K. H., R. A. Leonard, A. R. Bertrand, and S. Wilkinson. 1971. The metal complexing capacity and the nature of the chelating ligands of water extract of poultry litter. *Soil Sci. Soc. Am. Proc.* 35:265–269.

Tester, C. F. 1990. Organic amendment effects on physical and chemical properties of a sandy soil. *Soil Sci. Soc. Am. J.* 54:827–831.

Theng, B. K. G., K. R. Tate, and P. Sollins. 1989. Constituents of organic matter in temperate and tropical soils. In: *Dynamics of Soil Organic Matter in Tropical Ecosystems*, eds., D. C. Coleman et al., pp. 5–31. Honolulu, Hawaii: University of Hawaii Press.

Thomas, G. W. 1975. The relationship between organic matter content and exchangeable aluminum in acid soil. *Soil Sci. Soc. Am. Proc.* 39:591.

Tian, G., B. T. Kang, and L. Broussard. 1992. Biological effects of plant residues with contrasting chemical composition under humid tropical conditions: Decomposition and nutrient release. *Soil Biol. Biochem.* 24:1051–1060.

Tiessen, H. and J. Stewart. 1983. Particle-size fractions and their use in studies of soil organic matter. II. Cultivation effects on organic matter composition in size fractions. *Soil Sci. Soc. Am. J.* 47:509–514.

Tinker, P. B. 1982. Mycorrhizas: The present position. *Trans. Int. Congr. Soil Sci.*, 12th edition. 5:150–166.

Tirol-Padre, A. and J. K. Ladha. 2004. Assessing the reliability of permanent–oxidizable carbon as an index of soil labile carbon. *Soil Sci. Soc. Am. J.* 68:969–978.

Tisdall, J. M. 1994. Possible role of soil microorganisms in aggregation in soils. *Plant Soil* 159:115–121.

Tisdall, J. M. and J. M. Oades. 1982. Organic matter and water-stable aggregates in soils. *J. Soil Sci.* 33:141–163.

Tunney, H., O. T. Carton, P. C. Brookes, and A. E. Johnston. 1997. *Phosphorus Loss from Soil to Water*. Oxford, UK: CAN International.

Unger, P. W. and T. C. Kaspar. 1994. Soil compaction and root growth: A review. *Agron. J.* 86:759–766.

USEPA. 1996. *Environmental Indicators of Water Quality in the United States. EPA 841-R-96-002*. USEPA, Office of Water (4503F). Washington, DC: U. S. Government Printing Office.

Uusitalo, R., E. Turtola, T. Kauppila, and T. Lilja. 2001. Particulate phosphorus and sediment in surface runoff and drain flow from clayey soils. *J. Environ. Qual.* 30:589–595.

Van Veen, J. A. and P. J. Kuikman. 1990. Soil structural aspects of decomposition of organic matter by micro-organisms. *Biogeochemistry* 11:213–223.

Vance, E. D. 2000. Agricultural site productivity: Principles derived from long term experiments and their implications for intensively managed forests. *For. Ecol. Manage.* 138:369–396.

Vance, G. F., F. J. Stevenson, and F. J. Sikora. 1996. Environmental chemistry of aluminum–organic complexes. In: *The Environmental Chemistry of Aluminum*, ed., G. Spostio, pp. 169–220. New York: Lewis Publishers.

Vandenbygaart, A. J. and B. D. Kay. 2004. Persistence of soil organic carbon after plowing a long-term no-till field in southern Ontario, Canada. *Soil Sci. Soc. Am. J.* 68:1394–1402.

Ve, N. B., D. C. Olk, and K. G. Cassman. 2004. Characterization of humic acid fractions improves estimates of nitrogen mineralization kinetics for lowland rice soils. *Soil Sci. Soc. Am. J.* 68:1266–1277.

Vitousek, P. M., J. D. Aber, R. W. Howarth, G. E. Likens, P. A. Matson, D. W. Schindler, D. W. Schlesinger, and D. G. Tilman. 1997. Human alteration of the global nitrogen cycle: Sources and consequences. *Appl. Ecol.* 7:737–750.

Wander, M. M., M. G. Bidart, and S. Aref. 1998. Tillage impacts on depth distribution of total and particulate organic matter in the three Illinois soils. *Soil Sci. Soc. Am. J.* 62:1704–1711.

Wander, M. M., R. B. Dudley, S. J. Traina, S. J. Kaufman, B. R. Stinner, and G. K. Sims. 1996. Acetate fate in organic and conventionally managed soils. *Soil Sci. Soc. Am. J.* 60:1110–1116.

Wander, M. M. and S. J. Traina. 1996. Organic matter fractions from organically and conventionally managed soils: I. Carbon and nitrogen distribution. *Soil Sci. Soc. Am. J.* 60:1081–1087.

Wander, M. M., S. J. Traina, B. R. Stinner, and S. E. Peters. 1994. The effects of organic and conventional management on biological-active soil organic matter pools. *Soil Sci. Soc. Am. J.* 58:1130–1139.

Wani, S. P., W. B. McGill, K. L. Haugen-Kozyra, J. A. Robertson, and J. J. Thurston. 1994. Improved soil quality and barley yields with faba beans, manure, forages and crop rotation on a gray Luvisol. *Can. J. Soil Sci.* 74:75–84.

Webb, J., P. Bellamy, P. J. Loveland, and G. Goodlass. 2003. Crop residue returns and equilibrium soil organic carbon in England and Wales. *Soil Sci. Soc. Am. J.* 67:928–936.

Weaver, A. R., D. E. Kissel, F. Chen, L. T. West, W. Adkins, D. Rickman, and J. C. Luvall. 2004. Mapping soil pH buffering capacity of selected fields in the coastal plain. *Soil Sci. Soc. Am. J.* 68:662–668.

Weber, J. H. 1988. Binding and transport of metals by humic materials. In: *Humic Substances and Their Role in the Environmental*, eds., F. H. Frimmel, and R. F. Christman, pp. 165–178. Chichester, UK: John Wiley & Sons.

Weil, R. R. and F. Magdoff. 2004. Significance of soil organic matter to soil quality and health. In: *Soil Organic Matter in Sustainable Agriculture*, eds., F. Magdoff, and R. R. Weil, pp. 1–43. Boca Raton, Florida: CRC Press.

West, C. P., A. P. Mallarino, W. F. Wedin, and D. B. Marx. 1989. Spatial variability of soil chemical properties in grazed pastures. *Soil Sci. Soc. Am. J.* 53:84–789.

West, T. O. and W. M. Post. 2002. Soil organic carbon sequestration rates by tillage and crop rotation: A global data analysis. *Soil Sci. Soc. Am. J.* 66:1930–1946.

Wienhold, B. J. and A. D. Halvorson. 1999. Nitrogen mineralization responses to cropping, tillage, and nitrogen rate in the northern Great Plains. *Soil Sci. Soc. Am. J.* 63:192–196.

Wilcke, W. and J. Lilienfein. 2004. Soil carbon-13 natural abundance under native and managed vegetation in Brazil. *Soil Sci. Soc. Am. J.* 68:827–832.

Willert, F. J. V. and R. C. Stehouwer. 2003. Compost, limestone, and gypsum effects on calcium and aluminum transport in acidic mine spoil. *Soil Sci. Soc. Am. J.* 67:778–786.

Wilhelm, W. W., J. M. F. Johnson, J. L. Hatfield, W. B. Voorhees, and D. R. Linden. 2004. Crop and soil productivity response to corn residue removal: A literature review. *Agron. J.* 96:1–17.

Williams, S. M. and R. R. Weil. 2004. Crop cover root channels may alleviate soil compaction effects on soybean crop. *Soil Sci. Soc. Am. J.* 68:1403–1409.

Willis, R. J. 1985. The historical bases of the concept of allelopathy. *J. Hist. Biol.* 18:71–102.

Woods, L. E. and G. E. Schuman. 1988. Cultivation and slope position effects on soil organic matter. *Soil Sci. Soc. Am. J.* 52:1371–1376.

Wright, A. L. and F. M. Hons. 2004. Soil aggregation and carbon and nitrogen storage under soybean cropping sequences. *Soil Sci. Soc. Am. J.* 68:507–513.

Wright, A. L. and F. M. Hons. 2005. Carbon and nitrogen sequestration and soil aggregation under sorghum cropping sequences. *Biol. Fertil. Soils* 41:95–100.

Wright, S. F., J. L. Starr, and I. C. Paltineau. 1999. Changes in aggregate stability and concentration of glomalin during tillage management transition. *Soil Sci. Soc. Am. J.* 63:1825–1829.

Wuddivira, M. N., D. A. Robinson, I. Lebron, L. Brechet, M. Atewell, S. D. Caires, M. Oatham et al. 2012. Estimation of soil clay content from hygroscopic water content measurements. *Soil Sci. Soc. Am. J.* 76:1529–1535.

Xiang, H. and A. Banin. 1996. Solid-phase manganese fractionation changes in saturated arid-zone soils: Pathways and kinetics. *Soil Sci. Soc. Am. J.* 60:1072–1080.

Yamaguchi, N., S. Hiradate, M. Mizoguchi, and T. Miyazaki. 2004. Disappearance of aluminum tridecamer from hydroxyaluminum solution in the presence of humic acid. *Soil Sci. Soc. Am. J.* 68:1838–1843.

Yang, H. S. and B. H. Janssen. 1997. Analysis of impact of farmers practices on dynamics of soil organic matter in northern China. In: *Proceedings of the ESA-Congress*, Veldoven 1996, eds., M. van Ittersum, and S. C. van Geijn, pp. 267–275. Amsterdam: Elsevier.

Zaher, H., J. Caron, and B. Ouaki. 2005. Modeling aggregate internal pressure evolution following immersion to quantify mechanisms of structural stability. *Soil Sci. Soc. Am. J.* 69:1–12.

Zhang, H., K. H. Hartge, and H. Ringe. 1997. Effectiveness of organic matter incorporation in reducing soil compatibility. *Soil Sci. Soc. Am. J.* 61:239–245.

Zhang, M. K., Z. L. He, D. V. Calvert, P. J. Stoffella, X. E. Yang, and Y. C. Li. 2003a. Phosphorus and heavy metal attachment and release in sandy soil. *Soil Sci. Soc. Am. J.* 67:1158–1167.

Zhang, G., B. Liu, G. Liu, X. He, and M. A. Nearing. 2003b. Detachment of undisturbed soil by shallow flow. *Soil Sci. Soc. Am. J.* 67:713–719.

5 Nitrogen Use Efficiency in Crop Plants

5.1 INTRODUCTION

In the last six decades, nitrogen fertilization has been a powerful tool in increasing the grain yield of food crops, especially wheat, rice, and corn. However, in the current agricultural and economic environment farmers must optimize the application of nitrogen fertilizers to avoid pollution by nitrate and to preserve their economic margin (Hirel et al., 2001). In this context, nutrient use efficiency, including nitrogen in crop plants, has special importance due to economic and environmental implications. The importance of improving nutrient use efficiency in modern agriculture or crop production is higher because of the use of large amounts of chemical fertilizers. It has been estimated that almost 10^{11} kg of N per annum is applied as fertilizer worldwide, a 20-fold increase over the past 50 years, at a cost of 50 billion US dollars (Glass, 2003). The majority of crops except nitrogen-fixing legumes receive an application of N; the major requirements are for the production of seeds (Mengel et al., 2006) and forage (Kingston-Smith et al., 2006). However, crop plants are only able to convert 30–40% of this applied N into useful food products such as grain (Raun and Johnson, 1999). There is therefore extensive concern in relation to the N that is not used by the plant, which is lost by leaching of nitrate, denitrification from the soil, and the loss of ammonia to the atmosphere, all of which can have deleterious environmental effect (Vitousek et al., 1997; Glass, 2003). The possibility of more precise use of N in crop production is now being taken seriously (Day, 2005).

The widespread availability of N fertilizers from the 1950s onwards has enabled many farmers around the world to abandon exploitative, low-yielding agricultural practices that had minded the soil for macronutrients (Duan et al., 2011). Most of the food crops are highly responsive to N fertilization. According to FAO (2004), corn grain yields are highly responsive to N fertilization, leading to annual applications of an estimated 10 million Mg N fertilizer worldwide. Corn is not the only crop that requires N fertilization for higher yields but other crops such as rice, wheat, barley, and oats also need an adequate amount of N for maximizing yields. The use of chemical fertilizers in modern cropping systems is a prerequisite to maximize crop yields (Fageria, 1992). Nearly half or less of the N applied to soil, however, is unutilized by plants and is subject to leaching to groundwater or the emission of gaseous compounds into the atmosphere (Bundy and Andraski, 2005). This not only increases crop input costs but also threatens human health and the stability of natural ecosystems (Kaiser, 2001; Duan et al., 2011). Hence, this topic deserves ample discussion in order to provide information available in the literature on NUE in crop plants, which may be helpful in reducing the cost of crop production while at the same time reducing the risk of environmental pollution (air, water, and soil).

Global food security requires yield improvements or an expansion of land area used for agriculture. In addition, optimum resource use efficiency is a prerequisite for sustainability (Hawkesford, 2012). A major driver for yield, especially in intensive agriculture systems, is N fertilizer. Canopy growth requires N, and it is canopy photosynthesis that ultimately drives yield. The canopy also acts as a reservoir of N and other minerals, which are recycled into grain with potentially higher efficiency (Hawkesford, 2012). Inappropriate use of N can lead to a lower uptake of N, higher cost, and also environmental pollution. Well-adopted agronomic practices have a crucial role in optimizing fertilizer use to exploit the full potential of crop cultivars (Hawkesford, 2012).

In the twenty-first century, improving nutrient use efficiency in crop plants will be an important topic compared to the twentieth century, mainly due to limited land and water resources available for crop production, a higher cost of inorganic fertilizer inputs, declining trends in crop yields globally, and increasing environmental concerns. Furthermore, at least 60% of the world's arable lands have mineral deficiencies or elemental toxicity problems, and on such soils, fertilizers and lime amendments are essential to achieving improved crop yields. Fertilizer inputs are increasing the cost of production of farmers, and there is a major concern for environmental pollution due to excess fertilizer inputs. Higher demands for food and fiber by increasing world populations further enhance the importance of nutrient-efficient cultivars that are also higher producers. Nutrient-efficient plants are defined as those plants which produce higher yields per unit of nutrient, applied or absorbed, than other plants (standards) under similar agroecological conditions (Fageria et al., 2008).

During the last four decades, much research has been conducted to identify and/or breed nutrient-efficient plant species or genotypes/cultivars within species and to further understand the mechanisms of nutrient efficiency in crop plants. However, success in releasing nutrient-efficient cultivars has been limited. The main reasons for limited success are that the genetics of plant responses to nutrients and plant interactions with environmental variables are not well understood. Complexity of genes involved in nutrient use efficiency for macro- and micronutrients and limited collaborative efforts between breeders, soil scientists, physiologists, and agronomists to evaluate nutrient efficiency issues on a holistic basis have hampered progress in this area. Hence, during the twenty-first century, agricultural scientists have tremendous challenges, as well as opportunities, to develop nutrient-efficient crop plants and to develop best management practices that increase the plant efficiency for utilization of applied fertilizers (Fageria et al., 2008).

During the twentieth century, breeding for nutritional traits has been proposed as a strategy to improve the efficiency of fertilizer use or to obtain higher yields in low-input agricultural systems. This strategy should continue to receive top priority during the twenty-first century for developing nutrient-efficient crop genotypes, since an efficient use of N in plant production is an essential goal in crop management. This chapter overviews the importance of nutrient-efficient plants in increasing crop yields in modern agriculture. Further, the objective of this chapter is to discuss the latest available information on NUE in crop plants. This information may be helpful in improving NUE in crop plants and will produce significant economic and environmental benefits for worldwide agriculture (Fageria, 2009, 2013, 2014).

5.2 DEFINITIONS OF NUTRIENT-EFFICIENT PLANTS AND NITROGEN USE EFFICIENCY

Before discussing in detail the NUE in crop plants, it is important to discuss or define nutrient use efficiency and nutrient-efficient plants and finally discuss NUE. There are several definitions of nutrient use efficiency as well as nutrient-efficient plants. Large variations in defining nutrient-efficient plants and methods used in calculating nutrient use efficiency make it difficult to compare results of different studies (Fageria et al., 2008). The effort to measure yield response to an applied nutrient is further confounded by other factors, such as variable soil fertility levels, climatic conditions, crop rotations, and changes in production practices that affect nutrient use efficiency (Stewart et al., 2005). Shaviv and Mikkelsen (1993), Munoz et al. (2005), and Sato et al. (2012) also reported that the efficiency of fertilizers applied to crops varied due to diversity in production systems, crop type, growing season, fertilizer placement, irrigation, soil type, weather conditions, and chemical transformations of nutrients in the soil.

Since N is one of the most expensive nutrients to supply, one of the objectives of crop improvement programs is to measure and maximize NUE (Good et al., 2004; Castellanos et al., 2010). While the literature contains several definitions and methods for evaluating this index, it is essentially the ratio between crop output and nutrient inputs. In simple terms, efficiency is the ratio of output (economic yield) to input (fertilizers) for a process or complex system (Crop Science Society of America, 1992). Some of these definitions pertaining to two subjects (nutrient-efficient plants and nutrient use efficiency) reported in the literature are presented and discussed in Table 5.1. NUE in crop plants has been defined in several ways in the literature. Graham (1984) defined NUE as the ability of a genotype to produce superior grain yields under low soil N conditions in comparison with other genotypes. However, most of them denote the ability of a system to concert inputs into outputs (Fageria and Baligar, 2005).

Bandyopadhyay and Sarkar (2005) also reported that NUE can be described in different ways depending on whether the focus is on grain only or on the total biomass. Moll et al. (1982, 1987) defined NUE as the yield of grain per unit of available nitrogen in the soil. This NUE can be divided into two processes: uptake efficiency, the ability of the plant to remove N from the soil normally present as nitrate and ammonium ions; and the utilization efficiency (UE), the ability of the plant to transfer N to the grain, predominantly present as protein. There has recently been considerable interest in identifying the processes involved in regulating the N uptake and metabolism within the plant (Andrew et al., 2004; Gallais and Hirel, 2004; Lea and Azevedo, 2006).

TABLE 5.1
Definitions of Nutrient-Efficient Plants

Definition	Reference
Nutrient-efficient plant is defined as a plant that absorbs, translocates, or utilizes more of a specific nutrient than another plant under conditions of relatively low nutrient availability in the soil or growth media	Soil Science Society of America (2008)
The nutrient efficiency of genotypes (for each element separately) is defined as the ability to produce a high yield in a soil that is limiting in that element for a standard genotype	Graham (1984)
Nutrient efficiency of a genotype/cultivar is defined as the ability to acquire nutrients from a growth medium and/or to incorporate or utilize them in the production of shoot and root biomass or utilizable plant material (grain)	Blair (1993)
An efficient genotype is one that absorbs relatively high amounts of nutrients from the soil and fertilizer, produces a high grain yield per unit of absorbed nutrient, and stores relatively little nutrients in the straw	Isfan (1993)
Efficient plants are defined as those that produce more dry matter or have a greater increase in harvested portion per unit time, area, or applied nutrient, have fewer deficiency symptoms, or have greater incremental increases and higher concentrations of mineral nutrients than other plants grown under similar conditions or compared to a standard genotype	Clark (1990)
Efficient germplasm requires less nutrients than an inefficient one for normal metabolic processes	Gourley et al. (1994)
Efficient plant is defined as one that produces higher economic yield with a determined quantity of applied or absorbed nutrient compared to other or a standard plant under similar growing conditions	Fageria et al. (2008)

Source: Adapted from Graham, R. D. 1984. *Advances in Plant Nutrition*, Vol. 1, pp. 57–102. New York: Praeger Publisher; Clark, R. B. 1990. *Crops as Enhancers of Nutrient Use*, pp. 131–209. San Diego, California: Academic Press; Blair, G. 1993. *Genetic Aspects of Mineral Nutrition*, pp. 205–213. Dordrecht: Kluwer Academic Publishers; Isfan, D. 1993. *Plant Soil* 154:53–59; Gourley, C. J. P., D. L. Allan, and M. P. Russelle. 1994. *Plant Soil* 158:29–37; Fageria, N. K., V. C. Baligar, and Y. C. Li. 2008. *J. Plant Nutr.* 31:1121–1157; Soil Science Society of America. 2008. *Glossary of Soil Science Terms*. Madison, Wisconsin: Soil Science Society of America.

Low NUE in crop production is explained by N losses from the system, including plant senescence, denitrification, surface runoff, volatilization, leaching, and immobilization (Sato et al., 2012). Nitrogen in the $N-NO_3$ form is readily available to plants and also easily lost by leaching since its electrostatic adsorption is lower than of other anions present in the solid–liquid interface of the topsoil. Therefore, $N-NO_3^-$ is not absorbed to the soil, and remains in the soil solution from where it is easily lost by leaching to deeper soil layers out of reach of plant roots (Sartor et al., 2011).

There is a variation in NUE at lower and higher N rates. At high nitrogen input, variation in NUE was explained by variation in nitrogen uptake capabilities, whereas at low nitrogen input, variation in NUE was mainly due to differences in NUE defined as the ratio grain yield/nitrogen uptake (Hirel et al., 2001). These differences in the expression of genetic variability were further confirmed following the detection of specific quantitative trait *loci* (QTL) for a given level of fertilization (Bertin and Gallais, 2001). This suggests that several sets of genes are differentially expressed according to the amount of nitrogen provided to the plant (Bertin and Gallais, 2001).

In parallel with these agronomic studies, several investigators found that it is possible to detect genetic variation and select new genotypes that show increased or decreased activities of several enzymes involved in the nitrogen assimilatory pathway (Groat et al., 1984; Degenhart et al., 1992; Harrison et al., 2000; Hirel et al., 2001). In particular, in corn hybrids, several studies have been done to correlate the efficiency of primary nitrogen assimilation and nitrogen remobilization with yield and its components (Reed et al., 1980; Purcino et al., 1998). As a result of these studies, it was concluded that increases in grain yield observed during the last two decades were not due to additional enhancement in inorganic nitrogen assimilation but rather due to a better NUE as a result of a more efficient nitrogen remobilization (Hirel et al., 2001). In particular, leaf longevity was shown to be one of the main factors responsible for yield increase in modern corn hybrids (Tollenaar, 1991; Ma and Dwyer, 1998). Extension of leaf metabolism activity improved the ratio between the assimilate supply from source leaves and demand in sink leaves during grain filling and was independent of the level of fertilization in the soil (Racjan and Tollenaar, 1999a,b). During this metabolic process, the putative role of enzymes involved in inorganic nitrogen assimilation and recycling such as nitrate reductase (NR), cytosolic Gln synthetase (GS1), and Glu dehydrogenase (GDH) was suggested (Lea and Ireland, 1999; Hirel et al., 2001).

Moll et al. (1982) reported that efficiency in uptake and utilization of N in the production of grain requires that those processes associated with absorption, translocation, assimilation, and redistribution of N operate effectively. Taking into consideration these aspects of N use, definitions of NUEs have been grouped or classified as agronomic efficiency (AE), physiological efficiency (PE), agrophysiological efficiency (APE), apparent recovery efficiency (ARE), and UE (Fageria and Baligar, 2001, 2003, 2005; Fageria et al., 2003, 2011; Fageria, 2013, 2014). The determination of NUE in crop plants is an important approach to evaluating the fate of applied chemical fertilizers and their role in improving crop yields.

Raun and Johnson (1999) defined NUE as the percentage of fertilizer N removed in grain, and averages about 33% worldwide for cereal grain production. A low NUE is a result of fertilizer N losses through various physical, chemical, and biological pathways of the soil–plant–atmosphere systems (Zhang et al., 2012). These losses are proportional to the amount of excess fertilizer N applied (Johnson and Raun, 2003), especially for adverse climatic conditions. Hence, the most effective way to improve NUE is to reduce excess N throughout the entire growing season by synchronizing N supply with plant N demand. Given the dynamic nature of weather and other factors affecting plant N demand, N synchronization is best done using process-based, dynamic crop simulation models (Zhang et al., 2012).

Crop N requirement is a physiological component, which is directly related to the genetic potential of the crop and to plant growth conditions. This component is determined by the overall crop N uptake under optimum growing conditions (Zotarelli et al., 2009). The fertilizer uptake efficiency for a specific production system depends on factors, including environmental conditions, management, rate, timing, and source of nitrogen (Zotarelli et al., 2009).

5.2.1 AGRONOMIC EFFICIENCY

AE is defined as the economic production obtained per unit of N applied. It can be calculated with the help of the following equation:

$$AE(kg\ kg^{-1}) = \frac{GY_f\ in\ kg - GY_{uf}\ in\ kg}{N\ rate\ in\ kg}$$

where GY_f is the grain yield of fertilized plot and GY_{uf} is the grain yield of unfertilized plot. If the experiment is conducted under greenhouse or controlled conditions, the unit of AE is in mg mg^{-1} and grain yield and N rate should also be expressed in mg.

AE for lowland rice was calculated and the values are presented in Table 5.2. AE was higher at lower N rate and decreased with increasing N rates. This indicated that rice plants were unable to absorb N when applied in excess because their absorption mechanisms might have been saturated. Under these conditions, the possibility exists for more N being subject to loss by NH_3 volatilization, leaching, and denitrification. Decreases in N uptake efficiency at higher N rates have been reported by Kurtz et al. (1984). Similarly, Eagle et al. (2001) reported that NUE in rice, which has both physiological and soil N supply components, decreased with increases in soil N supply, indicating that some of the decrease in NUE may have been due to the increased soil N supply. Sowers et al. (1994) also reported that a reduction in N availability efficiency with an increasing N rate indicates a greater proportion of the N supplied was not recovered in the plant tissue or retained in the soil profile to the depth of rooting. This suggests N loss was a controlling factor in reducing the N uptake and use.

There is a general trend of decreasing NUE with increasing N fertilizer rates (Barbieri et al., 2008; Abbasi et al., 2012). Jokela and Randall (1997) reported that the N uptake in corn grain was increased by higher N rates in all cases, while the total N recovery ranged from 31% to 60% at the low N rates and decreased to 24–45% at the high rates. Fertilizer N recovery by the crop may sometimes be greater when N application is delayed as compared to application at planting (Russelle et al.,

---

TABLE 5.2
Agronomic Efficiency of Lowland Rice under Different N Levels

Nitrogen Rate (kg ha^{-1})	Agronomic Efficiency (kg kg^{-1})
30	35
60	32
90	22
120	22
150	18
180	16
210	13
Average	23
R^2	0.93**

Source: Adapted from Fageria, N. K., and V. C. Baligar. 2001. *Commun. Soil Sci. Plant Anal.* 32:1405–1429.

**Significant at the 1% probability level.

1983; Jokela and Randall, 1997). This is probably due to the greater exposure to N applied at plant-ing to a range of possible loss processes (immobilization, leaching, denitrification, and clay fixation) at a time when N uptake rates are relatively low (Scharf et al., 2002). It has also been reported that corn begins to rapidly take up N during the middle vegetative growth period, with the maximum rate of N uptake occurring near silking (Binder et al., 2000).

Maximum AE was 35 kg of grain produced per kg N applied at the 30 kg N ha^{-1} level and a minimum AE of 13 kg grain produced per kg N applied was obtained at the 210 kg N ha^{-1}. Overall, AE was 23 kg grain produced per kg of N applied. Bouldin (1986) reported that in lowland rice, the AE may range from 30 to 45 kg grain produced per kg of N applied. De Datta (1986) reported that the AE in lowland rice varied from 21 to 46 kg kg^{-1} depending on source, rate, and methods of split application. Similarly, Yoshida (1981) reported that AE in lowland rice varied from 15 to 25 kg rice grain produced per kg N applied. These results were confirmed by Prasad and De Datta (1979). De Datta (1986) reported that the NUE is about 20% higher in temperate regions than in the tropics. Hussain et al. (2000) also reported that AE of lowland rice in the Philippines was 18 kg of grain produced with the application of 1.0 kg of N. Thind et al. (2010) reported that the overall AE in lowland rice was 26 kg kg^{-1}.

The author studied AE in lowland and upland rice genotypes (Figures 5.1 and 5.2). There was a significant variation among lowland and upland rice genotypes in NUE. In lowland rice, it varied about 32–50 kg kg^{-1}, whereas in upland rice, the NUE varied from 10 to 28 kg kg^{-1} depending on the genotypes. Fageria et al. (2007) also reported that AE efficiency varied from 16 to 23 kg kg^{-1} among five lowland rice genotypes, with an average value of 10 kg kg^{-1}. Variations in NUE among lowland and upland rice genotypes have been reported by Fageria et al. (2003, 2007, 2010), and Fageria and Baligar (2005).

Novoa and Loomis (1981) reviewed the literature and reported that the AE of N for barley varied from 7.1 to 40 kg kg^{-1} and for wheat it varied from 15 to 45 kg kg^{-1}.

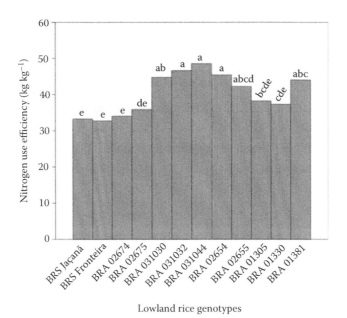

FIGURE 5.1 Agronomic efficiency of lowland rice genotypes. (From Fageria, N. K. 2014. *Mineral Nutrition of Rice*. Boca Raton, Florida: CRC Press. With permission.)

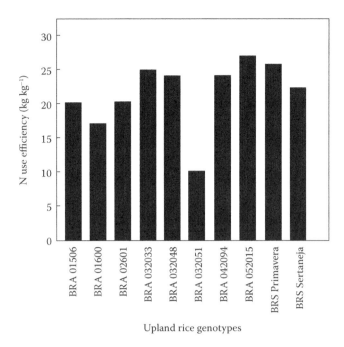

FIGURE 5.2 Agronomic efficiency of upland rice genotypes. (From Fageria, N. K. 2014. *Mineral Nutrition of Rice*. Boca Raton, Florida: CRC Press. With permission.)

5.2.2 Physiological Efficiency

PE is defined as the biological yield (grain plus straw) obtained per unit of N uptake by both grain and straw. It can be calculated by using the following equation:

$$PE(kg\ kg^{-1}) = \frac{BY_f\ in\ kg - BY_{uf}\ in\ kg}{N_{uf}\ in\ kg - N_{uuf}\ in\ kg}$$

where BY_f is the biological yield (grain plus straw) of the fertilized plot, BY_{uf} is the biological yield of the unfertilized plot, N_{uf} is the N uptake of the fertilized plot (grain plus straw), and N_{uuf} is the N uptake of the unfertilized plot (grain plus straw).

Fageria and Baligar (2001) calculated the PE of lowland rice under different N rates (Table 5.3). It varied from 113 to 182 kg kg^{-1}, with an average value of 146 kg kg^{-1}. There was a highly significant variation in PE when N was applied in the range of 30–210 kg ha^{-1}. The variation in PE was 62% with the addition of N in the range of 30–210 kg ha^{-1}. Fageria et al. (2007) reported that PE varied from 105 to 222 kg kg^{-1}, with an average value of 155 kg kg^{-1}.

5.2.3 Agrophysiological Efficiency

APE is defined as the economic production (grain yield in case of annual crops) per unit of N uptake in grain plus straw. APE can be calculated by using the following equation:

$$APE(kg\ kg^{-1}) = \frac{GY_f\ in\ kg - GY_{uf}\ in\ kg}{N_{uf}\ in\ kg - N_{uuf}\ in\ kg}$$

TABLE 5.3
Physiological Efficiency of Lowland Rice under Different N Levels

Nitrogen Rate (kg ha⁻¹)	Physiological Efficiency (kg kg⁻¹)
30	156
60	166
90	182
120	132
150	146
180	126
210	113
Average	146
R^2	0.62^{**}

Source: Adapted from Fageria, N. K. and V. C. Baligar. 2001. *Commun. Soil Sci. Plant Anal.* 32:1405–1429.
**Significant at the 5% probability level.

where GY_f is the grain yield of the fertilized plot, GY_{uf} is the grain yield of the unfertilized plot, N_{uf} is the N uptake of the fertilized plot (grain plus straw), and N_{uuf} is the N uptake of the unfertilized plot (grain plus straw).

Fageria and Baligar (2001) determined the APE of lowland rice under variable N levels (Table 5.4). The APE varied from 46 to 75 kg kg⁻¹, with an average value of 63 kg kg⁻¹. The APE was significantly influenced by N levels and the variation was 87% with the addition of N in the range of 30–210 kg ha⁻¹. Fageria et al. (2007) reported that APE among five lowland rice genotypes varied from 56 to 123 kg kg⁻¹, with an average value of 77 kg kg⁻¹. The Average APE (average of two locations and various N levels) in lowland rice reported by Thind et al. (2010) was 42 kg kg⁻¹. Gerik et al. (1994) reported that cotton requires 16–20 kg of N per 100 kg of lint. This translates into NUEs of 5.0–6.6 kg of lint per kg of N uptake.

TABLE 5.4
Agrophysiological Efficiency of Lowland Rice under Different N Levels

Nitrogen Rate (kg ha⁻¹)	Agronomic Efficiency (kg kg⁻¹)
30	72
60	73
90	75
120	66
150	57
180	51
210	46
Average	63
R^2	0.87^{**}

Source: Adapted from Fageria, N. K. and V. C. Baligar. 2001. *Commun. Soil Sci. Plant Anal.* 32:1405–1429.
**Significant at the 5% probability level.

5.2.4 APPARENT RECOVERY EFFICIENCY

Fertilizer N recovery efficiency also known as ARE is an important index to measure the fertilizer economic and environmental benefits (Xiong et al., 2013). It can be estimated by the ratio of increased plant N uptake that results from N application to the amount of applied N (Tian et al., 2011). ARE is defined as the quantity of N uptake per unit of N applied (Fageria et al., 2010). It can be calculated by using the following equation:

$$ARE(\%) = \frac{N_f - N_u}{\text{quantity of N applied}} \times 100$$

where N_f is the N accumulation by total biological yield (straw plus grain) in the fertilized plot (kg), N_u is the N accumulation by the total biological yield (straw plus grain) in the unfertilized plot (kg), and the quantity of N applied also in kg. This is related to field experimental data. If it is a greenhouse experiment, N uptake values can be expressed in mg and quantity of N applied should also be expressed in mg.

Nitrogen fertilizer has played an important role in increasing crop yields worldwide and the total consumption of N has increased gradually (Zhu and Chen, 2002; Singh et al., 2012; Yang et al., 2012) However, fertilizer NUE is low in most agroecological regions. For example, in rice, the N recovery efficiency is reported to be in the range of 25–45% and average about 35% (Dobermann and Cassman, 2002). More than half of the N fertilizer applied is lost and results not only in an environmental hazard but also in a substantial economic loss (Matsonet et al., 1997; Galloway, 1998; Choudhury and Kennedy, 2005; Li et al., 2009). Even though practices such as deep application (Roberts et al., 2009) and subsequent multiple topdressings of N fertilizer improve N fertilizer use efficiency, lack of application machinery and the rising cost of labor and the shortage of agricultural workers often limit the implementation of these practices (Zhang, 2008).

Fertilizer nutrients are not utilized efficiently in agriculture and the apparent recovery of fertilizer N in the soil–plant system seldom exceeds 50% of the N applied while the remaining is lost (Abbasi et al., 2005). Rice ARE in China ranged from 18% to 66% (Wang et al., 2001; Peng et al., 2006). Fageria and Baligar (2001) determined the ARE of lowland rice under different N rates (Table 5.5). It varied from 32% to 50%, with an average value of 39%. It was also significantly influenced by N levels and the variation was 82% due to the application of N rates. Nitrogen recovery

TABLE 5.5
Apparent Recovery Efficiency of Lowland Rice under Different N Levels

Nitrogen Rate (kg ha^{-1})	Apparent Recovery Efficiency (%)
30	49
60	50
90	37
120	38
150	34
180	33
210	32
Average	39
R^2	0.82**

Source: Adapted from Fageria, N. K. and V. C. Baligar. 2001. *Commun. Soil Sci. Plant Anal.* 32:1405–1429.

**Significant at the 5% probability level.

efficiency results reported in the literature varied from crop species to crop species and genotypes within species (Fageria et al., 2003). It was also influenced by crop management practices. Raun and Johnson (1999) reported that, worldwide, the ARE efficiency of N for cereal production (wheat, corn, rice, barley, sorghum, oat, millet, and rye) is approximately 33%. These authors also reported that the unaccounted 67% N loss represents a US$15.9 billion annual loss of N fertilizer. Plant N losses have accounted for 52–73% of the unaccounted N using ^{15}N in corn research (Francis et al., 1997), and between 21% (Harper et al., 1997) and 41% (Diagger et al., 1976) in winter wheat. Gases plant N loss excess of 45 kg N ha^{-1} year^{-1} has also been documented in soybean (Stutte et al., 1979). Bliss (1993) reported that in carefully conducted grain legume experiments, rarely is more than 30–50% of the applied N recovered in the plant.

Giambalvo et al. (2010) reported that the N recovery efficiency of wheat varied from 29.9% in the first year to 10.4% in the second year. These authors reported that the decrease observed in the second year may be due to the different characteristics of the soils in which the experiments were conducted. These authors also reported that their values of N recovery for wheat were comparable to values obtained in Tunisia by Sanaa et al. (1992), and in Syria by Pilbeam et al. (1997). Lopez-Bellido et al. (2006), in a study performed in Spain on durum wheat, reported values of labeled ^{15}N fertilizer recovery ranging from 12.7% when applied at sowing to 41.6% when applied at top dressing.

Fageria and Baligar (2001) reported that N recovery efficiency in irrigated or flooded rice is about 40%. De Datta and Buresh (1989) reported that Asia uses about 40% of the world's N fertilizer and about 60% of this is used for lowland rice production. According to De Datta and Buresh (1989), N recovery efficiency in lowland rice in Asia is around 40%. Similarly, Fageria et al. (2007) reported that N recovery efficiency use efficiency varied from 23% to 37% in irrigated or flooded rice among five flooded rice genotypes, with an average value of 29%. Apparent fertilizer N recovery efficiency in wheat was reported to be in the range of 21–59% depending on the year of cultivation, N rate, and soil preparation methods (Dalal et al., 2011). Ladha et al. (2005) reported that the global NUE for wheat is 34%. Similarly, Randall et al. (2003) reported that the global NUE for corn is about 31%. All these results suggest that a large part of applied N is lost in the soil–plant system.

Loss has been through ammonia volatilization, although leaching and denitrification also contribute to inefficient utilization. These authors further reported that because of the nature of these processes, attempts to reduce losses and improving recover efficiency have focused on water management. In the case of ammonia volatilization, loss is closely related to the concentration of ammonia in the floodwater. The estimates of the total losses of N as ammonia are between 36% and 44% (Fillery and De Datta, 1989). This loss was reduced to 22% by the use of urease inhibitors (Fillery and De Datta, 1989), but the cost is prohibitive (Buresh et al., 1988). Two approaches to reducing loss by reducing floodwater concentration of ammonia have been: (i) to drain the floodwater and apply the fertilizer to the soil surface; and (ii) to place the fertilizer below the soil surface. It was confirmed that losses were less when the fertilizer was applied to the drained soil surface than when it was broadcast on the floodwater (De Datta et al., 1987). It was shown that UE could be raised to more than 60% by placing the fertilizer below the surface of the soil (deep placement) (De Datta et al., 1989).

Nitrogen recovery efficiency as determined by the ^{15}N tracer technique in tropical lowland rice production has been reported to be approximately 30–50% (Bronsonet al., 2000; Eagle et al., 2001). Wilson et al. (1989) reported that, dependent on the application time, the rice plant had an observed total recovery of 53–74% of the applied N. Guindo et al. (1992) reported that N recovery efficiency of ^{15}N values, for drill-seeded, delayed flood rice in the range of 72–79% when ^{15}N fertilizer was applied a day before flood establishment and into the flood water at panicle differentiation. Dillon et al. (2012) reported that N recovery efficiency in flooded rice varied from 43% to 71% for multiple sources and multiple times between N fertilizer application and flooding across multiple environments.

The apparent N recovery efficiency of corn on a global basis varied from 25% to 50%, or an average of 33% (Raun and Johnson, 1999). The apparent NUE of corn in China is much lower (<25%) than in most developed countries (Ma et al., 2008; Ju et al., 2009). Cui et al. (2008) reported that due to boosting of agricultural production and the availability of chemical fertilizers to farmers since the 1990s, the apparent N recovery efficiency in cereal grain production decreased from 30% to 35% in the 1980s to <20% currently in China (Duan et al., 2011). Ju et al. (2009) demonstrated that the application rate of N fertilizer could be cut in half without the loss of yield if appropriate management practices are adopted. Apparent N recovery efficiency can be improved in crop plants with the adoption of improved management practices.

Benbi and Biswas (1996) reported that the apparent NUE of corn in India can be increased from 17% when N alone was applied to 33% by the application of balanced N and P fertilization together. Similarly, Duan et al. (2011) reported that, in China, NUE in corn can be increased from 20% to 45% by increasing the available P. Zhang et al. (2009) reported that the application of manure could significantly increase the total N content in the soil after 16 years of a wheat–corn cropping system in China.

Nitrogen recovery efficiency for flooded rice grown in Asia has been reported to range from 20% to 40% of applied N (De Datta et al., 1987, 1988; Schnier et al., 1990). These values were estimated using ^{15}N-labeled fertilizer and by differences in methods for determining N recovery efficiency values as calculated by Cassman et al. (1993). In some of these same studies, values ranged from 34% to 64%. Hussain et al. (2000) reported that N recovery efficiency in lowland rice grown in the Philippines was 36%. Bronson et al. (2000) reported that recovery efficiency in transplanted rice grown in Asia was higher (54%) when the difference method to calculate values was used than when the isotopic dilution method (44%) was used to calculate values. Thind et al. (2010 reported that ARE in lowland rice was 50%. Bijay-Singh and Yadvinder-Singh (2003) reported that despite considerable research to increase NUE in rice, the ARE of N fertilizer achieved by rice farmers ranges between 30% and 40%.

Even though agronomic and recovery efficiencies of N have been reported in the literature for various crops, other efficiencies (physiological, agrophysiological, and utilization) have rarely been reported. The N recovery efficiency in maize and sorghum has been reported to be about 25% from long-term plot research in Nebraska (Olson et al., 1986) and 40% for wheat in Oklahoma (Raun and Johnson, 1995). For cotton grown in Texas, N recovery efficiency has been reported to be in the range of 19–38% (Chua et al., 2003) depending on the management practices adopted. Randall et al. (2003) reported that apparent N recovery for corn ranged from 31% for total applied N at preplant to 44% for the split treatment (total 150 kg N ha^{-1}, 40% applied as preplant and 60% sidedress at V8 stage as defined by Ritchie and Hanway, 1984).

Halvorson et al. (2002) reported the fertilizer N recovery by onion (*Allium cepa* L.) to be 15% and unfertilized maize recovered 24% of the fertilizer applied to onion. Brown et al. (1988) reported N recovery efficiency by onion to be 19–26% depending on the rate and method of application. Huggins et al. (2001) reported the ARE of maize to be 44% when grown in monoculture and 50% when maize was grown in rotation with soybean (*Glycine max* L. Merr.).

Cassman et al. (2002) reported that N recovery efficiency in continuous lowland rice production systems in Asia to be about 31% and a somewhat higher efficiency of 37% for maize in the major maize-producing states of the United States. Errebhi et al. (1998) reported that during high rainfall and leaching events, an average of only 33% of the applied N was recovered by potato. While in the second year when growing season was characterized by less total rainfall and fewer leaching events, N recovery efficiency was 56%. Abbasi et al. (2005) reported that in the plots of grass sward, where NO_3^- was added as a source of N, N recovery efficiency was between 24% and 43%. In the same experiment, when N was added as NH_4^+, the N recovery efficiency was between 39% and 48%.

The Nitrogen recovery efficiency of fertilizer N by cotton plants has been reported to be low in field experiments (Stevens et al., 1996). Yasin (1991) reported that N recovery efficiency by cotton receiving 50, 100, and 150 kg N ha^{-1} was 34%, 38%, and 25%, respectively. Similarly, Stevens et al. (1996) reported that the N recovery efficiency by cotton was 38%, 28%, 19%, and 9% at rates of 45, 90, 135, and 180 kg N ha^{-1}. Constable and Rochester (1988) reported an N recovery efficiency of 30% for irrigated cotton grown on Vertisols. Wienhold et al. (1995) reported that corn grain utilized 35% and stover an additional 15% of the applied N fertilizer, while 30% of the N remained in the upper 0.6 m of the soil profile at the end of the growing season. Novoa and Loomis (1981) reported that, in calculating ARE, the majority of miscalculation is to do with basing the recovery estimate on grain N alone, ignoring the N accumulated into roots and straw residues, which recycles to soil humus and thus is available to subsequent crops. Fageria et al. (2011) compiled data from various sources and reported that AREs for major cereals in different continents are less than 50% (Table 5.6). In addition, the ARE was in the order of rice > wheat > corn = millet > sorghum.

Several authors have reported that current practices of N fertilizer management are often very inefficient as compared to natural systems, thus increasing the potential for N losses in crop production (Sanchez and Blackmer, 1988; Randall et al., 1997; Cambardella et al., 1999; Dinnes et al., 2002). As a result, first year recoveries of fertilizer N by corn have been reported to be 35% (Bijeriego et al., 1979), 14–65% (Meisinger et al., 1985), 23–45% (Kitur et al., 1984), 24–26% (Olson, 1980), 15–33% (Sanchez and Blackmer, 1988), and 45–59% (Reddy and Reddy, 1993). Approximately 28–55% of the N applied is taken up by the crop (Bundy and Andraski, 2005). Fertilizer recovery typically decreases with an increase in N application rate, and farmers incur economic loss by applying more N than is required to obtain a positive yield response (Macdonald et al., 1989).

Globally, a large land area is under pasture as compared to crop production. Hence, the knowledge of N recovery under pasture is very important. Schepers and Mosier (1991) reported that N recovery in pastures varies from 50% to 65% of the total N applied through N fertilizer. The N dynamics, at the site under grazing, with N application (urine) between 200 and 250 kg N ha^{-1} year^{-1} showed that one-third of the total applied N was retained by organic matter (25%) and by the plant root system (2–5%). Of the remaining N, a part is lost by ammonium volatilization after application (20%), ammonia volatilization from the spots of excreta deposition (20–30%), denitrification (6–12%), and by leaching or other types of losses, such as excretion outside the pasture (20–50%). Another fraction goes into animal production, be it meat (7–9%) or milk (14–18%) (Kimura and Kurashima, 1991; Whitehead, 2000).

TABLE 5.6

Average Apparent Recovery Efficiency (%) of Nitrogen for Five Major Cereals in Various Continents

Continent	Millet	Sorghum	Maize	Rice	Wheat
Europe			40 (3)		48 (34)
Africa	40 (25)	45 (6)	51 (11)	28 (30)	39 (4)
Asia	40 (4)	38 (9)	41 (12)	39 (66)	45 (22)
North America		18 (4)	29 (22)	53 (1)	51 (4)
South America		32 (9)	32 (42)	55 (17)	34 (32)
Australia		35 (9)	45 (3)	50 (9)	43 (12)
Average	40	34	40	45	43

Source: From Fageria, N. K., V. C. Baligar, and C. A. Jones. 2011. *Growth and Mineral Nutrition of Field Crops*, 3rd edition. Boca Raton, Florida: CRC Press. With permission.)

Note: Values in brackets are number of observations.

5.2.5 UTILIZATION EFFICIENCY

UE is defined as the product of physiological and apparent recovery efficiency. It can be calculated by using the following equation:

$$UE \ (kg \ kg^{-1}) = PE \times ARE$$

Fageria and Baligar (2001) studied the UE of N in lowland rice under different N levels (Table 5.7). It varied from 36 to 83 kg kg^{-1}, with an average value of 58 kg kg^{-1}. It decreased with increasing N rates and was significantly influenced by N levels. The variation in the UE was 90% due to the addition of N in the range of 30–210 kg ha^{-1}. Fageria et al. (2010) studied the UE and other NUE in upland rice under greenhouse conditions (Table 5.8).

AE and UE were significantly different among genotypes. PE, APE, and ARE were also varied among genotypes; however, the difference was statistically not significant. The AE varied from 12.8 to 26.7 mg grain produced per mg N applied with an average value of 21.4 mg grain produced per mg of N applied. The average value across the genotypes of PE was 86.6 mg dry matter production (grain plus straw) per mg N accumulated in grain and straw. The average value of APE was 45.2 mg grain produced per mg N accumulated in grain and straw. Fageria and Barbosa Filho (2001) determined the APE in eight lowland rice genotypes and reported an average value of 45.5 mg grain produced per mg N accumulated in the grain and straw. Hence, the APE of lowland and upland rice is comparable. The average ARE was 49.2%. The ARE in lowland rice is reported to be in the range of 31–40% in major rice-growing regions of the world (Cassman et al., 2002). Fageria and Baligar (2001) reported that the average ARE in lowland rice in Brazilian Inceptisol was 39%. This means that the ARE in upland rice is higher compared to lowland rice. The highest grain yield producing genotype CNAs 8993 had the highest AE, whereas the lowest yielding genotype CNA 8170 was having lowest AE. The genotype CNAs 8992 was also having highest PE and APE and reasonably good values of ARE and EU.

The author studied NUE in dry bean genotypes (Table 5.9). AE, PE, APE, ARE, and UE of N were significantly influenced by genotype treatment (Table 5.9). However, genotype means by Tukey's test were separated only for PE and APE. Further, the variation in NUEs among bean genotypes was reported to be related to genetic variability in N uptake and utilization under tropical conditions (Lynch and White, 1992; Lynch and Rodriguez, 1994).

TABLE 5.7
Utilization Efficiency of Lowland Rice under Different N Levels

Nitrogen Rate (kg ha^{-1})	Utilization Efficiency (kg kg^{-1})
30	76
60	83
90	67
120	50
150	50
180	42
210	36
Average	58
R^2	0.90**

Source: Adapted from Fageria, N. K. and V. C. Baligar. 2001. *Commun. Soil Sci. Plant Anal.* 32:1405–1429.

**Significant at the 5% probability level.

TABLE 5.8

Nitrogen Use Efficiency of 19 Upland Rice Genotypes

Genotype	Agronomic Efficiency (mg mg^{-1})	Physiological Efficiency (mg mg^{-1})	Agrophysiological Efficiency (mg mg^{-1})	Apparent Recovery Efficiency (%)	Utilization Efficiency (mg mg^{-1})
CRO 97505	23.5ab	100.2a	55.4a	43.5a	42.5ab
CNAs 8993	26.7a	101.4a	59.0a	47.8a	45.5ab
CNAs 8812	24.2ab	80.5a	41.3a	58.2a	47.0ab
CNAs 8938	21.8ab	87.8a	46.3a	48.7a	40.9ab
CNAs 8960	21.7ab	79.3a	46.1a	48.3a	37.3ab
CNAs 8989	24.8aab	72.3a	42.3a	59.3a	42.4ab
CNAs 8824	17.0ab	73.7a	36.7a	47.0a	33.5b
CNAs 8957	22.4ab	72.3a	46.3a	50.7a	35.2ab
CRO 97422	19.8ab	85.7a	47.1a	45.0a	36.6ab
CNAs 8817	17.9ab	72.7a	33.6a	53.2a	38.3ab
CNAs 8934	20.3ab	83.3a	40.5a	50.2a	41.8ab
CNAs 9852	22.9ab	81.7a	48.9a	47.7a	38.4ab
CNAs 8950	21.0ab	88.5a	49.9a	44.7a	37.6ab
CNA 8540	22.2ab	89.7a	45.2a	53.5a	44.1ab
CNA 8711	19.4ab	79.2a	40.0a	48.1a	38.0ab
CNA 8170	12.8b	92.8a	28.0a	45.4a	42.0ab
BRS Primavera	22.6ab	89.8a	51.2a	45.8a	39.9ab
BRS Canastra	21.7ab	125.7a	57.7a	40.6a	47.4ab
BRS Carisma	24.2ab	89.6a	42.8a	56.2a	50.5a
Average	21.4	86.6	45.2	49.2	41.0
F-test					
Genotype	*	NS	NS	NS	**
CV (%)	17	23	26	24	12

Source: From Fageria, N. K., O. P. Morais, and A. B. Santos. 2010. *J. Plant Nutr.* 33:1696–1711. With permission.
*,**, NS: Significant at the 5% and 1% probability level and nonsignificant, respectively. Means followed by the same letter in the same column are not significantly different at the 5% probability level by Tukey's test.

5.2.6 NUTRIENT EFFICIENCY RATIO

Nitrogen or nutrient use efficiency is also expressed in nutrient efficiency ratio (NER), which was suggested by Gerlof and Gabelman (1983) to differentiate genotypes into efficient and inefficient nutrient utilizers. The equation to calculate this efficiency ratio is as follows (Baligar et al., 2001):

$$\text{NER (kg kg}^{-1}) = \frac{\text{units of yield in kg}}{\text{units of element in tissue in kg}}$$

The above equation is used for expressing the results of field experimentation. If the experiment is conducted under greenhouse conditions, the unit of result expression is mg mg^{-1}.

The author studied the nitrogen efficiency ratio of cover crops under different pH levels (Table 5.10). Acidity and cover crop species interaction was significant for NUE, indicating variations in cover crop species to acidity levels for N uptake and utilization. Hence, the results of NUE under three acidity levels are presented (Table 5.10). The NUE varied from 27.43 mg dry matter produced with the accumulation of 1 mg of N in the shoot of Brazilian lucerne to 57 mg dry matter produced per mg of N uptake in the shoot of Black mucuna bean at high acidity level. The average

TABLE 5.9
Nitrogen Use Efficiency by Dry Bean Genotypes

Genotype	AE (mg mg⁻¹)	PE (mg mg⁻¹)	APE (mg mg⁻¹)	ARE (%)	UE (mg mg⁻¹)
Pérola	5.9	22.0efg	13.3ab	47.5	10.1
BRS Valente	3.7	26.5cdefg	12.7ab	27.0	7.3
CNFM 6911	4.7	28.7bcdef	13.9ab	30.3	8.7
CNFR 7552	8.4	48.3a	21.9ab	38.7	18.7
BRS Radiante	9.0	33.9abcdef	18.1ab	53.6	17.9
Jalo Precoce	6.3	30.7bcdef	14.8ab	40.9	13.1
Diamante Negro	7.1	24.7defg	12.6ab	48.1	13.1
CNFP 7624	11.5	42.0ab	22.8ab	50.4	21.2
CNFR 7847	9.0	27.8bcdefg	16.4ab	43.3	12.4
CNFR 7866	6.1	13.0 g	10.2b	60.5	8.2
CNFR 7865	6.2	27.6bcdefg	12.5ab	47.6	12.4
CNFM 7875	11.2	36.0abcde	21.5ab	52.3	18.8
CNFM 7886	10.6	35.7abcde	19.3ab	54.7	19.5
CNFC 7813	10.7	35.0abcde	20.6ab	51.6	18.1
CNFC 7827	11.8	40.1abc	24.4a	49.0	19.5
CNFC 7806	8.1	26.9bcdefg	15.3ab	48.7	14.0
CNFP 7677	10.7	35.6abcde	21.2ab	50.5	18.0
CNFP 7775	10.7	39.9abcd	23.2ab	46.2	18.3
CNFP 7777	8.1	30.8bcdef	15.7ab	51.4	15.9
CNFP 7792	3.9	19.2fg	10.5ab	37.2	7.4
Average	8.2	31.2	17.0	46.5	14.6
F-test	**	**	**	NS	**

Source: From Fageria, N. K., L. C. Melo, and J. P. Oliveira. 2013. *J. Plant Nutr.* 36:2179–2190. With permission.

Note: AE = agronomic efficiency, PE = physiological efficiency, APE = agrophysiological efficiency, ARE = apparent recovery efficiency, and UE = utilization efficiency.

**, NS: Significant at the 1% probability level and nonsignificant, respectively. Within the same column, means followed by the same letter do not differ significantly at 5% probability level by Tukey's test.

value of NUE at high acidity was 36.88 mg shoot dry weight per mg of N uptake. At the medium acidity level, NUE varied from 23.77 to 48.42 mg mg⁻¹, with an average value of 34.36 mg mg⁻¹. Similarly, at the low acidity level, maximum NUE of 40.44 mg kg⁻¹ was of Mucuna bean ana and a minimum NUE of 24.97 mg mg⁻¹ was of the white jack bean. The average NUE at low acidity was 31.54 mg mg⁻¹. The decrease in NUE at medium acidity level was about 7% as compared to the high acidity level. Similarly, the decrease in NUE at the low acidity level was 17% as compared to the high acidity level. The decrease in NUE with decreasing acidity level may be associated with a decrease in the shoot dry weight of cover crop species (Fageria et al., 2008). The decrease in the shoot dry weight with decreasing soil acidity was associated with tolerance of tropical cover crop species to soil acidity (Fageria et al., 2008).

5.2.7 NUE versus Crop Yield

As mentioned earlier, NUE is defined in several ways in the literature. The question is how the NUEs are related to crop yields. Nitrogen is the principal constituent of numerous organic compounds such as amino acids, proteins, nucleic acids, and compounds of secondary plant metabolism such as alkaloids (Mengel et al., 2001). The efficiency of N uptake and use relative to the production of grain requires that the processes associated with absorption, translocation, assimilation, and redistribution of N operate effectively (Moll et al., 1982). Among these processes, the uptake of N

TABLE 5.10

Nitrogen Use Ratio (NUR) of Tropical Legume Cover Crops as Influenced by Soil pH Levels

Cover Crop Species	NER (mg mg^{-1})		
	High Acidity	Medium Acidity	Low Acidity
Short-flowered crotalaria	41.84abc	33.04abcd	30.58a
Sunnhemp	36.90abc	32.34abcd	37.03a
Smooth crotalaria	30.38c	22.42d	25.38a
Showy crotalaria	30.88c	27.33bcd	28.42a
Crotalaria	28.00c	23.77cd	30.60a
Calapo	36.78abc	40.21abcd	31.46a
Black Jack bean	28.56c	27.94bcd	30.29a
Bicolor pigeon pea	26.86c	25.79bcd	32.23a
Black pigeon pea	33.35c	31.02abcd	35.01a
Mulato pigeon pea	37.29abc	41.62abc	31.72a
Lablab	37.40abc	37.50abcd	32.18a
Mucuna bean ana	56.18ab	47.01a	40.44a
Black mucuna bean	57.00a	48.42a	28.88a
Gray mucuna bean	46.48abc	43.34ab	34.01a
White Jack bean	34.69bc	34.51abcd	24.97a
Brazilian Lucerne	27.43c	33.58abcd	31.48a
Average	36.88	34.36	31.54
F-test			
Acidity level (A)	*		
Crop species (C)	**		
A × C	**		

Note: The acidity treatments were created by applying dolomitic lime at the rate of 0 g kg^{-1} (high acidity), 3.3 g kg^{-1} (medium acidity), and 8.3 g kg^{-1} of soil (low acidity).

*,**Significant at the 5% and 1% probability levels, respectively. Means within the same column followed by the same letter do not differ significant at the 5% probability level by Tukey's test.

in higher amounts by plants and its translocation to grain is crucial for increasing yields. A relationship was determined between grain yield and NUE of 19 upland rice genotypes tested under two N rates (zero, low, and high, 400 mg) (Figure 5.3). AE, APE, ARE, and UE had significant positive associations with grain yield.

Similarly, in a field experiment, NUE of lowland rice genotypes grown in field experiments had significant positive quadratic association with grain yield (Fageria and Baligar, 2005). López-Bellido et al. (2003) also reported that NUE in the faba bean was higher in the years with higher seed yield and higher N uptake. However, plant N accumulation and grain yield generally had positive quadratic relationships (Fageria and Baligar, 2001; Cassman et al., 2002), which means that diminishing returns appear in the conversion of plant N to grain as yields approach yield potential ceilings.

The author studied the relationship between NUEs and dry bean yield (Table 5.11). All the five NUEs were having highly significant ($P < 0.01$) association with grain yield (Table 5.11). AE was having higher coefficient of determination ($R^2 = 0.57*$), followed by UE ($R^2 = 0.47*$), APE ($R^2 = 0.37*$), ARE ($R^2 = 0.34*$), and PE ($R^2 = 0.27*$). Significant positive association between grain yield and NUEs indicates that improving NUE by using N-efficient crop genotypes along with improved N management practices can improve grain yield as well as reduce environmental pollution (Wiesler et al., 2001). Good et al. (2004) reported that N fertilization is correlated closely with crop yields. Farmers tend to use large amounts to maximize the quality and quantity and, so,

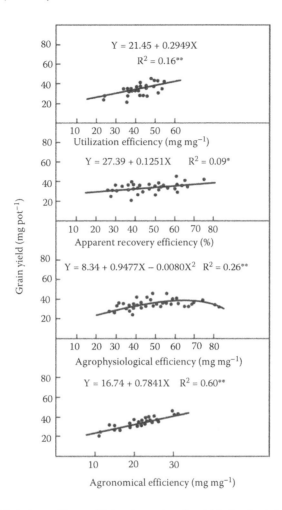

FIGURE 5.3 Relationship between N use efficiencies and grain yield of upland rice. (From Fageria, N. K., O. P. Morais, and A. B. Santos. 2010. *J. Plant Nutr.* 33:1696–1711. With permission.)

TABLE 5.11

Relationship between N Use Efficiency and Grain Yield of Dry Bean

Variable	Regression Equation	R^2
AE (X) versus grain yield (Y)	$Y = 4.1500 + 1.7022X - 0.0554X^2$	0.578**
PE (X) versus grain yield (Y)	$Y = -1.2031 + 0.8366X - 0.0106X^2$	0.27**
APE (X) versus grain yield (Y)	$Y = 1.7855 + 1.1349X - 0.0228X^2$	0.37**
ARE (X) versus grain yield (Y)	$Y = 0.8856 + 0.4415X - 0.0032X^2$	0.34**
EU (X) versus grain yield (Y)	$Y = 3.0899 + 1.2579X - 0.0311X^2$	0.47**

Source: From Fageria, N. K., L. C. Melo, and J. P. Oliveira. 2013. *J. Plant Nutr.* 36:2179–2190. With permission.
Note: AE = agronomic efficiency, PE = physiological efficiency, APE = agrophysiological efficiency, ARE = apparent recovery efficiency, and UE = utilization efficiency. With permission.
**Significant at the 1% probability level.

commercial fertilizers constitute a major production cost. This overdose of N clashes with one of the key aims of today's agriculture, that is, sustainability.

5.2.8 Mechanisms Responsible for Variation in NUE in Crop Plants

It is widely reported in the literature that crop species and genotypes within species differ significantly in NUE (Fageria and Baligar, 2003, 2005; Fageria et al., 2003, 2008, 2011; Fageria, 2013, 2014). Genetic differences in N uptake and/or grain per unit of N applied has also been reported in different crops, including wheat, rice, corn, sorghum, and barley (Ortiz-Monasterio et al., 1997; Muchow, 1998; Presterl et al., 2003; Anbessa et al., 2009). Several reasons have been cited why some genotypes are more efficient in N utilization as compared to others (Thomason et al., 2002). Moll et al. (1982) reported that NUE differences among corn hybrids were due to differing utilization of N already accumulated in the plant prior to anthesis, especially at low N levels. Eghball and Maranville (1991) reported that NUE generally parallels water use efficiency in corn. Hence, both NUE and water use efficiency traits might be selected simultaneously where such parallels exist.

Kanampiu et al. (1997) reported that wheat cultivars with a higher grain harvest indexes had higher NUE. Cox et al. (1985) reported that wheat cultivars that accumulate large amounts of N early in the growing season do not necessarily have high NUE. Plants must convert this accumulated N into grain N and must assimilate N after anthesis to produce high NUE. Forms of N uptake (NH_4^+ vs. NO_3^-) may also have effects on NUE (Thomason et al., 2002). Plants with a preferential uptake of NH_4^+ during grain fill may provide increased NUE over plants without this preference (Tsai et al., 1992). Ammonium-N supplied to high-yielding corn genotypes increased yield over plants supplied with NO_3^- during critical ear development (Pan et al., 1984). Salsac et al. (1987) reported that NH_4^+ assimilation processes require 5 ATP (adenosine triphosphate) mol^{-1} of NH_4^+, whereas NO_3^- assimilation processes require 20 ATP mol^{-1} NO_3^-. This energy-saving mechanism may be responsible for higher NUE in NH_4^+–N.

In addition to the above-mentioned reasons, Table 5.12 summarizes various soil and plant mechanisms and processes and other factors that influence genotypic differences in plant nutrient efficiency. No attempt has been made to discuss these mechanisms or processes in details. For extensive reviews related to nutrient flux and mechanisms of uptake and utilization in soil–plant systems, see Mengel et al. (2001), Barber (1995), Marschner (1995), Fageria et al. (2011), and Baligar et al. (2001).

Regarding genotypic variability for NUE, Rosielle and Hamblin (1981) reported that the heritability for grain yield is usually lower for plants grown under low N versus high N. Thus, potential progress would be lower for plants grown with low N as compared to high N target environments. Banziger and Lafitte (1997) reported that the heritability of grain yield usually decreases for plants grown under low N. Banziger et al. (1997) reported that secondary traits (ears per plant, leaf senescence, and leaf chlorophyll concentration) are valuable for increasing the efficiency of selection for grain yield when the broad-sense heritability of grain yield is low under low-N environments.

5.2.8.1 Physiological and Molecular Mechanisms Associated with NUE

There are two components of NUE in the plants. These are known as uptake and assimilation efficiency (N is stored and assimilated into amino acids and other important nitrogenous compounds in the leaves, roots, and young organs) and retranslocation or distribution and utilization in the grain formation. During the growth and development process, the stored amino acids are further utilized in the synthesis of proteins and enzymes involved in different biochemical pathways and the photosynthetic machinery governing plant growth, architecture, and development (Kant et al., 2010). Both the components of NUE are important in increasing NUE in crop plants. Both the components of NUE are influenced by environmental factors. In addition, several physiological and biochemical changes occur in plants as adaptive responses to N limitations, including an increase in N uptake by high-affinity transporters, remobilization of N from older to younger leaves and reproductive

TABLE 5.12

Soil and Plant Mechanisms and Processes and Other Factors Influencing Crop Species/ Genotypic Differences in Nutrient Use Efficiency in Plants

Nutrient Acquisition

Diffusion and mass flow in soil: buffer capacity, ionic concentration and properties, tortuosity, moisture, bulk density, temperature

Root morphological factors: number, length, extension, density, root hair density

Physiological: root/shoot ratio, root microorganisms (rhizobia, azotobacter, mycorrhizae), nutrient status, water uptake, nutrient influx and efflux, nutrient transport rates, affinity for uptake (Km), threshold concentration ($Cmin$)

Biochemical: enzyme secretion (phosphatases), chelating compounds, phytosiderophores, proton exudates, organic acid exudates (citric, malic, *trans*-aconitic)

Nutrient Movement in Root

Transfer across endodermal cells and transport in roots

Compartmentalization/binding within roots

Rate of nutrient release to xylem

Nutrient Accumulation and Remobilization in Shoot

Demand at cellular level and storage in vacuoles

Retransport from older to younger leaves and from vegetative to reproductive tissues

Rate of chelation in xylem transport

Nutrient Utilization and Growth

Nutrient metabolism at reduced tissue concentrations

Lower element concentrations in supporting structure, particularly stems

Elemental substitution (Fe for Mn, Mo for P, Co for Ni)

Biochemical: peroxidase for Fe efficiency, ascorbic acid oxidase for Cu, carbonic anhydrase for Zn, metallothionein for metal toxicities

Other Factors

Soil factors

Soil solution: ionic equilibria, solubility, precipitation, competing ions, organic ions, pH, phytotoxic ions

Physiochemical properties: organic matter, pH, aeration, structure, texture, compaction, moisture

Environmental effects

Intensity and quality of light (solar radiation)

Temperature

Moisture (rainfall, humidity, drought)

Plant diseases, insects, and allelopathy

Source: Adapted from Baligar, V. C., N. K. Fageria, and Z. L. He. 2001. *Commun. Soil Sci. Plant Anal.* 32:921–950; Fageria, N. K. and V. C. Baligar. 2003. *J. Plant Nutr.* 26:1315–1333.; Fageria, N. K., V. C. Baligar, and C. A. Jones. 2011. *Growth and Mineral Nutrition of Field Crops*, 3rd edition. Boca Raton, Florida: CRC Press.

parts, retardation of growth and photosynthesis, and increased anthocyanin accumulation (Ono et al., 1996; Chalker-Scott, 1999; Ding et al., 2005; Diaz et al., 2006).

5.2.8.2 Genetic Mechanisms for Nitrogen Uptake and Assimilation

Nitrogen uptake by plants is mainly in two forms, that is, nitrate (NO_3^-) and ammonia (NH_4^+). However, nitrate is the dominant form in most agriculture soils (Fageria, 2014). Nitrate is not only the predominant source of N supply to plants, but also acts as an important signal for several development processes. This regulation includes a rapid change in expression pattern of genes involved in carbon and N metabolism and other metabolic pathways (Kant et al., 2010). In addition, nitrate is reported to be responsible for the growth of root system and also root-to-shoot ratio. The function of several structural genes involved in N uptake and assimilation have been studied extensively in the past decade (Kant et al., 2010). In *Arabidopsis*, there are three families of nitrate transporters, that is, NRT1, NRT2, and CLC with 53 NRT1, 7 NRT2, and 7 CLC genes identified (Kant et al., 2010).

The NRT2 are high-affinity nitrate transporters while most of the NRT1 family members characterized so far are low-affinity nitrate transporters, except NRT1.1, which is a dual-affinity nitrate transporter. The NRT1.1, NRT1.2, NRT2.1, and NRT2.2 are involved primarily in nitrate uptake from the external environment (Miller et al., 2007; Tsay et al., 2007; Ho et al., 2009). Among the CLC family members, CLCa is known to mediate nitrate accumulation in the plant vacuole (De Angeli et al., 2006). Nitrate, after entering the plant cell, is reduced to nitrite by NR and further to ammonium by nitrite reductase. The ammonium derived from nitrate or from direct ammonium uptake by AMT transporters is further assimilated into amino acids via the glutamine synthetase (GS)/glutamate synthase (GOGAT) cycle (Kant et al., 2010).

5.2.8.3 Genetic Mechanisms for Nitrogen Remobilization

The N accumulated in the leaves is the main source for retranslocation to the seeds during the ripening growth stage of the plants. The majority of this remobilization occurs during senescence and N is mainly transported via amino acids. Up to 80% of the grain N contents are derived from leaves in rice and wheat (Tabuchi et al., 2005; Kichey et al., 2007). Plants have developed efficient methods and mechanisms that release tied-up N from source tissues via protease activities during leaf senescence (Kant et al., 2010). Approximately 80% of total N is located in the chloroplasts mainly in the form of proteins and this is an important N pool for remobilization (Adam et al., 2001). Among chloroplastic proteins, Rubisco (about 50% of total cellular proteins in C_3 and about 20% in C_4 plants) seems to serve as the major protein subjected to proteolysis and is responsible for most N remobilization during leaf senescence for grain filling (Mae et al., 1993). All plant species harbor ATG (AuTophaGy) genes designed to carry out this important recycling process and several ATG genes have been identified in plants. Among these, ATG8s participate in tagging proteins for degradation (Slavikova et al., 2005). A detailed discussion of genetic mechanism for N remobilization is given by Kant et al. (2010) and readers may consult this article for additional information.

5.3 CONCLUSIONS

Fertilizer use efficiency is an important factor that needs to be taken into consideration. In agricultural production systems, inefficient use of fertilizer inputs represents not only an environmental hazard but also a substantial economic loss. In general, efficiency is output divided by input. The higher the means value, the higher the efficiency or better use of applied resource. NUE in crop plants are classified as AE, PE, APE, ARE, and UE. AE is defined as the economic production obtained per unit of N applied. PE is defined as the biological yield obtained per unit of N uptake. APE is defined as the economic production (grain yield in case of annual crops) obtained per unit of N uptake. ARE is defined as the quantity of N uptake per unit of N applied. Nitrogen utilization efficiency is the product of PE and ARE. In addition, nutrient use efficiency or NUE can also be expressed in NER. NER is defined as the yield produced per unit element in the plant tissue. Nitrogen recovery efficiency is less than 50% in most crop plants, indicating large loss of this element in soil–plant systems. The losses of N may be related to leaching, denitrification, volatilization, and surface runoff. Hence, there is significant opportunity and challenge to agricultural scientists to improve NUE and reduce the cost of crop production as well as environmental pollution. Generally, NUE has positive quadratic association with grain yield in crop pants. Hence, improving NUE improves grain yield in crops. Climatic, soil, and plant factors are principal determinants of NUE. A detailed knowledge of these factors is fundamental in improving NUE in crop plants. In addition, the components of NUE interact in multiple and complex ways with other metabolic pathways.

REFERENCES

Abbasi, M. K., M. Kazmi, and F. U. Hussan. 2005. Nitrogen use efficiency and herbage production of an established grass sward in relation to moisture and nitrogen fertilization. *J. Plant Nutr.* 28:1693–1708.

Abbasi, M. K., M. M. Tahir, A. Sadiq, M. Iqbal, and M. Zafar. 2012. Yield and nitrogen use efficiency of rainfed maize response to splitting and nitrogen rates in Kashmir, Pakistan. *Agron. J.* 104:448–457.

Adam, Z., I. Adamska, and K. Nakabayashi. 2001. Chloroplast and mitochondrial proteases in *Arabidopsis*. A proposed nomenclature. *Plant Physiol.* 125:1912–1918.

Anbessa, Y., P. Juskiw, A. Good, J. Nyachiro, and J. Helm. 2009. Genetic variability in nitrogen use efficiency of spring barley. *Crop Sci.* 49:1259–1269.

Andrews, M., P. J. Lea, J. A. Raven, and K. Lindsey. 2004. Can genetic manipulation of plant nitrogen assimilation enzymes result in increased crop yield and greater N-use efficiency? An assessment. *Ann. Appl. Bot.* 145:25–40.

Baligar, V. C., N. K. Fageria, and Z. L. He. 2001. Nutrient use efficiency in plants. *Commun. Soil Sci. Plant Anal.* 32:921–950.

Bandyopadhyay, K. K., and M. C. Sarkar. 2005. Nitrogen use efficiency, 15N balance, and nitrogen losses in flooded rice in an Inceptisol. *Commun. Soil Sci. Plant Anal.* 36:1661–1679.

Banziger, M. and H. R. Lafitte. 1997. Efficiency of secondary traits for improving maize for low nitrogen target environments. *Crop Sci.* 37:1110–1117.

Banziger, M., F. J. Betran, and H. R. Lafitte. 1997. Efficiency of high nitrogen selection environments for improving maize low-nitrogen target environment. *Crop Sci.* 37:1103–1109.

Barber, S. A. 1995. *Soil Nutrient Bioavailability: A Mechanistic Approach*, 2nd edition. New York: John Wiley.

Barbieri, P. A., H. E. Echeverria, H. R. S. Rozas, and F. H. Andrade. 2008. Nitrogen use efficiency in maize as affected by nitrogen availability and row spacing. *Agron. J.* 100:1094–1100.

Benbi, D. K. and C. R. Biswas. 1996. Nitrogen balance and N recovery after 22 years of maize-wheat-cowpea cropping in a long term experiment. *Nutr. Cycling Agroecosyst.* 47:107–114.

Bertin, P. and A. Gallais. 2001. Physiological and genetic basis of nitrogen use efficiency. II. QTL detection and coincidences. *Maydica* 46:53–68.

Bijay-Singh, and Yadvinder-Singh. 2003. Efficient nitrogen management in rice-wheat system in the Indo-Gangetic plains. In: *Nutrient Management for Sustainable Rice-Wheat Cropping System*, eds., Yadvinder-Singh, Bijay-Singh, V. K. Nayyar, and J. Singh, pp. 99–114. New Delhi: Indian Council of Agricultural Research.

Bijeriego, M., R. D. Hauck, and R. A. Olson. 1979. Uptake, translocation and utilization of 15N depleted fertilizer in irrigated corn. *Soil Sci. Soc. Am. J.* 43:528–533.

Binder, D. L., D. H. Sander, and D. T. Walters. 2000. Maize response to time of nitrogen application as affected by level of nitrogen deficiency. *Agron. J.* 92:1228–1236.

Blair, G. 1993. Nutrient efficiency—What do we really mean. In: *Genetic Aspects of Mineral Nutrition*, eds., P. J. Randall, E. Delhaize, R. A. Richards, and R. Munns, pp. 205–213. Dordrecht: Kluwer Academic Publishers.

Bliss, F. A. 1993. Utilizing the potential for increased fixation in common bean. *Plant Soil* 152:157–160.

Bouldin, D. R. 1986. The chemistry and biology of flooded soils in relation to the nitrogen economy. *Fertilizer Res.* 9:1–14.

Bronson, K. F., F. Hussain, E. Pasuquin, and J. K. Ladha. 2000. Use of 15N labeled soil in measuring fertilizer recovery efficiency in transplanted rice. *Soil Sci. Soc. Am. J.* 64:235–239.

Brown, B. D., A. J. Hornbacher, and D. V. Naylor. 1988. Sulfur coated urea as a slow release nitrogen source for onions. Bulletin 683. Coop. Ext. Serv., University of Idaho, Moscow.

Bundy, L. G. and T. W. Andraski. 2005. Recovery of fertilizer nitrogen in crop residues and cover crops on an irrigated sandy soil. *Soil Sci. Soc. Am. J.* 69:640–648.

Buresh, R. J., S. K. De Datta, J. L. Padilla, and M. I. Samson. 1988. Effect of two urease inhibitors on floodwater ammonia following urea application to lowland rice. *Soil Sci. Soc. Am. J.* 52:856–861.

Cambardella, C. A., T. B. Moorman, D. B. Jaynes, T. B. Parkin, and D. L. Karlen. 1999. Water quality in walnut creek watershed: Nitrate nitrogen in soils, subsurface drainage water and shallow groundwater. *J. Environ. Qual.* 28:25–34.

Cassman, K. G., A. Dobermann, and D. T. Walters. 2002. Agroecosystems, nitrogen-use efficiency, and nitrogen management. *AMBIO* 31:132–140.

Cassman, K. G., M. J. Kropff, J. Gaunt, and S. Peng. 1993. Nitrogen use efficiency of rice reconsidered: What are the key constraints? *Plant Soil* 155:359–362.

Castellanos, M. T., M. C. Cartagena, F. Ribas, M. J. Cabello, A. Arce, and A. M. Tarquis. 2010. Efficiency indexes for melon crop optimization. *Agron. J.* 102:716–722.

Chalker-Scott, L. 1999. Environmental significance of anthocyanins in plant stress responses. *Photochem. Photobiol.* 70:1–9.

Choudhury, A. T. M. A. and I. R. Kennedy. 2005. Nitrogen fertilizer losses from rice soils and control of environmental pollution problems. *Commun. Soil Sci. Plant Anal.* 36:1625–1639.

Chua, T. T., K. F. Bronson, J. D. Booker, J. W. Keeling, A. R. Mosier, J. P. Bordovsky, R. J. Lascano, C. J. Green, and E. Segarra. 2003. In-season nitrogen status sensing in irrigated cotton. I. Yields and nitrogen-15 recovery. *Soil Sci. Soc. Am. J.* 67:1428–1438.

Clark, R. B. 1990. Physiology of cereals for mineral nutrient uptake, use, and efficiency. In: *Crops as Enhancers of Nutrient Use*, eds., V. C. Baligar, and R. R. Duncan, pp. 131–209. San Diego, California: Academic Press.

Constable, G. A. and I. J. Rochester. 1988. Nitrogen application to cotton on clay soil: Timing and soil testing. *Agron. J.* 80:498–502.

Cox, M. C., C. O. Qualset, and D. W. Rains. 1985. Genetic variation for nitrogen assimilation and translocation in wheat. II. Nitrogen assimilation in relation to grain yield and protein. *Crop Sci.* 25:435–440.

Crop Science Society of America. 1992. *Glossary of Crop Science Terms*. Madison, Wisconsin: Crop Science Society of America.

Cui, Z. L., X. P. Chen, Y. X. Miao, F. S. Zhang, Q. P. Sun, and J. Schroder. 2008. On farm evaluation of the improved soil Nmin-based nitrogen management for summer maize in North China Plain. *Agron. J.* 100:517–525.

Dalal, R. C., W. Wang, D. E. Allen, S. Reeves, and N. W. Menzies. 2011. Soil nitrogen and nitrogen use efficiency under long term no-tillage practice. *Soil Sci. Soc. Am. J.* 75:2251–2261.

Day, W. 2005. Engineering precision into variable biological systems. *Ann. Appl. Biol.* 146:155–162.

De Angeli, A., D. Monachello, G. Ephritikhine, J. M. Frachisse, S. Thomine, F. Gambale, and H. Barbier-Brygoo. 2006. The nitrate/proton antiporter atCKCa mediates nitrate accumulation in plant vacuoles. *Nature* 442:939–942.

De Datta, S. K. 1986. Improving nitrogen fertilizer efficiency in lowland rice. *Fertilizer Res.* 9:171–186.

De Datta, S. K. and R. J. Buresh. 1989. Integrated nitrogen management in irrigated rice. *Adv. Soil Sci.* 10:143–169.

De Datta, S. K., R. J. Buresh, M. I. Samson, and W. Kai-Rong. 1988. Nitrogen use efficiency and nitrogen-15 balances in broadcast-seeded flooded and transplanted rice. *Soil Sci. Soc. Am. J.* 52:849–855.

De Datta, S. K., I. R. P. Fillery, W. N. Obcemea, and R. C. Evangelista. 1987. Floodwater properties, nitrogen utilization, and nitrogen-15 balance in a calcareous lowland rice soil. *Soil Sci. Soc. Am. J.* 51:1355–1362.

De Datta, S. K., W. N. Obcemea, R. Y. Chen, J. C. Calabio, and R. C. Evangelista. 1987. Effect of water depth on nitrogen use efficiency and nitrogen-15 balance in lowland rice. *Agron. J.* 79:210–216.

De Datta, S. K., A. C. F. Trevitt, J. R. Freney, W. N. Obcemea, J. G. Real, and J. R. Simpson. 1989. Measuring nitrogen losses from lowland rice using bulk aerodynamic and nitrogen-15 balance methods. *Soil Sci. Soc. Am. J.* 53:1275–1281.

Degenhart, N. R., D. K. Barnes, and C. P. Vance. 1992. Divergent selection for nodule aspartate aminotransferase and asparagine synthetase activities in alfalfa. *Crop Sci.* 32:313–317.

Diagger, L. A., D. H. Sander, and G. A. Peterson. 1976. Nitrogen content of winter wheat during growth and maturation. *Agron. J.* 68:815–818.

Diaz, C., V. Saliba-Colombani, O. Loudet, P. Belluomo, L. Moreau, F. Daniel-Vedele, J. F. Moroto-Gaudry, and C. Msclaux-Daubresse. 2006. Leaf yellowing and anthocyanin accumulation are two genetically independent strategies in response to nitrogen limitation in *Arabidopsis thaliana*. *Plant Cell Physiol.* 47:74–83.

Dillon, K. A., T. W. Walker, D. L. Harrell, L. J. Krutz, J. J. Varco, C. H. Koger, and M. S. Cox. 2012. Nitrogen sources and timing effects on nitrogen loss and uptake in delayed fold rice. *Agron. J.* 104:466–472.

Ding, L., K. J. Wang, G. M. Jiang, D. K. Biswas, H. Xu, L. F. Li, and Y. H. Li. 2005. Effects of nitrogen deficiency on photosynthetic traits of maize hybrids released in different years. *Ann. Bot.* 96:925–930.

Dinnes, D. L., D. L. Karlen, D. B. Jaynes, T. C. Kaspar, J. L. Hatfield, T. S. Colvin, and C. A. Cambardella. 2002. Nitrogen management strategies to reduce nitrate leaching in tile-drained Midwestern soils. *Agron. J.* 94:153–171.

Dobermann, A. and K. G. Cassman. 2002. Plant nutrient management for enhanced productivity in intensive grain production systems of the United States and Asia. *Plant Soil* 247:153–175.

Duan, Y., M. Xu, B. Wang, X. Yang, S. Huang, and S. Gao. 2011. Long-term evaluation of manure application on maize yield and nitrogen use efficiency in China. *Soil Sci. Soc. Am. J.* 75:1562–1573.

Eagle, A. J., J. A. Bird, J. E. Hill, W. R. Horwath, and C. V. Kessel. 2001. Nitrogen dynamics and fertilizer use efficiency in rice following straw incorporation and winter flooding. *Agron. J.* 93:1346–1354.

Eghball, B. and J. W. Maranville. 1991. Interactive effects of water and nitrogen stresses on nitrogen utilization efficiency, leaf water status and yield of corn genotypes. *Commun. Soil Sci. Plant Anal.* 22, 1367–1382.

Errebhi, M., C. J. Rosen, S. C. Gupta, and D. E. Birong. 1998. Potato yield response and nitrate leaching as influenced by nitrogen management. *Agron. J.* 90:10–15.

Fageria, N. K. 1992. *Maximizing Crop Yields*. New York: Marcel Dekker.

Fageria, N. K. 2009. *The Use of Nutrients in Crop Plants*. Boca Raton, Florida: CRC Press.

Fageria, N. K. 2013. *The Role of Plant Roots in Crop Production*. Boca Raton, Florida: CRC Press.

Fageria, N. K. 2014. *Mineral Nutrition of Rice*. Boca Raton, Florida: CRC Press.

Fageria, N. K. and V. C. Baligar. 2001. Lowland rice response to nitrogen fertilization. *Commun. Soil Sci. Plant Anal.* 32:1405–1429.

Fageria, N. K. and V. C. Baligar. 2003. Methodology for evaluation of lowland rice genotypes for nitrogen use efficiency. *J. Plant Nutr.* 26:1315–1333.

Fageria, N. K. and V. C. Baligar. 2005. Enhancing nitrogen use efficiency in crop plants. *Adv. Agron.* 88:97–185.

Fageria, N. K. and M. P. Barbosa Filho. 2001. Nitrogen use efficiency in lowland rice genotypes. *Commun. Soil Sci. Plant Anal.* 32:2079–2089.

Fageria, N. K., V. C. Baligar, and C. A. Jones. 2011. *Growth and Mineral Nutrition of Field Crops*, 3rd edition. Boca Raton, Florida: CRC Press.

Fageria, N. K., V. C. Baligar, and Y. C. Li. 2008. The role of nutrient efficient plants in improving crop yields in the twenty first century. *J. Plant Nutr.* 31:1121–1157.

Fageria, N. K., L. C. Melo, and J. P. Oliveira. 2013. Nitrogen use efficiency in dry bean genotypes. *J. Plant Nutr.* 36:2179–2190.

Fageria, N. K., O. P. Morais, and A. B. Santos. 2010. Nitrogen use efficiency in upland rice. *J. Plant Nutr.* 33:1696–1711.

Fagria, N. K., N. A. Slaton, and V. C. Baligar. 2003. Nutrient management for improving lowland rice productivity and sustainability. *Adv. Agron.* 80:63–152.

Fageria, N. K., A. B. Santos, and V. A. Cutrim. 2007. Yield and nitrogen use efficiency of lowland rice genotypes as influenced by nitrogen fertilization. *Pesq. Agropec. Bras.* 42:1029–1034.

FAO. 2004. *Current World Fertilizer Trends and Outlook to 2008/09*. Rome: FAO.

Fillery, I. R. P. and S. K. De Datta. 1989. Ammonia volatilization from nitrogen source applied to rice fields. I. Methodology, ammonia fluxes and nitrogen-15 loss. *Soil Sci. Soc. Am. J.* 50:80–86.

Francis, D. D., J. S. Schepers, and A. L. Sims. 1997. Ammonium exchange from corn foliage during reproductive growth. *Agron. J.* 89:941–946.

Gallais, A. and B. Hirel. 2004. An approach to the genetics of nitrogen use efficiency in maize. *J. Exp. Bot.* 55:295–306.

Galloway, J. N. 1998. The global nitrogen cycle: Changes and consequences. *Environ. Pollut.* 102:15–24.

Gerik, T. J., B. S. Jackson, C. O. Stockle, and W. D. Rosenthal. 1994. Plant nitrogen status and boll load of cotton. *Agron. J.* 86:514–518.

Gerloff, G. C. and W. H. Gabelman. 1983. Genetic basis of inorganic nutrition. In: *Inorganic Plant Nutrition. Encyclopedia and Plant Physiology New Series, Volume 15B*, eds., A. Lauchli, and R. L. Bieleski, pp. 453–480. New York: Springer Verlag.

Giambalvo, D., P. Ruisi, G. D. Miceli, A. S. Frends, and G. Amato. 2010. Nitrogen use efficiency and nitrogen fertilizer recovery of durum wheat genotypes as affected by interspecific competition. *Agron. J.* 102:707–715.

Glass, A. D. M. 2003. Nitrogen use efficiency of crop plants: Physiological constraints upon nitrogen absorption. *Crit. Rev. Plant Sci.* 22:453–470.

Good, A., A. Shrawat, and D. Muench. 2004. Can less yield more? Is reducing nutrient input into the environment compatible with maintaining crop production? *Trends Plant Sci.* 9:597–605.

Gourley, C. J. P., D. L. Allan, and M. P. Russelle. 1994. Plant nutrition efficiency: A comparison of definitions and suggested improvement. *Plant Soil* 158:29–37.

Graham, R. D. 1984. Breeding for nutritional characteristics in cereals. In: *Advances in Plant Nutrition*, Vol. 1, eds., P. B. Tinker, and A. Lauchi, pp. 57–102. New York: Praeger Publisher.

Groat, R. C., C. P. Vance, and D. K. Barnes. 1984. Host plant nodule enzymes associated with selection for increased N_2 fixation in alfalfa. *Crop Sci.* 24:895–898.

Guindo, D., B. R. Wells, C. E. Wilson, and R. J. Norman. 1992. Seasonal accumulation and partitioning of nitrogen-15 in rice. *Soil Sci. Soc. Am. J.* 56:1521–1526.

Halvorson, A. D., R. F. Follett, M. E. Bartolo, and F. C. Schweissing. 2002. Nitrogen fertilizer use efficiency of furrow-irrigated onion and corn. *Agron. J.* 94:442–449.

Harper, L. A., R. R. Sharpe, G. W. Langdale, and J. E. Giddens. 1997. Nitrogen cycling in a wheat crop: Soil, plant, and aerial nitrogen transport. *Agron. J.* 79:965–973.

Harrison, J., N. Brugiere, B. Phillipson, S. Ferrario, T. Becker, A. Limami, and B. Hirel. 2000. Manipulating the pathway of ammonium assimilation through genetic manipulation and breeding: Consequences on plant physiology and development. *Plant Soil* 221:81–93.

Hawkesford, M. J. 2012. The diversity of nitrogen use efficiency for wheat varieties and the potential for crop improvement. *Better Crops Plant Food* 96:10–12.

Hirel, B., P. Bertin, I. Quillere, W. Bourdoncle, C. Attagnant, C. Dellay, A. Gouy et al. 2001. Towards a better understanding of the genetic and physiological basis for nitrogen use efficiency in maize. *Plant Physiol.* 125:1258–1270.

Ho, C. H., S. H. Lin, H. C. Hu, and Y. F. Tsay. 2009. CHL1 functions as a nitrate sensor in plants. *Cell* 138:1184–1194.

Huggins, D. R., G. W. Randall, and M. P. Russelle. 2001. Subsurface drain losses of water and nitrate following conversion of perennials to row crops. *Agron. J.* 93:477–486.

Hussain, F., K. F. Bronson, Y. Singh, B. Singh, and S. Peng. 2000. Use of chlorophyll meter sufficiency indices for nitrogen management of irrigated rice in Asia. *Agron. J.* 93:477–486.

Isfan, D. 1993. Genotypic variability for physiological efficiency index of nitrogen in oats. *Plant Soil* 154:53–59.

Johnson, G. V. and W. R. Raun. 2003. Nitrogen response index as a guide to fertilizer management. *J. Plant Nutr.* 26:249–262.

Jokela, W. E. and G. W. Randall. 1997. Fate of fertilizer nitrogen as affected by time and rate of application on corn. *Soil Sci. Soc. Am. J.* 61:1695–1703.

Ju, X. T., G. X. Xing, X. P. Chen, S. L. Zhang, L. J. Zhang, X. J. Liu, Z. L. Cui et al. 2009. Reducing environmental risk by improving N management in intensive Chinese agricultural systems. *Proc. Natl. Acad. Sci.* 106:3041–3046.

Kaiser, J. 2001. The other global pollutant; Nitrogen proves tough to curb. *Science* 294:1268–1269.

Kanampiu, F. K., W. R. Raun, and G. V. Johnson. 1997. Effect of nitrogen rate on plant nitrogen loss in winter wheat varieties. *J. Plant Nutr.* 20:389–404.

Kant, S., Y. M. Bi, and S. J. Rothstein. 2010. Understanding plant response to nitrogen limitation for the improvement of crop nitrogen use efficiency. *J. Exp. Bot.* 6:1–11.

Kichey, T., B. Hirel, E. Heumez, F. Dubois, and J. L. Gouis. 2007. In winter wheat (*Triticum aestivum* L.), post-anthesis nitrogen uptake and remobilization to the grain correlate with agronomic traits and nitrogen physiological markers. *Field Crops Res.* 102:22–32.

Kimura, T. and K. Kurashima. 1991. Quantitative estimates of the budgets of nitrogen applied as fertilizer, urine and feces in a soil. *Grass. System.* 25:101–107.

Kingston-Smith, A. H., A. L. Bollard, and F. R. Minchin. 2006. The effect of nitrogen status on the regulation of plant-mediated proteolysis in ingested forage, an assessment using non-nodulating white clover. *Ann. Appl. Bot.* 149:35–41.

Kitur, B. K., M. S. Smith, R. L. Blevins, and W. W. Frye. 1984. Fate of ^{15}N depleted ammonium nitrate applied to no-tillage and conventional tillage corn. *Agron. J.* 76:240–242.

Kurtz, L. T., L. V. Boone, T. R. Peck, and R. G. Hoeft. 1984. Crop rotations for efficient nitrogen use. In: *Nitrogen in Crop Production*, ed., R. D. Hauck, pp. 295–306. Madison, Wisconsin: ASA, CSSA, and SSSA.

Ladha, J. K., H. Pathak, T. J. Krupnik, J. Six, and C. V. Kessel. 2005. Efficiency of fertilizer nitrogen in cereal production: Retrospects and prospects. *Adv. Agron.* 87:85–156.

Lea, P. J. and R. A. Azevedo. 2006. Nitrogen use efficiency: Uptake of nitrogen from the soil. *Ann. Appl. Biol.* 149:243–247.

Lea, P. J. and R. J. Ireland. 1999. Plant amino acids. In: *Nitrogen Metabolism in Higher Plants*, ed., B. K. Singh, pp. 1–47. New York: Marcel Dekker.

Li, H., X. Q. Liang, Y. F. Lian, L. Xu, and Y. X. Chen. 2009. Reduction of ammonia volatilization from urea by a floating duckweed in flooded rice fields. *Soil Sci. Soc. Am. J.* 73:1890–1895.

Lopez-Bellido, L., R. J. Lopez-Bellido, and F. J. Lopez-Bellido. 2006. Fertilizer nitrogen efficiency in durum wheat under rainfed Mediterranean conditions: Effects of split application. *Agron. J.* 98:55–62.

López-Bellido, R. J., L. López-Bellido, F. J. López-Bellido, and J. E. Castillo. 2003. Faba bean (*Vicia faba* L.) response to tillage and soil residual nitrogen in a continuous rotation with wheat (*Triticum aestivum* L.) under rainfed Mediterranean conditions. *Agron. J.* 95:1253–1261.

Lynch, J. and J. W. White. 1992. Shoot nitrogen dynamics in tropical common bean. *Crop Sci.* 32:392–397.

Lynch, J. and N. S. Rodriguez. 1994. Photosynthetic nitrogen use efficiency in relation to leaf longevity in common bean. *Crop Sci.* 34:1284–1290.

Ma, B. L. and M. L. Dwyer. 1998. Nitrogen uptake and use in two contrasting maize hybrids differing in leaf senescence. *Plant Soil* 199:283–291.

Ma, W., J. Li, L. Ma, F. Wang, I. Sisak, G. Cushman, and F. Zhang. 2008. Nitrogen flow and use efficiency in production and utilization of wheat, rice, and maize in China. *Agric. Syst.* 99:53–63.

Macdonald, A. J., D. S. Powlson, P. R. Poulton, and D. S. Jenkinson. 1989. Unused fertilizer nitrogen in arable soils-Its contribution to nitrate leaching. *J. Sci. Food Agric.* 46:407–419.

Mae, T., H. Thomas, A. P. Gay, A. Makino, and J. Hidema. 1993. Leaf development in *Lolium temulentum*: Photosynthesis and photosynthetic proteins in leaves senescing under different irridance. *Plant Cell Physiol.* 34:391–399.

Marschner, H. H. 1995. *Mineral Nutrition of Higher Plants*. New York: Academic Press.

Matson, P. A., W. J. Parton, A. G. Power, and M. J. Swift. 1997. Agricultural intensification and ecosystem properties. *Science* 277:504–509.

Meisinger, J. J., V. A. Bandel, G. Stanford, and J. O. Legg. 1985. Nitrogen utilization of corn under minimum tillage and moldboard plow tillage: I. Four year results using labeled fertilizer on an Atlantic Coastal Plain soil. *Agron. J.* 77:602–611.

Mengel, K., B. Hutsch, and Y. Kane. 2006. Nitrogen fertilizer application rates on cereal crops according to available mineral and organic soil nitrogen. *Eur. J. Agron.* 24:343–348.

Mengel, K., E. A. Kirkbay, H. Kosegarten, and T. Appel. 2001. *Principles of Plant Nutrition,* 5th edition. Dordrecht: Kluwer Academic Publishers.

Miller, A. J., X. R. Fan, M. Orsel, S. J. Smith, and D. M. Wells. 2007. Nitrate transport and signaling. *J. Exp. Bot.* 58:2297–2306.

Moll, R. H., E. J. Kamprath, and W. A. Jackson. 1982. Analysis and interpretation of factors which contribute to efficiency of nitrogen utilization. *Agron. J.* 74:562–564.

Moll, R. H., E. J. Kamprath, and W. A. Jackson. 1987. Development of nitrogen efficient prolific hybrids of maize. *Crop Sci.* 27:181–186.

Muchow, R. C. 1998. Nitrogen utilization efficiency in maize and grain sorghum. *Field Crops Res.* 56:209–216.

Munoz, F., R. S. Mylavarapu, and C. M. Hutchinson. 2005. Environmentally responsible potato production systems: A review. *J. Plant Nutr.* 28:1287–1309.

Novoa, R. and R. S. Loomis. 1981. Nitrogen and plant production. *Plant Soil* 58:177–204.

Olson, R. A., W. R. Raun, S. C. Yang, and J. Skopp. 1986. Nitrogen management and interseeding effects on irrigated corn and sorghum and on soil strength. *Agron. J.* 78: 856–862.

Olson, R. V. 1980. Fate of tagged nitrogen fertilizer applied to irrigated corn. *Soil Sci. Soc. Am. J.* 55:1616–1621.

Ono, K., I. Terashima, and A. Watanabe. 1996. Interaction between nitrogen deficit of a plant and nitrogen content in the old leaves. *Plant Cell Physiol.* 37:1083–1089.

Ortiz-Monasteiro, J. I., K. D. Sayre, S. Rajaram, and M. McMahon. 1997. Genetic progress in wheat yield and nitrogen use efficiency under four nitrogen rates. *Crop Sci.* 37:989–904.

Pan, W. L., E. J. Kamprath, R. H. Moll, and W. A. Jackson. 1984. Prolificacy in corn: Its effects on nitrate and ammonium uptake and utilization. *Soil Sci. Soc. Am. J.* 48:1101–1106.

Peng, S. B., R. J. Buresh, J. L. Huang, J. C. Yang, Y. B. Zou, and X. H. Zhong. 2006. Strategies for overcoming low agronomic nitrogen use efficiency in irrigated rice systems in China. *Field Crops Res.* 96:37–47.

Pilbeam, C. J., A. M. McNeill, H. C. Harris, and R. S. Swift. 1997. Effect of fertilizer rate and form on the recovery of ^{15}N-labeled fertilizer applied to wheat in Syria. *J. Agric. Sci.* 128:415–424.

Prasad, R. and De Datta, S. K. 1979. Increasing fertilizer nitrogen efficiency in wetland rice. In: *Nitrogen and Rice*, ed., IRRI, pp. 465–484. Los Baños, Philippines: IRRI.

Presterl, T., G. Seitz, M. Landbeck, E. M. Thiemt, W. Schmidt, and H. H. Geiger. 2003. Improving nitrogen use efficiency in European maize: Estimation of quantitative genetic parameters. *Crop Sci.* 43:1259–1265.

Purcino, A. A. C., C. Arellano, G. S. Athwal, and S. C. Huber. 1998. Nitrate effect on carbon and nitrogen assimilating enzymes of maize hybrids representing seven eras of breeding. *Maydica* 43:83–94.

Racjan, I. and M. Tollenaar. 1999a. Source:sink ratio and leaf senescence in maize: I. Dry matter accumulation and partitioning during grain filling. *Field Crops Res.* 60:245–253.

Racjan, I. and M. Tollenaar. 1999b. Source:sink ratio and leaf senescence in maize: II Nitrogen metabolism during grain filling. *Field Crops Res.* 60:255–265.

Randall, G. W., D. R. Huggins, M. P. Russelle, D. J. Fuchs, W. W. Nelson, and J. L. Anderson. 1997. Nitrate losses through subsurface tile drainage in conservation reserve program, alfalfa and row crop systems. *J. Environ. Qual.* 26:1240–1247.

Randall, G. W., J. A. Vetsch, and J. R. Huffman. 2003. Corn production on a subsurface drained Mollisols as affected by time of nitrogen application and nitrapyrin. *Agron. J.* 95:1213–1219.

Raun, W. R. and G. V. Johnson. 1995. Soil-plant buffering of organic nitrogen in continuous winter wheat. *Agron. J.* 87:827–834.

Raun, W. R. and G. V. Johnson. 1999. Improving nitrogen use efficiency for cereal production. *Agron. J.* 91:357–363.

Reddy, G. B. and K. R. Reddy. 1993. Fate of nitrogen-15 enriched ammonium nitrate applied to corn. *Soil Sci. Soc. Am. J.* 57:111–115.

Reed, A. J., F. E. Below, and R. H. Hageman. 1980. Grain protein accumulation and the relationship between leaf nitrate reductase and protease activities during grain development in maize: I. Variation between genotypes. *Plant Physiol.* 66:164–170.

Ritchie, S. W. and J. J. Hanway. 1984. How a corn plant develop. Spec. Rep. 48. Iowa State University Coop. Ext. Serv., Ames, IA.

Roberts, T. L., R. J. Norman, N. A. Slaton, and C. E. Wilson. 2009. Changes in alkaline hydrolysable nitrogen distribution with soil depth; Fertilizer correlation and calibration implications. *Soil Sci. Soc. Am. J.* 73:2151–2158.

Rosielle, A. A. and J. Hamblin. 1981. Theoretical aspects of selection for yield in stress and non-stress environments. *Crop Sci.* 21:943–946.

Russelle, M. P., R. D. Hauck, and R. A. Olson. 1983. Nitrogen accumulation rates of irrigated corn. *Agron. J.* 75:593–598.

Salsac, L., S. Chaillou, J. F. Morot-Gaudry, C. Lesaint, and E. Jolivoe. 1987. Nitrate and ammonium nutrition in plants. *Plant Physiol. Biochem.* 25:805–812.

Sanaa, M., O. V. Clemput, L. Baert, and A. Mhiri. 1992. Field study of the fate of labeled fertilizer nitrogen applied to wheat on calcareous Tunisian soils. *Pedologie* 42:245–255.

Sanchez, C. A. and A. M. Blackmer. 1988. Recovery of anhydrous ammonium-derived nitrogen-15 during three years of corn production in Iowa. *Agron. J.* 80:102–108.

Sartor, L. R., T. S. Assmann, A. B. Soares, P. F. Adami, A. L. Assmann, and C. S. R. Pitta. 2011. Nitrogen fertilizer use efficiency, recovery and leaching of an alexandergrass pasture. *R. Bras. Ci. Solo* 35:899–906.

Sato, S., K. T. Morgan, M. Ozores-Hampton, K. Mahmoud, and E. H. Simonne. 2012. Nutrient balance and use efficiency in sandy soils cropped with tomato underseepage irrigation. *Soil Sci. Soc. Am. J.* 76:1867–1876.

Scharf, P. C., W. J. Wiebold, and J. A. Lory. 2002. Corn yield response to nitrogen fertilizer timing and deficiency level. *Agron. J.* 94:435–441.

Schepers, J. S. and A. R. Mosier. 1991. Accounting for nitrogen in nonequilibrium soil-crop systems. In: *Managing Nitrogen for Groundwater Quality and Farm Profitability*, eds., R. F. Follett, D. R. Keeney, and R. M. Cruse, pp. 125–138.

Schnier, H. F., M. Dingkuhn, S. K. De Datta, E. P. Marquesses, and J. E. Faronilo. 1990. Nitrogen-15 balance in transplanted and direct seeded flooded rice as affected by different methods of urea application. *Biol. Fertl. Soils* 10:89–96.

Shaviv, A. and R. L. Mikkelsen. 1993. Controlled-release fertilizers to increase efficiency of nutrient use and minimize environmentally degradation: A review. *Fert. Res.* 35:1–12.

Sing, B., V. Singh, Y. Singh, H. S. Thind, A. Kumar, and R. K. Gupta. 2012. Fixed time adjustable dose site-specific fertilizer nitrogen management in transplanted irrigated rice (*Oryza sativa* L.) in South Asia. *Field Crops Res.* 126:63–69.

Slavikova, S., G. Shy, Y. L. Yao, R. Giozman, H. Levanony, S. Peitrokovski, Z. Elazar, and G. Galili. 2005. The autophagy-associated Atg8 gene family operates both under favorable growth conditions and under starvation stresses in *Arabidopsis* plants. *J. Exp. Bot.* 56:2839–2849.

Soil Science Society of America. 2008. *Glossary of Soil Science Terms*. Madison, Wisconsin: Soil Science Society of America.

Sowers, K. E., W. L. Pan, B. C. Miller, and J. L. Smith. 1994. Nitrogen use efficiency of split nitrogen application in soft white inter wheat. *Agron. J.* 86:942–948.

Stevens, W. E., J. J. Varco, and J. R. Johnson. 1996. Evaluating cotton nitrogen dynamics in the GOSSYM simulation model. *Agron. J.* 88:127–132.

Stewart, W. M., D. W. Dibb, A. E. Johnston, and T. J. Smith. 2005. The contribution of commercial fertilizer nutrients to food production. *Agron. J.* 97:1–6.

Stutte, C. A., R. T. Weiland, and A. R. Blem. 1979. Gaseous nitrogen loss from soybean foliage. *Agron. J.* 71:95–97.

Tabuchi, M., T. K. Sugiyama, K. Ishiyama, E. Inoue, T. Sato, H. Takahashi, and T. Yamaya. 2005. Severe reduction in growth rate and grain filling of rice mutants lacking OsGS1:1, a cytosolic glutamine synthetase 1:1. *Plant J.* 42:641–651.

Thind, H. S., B. Singh, R. P. S. Pannu, Y. Singh, V. Singh, R. K. Gupta, M. Vashistha, J. Singh, and A. Kumar. 2010. Relative performance of need (*Azadirachta indica*) coated urea vis-à-vis ordinary urea applied to rice on the basis of soil test or following need based nitrogen management using leaf color chart. *Nutr. Cycl. Agroecosyst.* 87:1–8.

Thomason, W. E., W. R. Raun, G. V. Johnson, K. W. Freeman, K. J. Wynn, and R. W. Mullen. 2002. Production system techniques to increase nitrogen use efficiency in winter wheat. *J. Plant Nutr.* 25:2261–2283.

Tian, C. Y., Z. A. Lin, Y. B. Zuo, W. Y. Sun, S. G. Che, M. F. Cheng, and B. Q. Zhao. 2011. Review on several concepts on fertilizer nitrogen recovery rate and its calculation. *Chin. J. Soil Sci.* 42:1530–1536.

Tollenaar, M. 1991. Physiological basis of genetic improvement of maize hybrids in Ontario from 1959 to 1988. *Crop Sci.* 31:119–124.

Tsai, C. Y., I. Dweikat, D. M. Huber, and H. L. Warren. 1992. Interrelationship of nitrogen nutrition with maize (*Zea mays* L.) grain yield, nitrogen use efficiency and grain quality. *J. Sci. Food Agric.* 58:1–8.

Tsay, Y. F., C. C. Chiu, C. B. Tsai, C. H. Ho, and P. K. Hsu. 2007. Nitrate transporter and peptide transporters. *FEBS Lett.* 581:2290–2300.

Vitousek, P. M., J. D. Aber, R. W. Howarth, G. E. Likens, P. A. Matson, D. W. Schindler, W. H. Schlesinger, and D. G. Tilman. 1997. Human alteration of the global nitrogen cycle: Sources and consequences. *Ecol. Appl.* 7:737–750.

Wang, G. H., A. Dobermann, C. Witt, Q. Z. Sun, and R. X. Fu. 2001. Performance of site-specific nutrient management for irrigated rice in southeast China. *Agron. J.* 93:869–878.

Whitehead, D. C. 2000. *Nutrient Elements in Grasslands: Soil-Plant-Animal Relationships.* Wallingford, UK: CAB.

Wienhold, B. J., T. P. Trooien, and G. A. Reichman. 1995. Yield and nitrogen use efficiency of irrigated corn in the northern Great Plains. *Agron. J.* 87:842–846.

Wiesler, F., T. Behrens, and W. J. Horst. 2001. The role of nitrogen efficient cultivars in sustainable agriculture. *Scientific World* 1:61–69.

Wilson, C. E. Jr., R. J. Norman, and B. R. Wells. 1989. Seasonal uptake patterns of fertilizer nitrogen applied in split application to rice. *Soil Sci. Soc. Am. J.* 53:1884–1887.

Xiong, J., C. Q. Ding, G. B. Wei, Y. F. Ding, and S. H. Wang. 2013. Characteristic of dry matter accumulation and nitrogen uptake of super-high yielding early rice in China. *Agron. J.* 105:1142–1150.

Yang, Y., M. Zhang, Y. C. Li, X. Fan, and Y. Ceng. 2012. Controlled released urea improved nitrogen use efficiency, activities of leaf enzymes, and rice yield. *Soil Sci. Soc. Am. J.* 76:2307–2317.

Yasin, I. 1991. No-tillage cotton response to winter cover management and fertilizer nitrogen. M.Sc. Thesis. Mississippi State, MS: Mississippi State University.

Yoshida, S. 1981. *Fundamentals of Rice Crop Science.* Los Baños, the Philippines: IRRI.

Zhang, F. S. 2008. *Strategy of Chinese Fertilizer Industry and Scientific Application.* Beijing: Chinese Agricultural University Press.

Zhang, W., M. Xu, B. Wang, and X. Wang. 2009. Soil organic carbon, total nitrogen and grain yields under long-term fertilizations in the upland red soil of southern China. *Nutr. Cycling Agroecosyst.* 84:59–69.

Zhang, X. C., C. T. Mackown, J. D. Garbrecht, H. Zhang, and J. T. Edwards. 2012. Variable environment and market affect optimal nitrogen management in wheat and cattle production systems. *Agron. J.* 104:1136–1148.

Zhu, Z. L. and D. L. Chen. 2002. Nitrogen fertilizer use in China contributions to food production, impacts on the environment and best management strategies. Nutrient cycling in impacts on the environment and best management strategies. *Nutr. Cycling Agroecosyst.* 63:117–127.

Zotarelli, L., L. Avila, J. M. S. Scholberg, and B. J. R. Alves. 2009. Benefits of vetch and rye cover crops to sweet corn under no-tillage. *Agron. J.* 101:252–260.

6 Nitrogen Interaction with Other Nutrients

6.1 INTRODUCTION

Among the 17 essential plant nutrients, nitrogen plays the most important role in augmenting agricultural production and potential environmental risks and impacting human and animal health (Aulakh and Malhi, 2005). Nitrogen fertilization significantly influences crop yields in low-fertility soils, especially in soils having low organic matter content and low cation exchange capacity. The positive influence of N on crop growth also influences the uptake of other essential nutrients, known as nutrient interaction. Interactions among nutrients occur when the supply of one nutrient affects the absorption, distribution, or function of another nutrient in crop plants (Robson and Pitman, 1983). In crop production, nutrient interactions assume added significance by affecting the crop productivity and returns from investments made by farmers in fertilizers (Aulakh and Malhi, 2005).

Nutrient interaction in crop plants is probably one of the most important factors affecting yields of annual crops. Nutrient interaction may be positive (synergistic), negative (antagonistic), and no interaction (Fageria, 1983; Fageria et al., 2011a). Aulakh and Malhi (2005) stated that nutrient interactions have a role to play in determining the course and outcome of two major issues of interest in fertilizer management, namely, balanced fertilizer input and efficient fertilizer use. Wilkinson et al. (2000) reviewed the literature on nutrient interactions in soil and plant and stated that if two nutrients are limiting, or nearly limiting growth or concentration in the plant tissue, where adding only one of the nutrients has little effect while adding both gives a considerable effect, the effect is said to be a positive interaction. It can be measured in terms of crop growth and nutrient concentrations in the plant tissue. These authors further reported that if adding the two together has less effect than when each is added separately, the effect is said to be a negative interaction. When the factors acting in concert result in a positive growth response that is greater than the sum of their individual effects, the interaction is positive or synergistic; when it is less than the sum of the individual effects, the interaction is negative or antagonistic; and when it is the same, there is no interaction.

Soil, plant, and climatic factors can influence interaction. In the nutrient interaction studies, all other factors should be at an optimum level except the variation in the level of the nutrient under investigation. Hence, nutrient interaction results are valid only when other environmental factors are at optimum levels, except the variation in the concentrations of the two nutrients under investigation. Wallace (1990) proposed the law of maximum in nutrient interaction studies rather than Liebig's law of minimum. The law of maximum states that, when the need is fully satisfied for every factor involved in the process, the rate of the process can be at its maximum potential, which is greater than the sum of its parts because of the sequentially additive interaction (Wallace, 1990).

Nutrient interaction can occur at the root surface or within the plant. Interactions at the root surface are due to the formation of chemical bonds by ions and precipitation or complexes. One example of this type of interaction is that the liming of acid soils decreases the concentration of almost all the micronutrients except molybdenum (Fageria and Zimmermannn, 1998; Fageria, 2000). The second type of interaction is between ions whose chemical properties are sufficiently similar that they compete for the site of absorption, transport, and function on the plant root surface or within plant tissues. Such interactions are more common between nutrients of similar size, charge, and geometry of coordination and electronic configuration (Robson and Pitman, 1983).

Interactions occur when the supply of one nutrient affects the absorption and utilization of another nutrient (Robson and Pitman, 1983; Wilkinson et al., 2000). Nutrient interactions affect plant growth and development only when the supply of a determined nutrient is too low compared to the applied ones. In other words, yield decrease occurs only when the supply of some nutrients falls below the critical level. If the soil or growth medium has sufficient supply of other essential nutrients compared to the added one, plant growth will not be affected adversely, even though the uptake of some nutrients may decrease. Hence, plant growth or yield is considered a better criterion for evaluating nutrient interactions in crop plants.

The interaction varied from nutrient to nutrient and from crop species to species and sometimes among cultivars of the same species. Therefore, this is a very complex issue in mineral nutrition that is not well understood in annual crops grown on many soil orders. Most of the research data available in the literature related to nutrient interactions were generated in solution culture or sand culture (Marschner, 1995). Solution or sand culture results may not be applicable to field conditions. Pan (2012) reported that nutrient interactions will be delineated as to their specificity, specific or primary interactions being those in which two nutrients directly react in a chemical or biological process. Nonspecific or secondary nutrient interactions occur when the uptake of one nutrient is indirectly affected by the activity of another nutrient through a series of intermediate plant processes (Pan, 2012). The objective of this chapter is to discuss nitrogen interaction with other plant essential nutrients under field conditions or conducted in a soil medium under controlled conditions. This information may help in understanding the increasing, decreasing, or no effect of nitrogen on the uptake of other nutrients and vice versa. Finally, it may help in better management of N for crop production.

6.2   INTERACTION OF NITROGEN WITH MACRONUTRIENTS

Macronutrients that are essential for higher plants are nitrogen, phosphorus, potassium, calcium, magnesium, and sulfur. They are required by plants in higher amounts and are known as macronutrients. Nutrients are absorbed by plants from soil solution, some as cations and others as anions, which interact with each other in the absorption as well as utilization processes in plants. Since nitrogen is absorbed or utilized by plants in higher amounts, its interaction with other nutrients is important in determining nutritional balance and, consequently, crop yields. Hence, the knowledge of nitrogen interaction with other nutrients is fundamental in nutrient management for obtaining the maximum economic yield of crops.

6.2.1   NITROGEN VERSUS PHOSPHORUS

The majority of agricultural soils are deficient in available N and are either low or medium in available P (Aulakh and Malhi, 2005). In addition, nitrogen and phosphorus are required by crop plants in higher amounts for maximizing yields. Aulakh and Malhi (2005) reported that the N × P interaction is one of the most important nutrient interactions in crop production. A positive interaction of N with P has been reported (Terman et al., 1977; Wilkinson et al., 2000; Fageria and Baligar, 2005). Schulthess et al. (1997) reported that the accumulation of N and P in the shoot and grain of wheat was positively associated. Similarly, Fageria and Baligar (2005) reported a positive association between N and P in dry bean production. The positive interaction of N with P may be associated with improved yield with the addition of N (Table 6.1). Pederson et al. (2002) reported that N concentration was highly correlated with P concentration in aboveground plant parts of ryegrass (*Lolium multiflorum* Lam.). The improvement in the uptake of P with the addition may also be related to the increase in root hairs, chemical changes in the rhizosphere, and physiological changes stimulated by N, which influence the transport of P (Marschner, 1995; Baligar et al., 2001). Nitrogen has a synergistic relationship with P in crop plants (Black, 1993). Kaiser and Kim (2013) also reported the synergistic effects of N and P in

TABLE 6.1

Dry Weight of Shoot and P Uptake in the Shoot of 60-Day-Old Dry Bean Plants as Influenced by Nitrogen Rates

N Rate (kg ha^{-1})	Shoot Dry Weight (kg ha^{-1})	P Uptake (kg ha^{-1})
0	220	0.6
40	494	1.4
80	814	1.6
120	828	2.1
160	1260	2.9
200	1659	3.4
R^2	0.83**	0.81**

** Significant at the 1% probability level.

soybean production. Abbasi et al. (2012) also reported that the application of P significantly increased the uptake of N in soybean.

The author studied the uptake of N under different P levels in 14 cover crop species (Table 6.2). Uptake (concentration × dry matter) was significantly affected by P levels as well as cover crop treatments (Table 6.2). Similarly, P × cover crop interactions were also significant for N uptake,

TABLE 6.2

Uptake of N (mg Plant^{-1}) in the Tops of 14 Tropical Cover Crops under Three P Levels

Cover Crops	P Levels (mg kg^{-1})		
	0 (Low)	100 (Medium)	200 (High)
Crotalaria	5.95f	21.59e	25.47ef
Sunnhemp	58.18de	152.02bcd	176.90cd
Crotalaria	11.11f	35.34e	36.51ef
Crotalaria	13.98def	42.80e	54.27ef
Crotalaria	13.89def	68.16de	109.22de
Calapogonio	11.75ef	38.37e	7.32f
Pueraria	7.31f	27.50e	35.63ef
Pigeon pea (black)	37.90def	97.21cde	87.55def
Pigeon pea (mixed color)	16.58def	76.73de	79.67def
Lablab	48.32def	192.86bc	237.84bc
Mucuna bean ana	123.10c	190.81b	220.79bc
Black mucuna bean	57.63d	194.44b	265.75ab
Gray mucuna bean	190.41b	199.05bc	304.37ab
White jack bean	282.48a	350.79a	376.39a
Average	62.76	120.55	144.12
F-test			
P levels (P)	**		
Cover crops (C)	**		
P × C	**		
CV (%)	24.58		

**Significant at the 1% probability level. Means followed by the same letters in the same column are not significant at the 5% probability level by Tukey's test.

indicating variable responses of cover crops in N uptake with changing P levels. The uptake of N increased with increasing P levels, indicating a positive interaction between these two elements.

Nitrogen can increase P concentration in plants by increasing root growth and by increasing the ability of roots to absorb and translocate P (Wilkinson et al., 2000). One of the important aspects of improving P uptake with the addition of N is vigorous root growth. Data in Table 6.3 show the influence of N levels on root growth of upland rice grown on a Brazilian Oxisol. Nitrogen × genotype interaction was significant for root length and root dry weight; therefore, data are reported separately for the two N rates (Table 6.3). Root length varied from 27 cm produced by genotype BRA01600 to 43 cm produced by genotype BRA032039, with an average value of 30.49 cm at a low N rate (0 mg N kg^{-1}). Similarly, at a high N rate (300 mg N kg^{-1}), the root length ranged from 21 cm produced by genotype BRA01506 to 40.33 cm produced by genotype BRS Sertaneja, with an average value of 31.99 cm. Overall, the root length was 5% higher at the higher N rate compared to the lower N rate. However, 35% genotypes produced a lower root length at the higher N rate compared

TABLE 6.3

Root Length and Root Dry Weight of 20 Upland Rice Genotypes as Influenced by Nitrogen Fertilization

Genotype	Root Length (cm)		Root Dry Weight (g Plant^{-1})	
	0 mg N kg^{-1}	300 mg N kg^{-1}	0 mg N kg^{-1}	300 mg N kg^{-1}
BRA01506	34.67ab	21.00cd	0.92a	0.40f
BRA01596	30.00b	15.67d	0.87a	0.45f
BRA01600	27.00ab	25.33bcd	1.14a	1.03ef
BRA02535	31.67ab	30.00abcd	1.33a	3.25abcd
BRA02601	28.00b	32.67abc	1.11a	3.73ab
BRA032033	30.00a	28.00abcd	1.12a	2.41cd
BRA032039	43.00a	34.67abc	1.24a	3.77ab
BRA032048	28.50b	32.00abc	1.05a	3.62abc
BRA032051	30.67ab	35.67ab	1.31a	2.82bcd
BRA042094	30.33b	38.00ab	1.78a	2.31de
BRA042156	29.33b	33.00abc	1.19a	2.84bcd
BRA042160	29.67b	32.50abc	1.51a	3.33abcd
BRA052015	31.00ab	35.00abc	1.67a	2.83bcd
BRA052023	29.67b	26.50abcd	1.48a	4.14a
BRA052033	30.33b	31.33abc	1.56a	3.38abcd
BRA052034	29.00b	37.67ab	1.78a	2.58bcd
BRA052045	28.67b	37.67ab	1.75a	3.66abc
BRA052053	29.00b	36.50ab	1.65a	2.72bcd
BRS Primavera	29.33b	36.33ab	1.68a	2.49bcd
BRS Sertaneja	30.00b	40.33a	1.45a	2.68bcd
Average	30.49	31.99	1.38	2.72
F-test				
N rate (N)	NS		*	
Genotype (G)	**		**	
N × G	**		**	
CV (%)	14.25		15.86	

Source: Adapted from Fageria, N. K. 2010. Root growth of upland rice genotypes as influenced by nitrogen fertilization. Paper presented at the 19th World Soil Science Congress, Brisbane, Australia. 1–6 August, 2010.

*, **, NS: Significant at the 5% and 1% probability levels and not significant, respectively. Means followed by the same letter in the same column are not significant at the 5% probability level by Tukey's test.

FIGURE 6.1 Root growth of upland rice genotype BRS Primavera at two N levels.

to the lower N rate. Fageria (1992) reported a higher root length of rice at the low N rate compared to the high N rate in nutrient solution. Fageria (1992) also reported that at nutrient deficient levels, the root length is higher compared to the high nutrient levels because of the tendency of plants to tap nutrients from deeper soil layers.

Root dry weight varied from 0.87 g plant^{-1} produced by genotype BRA01596 to 1.78 g plant^{-1} produced by genotype BRA052034, with an average value of 1.38 g plant^{-1} at the lower N rate (0 mg N kg^{-1}). At the higher N rate (300 mg N kg^{-1}), the root dry weight ranged from 0.40 g plant^{-1} produced by genotype BRA01506 to 4.14 g plant^{-1} produced by genotype BRA052023, with an average value of 2.72 g plant^{-1}. Overall, the root dry weight was 97% higher at the higher N rate compared to the lower N rate. Fageria and Baligar (2005) and Fageria (2009) reported that N fertilization improved the root dry weight in crop plants, including upland rice. The positive effect of N on root dry matter has been previously documented (Fageria, 2009). Figures 6.1 through 6.3 show the influence of N on the root growth of upland rice genotypes.

NUE was determined in selected field crops without N and with N + P by several authors and the results are presented in Table 6.4. In corn, in one case, it was 31% higher, and in another case, it was 23% higher with the addition of N + P compared to the treatment that did not receive P. In case of wheat, the increase was 28% in the treatment that received N + P in comparison to the treatment

FIGURE 6.2 Root growth of upland rice genotype BRA 02535 at two N levels.

FIGURE 6.3   Root growth of upland rice genotype BRA 02601 at two N levels.

received only N but no P. In rice, NUE increased 14% with the addition of P together with N compared to N alone. In sunflower and field pea, NUE increased 59% and 46%, respectively, with the addition of P and N together compared to N addition alone. These results clearly showed the positive influence of the balanced supply of N and P in improving NUE in crop plants.

For nonirrigated crops, better root growth as a result of adequate P supply could enable the plants to absorb water from deeper soil layers during droughty spells, thereby increasing NUE as well (Aulakh and Malhi, 2005). In addition, when N is provided as an ammonium or ammonium-producing fertilizer, the acidifying effect could enhance N concentration in plants (Malhi et al., 1988) and P solubility in soil (Prasad and Power, 1997), thus providing a positive interaction. However, in those few soils that tested high in available N and P, the addition of single or both element fertilizers may not provide a grain yield advantage, regardless of the type of the crop, fertilizer, or placement method (Buah et al., 2000). Aulakh and Malhi (2005) concluded that, in a situation where a farmer cannot afford to apply both N and P in optimum amounts, it is better

TABLE 6.4
Influence of Nitrogen × P Interaction on NUE in Principal Field Crops

	NUE (kg Grain kg⁻¹ N)		Reference
Crop Species	**Without N**	**With N + P**	
Corn	8.8	11.5	Singh (1991)
Corn	32.4	40.0	Satyanarayan et al. (1978)
Wheat	20.3	25.9	Dwivedi et al. (2003)
Rice	22.4	25.5	Dwivedi et al. (2003)
Sunflower	8.7	13.8	Aulakh and Pasricha (1990)
Field peas	10.3	15.0	Pasricha et al. (1987)

Source: Adapted form Aulakh, M. S., N. S. Pasricha, and A. S. Azad. 1990. *Soil Sci.* 150:705–709; Dwivedi, B. S. et al. 2003. *Field Crops Res.* 80:167–193; Pasricha, N. S. et al. 1987. In: *Research Bulletin* 15, p. 92. Department of Soils, Ludhiana, India: Punjab Agricultural University; Satyanarayana, T., V. P. Badanur, and G. V. Havanagi. 1978. *Indian J. Agron.* 23:49–51; Singh, A. K. 1991. *Indian J. Agron.* 36:508–510.

to apply lower amounts of both N and P instead of a large amount of N alone. These authors also concluded that the N × P interaction varied among crop species and also within genotypes of the same species.

In a field study conducted for 5 years in a loamy sand soil in the state of Punjab in India, the interaction effect of N and P on the yield and protein content of field peas was significant (Pasricha et al., 1987) for harnessing the optimum yield potential, where the interaction impact was 23%. In dry beans (*Phaseolus vulgaris* L.), while N alone was beneficial only up to 30 kg N ha^{-1}, the crop made effective use of 60 kg N ha^{-1} when this was combined with 100 kg P_2O_5 ha^{-1} (Srinivas and Rao, 1984). Compared to the control plots, dry bean yields could be increased by more than five times by a judicious N + P combination, of which 59% was due to the interaction effect. Phosphorus application can create a more favorable environment for biological nitrogen fixation in legumes compared to N application alone. A balanced application of N and P maintains nitrogenase activity at high levels in field peas (Pasricha et al., 1987). Hence, the N × P interaction is favorable for biological nitrogen fixation in legumes enhancing N fixation as was also observed by Muller et al. (1993) and Amanuel et al. (2000) in faba beans (*Vicia faba* L.).

In a field study in Argentina, Zubillaga et al. (2002) reported that the yield of sunflower (*Helianthus annuus* L.) can be increased by 20% with the application of N and P together as compared to N alone. They concluded that P fertilizer provides a more efficient use of N by producing greater and consistent effects on crop performance most likely due to early root development. Available data on other oilseeds revealed that the interaction was positive in sesame (*Sesamum indicum* L.) (Daulay and Singh, 1982). Application of P to rapeseed (*Brassica napus* L.) and mustard (*Brassica juncea* L.) was more effective when combined with N, and, as a general guideline, N and P are recommended in a 2:1 ratio (Pasricha et al., 1991). The application of excessive N could increase aphid infestation in rapeseed, whereas a combined application of N + P suppresses its attack and increases the yield significantly (Khattak et al., 1996). In cotton (*Gossypium hirsutum* L.), the interaction between N and P was synergistic and accounted for 15% of the response to N + P in the first year and 29% in the second year (Raghuvanshi et al., 1989). Aulakh and Malhi (2005) concluded that the positive N × P interaction in cereals, legumes and nonlegume oilseeds, and other crops is responsible for improved N and P use efficiencies. However, the magnitude of this interaction can be modified by the climatic condition, crop species and genotypes within species, and level of available soil P and N.

Nonspecific N × P interactions are commonly observed if for no other reason that they are typically the two most limiting nutrients in crop production and therefore demonstrate additive or synergistic yield effects in bringing both nutrients to sufficiency levels of availability (Pan, 2012). Mechanisms of N × P interactions have been ascribed to all stages of the soil–plant continuum (Miller, 1974). Brennan and Bolland (2009) demonstrated classic Liebig-type responses to P at varying levels of N, suggesting that P was critically more limiting than N in these soils (Pan, 2012).

6.2.2 NITROGEN VERSUS POTASSIUM

These two nutrients are required in large amounts by crop plants. Hence, their sufficiency as well as deficiency has significant effects on the growth and development of plants. In addition, interactions having economic significance occur when one of these two nutrients is present at near deficiency levels, and the other at high or toxic levels (Wilkinson et al., 2000). Aulakh and Malhi (2005) reported that after N × P interactions, N × K interactions are the second most important interaction in crop production. The significance of the N × K interaction and its optimum management is increasing due to the increasing cropping intensity, higher crop yield, and greater depletion of soil K. Crops with a high requirement of K, such as corn and rice, often show strong N × K interaction (Loue, 1979; Singh, 1992; Fageria, 2014). Application of K was having a significant positive influence on the uptake of N in soybean (Abbasi et al., 2012).

Generally, N has a positive interaction with potassium. Increasing the levels of N improved K uptake if K was present in sufficient amounts in the growth medium. The influence of K on the absorption of NO_3-N by wheat plants was stronger than on the absorption of NH_4-N (Ali et al., 1987). These authors reported that the absorption of NO_3-N increased with the addition of K in the K-deficient plants while the absorption of NH_4-N still remained at a lower rate in spite of the addition of K (Ali et al., 1987). The influence of K was stronger on the translocation of N from roots to shoots and the translocation of NO_3-N was much higher compared to NH_4-N (Ali et al., 1987). Although K^+ is a cation, it does not compete in the absorption of the NH_4^+ ion; rather, it increases NH_4^+ assimilation in plants and avoids possible NH_4^+ toxicity (Aulakh and Malhi, 2005). Mengel et al. (1976) concluded that it was unlikely that K^+ competes with NH_4^+ for selective binding sites in the absorption process. Rice yield was significantly increased when N, P, and K were added in adequate amounts in comparison to control as well as N and P alone (Aulakh and Malhi, 2005).

Macleod (1969) reported that the optimum supply of K was important in promoting barley grain and straw yield, as deficient K levels had a depressing effect, especially when N was supplied at high rates. Johnson and Reetz (1975) observed that adequate soil test K levels are critical to realize the full benefits of applied N for harnessing optimum corn yields and NUE in Ohio. Also, more of the applied N was left in the soil after harvest, resulting in lower profitability and creating a greater potential for a negative environmental impact.

Potassium is one of the most important nutrients affecting the nitrogen metabolism of rice plants (Ali et al., 1985). Ali et al. (1985) reported that the N metabolism of rice plants was impaired by the limited supply of K when plants were fed with NO_3-N, but not when they were fed with NH_4-N. Data in Table 6.5 show the uptake of K at different N levels in the shoot of dry bean. There was a significant quadratic uptake of K when N levels were in the range of 0–200 kg ha^{-1}. Variation in K uptake due to N application was 71%. The positive interaction of N with K was related to the improvement in yield with the addition of N (Fageria and Baligar, 2005).

A positive interaction of N with K has been reported by Dibb and Thomson (1985). Dibb and Welch (1976) reported that the increased K allowed for rapid assimilation of absorbed NH_4^+ ions in the plant, as well as maintenance of a low, nontoxic level of NH_3. The increased yield of crops with the addition of N and P requires a higher level of K in the soil (Dibb and Thomson, 1985; Fageria et al., 2011a). A positive interaction between N and K has been reported by Kemp (1983) in barley. Barley yield was significantly increased with increasing N and K levels simultaneously.

Tropical soils such as Oxisols and Ultisols are poor in available P and K (Fageria and Baligar, 2008) and data of field experiments conducted on these soils show positive interactions among N × P × K (Aulakh and Malhi, 2005). Data from Brazil show a positive N × K interaction in rice

TABLE 6.5
Influence of N on the Uptake of K in the Shoot of 60-Day-Old Dry Bean Plants

N Rate (kg ha^{-1})	K Uptake (kg ha^{-1})
0	4.4
40	9.4
80	17.2
120	16.7
160	28.6
200	33.2
R^2	0.71*

*Significant at the 1% probability level.

where a good response to K was obtained only when adequate N (90 kg ha^{-1}) was applied (PPI, 1988). Also, the response to N increased as the level of K was increased; the highest rice yield as well as NUE and K use efficiency was obtained when both N and K were applied (PPI, 1988).

The author studied the interaction among N, P, and K in upland rice growth, yield, and yield components (Table 6.6). There was a significant N × P × K interaction for plant height, shoot dry

TABLE 6.6
Influence of N, P, and K Treatments on Plant Height, Shoot Dry Weight, Grain Yield, and Grain Harvest Index of Upland Rice

N, P, and K Treatments	Plant Height (cm)	Shoot Dry Weight (g Plant^{-1})	Grain Yield (g Plant^{-1})	Grain Harvest Index
$N_0P_0K_0$	24.33i	0.30i	0.00j	0.00g
$N_0P_0K_1$	26.00i	0.40hi	0.00j	0.00g
$N_0P_0K_2$	46.83f	0.86hi	0.00j	0.00g
$N_0P_1K_0$	75.00e	3.82gh	2.32ij	0.38abcde
$N_0P_1K_1$	75.17de	8.55ef	3.05ij	0.26f
$N_0P_1K_2$	91.92bc	7.30ef	3.75i	0.34bcdef
$N_0P_2K_0$	78.42de	8.56ef	3.20ij	0.27f
$N_0P_2K_1$	79.92de	8.17ef	3.34ij	0.29ef
$N_0P_2K_2$	83.33cde	8.86e	4.05hi	0.31def
$N_1P_0K_0$	39.21fgh	0.52hi	0.00j	0.00g
$N_1P_0K_1$	43.33fg	0.49hi	0.00j	0.00g
$N_1P_0K_2$	35.08ghi	0.54hi	0.00j	0.00g
$N_1P_1K_0$	85.25cde	12.75d	5.93ghi	0.32cdef
$N_1P_1K_1$	106.83a	16.32c	10.71ef	0.39abcde
$N_1P_1K_2$	111.58a	17.69c	12.64de	0.42abc
$N_1P_2K_0$	93.67bc	15.35cd	10.21ef	0.40abcd
$N_1P_2K_1$	111.08a	18.66c	12.35cd	0.43ab
$N_1P_2K_2$	109.08a	22.19b	16.64bc	0.43ab
$N_2P_0K_0$	27.75hi	0.35hi	0.00j	0.00g
$N_2P_0K_1$	30.75hi	0.44hi	0.00j	0.00g
$N_2P_0K_2$	35.83fghi	5.24fg	0.00j	0.00g
$N_2P_1K_0$	86.67cd	15.29cd	7.56fgh	0.33bcdef
$N_2P_1K_1$	110.67a	22.38b	16.20bcd	0.42abc
$N_2P_1K_2$	110.58a	22.89b	18.63b	0.45a
$N_2P_2K_0$	92.75bc	16.73c	8.82fg	0.35abcdef
$N_2P_2K_1$	103.33ab	24.28b	18.75b	0.44ab
$N_2P_2K_2$	108.42a	28.66a	23.01a	0.45a
Average	74.92	10.65	6.78	0.25
F-test				
N	**	**	**	**
P	**	**	**	**
K	**	**	**	**
N × P	**	**	**	**
N × K	**	**	**	**
P × K	**	**	**	**
N × P × K	**	**	**	**
CV (%)	4.69	9.98	16.31	9.66

Note: Means within the same column followed by the same letter do not differ significantly at the 5% probability level by Tukey's test.

**Significant at the 1% probability level.

weight, grain yield, and grain harvest index (GHI) (Table 6.6). Hence, the response of these plant characteristics is associated with an adequate rate of N, P, and K fertilization. Plant height varied from 24.33 cm in the $N_0P_0K_0$ treatment to 111.58 cm in the $N_1P_1K_2$ (N = 150 mg kg^{-1}, P = 100 mg kg^{-1}, and K = 200 mg kg^{-1}) treatment, with an average value of 74.92 cm. Shoot dry weight varied from 0.30 g plant^{-1} in the $N_0P_0K_0$ treatment to 28.66 g plant^{-1} in the $N_2P_2K_2$ (N = 300 mg kg^{-1}, P = 200 mg kg^{-1}, and K = 200 mg kg^{-1}) treatment, with an average value of 10.65 g plant^{-1}. Fageria and Baligar (1997) reported a significant increase in rice plant height and shoot dry weight with the addition of N, P, and K fertilization in the Brazilian Oxisol.

Grain yield varied from 0 to 23.01 g plant^{-1} in the $N_0P_0K_0$ and $N_2P_2K_2$ treatments, respectively, with an average value of 6.78 g plant^{-1} (Table 6.6). Plants that did not receive P fertilization but received adequate rates of N and K did not produce panicles or grains. Hence, it can be concluded that P is the most yield-limiting nutrient in the highly weathered Brazilian Oxisol. Fageria and Baligar (1997, 2001) have reported similar results. Grain yield results also showed that there is a strong positive interaction among N, P, and K fertilization in upland production. This type of interaction is widely reported in the literature (Wilkinson et al., 2000). These authors also reported that increasing the N rate increases the demand for other nutrients, especially P and K, and higher yields were obtained at the highest rates of N, P, and K. Wilson (1993) also confirmed the generalization that the response to one nutrient depends on the sufficiency level of other nutrients. Yield reductions were found when high levels of one nutrient were combined with low levels of other nutrients (Wilkinson et al., 2000). Alleviating the yield-depressing effect of excessive macronutrient supply involved removing the limitation of a low supply of other nutrients.

GHI varied from 0 in the treatment that did not receive P to 0.45 in the treatment $N_2P_2K_2$, with an average value of 0.25 (Table 6.6). GHI is an important index in determining the partitioning of dry matter between shoot and grain. Fageria (2009) reported that rice GHI is influenced by environmental factors, including mineral nutrition. Fageria and Baligar (2005) reported that variation in rice GHI is from 0.23 to 0.50. However, Kiniry et al. (2001) reported that rice GHI values varied greatly among cultivars, locations, seasons, and ecosystems, ranging from 0.35 to 0.62. The GHI values of modern crop cultivars are commonly higher than for old traditional cultivars for major field crops (Ludlow and Muchow, 1990). The limit to which GHI can be increased is considered to be about 0.60 (Austin et al., 1980). Hence, cultivars with low harvest indices would indicate that further improvement in the partitioning of a biomass would be possible (Fageria and Baligar, 2005). On the other hand, cultivars with harvest indices between 0.50 and 0.60 would probably not benefit by increasing the harvest index (Sharma and Smith, 1986).

Average response values of plant height, shoot dry weight, grain yield, and GHI with the application of N, P, and K nutrients are presented in Table 6.7. Maximum plant height was achieved at the N_1 level (150 mg N kg^{-1} of soil), P_2 level (100 mg P kg^{-1} of soil), and K_2 level (200 mg K kg^{-1} of soil). Similarly, shoot dry weight, grain yield, and grain harvest values were maximum at the highest levels of N_2 (300 mg kg^{-1} of soil), P_2 (200 mg P kg^{-1} of soil), and K_2 (200 mg K kg^{-1} of soil). The increase in plant height was 25% at the N_2 level compared to the N_0 level. Similarly, the increase in plant height at the P_2 level was 179% compared to the P_0 level and 25% at the K_2 level compared to the K_0 level. Shoot dry weight increase was 191% at the highest N level compared to the lowest N level; in case of P, the increase at the highest level was 1550% compared to the lowest P level. Similarly, the increase in shoot dry weight at the highest K level was 55% compared to the zero K level. Grain yield increase was 372% at 300 mg N kg^{-1} compared to control treatment. At the highest K level (200 mg K kg^{-1}), the increase was 107% compared to control treatment. In case of P at the zero P level, plants did not produce grain. The increase in harvest index at the highest N, P, and K level followed the same pattern as the plant height, shoot dry weight, and grain yield. Hence, it can be concluded that P was the most yield-limiting nutrient, followed by N and K in upland rice production in Brazilian Oxisol. Fageria and Baligar (1997) and Fageria et al. (2010) reported similar conclusions. The low availability of P in Brazilian Oxisol is associated with the

TABLE 6.7

Average Values of Plant Height, Shoot Dry Weight, Grain Yield, and Grain Harvest Index across N, P, and K Levels

N, P, and K Treatments	Plant Height (cm)	Shoot Dry Weight (g Plant^{-1})	Grain Yield (g Plant^{-1})	Grain Harvest Index
N_0	65.55b	5.20c	2.19c	0.21b
N_1	81.68a	11.61b	7.83b	0.27a
N_2	78.53a	15.12a	10.33a	0.27a
Average	74.92	10.65	6.78	0.25
P_0	34.35b	1.02c	0.00c	0.00b
P_1	94.85a	12.11b	8.98b	0.37a
P_2	95.56a	16.83a	11.37a	0.37a
Average	74.92	10.65	6.78	0.25
K_0	67.00b	8.18b	4.23b	0.23a
K_1	76.34a	11.08a	7.38a	0.25a
K_2	81.41a	12.69a	8.75a	0.27a
Average	74.92	10.65	6.78	0.25

Note: Means within the same column and same nutrient levels followed by the same letter do not differ significantly at the 5% probability level by Tukey's test.

natural low level of this element and higher P immobilization capacity (Fageria, 1989; Fageria and Baligar, 2008).

Panicles per plant were significantly influenced by N, P, and K treatments and their interactions (Table 6.8). The 1000 grain weight was also influenced by N, P, and K treatments and N × P and N × K interactions. Root dry weight was influenced by N, P, and K treatments and N × P, N × K, and P × K interactions. This means that for obtaining the maximum panicle number, 1000 grain weight, and root dry weight, there is need for an adequate level of N, P, and K in the growth medium. Average analysis of N, P, and K showed the maximum effect of P on the panicle number, followed by N and K (Table 6.9). Root dry weight and maximum root length were influenced by N and P treatments. The K application improves these parameters but the effect was not significant. Spikelet sterility was reducing with the application of K compared to the control treatment. The improvement in panicle number, 1000 grain weight, and root growth with the application of N, P, and K has been reported by Fageria et al. (2010) and Fageria (2009).

The N, P, and K interactions were also observed in the growth of rice plants. Plants that did not receive N, P, and K fertilization ($N_0P_0K_0$) had a significantly lower height, did not produce tillers, and also did not produce grain (Figure 6.4). Similarly, plants that did not receive N and P but only K also did not produce grain or tillers, and the height was significantly reduced (Figure 6.4). Rice growth was significantly improved with increasing N rates from 0 to 300 mg N kg^{-1} along with P and K 200 mg kg^{-1} of soil (Figure 6.5). Plants with adequate P_{200} and K_{200} but which did not receive N were yellow in color, with reduced tillering, and had few panicles (Figure 6.5). Increasing P levels from 0 to 200 mg kg^{-1} significantly improved the rice growth, tillering, and panicle number (Figure 6.6). However, plants without P did not produce panicles or grains (Figure 6.6). The plants that received adequate amounts of N and K but did not receive P had leaves that were dark green and straight (Figure 6.6), indicating that P deficiency is the most yield-limiting factor in Brazilian Oxisols. The K-deficient plants produced yellow leaves whose margins and tips were dry (Figure 6.7). Potassium symptoms first started on older leaves, and with time the whole plant was showing K deficiency symptoms. Maturity was delayed by about 10 days in pots that did not receive K but

TABLE 6.8

Influence of Nitrogen, Phosphorus, and Potassium Levels on Panicle Number, 1000 Grain Weight, and Root Dry Weight

N, P, and K Treatments	Panicle Number (Plant^{-1})	1000 Grain Weight (g)	Root Dry Weight (g Plant^{-1})
1. N_0P_0	0.00e	0.00e	0.18d
2. N_0P_1	2.42d	29.67ab	1.59cd
3. N_0P_2	2.66d	30.29a	1.53cd
4. N_1P_0	0.0e	0.00e	0.22d
5. N_1P_1	3.89c	27.92bc	2.77bc
6. N_1P_2	5.58b	26.45cd	4.32ab
7. N_2P_0	0.0e	0.00e	0.28b
8. N_2P_1	5.66b	26.27cd	4.09ab
9. N_2P_2	7.27a	25.65d	5.11a
1. N_0K_0	1.44a	20.13a	1.04c
2. N_0K_1	1.44a	20.12a	1.03c
3. N_0K_2	2.19a	19.71ab	1.23c
4. N_1K_0	2.55a	17.73bcd	2.15bc
5. N_1K_1	3.11a	18.79abc	2.04bc
6. N_1K_2	3.81a	17.89bcd	3.12ab
7. N_2K_0	3.25a	16.11d	1.68bc
8. N_2K_1	4.53a	17.41cd	3.21ab
9. N_2K_2	5.17a	18.39abc	4.58a
1. P_0K_0	0.00d	0.00a	0.14c
2. P_0K_1	0.00d	0.00a	0.26c
3. P_0K_2	0.00d	0.00a	0.29c
4. P_1K_0	2.58c	27.66a	2.31b
5. P_1K_1	4.42b	28.11a	2.43b
6. P_1K_2	4.97b	28.10a	3.73ab
7. P_2K_0	4.67b	26.31a	2.44b
8. P_2K_1	4.66b	28.22a	3.61ab
9. P_2K_2	6.19a	27.86a	4.91a
F-test			
N	**	**	**
P	**	**	**
K	**	**	**
N × P	**	**	**
N × K	NS	**	**
P × K	**	NS	**
N × P × K	**	NS	NS
CV (%)	22.40	7.20	47.36

Note: Means within the same column and interaction N × P, N × K, and P × K nutrient levels followed by the same letter do not differ significantly at the 5% probability level by Tukey's test.

**, NS: Significant at the 1% probability level and not significant, respectively.

received adequate amounts of N and P (Figure 6.7). The root system was also adversely affected in the absence of N, P, and K fertilization. Figure 6.8 shows that root growth increased with increasing K fertilization. Fageria (1984, 1989) and Fageria and Baligar (1997) presented similar results in relation to N, P, and K deficiency symptoms, as well as root and shoot growth of rice grown on Brazilian Oxisol.

TABLE 6.9

Average Values of Panicle Number, 1000 Grain Weight, and Root Dry Weight across N, P, and K Levels

N, P, and K Treatments	Panicle Number (Plant^{-1})	1000 Grain Weight (g)	Root Dry Weight (g Plant^{-1})	Maximum Root Length (cm)	Spikelet Sterility (%)
N_0	1.69c	19.99a	1.10a	29.56b	9.17b
N_1	3.16b	18.13b	2.44ab	32.93a	11.75a
N_2	4.31a	17.31b	3.16a	34.63a	9.98ab
Average	3.06	18.48	2.23	32.37	10.30
P_0	0.00c	0.00b	0.23b	24.70c	0.00b
P_1	3.99b	27.96a	2.82a	33.89b	14.25a
P_2	5.18a	27.47a	3.66a	38.52a	16.64a
Average	3.06	18.48	2.23	32.37	10.30
K_0	2.42a	17.99a	1.63a	32.37a	12.27a
K_1	3.03ab	18.78a	2.10a	32.33a	8.92b
K_2	3.72a	18.65a	2.98a	32.41a	9.70b
Average	3.06	18.48	2.23	32.37	10.30

Note: Means within the same column and same nutrient levels followed by the same letter do not differ significantly at the 5% probability level by Tukey's test.

6.2.3 NITROGEN VERSUS CALCIUM

Nitrogen and calcium interaction in crop plants is important, especially in legumes where both these nutrients are required in large amounts. Positive interactions have been reported in most of the studies related to N and Ca interactions (Fageria and Baligar, 2005). Data in Table 6.10 show the influence on N on the uptake of Ca in the dry bean shoot. Calcium uptake was significantly and quadratically increased with increasing N levels from 0 to 200 kg ha^{-1}. The increase in Ca uptake was related to the increase in the dry matter of shoot with the addition of N (Fageria and Baligra,

FIGURE 6.4 Upland rice plants without N, P, and K fertilization (left pot), without N and P and with 100 mg K kg^{-1} soil (middle pot), and without N and P and with 200 mg K kg^{-1} soil (right pot).

FIGURE 6.5 Upland rice plants without N and with 200 mg P and K kg⁻¹ of soil (left pot), with 150 mg N and 200 mg P and K kg⁻¹ of soil (middle pot), and with 300 mg N and 200 mg P and K kg⁻¹ of soil (right pot).

FIGURE 6.6 Upland rice plants without P and with 300 mg N and 200 mg K kg⁻¹ soil (left pot), with 100 mg P and with 300 mg N and 200 mg K kg⁻¹ soil (center pot), and with 200 mg P and with 300 mg N and 200 mg K kg⁻¹ soil (right pot).

FIGURE 6.7 Upland rice plants without K and with 300 mg N and 200 mg P kg⁻¹ soil (left pot), with 100 mg K and with 300 mg N and 200 mg K kg⁻¹ soil (center pot), and with 200 mg K and with 300 mg N and 200 mg P kg⁻¹ soil (right pot).

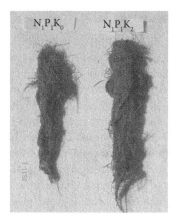

FIGURE 6.8 Root growth of upland rice at different N, P, and K levels. ($N_1P_1K_0 = N_1$, 150 mg N kg^{-1}; P_1, 100 mg P kg^{-1}; K_0, 0 mg K kg^{-1}; and K_2, 200 mg K kg^{-1}.)

2005). Wilkinson et al. (2000) reported that plants supplied with NO_3-N accumulate more cations (K, Ca, and Mg), while those supplied with NH_4^+ accumulated more anions (SO_4^{2-}, $H_2PO_4^-$, Cl$^-$). Wilkinson et al. (2000) also reported that when N is supplied as NH_4^+, it may decrease the uptake of cations such as Ca^{2+}. However, in well-drained or oxidized soils, NO_3-N is the major form of N uptake. In flooded rice, nitrogen may be present in NH_4^+ form in the reduced soil layer due to lack of oxygen, which inhibits nitrification. Under this situation, there may be a negative interaction between N and Ca^{2+}.

The use of lime and farmyard manures significantly increased water-soluble N and fixed NH_4^+ in acid soils of India, leading to increased N uptake by soybean and wheat (Bishnoi et al., 1984; Prasad et al., 1986). In a highly acidic soil (pH 4.5), a substantially higher rice yield obtained with the combined application of lime and NPK than NPK or lime alone indicated that soil acidity is the main constraint in the utilization of nutrients by the crop (Fageria and Baligar, 2001). Once acidity is corrected, the uptake of soil N, Ca, and some other nutrients registered a multifold increase. Other associated problems, such as high concentrations of Al and Mn, reduced biological nitrogen fixation in legumes, and decreased root growth, may lead to a decline in water and nutrient use efficiencies (Aulakh and Malhi, 2005). Malhi et al. (1995) reported that with the application of lime with an adequate rate of N, there was a significant yield increase of barley and also improved NUE.

TABLE 6.10
Influence of N on the Uptake of Ca in the Shoot
of 60-Day-Old Dry Bean Plants

N Rate (kg ha^{-1})	Ca Uptake (kg ha^{-1})
0	4.1
40	8.2
80	12.8
120	12.9
160	18.4
200	24.9
R^2	0.79**

**Significant at the 1% probability level.

6.2.4 Nitrogen versus Magnesium

Magnesium is a macronutrient essential for all plant growth and development. The adequacy of its level in the soil is important for producing maximum economic yields. Like calcium (Ca^{2+}), magnesium (Mg^{2+}) deficiency in crop production is more common in highly weathered acid soils (Fageria and Souza, 1991). A deficiency of Mg^{2+} may also occur in coarse-textured soils of humid regions with low cation exchange capacities. The functions of Mg^{2+} in plants are many and the most important are the enzyme activator and also part of the chlorophyll molecule. Magnesium is a mineral constituent of plant chlorophyll, so it is actively involved in photosynthesis. Magnesium also aids in phosphate metabolism, plant respiration, and the activation of several enzyme systems involved in energy metabolism (Fageria and Gheyi, 1999). Magnesium aids in the formation of sugars, oils, and fats. It also activates the formation of polypeptide chains from amino acids (Tisdale et al., 1985). Magnesium is also an essential element for microbial growth and was implicated in microbial ecology in the early studies of soil microbiology, since magnesium carbonate applied to certain soils increases the reproduction of soil bacteria (Jones and Huber, 2007).

Looking into the importance of Mg, the knowledge of factors affecting its uptake is important in the management of this element for higher or sustainable crop yields. Nitrogen has a positive interaction in the uptake of magnesium (Fageria and Baligar, 2005). Data in Table 6.11 show that Mg uptake in dry bean plant shoots was significantly and quadratically increased with increasing N levels in the range of 0–200 kg ha^{-1}. The variation in the uptake of Mg was 79% with the addition of N.

6.2.5 Nitrogen versus Sulfur

Sulfur is the fourth major plant nutrient along with N, P, and K. Deficiency of sulfur has been reported in many soils and crops in many parts of the world (Scherer, 2001). Among different regions, Asia represents the region with the highest S fertilizer requirement (Aulakh and Malhi, 2005). Sulfur plays many important roles in the growth and development of plants. Fageria and Gheyi (1999) have summarized the important functions of sulfur in plants: (i) Sulfur is an important component of two amino acids, cysteine, and methionine, which are essential for protein formation. Since animals cannot reduce sulfate, plants play a vital role in supplying essential S-containing amino acids to them (Fageria and Gheyi, 1999). (ii) It plays an important role in enzyme activation. (iii) It promotes nodule formation in legumes. (iv) It is necessary in chlorophyll formation, although it is not a constituent of chlorophyll. (v) The maturity of seeds and fruits is delayed in the absence of adequate sulfur. (vi) Sulfur is required by plants in the formation of nitrogenase. (vii) It increases the crude protein content of forages. (viii) It improves the quality of the cereal crop for milling and baking. (ix) It increases the oil content of oilseed crops. (x) It increases winter hardiness in plants. (xi) It

TABLE 6.11
Influence of N on the Uptake of Mg in the Shoot of 60-Days-Old Dry Bean Plants

N Rate (kg ha^{-1})	Mg Uptake (kg ha^{-1})
0	1.0
40	2.1
80	3.2
120	3.3
160	5.1
200	6.5
R^2	0.79**

**Significant at the 1% probability level.

increases drought tolerance in plants. (xii) It controls certain soil-borne diseases. (xiii) It helps in the formation of glycosides that give characteristic odors and flavors to onion, garlic, and mustard. (xiv) It is necessary for the formation of vitamins and the synthesis of some hormones and glutathione. (xv) It is involved in oxidation–reduction reactions. (xvi) It improves the tolerance to heavy metal toxicity in plants. (xvii) It is a component of sulfur-containing sulfolipids. (xviii) Organic sulfates may serve to enhance the water solubility of organic compounds, which may be important in dealing with salinity stress (Clarkson and Hanson, 1980). (xix) Fertilization with soil applied S in sulfate form decreases fungal diseases in many crops (Klikocka et al., 2005; Haneklaus et al., 2007).

Since S has many important functions in the plant, improvement in its uptake is important to achieve maximum economic yields. Nitrogen fertilization has positive interactions in crop plants (Zhao et al., 1997). Nitrogen has a strong influence on S assimilation and vice versa (Fageria and Gheyi, 1999). Increasing N levels along with S improves crop yields quadratically (Fageria et al., 2011a). Jackson (2000) reported that canola's (*Brassica napus* L.) response to N fertilization reached a plateau at about 200 kg N ha^{-1} without S addition. However, canola responds almost linearly to N fertilization up to 250 kg N ha^{-1} when 22 kg S ha^{-1} was added.

When there is S deficiency in the soil, the growth of most crop plants reduces significantly. While N directly affects the photosynthesis efficiency of plants, S affects the photosynthesis efficiency indirectly by improving the NUE of plants, as evident from the relationship between N content and photosynthesis rate in the leaves with S and without S-treated *Brassica* plants (Ahmad and Abdin, 2000). In without S plants, photosynthesis was related to leaf N content only up to 1.5 g m^{-2}, whereas the relationship was linear even beyond 1.5 g m^{-2} in S-treated plants. Rapeseed plants grown on S-limiting soils suppress the development of reproductive growth and could even lead to poor seed set (Nuttall et al., 1987) or pod absorption (Fismes et al., 2000).

Oilseeds and legumes require a large amount of S compared to cereals and grasses (Aulakh and Chhibba, 1992). Malhi and Gill (2002) reported that the application of N in association with S improved grain yield, oil content, and S uptake in canola compared to no fertilizer and N-alone treatments significantly. These authors also reported that NUE was 2.0 kg seed kg^{-1} N when N fertilizer was applied alone and it increased more than five times (10.2 kg seed kg^{-1} N) when both N and S fertilizers were applied. Mcgrath and Zhao (1996) reported an increase of 42–267% in the seed yield of *Brassica napus* with the application of 40 kg S ha^{-1} along with 180 and 230 kg N ha^{-1}. Without S application, the seed yield declined drastically due to S deficiency when the N fertilization rate increased from 180 to 230 kg N ha^{-1}. Such severe negative impacts when N alone was applied to S-deficient soils on seed yield, oil content and production, protein content, and NUE in rapeseed and mustard crops have been reported in several other studies from Canada (Janzen and Bettany, 1984; Nuttall et al., 1987), India (Abdin et al., 2003), and Europe (Fismes et al., 2000; Walker and Booth, 2003).

Several field experiments with pulses such as chickpea (*Cicer arietinum* L.), lentil (*Lens culinaris* Medik), mungbean (*Vigna radiata*), black gram (*Vigna mungo*), pigeon pea (*Cajanus cajan* L. Millsp.), and cowpea (*Vigna unguiculata* L. Walp) showed a significant increase in grain yield due to balanced N, P, and S fertilization (Aulakh and Malhi, 2005). A synergistic interaction of N × S has been reported by Tandon (1992) in several crop species in relation to yield, protein content, and nutrient uptake. The positive interaction of N × S in grain yield and seed oil content has been reported by Aulakh et al. (1990). The yield of wheat, grown in the coastal plain of Virginia, increased linearly with N + S application (Reneau et al., 1986). In four different studies in India, the application of S with N and P produced an additional yield of 700–1300 kg and 400 kg ha^{-1} of wheat and corn, respectively (Aulakh and Chhibba, 1992). Dev et al. (1979) reported a significant N × S interaction in corn resulting in higher S uptake in the leaves and stems with increasing levels of N and S in the growth medium. Mixing urea with elemental S in a 4:1 ratio prior to its surface application onto a calcareous soil enhanced the NUE of pearl millets from 15% to 48% while reducing the NH$_3$ volatilization by about 50% (Aggarwal et al., 1987). Sexton et al. (1998) reported that N and S affect the level and composition of seed storage proteins in soybean. Kim et al. (1999) reported that SO$_4^{2-}$ and NO$_3^-$ are related to protein synthesis during N and S assimilation.

Because of the importance of both S and N in protein synthesis, these nutrients are intimately linked and are often considered to be colimiting. It has been established that for every 15 parts of N in protein, there is approximately 1 part of S (i.e., 15:1 ratio of N:S) (Norton et al., 2013). However, this general guide will vary for different crops. For example, wheat grain has an N:S ratio of around 16:1, while the N:S ratio for canola seed is around 6:a (Norton et al., 2013). An inadequate S supply will not only reduce yield and crop quality but it will also decrease NUE and enhance the risk of N loss to the environment. Studies have demonstrated that supplying S to deficient pastures increased yields and NUE, and lowered N losses from the soil (Norton et al., 2013). Owing to the close linkage between S and N, Norton et al. (2013) reported that one unit of S deficit to meet plant demand can result in 15 units of N that are potentially lost to the environment.

Nitrogen and sulfur interaction is synergistic to affect crop grain yield (Wang et al., 1976; Randall et al., 1981), grain protein and grain quality (Randall et al., 1990), and N utilization efficiency (Malhi and Gill, 2002, Salvagiotti and Miralles, 2008). Optimizing S nutrition increases NUE, mainly by increasing N uptake efficiency in grass species (Brown et al., 2000; Salvagiotti et al., 2009). In brassica crops, yield and oil content (Ahmad et al., 1999), glucosinolate (Kim et al., 2002; Schonhof et al., 2007), and isothiocyanate (Gerendas et al., 2008) concentrations are influenced by relative supplies of N and S (Pan, 2012). The field observation of N and S fertility interactions are well reflected at the cellular and root levels (Reuveny et al., 1980; Hesse et al., 2004). Mineral nutrition studies have revealed that the uptake and assimilation of N and S are coregulated by the substrate ions and their assimilatory products (Clarkson et al., 1989; Koprivova et al., 2000; Hesse et al., 2004; Pan, 2012). Sulfate uptake and assimilation are regulated by O-acetylserine, a cysteine precursor that is in itself regulated by N availability and assimilation (Koprivova et al., 2000; Pan, 2012). Excess cysteine production when S is high or N is limiting will repress S uptake and assimilation (Zhao et al., 1999). Conversely, N uptake and assimilation is depressed during S starvation (Clarkson et al., 1989; Prosser et al., 2001) as arginine and asparagine accumulated with reduced cysteine and methionine production (Thomas et al., 2000; Prosser et al., 2001; Pan, 2012).

6.3   INTERACTION OF NITROGEN WITH MICRONUTRIENTS

Micronutrients are required by plants in small amounts compared to macronutrients. Overall, the concentration of macronutrients (except C, H, and O) in plant dry matter at maturity varied from 1 to 12 g kg^{-1} (0.1–1.2%) (Fageria et al., 2011a). Micronutrients essential for plant growth are Zn, Fe, Mn, Bo, Cu, Mo, Ni, and Cl. The deficiency of micronutrients is not as widespread as that of macronutrients. However, whenever it occurs, it can result in a significant reduction in the yield and quality of crops and utilization efficiency of other nutrients and water (Aulakh and Malhi, 2005). For example, the deficiency of Zn is very common in upland rice in the central part of Brazil locally known as the "Cerrado region." Addition of about 10 kg Zn ha^{-1} with zinc sulfate can correct the deficiency and improve the yield and nutrient use efficiency (Fageria, 2009, 2013; Fageria et al., 2011a).

6.3.1   NITROGEN VERSUS ZINC

Zinc deficiency is one of the main problems in crop production around the world (Fageria et al., 2002, 2012; Alloway, 2008). Graham (2008) reported that half of the world's soils are intrinsically deficient in Zn. Zinc deficiency in annual crops is reported in Brazil (Fageria and Stone, 2008), Australia (Graham, 2008), India (Singh, 2008), China (Zou et al., 2008), Turkey (Cakmak, 2008), Europe (Sinclair and Edwards, 2008), the United States (Brown, 2008), and Africa (Waals and Laker, 2008). Fageria and Baligra (1999) reported that liming of acid soils is one of the main factors creating Zn deficiency in crop plants (Figure 6.9). Hence, it can be concluded that the knowledge of factors affecting Zn uptake (including interactions with other nutrients) is important to its adequate management.

Nitrogen application improved crop growth, which may create Zn deficiency in soils having low Zn content. Such interaction of N × Zn has been observed by Fageria (2013) in upland rice grown on a

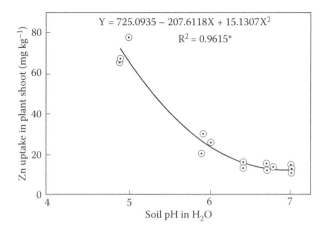

FIGURE 6.9 Influence of soil pH on the uptake of Zn in the shoot of dry bean plants. *—Significant at the 5% probability level. (Adapted from Fageria, N. K. and V. C. Baligar. 1999. *J. Plant Nutr.* 22:23–32.)

Brazilian Oxisol. Singh and Singh (1985) also reported that as the N supply to rice increased, Zn deficiency was observed. Savithri and Ramnathan (1990) reported that the application of $ZnSO_4$ increased the response of rice to urea–N by 400–600 kg grain ha^{-1}. It is widely reported in the literature that $N \times Zn$ interaction is very common in crops that respond to N fertilization (Bajwa and Paul, 1978; Kene and Deshpande, 1980; Kumar et al., 1985; Sakal et al., 1988; Verma and Bhagat, 1990). From these studies, it was concluded that N and Zn having synergistic interaction and maximum yield could be obtained with optimum balance between these two nutrients (Aulakh and Malhi, 2005). Data in Table 6.12 show that the maximum yield of wheat was obtained when N, P, K, and Zn were at adequate levels. Synergistic $N \times Zn$ interactions have been reported to increase the N concentration in different crop species (Singh and Tripathi, 1974; Hulagur and Dangarwala, 1983; Kene and Deshpande, 1980). The use of balanced N and Zn improved biological nitrogen fixation in legumes and also amino acids and proteins in cereals (Dwivedi and Randhawa, 1973; Kene and Deshpande, 1980).

Kutman et al. (2011) reported that Zn uptake by wheat plants enhanced up to fourfold by high N supply while the increase in plant growth by high N supply was much less. They further reported that when both the Zn and N supplies were high, approximately 50% of grain Zn was provided by postanthesis shoot uptake, indicating that the contribution of remobilization to grain accumulation was higher for Zn. Kutman et al. (2010) and Shi et al. (2010) also reported that the grain concentration of Zn can be enhanced by increasing the N supply and that Zn and N applications have a synergistic effect on the grain Zn concentration of durum wheat. Nitrogen nutrition of plants appears to be a critical component for an effective biofortification of food crops with Zn due to several physiological and molecular mechanisms, which are under the influence of N nutritional status (Cakmak et al., 2010).

TABLE 6.12
Grain Yield of Wheat (kg ha^{-1}) as Influenced by N, P, K, and Zn Levels

N (kg ha^{-1})	P (kg ha^{-1})	K (kg ha^{-1})	0 kg Zn ha^{-1}	5 kg Zn ha^{-1}	10 kg Zn ha^{-1}
0	0	0	1450	1580	1640
50	30	25	2730	2880	3030
100	60	50	3530	3840	4040
LSD (5%)				220	

Source: Adapted from Sakal, R., A. P. Singh, and R. B. Sinha. 1988. *J. Indian Soc. Soil Sci.* 36:125–127.

TABLE 6.13
Influence of N on the Uptake of Zn in the Shoot
of 60-Day-Old Dry Bean Plants

N Rate (kg ha^{-1})	Zn Uptake (g ha^{-1})
0	14.0
40	27.7
80	38.4
120	42.1
160	62.0
200	88.5
R^2	0.79**

**Significant at the 1% probability level.

The trafficking of Zn from the rhizosphere into grains is dependent on various protein and other nitrogenous compounds, including amino acids and peptides (Kutman et al., 2011). Kutman et al. (2010) reported that grain protein is a sink of Zn. High N can increase the grain Zn concentration by enhancing the grain protein concentration and thereby the sink strength of the grain for Zn. Significant positive correlations between seed protein and Zn have been documented in various studies (Peleg et al., 2008; Zhao et al., 2009). Although both the uptake and remobilization of Zn in wheat plants are positively affected by N nutrition (Erenoglu et al., 2001), the share of concurrent uptake during grain filling in grain Zn deposition is increased by higher N supply (Kutman et al., 2011, 2012).

Fageria and Baligar (2005) studied influence of N on the uptake of Zn in the shoot of dry bean (Table 6.13). Nitrogen fertilization significantly increased Zn uptake in the plant tissue with the increasing N levels in the range of 0 to 200 kg N ha^{-1}. Nitrogen application was responsible for 79% variation in the uptake of Zn. Hence, it can be concluded that N is one of the main factors in increasing Zn uptake. These authors reported that increase in Zn uptake was related to increase in dry matter of dry bean plants. Pederson et al. (2002) also reported that N concentration was highly correlated with Zn concentration in aboveground plant parts of ryegrass. Nitrogen application has been reported to influence Zn absorption by plants and vice versa (Aulakh and Malhi, 2005). In corn, the Zn concentration in shoots was higher when both N and Zn were applied together followed by the application of Zn and N alone (Dev and Shukla, 1980).

6.3.2 NITROGEN VERSUS COPPER

Copper deficiency has been reported in many parts of the world in crop plants. According to Sillanpaa (1990), 14% of the world soils are Cu deficient. Copper deficiency has been reported in Brazilian Oxisols (Goedert, 1983; Hitsuda et al., 2010). Nearly 70–80% of Cerrado Oxisols are deficient in Zn, Cu, or Mn (Lopes and Cox, 1977). Micronutrient deficiencies in crop plants are widespread because of (i) increased micronutrient demands from intensive cropping practices and adaptation of high-yielding cultivars, which may have higher micronutrient demand, (ii) enhanced production of crops on marginal soils that contain low levels of essential nutrients, (iii) increased use of high analysis fertilizers with low amounts of micronutrients, (iv) decreased use of animal manures, composts, and crop residues, (v) use of many soils that are inherently low in micronutrient reserves, (vi) involvement of natural and anthropogenic factors that limit adequate supplies and create element imbalances (Fageria et al., 2002), and (vii) liming acid soils (Figure 6.10). Fageria and Baligar (1997) reported that cereals and legumes grown on Oxisols responded significantly to macro- and micronutrient fertilization.

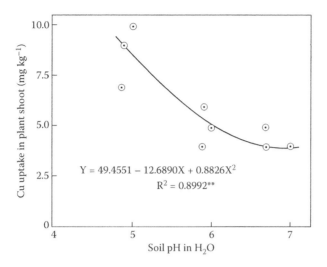

FIGURE 6.10 Influence of soil pH on the uptake of Cu in the shoot of dry bean plants. **—Significant at the 1% probability level. (Adapted from Fageria, N. K. and V. C. Baligar. 1999. *J. Plant Nutr.* 22:23–32.)

Fageria and Baligar (2005) studied the influence of N on the uptake of Cu in the shoots of dry bean plants (Table 6.14). The uptake of Cu was significantly increased in a quadratic fashion with increasing N rates in the range of 0–200 kg ha^{-1}. The variation in Cu uptake was 78% due to N fertilizer application. Pederson et al. (2002) also reported that N concentration was highly correlated with Zn concentration in the aboveground plant parts of ryegrass.

6.3.3 Nitrogen versus Manganese

Manganese plays an important role in the growth and development of plants. It plays an important role in the photolysis of water in chloroplasts, regulation of enzyme activities, and protection against oxidative damage of membranes (Alloway, 2008). Sillanpaa (1990) reported that Mn deficiency is present in about 10% soils in 15 different countries. In acid soils, liming is the main factor for Mn deficiency (Fageria et al., 2002; Fageria and Stone, 2008). Figure 6.11 shows that Mn uptake decreased significantly in dry bean plants when the soil pH was raised from 4.8 to 7.0 in a Brazilian Inceptisol.

TABLE 6.14
Influence of N on the Uptake of Cu in the Shoot
of 60-Day-Old Dry Bean Plants

N Rate (kg ha^{-1})	Cu Uptake (g ha^{-1})
0	3.0
40	5.2
80	8.9
120	9.6
160	15.0
200	18.7
R^2	0.78**

**Significant at the 1% probability level.

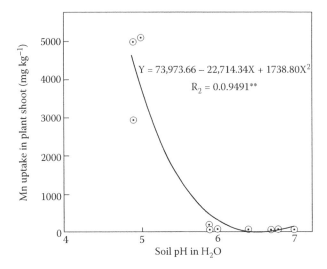

FIGURE 6.11   Influence of soil pH on the uptake of Mn in the shoot of dry bean plants. **—Significant at the 1% probability level. (Adapted from Fageria, N. K. and V. C. Baligar. 1999. *J. Plant Nutr.* 22:23–32.)

Fageria and Baligar (2005) studied the influence of N on the uptake of Mn in the shoots of dry bean plants (Table 6.15). The uptake of Mn was significantly increased with increasing N rates in the range of 0–200 kg ha^{-1}. Nitrogen application was responsible for 88% variation in the Mn uptake. The positive association of N with Mn was related to the improvement in dry matter production of dry bean (Fageria and Baligar, 2005). Xue et al. (2012) reported that applying N at an optimal rate (198 kg ha^{-1}) to wheat resulted in a significantly higher grain Zn concentration compared to the control treatment. For example, grain Zn concentration increased from 21.5 mg kg^{-1} in the control to 30.9 mg kg^{-1} with the optimum N supply.

6.3.4   NITROGEN VERSUS IRON

Among micronutrients, except chlorine, iron is required in higher amounts for the growth and development of plants. Iron stress (deficiency or toxicity) in crop plants often represents a serious constraint for stabilizing and/or increasing crop yields. Any factor that decreases the availability of Fe in a soil or competes in the plant absorption process contributes to Fe deficiency. Iron deficiency

TABLE 6.15
Influence of N on the Uptake of Mn in the Shoot of 60-Day-Old Dry Bean Plants

N Rate (kg ha^{-1})	Mn Uptake (g ha^{-1})
0	43.3
40	115.9
80	156.9
120	153.3
160	222.1
200	361.8
R^2	0.88**

**Significant at the 1% probability level.

has been observed in important crops such as corn, sorghum, peanuts, soybeans, common bean, oats, barley, and upland rice (Fageria et al., 2006), whereas Fe toxicity is mostly restricted to flooded or lowland rice (Fageria et al., 1990). In general, monocotyledonous species are less Fe efficient than dicotyledonous species. Plants are classified as Fe efficient if they respond to Fe deficiency stress by inducing biochemical reactions that make Fe available in a useful form and Fe inefficient if they do not. These induced reactions or compounds are: release of hydrogen ions from their roots, release of reducing compounds from their roots, reduction of Fe^{3+} to Fe^{2+} at their roots, and increases in organic acids in their roots (Brown, 1978).

Iron is required in higher plants for chlorophyll synthesis. Iron is a component of many enzymes that catalyze the metabolism of plants. In general, enzymes containing metal are divided into two groups. One group is known as a metal activator and the other is known as a metalloenzyme. In the former, Fe acts as a temporary link between the enzyme and the substrate during biochemical reactions, hence activating a number of oxidases. However, the majority of Fe enzymes belong to the metalloenzyme group, in which Fe is firmly bound to a protein. Frequently, Fe is chelated by or attached to a small molecule called a prosthetic group; peroxidase, catalase, and cytochrome oxidase all contain Fe bound in the heme group. This means that Fe nutrition has a close relationship with these oxidase activities (Okajima et al., 1975). Among the Fe enzymes, the ferredoxins are of special interest because of their importance in photosynthesis. Ferredoxins are small protein molecules containing Fe and labile sulfides. They catalyze the phosphorylation of ADP in the presence of light.

Iron is immobile in plant tissues; therefore, its deficiency first appears in the young leaves as an interveinal chlorosis. At an advanced stage, the entire leaf blade may become yellow or white. The leaf veins are the last to lose chlorophyll. In green leaves, approximately 80% of the iron is located in the chloroplasts (Terry, 1980); thus, iron deficiency affects processes located in the chloroplasts. In higher plants, one of the most obvious effects of iron deficiency is the development of chlorotic leaves. Chlorosis is associated with a loss of not only chlorophyll but also all thylakoid constituents. The reduction in thylakoid membranes during iron deficiency is accompanied by decreases in all photosynthetic pigments (Terry and Abadia, 1986; Monge et al., 1987). Some of the environmental factors that can induce iron deficiency in crop plants are (i) low iron content of the soil, (ii) high level of lime application or lime-induced chlorosis, (iii) poor aeration, (iv) high P concentration, (v) high levels of Mn, Zn, and Cu, (vi) high light intensity, (vii) high level of nitrate, (viii) high or low temperatures, (ix) unbalanced cation ratios, (x) addition of organic matter to soil, (xi) virus infection (Hale and Orcutt, 1987), and (xii) liming acid soils.

Tissue analysis as ordinarily performed does not distinguish between functional iron and that inactivated by precipitation or complexing. Active iron is that which is extracted by normal hydrochloric acid from grounded dry plant tissue (Hale and Orcutt, 1987). Iron uptake reduced significantly with increasing soil pH (Figure 6.12). At pH 4.9, the value of Fe uptake by dry bean plants was 328 mg kg^{-1} shoot dry weight, while at pH 7.0, the Fe uptake value dropped to 96 mg Fe kg^{-1}. This means that the decrease at the highest pH value was 242% as compared to the lowest pH value.

Plants have different strategies for solubilization and uptake of iron. Graminaceous plants exude phytosiderophores or Fe bearers, for example, mugeneic or avenic acids that carry Fe to the roots (Romheld, 1987). Phytosiderophores have been defined by Takagi et al. (1984) as a group of root exudates exhibiting strong complexing properties with respect to ferric Fe and identified as nonproteinogenic amino acids, such as mugineic acid and its derivatives. In this respect, they are analogs of microbial siderophores, which are literally iron bearers (Hinsinger, 1998). The literature on this topic has been extensively reviewed by Romheld and Marschner (1986), Marschner et al. (1989), and Romheld (1991). The synthesis and release of phytosiderophores in the rhizosphere are stimulated by Fe deficiency (Romheld, 1991) and have been described as strategy II for Fe acquisition as developed exclusively by graminaceous species (Marschner, 1995). Graminaceae species differ widely in their ability to produce phytosiderophores, both quantitatively and qualitatively. Most remarkably, among the range of graminaceous species studied by Marschner et al. (1989), the enhancement of

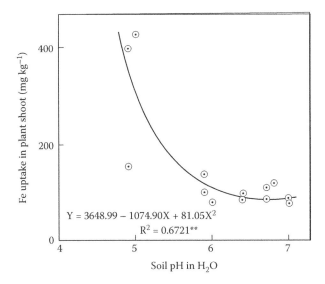

FIGURE 6.12 Influence of soil pH on the uptake of Fe in the shoot of dry bean plants. (Adapted from Fageria, N. K. and V. C. Baligar. 1999. *J. Plant Nutr.* 22:23–32.)

the release of phytosiderophores by Fe deficiency was reported to increase according to the resistance of the species to lime-induced chlorosis (Fageria et al., 2006).

Iron in adequate amounts developed N_2-fixing nodules in dry bean, an increase in yield, and improved NUE (Garg, 1987). The use of ammoniacal N fertilizers lowers soil pH and improves Fe uptake in crop plants. The author studied the influence of N on the uptake of Fe in the shoots of dry bean plants (Table 6.16). The uptake of Fe was significantly increased with the increasing N rates in the range of 0–200 kg ha^{-1}. Nitrogen application was responsible for 83% variation in the Fe uptake. The positive association of N with Mn was related to the improvement in dry matter production of dry bean (Fageria and Baligar, 2005). Fageria (2009) reported that N uptake in the form of NO_3^- can increase soil pH due to the liberation of OH^- ions, thus decreasing the iron solubility and uptake. If NH_4^+ N is absorbed, the reverse will occur: a decrease in soil pH may solubilize iron in the soil, and its uptake may increase.

Kutman et al. (2011) reported that N enhanced the uptake of iron by wheat plants. These authors also reported that N nutrition is a critical factor in both the acquisition and grain allocation of Fe in wheat. Kutman et al. (2011) reported that the positive effect of N on the uptake of Fe in the grain of

TABLE 6.16
Influence of N on the Uptake of Fe in the Shoot
of 60-Day-Old Dry Bean Plants

N Rate (kg ha^{-1})	Fe Uptake (g ha^{-1})
0	57.9
40	142.8
80	225.4
120	219.0
160	365.1
200	519.1
R^2	0.83**

**Significant at the 1% probability level.

wheat has important implications in terms of human nutrition, and should be considered in agronomic and genetic biofortification of wheat with Fe. Various genetic studies revealed the existence of a close linkage between the remobilization of N and remobilization of Fe from senescing leaf tissue into grain in durum wheat (Uauy et al., 2006a,b; Distelfeld et al., 2007; Waters et al., 2009). Kutman et al. (2011) reported that the Fe harvest index (Fe in the grain/Fe in grain plus straw) strongly depends on N supply in wheat plants.

6.3.5 Nitrogen versus Boron

Boron deficiency in crop plants has been reported throughout the world (Gupta, 1979; Blevins and Lukaszewski, 1998). The deficiency of B is generally common for plants grown on light-textured soils in humid climates where B is readily leached from the soil. Boron in soil solutions usually occurs as the undissociated boric acid (H_3BO_3). Adequate B levels are essential for higher crop yields and quality. Boron improved the root development in common bean, soybean, and wheat grown on an Oxisol of central Brazil (Baligar et al., 1998). Boron has been reported to have counteracting toxic effects of Al on the root growth of dicotyledonous plants (Blevins and Lukaszewski, 1998). Boron requirements for reproductive growth are much higher than for vegetative growth of most crop species (Loomis and Durst, 1992).

Boron has also been reported to be essential for N_2 fixation (Mateo et al., 1986). The application of B increased the N concentration in chickpea (Yadav and Manchanda, 1979), lentil (Singh and Singh, 1983), and peanut (Patel and Golakia, 1986), presumably due to the favorable effect of B on nodulation as nodule counts were found to increase by 37% over no-B control (Patel and Golakia, 1986). Nitrogen in adequate amounts reduces B toxicity in crop plants (Willett et al., 1985).

Warington (1923) is credited with the first definitive proof that B is required by higher plants. Later, Sommer and Lipman (1926) established B requirements for six nonleguminous dicots and for one graminaceous plant (barley). However, B requirements are highly variable, and optimum quantities for one plant species could be either toxic or insufficient for other plant species. Based on the B requirements, plants can be divided into three general groups: (i) graminaceous plants that have the lowest B demands; (ii) the remaining monocots and most dicots with intermediate B requirements; and (iii) latex-forming plants with the highest B requirements (Mengel et al., 2001).

Boron deficiency affects the development of meristems or actively growing tissues so that deficiency symptoms are death of growing points of shoots and roots, failure of flower buds to develop, and ultimately blackening and death of these tissues. Boron uptake normally decreases with increasing soil pH (Figure 6.13). In common bean grown on an Oxisol, decreases were 120% when the soil pH was raised from 4.9 to 7.0. This decrease in B uptake may have been related to adsorption processes as the soil pH increases. Boron adsorption increased as the pH increased above 4 and reached a maximum at pH 8–9 before decreasing at higher pH values (Barber, 1995). Soil texture also affects B adsorption; it is higher in heavy-textured soils compared to light-textured soils.

Positive relations have been noted between B and N fertilizers for improving crop yields (Hill and Morrill, 1975; Moraghan and Mascagni, 1991; Fageria et al., 2002). Miley et al. (1969) reported that B fertilization in adequate amounts enhances utilization of applied N in cotton by increasing the translocation of N compounds into the boll. Similarly, Smith and Heathcote (1976) found that, when B deficiency occurred in cotton, the application of 250 kg N ha^{-1} depressed yield. However, this rate of N produced a yield increase when B was applied. Inal and Tarakcioglu (2001) and Lopez-Lefebre et al. (2002) reported the positive effects of B on uptake and metabolism.

6.3.6 Nitrogen versus Molybdenum

Arnon and Stout (1939) established that Mo was an essential nutrient for higher plants using tomato as a test plant. Molybdenum is indispensable in the nitrogen fixation process to aerobic as well as

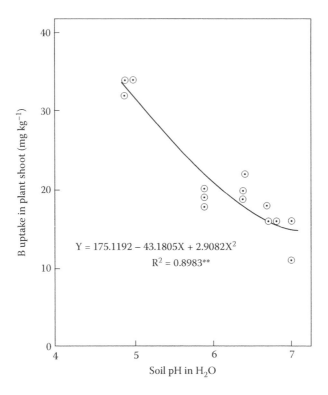

$$Y = 175.1192 - 43.1805X + 2.9082X^2$$
$$R^2 = 0.8983**$$

FIGURE 6.13   Influence of soil pH on the uptake of N in the shoot of dry bean plants. **—Significant at the 1% probability level. (Adapted from Fageria, N. K. and V. C. Baligar. 1999. *J. Plant Nutr.* 22:23–32.)

anaerobic N fixers (*Azotobacter* and *Clostridium*) and also to the symbiotic *Rhizobium*. The most important soil factor affecting the availability of Mo to the plant is the soil pH. As an acid former, Mo appears to increase its solubility with the increase in soil pH. Soil solution molybdenum levels increase 10-fold for each unit increase in soil pH. Where Mo has been added to the soil, the amount adsorbed will decrease as the pH increases (Barber, 1995). Generally, liming of an acid soil is sufficient to correct Mo deficiency. However, if parent materials are too low in Mo, this statement is not valid. Soils on which deficiencies were found were all acidic, while excess Mo occurred on alkaline soils (Okajima et al., 1975).

The plant requirement for Mo is lower than for any other mineral element. The critical deficiency levels range from 0.1 to 1 mg kg^{-1} plant dry weight. The molybdenum requirement is much greater for legumes, because it is used in the nodule for nitrogen fixation, and levels needed for plant growth are much lower than levels needed to supply the root nodule with sufficient molybdenum for nitrogen fixation. An antagonistic relationship between Mo and Cu and sulfate was found, whereas phosphate promotes the absorption of Mo. Molybdenum is an essential component of two major enzymes, nitrogenase and nitrate reductase, both of which depend on an Mo valency change for their function (Shuman, 1994). One could generalize by saying that this element is closely associated with N metabolism in most crop plants (Blevins, 1994). The function of Mo in plant metabolism is to reduce nitrate N to nitrite N; thus, this element functions in a way that is different from other elements. If a plant absorbs N in the nitrate form, Mo is indispensable for normal growth. On the other hand, if N has been supplied in the ammonium form, the presence of Mo will not be necessary (Ishizuka, 1978). Molybdenum is immobile in the plant and hence Mo deficiency first appears in the younger leaves. The molybdenum deficiency symptoms are very similar to the symptoms of iron deficiency.

6.3.7 NITROGEN VERSUS CHLORINE

In 1954, the essentiality of chlorine for plant growth was confirmed (Broyer et al., 1954). For more than 20 years, it was generally assumed that field-grown crops would not benefit from the application of Cl-containing fertilizers because it is ubiquitous in the environment (Fixen, 1993). Pacific Northwest studies in the United States (Taylor et al., 1981; Christensen et al., 1981) in the early 1980s provided the first evidence that field-grown wheat could benefit from chlorine fertilization. Engel et al. (1998) reported that wheat grain yield increased at an average of 417 kg ha^{-1} (9.7%) in 86 cases where significant responses to Cl$^-$ were measured. The largest grain yield responses (>800 kg ha^{-1}) occurred at sites with the lowest plant Cl$^-$ concentration (<0.50 g kg^{-1}). Kernel size was the most important yield component affected by applied Cl$^-$. Approximately 73% of the yield response to applied Cl$^-$ could be accounted for by the larger kernel size. Biological functions of Cl$^-$ in plants are presumed to require a concentration of no more than 0.10 g kg^{-1} (Fixen, 1993). The beneficial effects of chlorine are more likely due to its osmoregulatory role in the plant (Flowers, 1988). The importance of this function on plant growth and grain yield should be highly dependent on the growing environment (e.g., water and temperature).

Among micronutrients, chlorine is absorbed in maximum quantity by crop plants. Chlorine takes part in the capture and storage of light energy through its involvement in photophosphorylation reactions in photosynthesis. It is not present in the plant as a true metabolite but as a mobile anion. It is involved with K in the regulation of osmotic pressure, acting as an anion in counterbalance to cations. General plant symptoms are chlorosis in younger leaves and overall wilting. Chlorine is ubiquitous in the environment and appears to be involved in several plant processes, including photosynthesis, sugar translocation, and maintaining or increasing water potential (Voss, 1993).

Nitrogen and chlorine interact via several mechanisms including both soil and plant processes. The rates of some steps in the mineralization of soil organic matter are affected by Cl$^-$ (Fixen, 1993). This can influence the form of N absorbed by crop plants. At the root surface, the NO_3^- and Cl$^-$ ions are known to compete with each other in the uptake process (Fixen, 1993). Increasing the supply of either one tends to reduce the tissue concentration of the other. Nitrate and Cl^{-1} compete with each other for uptake in many species (Glass and Siddiqi, 1985; Christensen and Brett, 1985; Goos et al., 1987). However, under some conditions, positive interactions between NO_3^- and Cl$^-$ have been reported (Murarka et al., 1973). At soil Cl$^-$ levels greater than 19 mg kg^{-1} soil, spring wheat Cl$^-$ concentrations in South Dakota increased with increasing soil NO_3^- levels (Fixen et al., 1987). Below 19 mg Cl$^-$ kg^{-1} soil, they decreased with increasing soil NO_3^-.

6.3.8 NITROGEN VERSUS NICKEL

The essentiality of nickel (Ni) for higher plants is the most recent addition to the list of micronutrients. Nickel was suspected of being an essential plant nutrient in the early twentieth century, when it was discovered to be a constituent of plant ash (Wood and Reilly, 2007). Responses of crops in the field of foliar Ni sprays were noted to increase yields of wheat, potatoes, and broad beans as early as 1946 (Roach and Barclay, 1946), and responses in other crops were noted in subsequent years (Dixon et al., 1975; Welch, 1981). Its essentiality for higher plants was established in the 1980s (Welch, 1981; Eskew et al., 1983) using soybean as a test plant. Soybean plants were grown in a highly purified nutrient solution (without Ni) and urea accumulated in the toxic level in the tips of leaflets, which become necrotic. When Ni was supplied to soybean plants in the concentration of 1 µg L^{-1}, no excess urea was accumulated in the leaf tips and necrosis was also absent (Epstein and Bloom, 2005).

Subsequent research by Brown et al. (1987a,b) established the essentiality of Ni for other crop species such as barley. These authors reported that Ni is essential for plants supplied with urea. The Ni-deficient plants accumulate toxic levels of urea in leaf tips, because of reduced urease

FIGURE 6.14 Response of upland rice to N fertilization applied with ammonium sulfate.

activity (Daroub and Snyder, 2007). Daroub and Snyder (2007) also reported that the essentiality of Ni as an essential nutrient was established in 1987. In 1992, the United States Department of Agriculture, Agriculture Research Service added Ni to its list of essential plant nutrient elements (Wood and Reilly, 2007). It essentially was also recognized by the American Association of Plant Food Control Officials (AAPFCO), the umbrella organization that functions to govern and influence regulation, labeling, ingredients, and amounts of elements in fertilizer products in the United States (Terry, 2004). Ni is now listed on fertilizer labels in the United States, and commercial Ni fertilizer products are now marketed (Wood and Reilly, 2007). The deficiency of Ni in crop plants is rarely observed under field conditions. Even in controlled conditions, creating deficiency symptoms is difficult (Fageria, 2009).

Nickel deficiency causes severe disruption in N metabolism and other metabolic process (Brown et al., 1990). Nickel stimulates proline biosynthesis in plants, which is responsible for osmotic balance in plant tissues (Salt et al., 1995; Singh et al., 2004). Nickel function in the utilization of nitrogen translocated from roots to tops via guanidines or ureides that are subsequently used for anabolic reactions in growing tissues (Welch, 1981). Simultaneous supplies of NO_3-N and NH_4-N reduced Ni toxicity in sunflower, and growth was enhanced from added Ni (Zornoza et al., 1999). Low-Ni plants became N deficient from lack of urease activity with a high accumulation of urea but low tissue N (Gerendas and Sattelmacher, 1997). Figures 6.14 and 6.15 show upland rice growth

FIGURE 6.15 Response of upland rice to N fertilization applied with urea.

at different N levels. In Figure 6.14, plant growth increased with increasing N levels when the N source was ammonium sulfate. However, in Figure 6.15, plant growth was reduced when the N rate was 400 mg kg^{-1} supplied with urea. In this case, nickel deficiency may be suspected for reduction in rice growth at a higher urea level because, in addition to reduced growth, tips of rice leaves were showing necrotic or toxic symptoms of urea as described by Fageria (2009).

6.4 NITROGEN VERSUS SOIL SALINITY AND ALKALINITY

Salt-affected soils limit crop yields around the world. It is especially prevalent in irrigated agriculture and in marginal lands associated with poor drainage or high water tables. Estimates for the extent of salinity damage vary from 25% to 50% of the world's irrigated land (Adams and Hughes, 1990). The knowledge of how nutrient availability is affected in plants growing on salt-affected soils is important to adopt appropriate management practices to satisfy plants' nutritional requirements and improve yields to meet food demands of increasing world populations. In the salt-affected environment, plants are required to absorb essential nutrients from a dilute source in the presence of highly concentrated nonessential nutrients (Fageria et al., 2011b). Further, among nutrients, N is one of the most widely limiting elements for crop production, when plants are subjected to salt stress. The uptake of N by rice was inhibited under high NaCl and Na_2SO_4 concentration of the root medium, and the excess amount of absorbed Na depressed NH_4^+ absorption (Palfi, 1965; Mahajan and Sonar, 1980). Reduction in root permeability and the consequent decrease in water and nutrient uptake under high salt concentrations (Frota and Tucker, 1978) have been associated with impaired N absorption by plant under salt stress conditions.

Grattan and Grieve (1999) reviewed the literature on the interaction between salinity and N accumulation in crop plants and concluded that salinity can reduce N accumulation. These authors further reported that this is not surprising since an increase in Cl$^-$ uptake and accumulation is often accompanied by a decrease in shoot NO_3^- of many crops (Khan and Srivastava, 1998; Kaya and Higgs, 2003). This may be associated with the competition between NO_3^- and Cl$^-$ ions. Kafkafi et al. (1992) reported that the nitrate influx rate or interaction between NO_3^- and Cl$^-$ might be related to the salt tolerance of the cultivar under investigation. These authors found that salt tolerance cultivars had higher NO_3^- influx rates than the more sensitive cultivars. Forms of N also influence the interaction of salinity with N. The NH_4^+ feed plants were more sensitive to salinity than NO_3^- feed plants in nutrient solutions (Grattan and Grieve, 1999).

Soil salinity may disrupt symbiotic N_2 fixation systems in several ways. Salts can limit nodule formation by reducing the population of *Rhizobium* in the soil or by impairing their ability to infect root hairs. The direct effects of salinity on the host plant can limit N fixation, independently of the effects of salinity on the *Rhizobium* bacteria and the nodulation process (Keck et al., 1984; Fageria, 1992). Stunted growth of the host plant may reduce the supply of photosynthate to the root nodules. Since photosynthate supply is a major limiting factor in N_2 fixation, this indirect effect can be quite important (Fageria, 1992).

Sodic solonetzic soils are also deficient in calcium for optimum plant growth and can be reclaimed by applying gypsum to replace Na$^+$ with Ca^{2+} on the cation exchange capacity (Aulakh and Malhi, 2005). In a black solonetzic soil in Alberta, Canada, an application of gypsum increased the concentration of extractable Ca and reduced the sodium adsorption ratio (Malhi et al., 1992). In this experiment, an N × Ca interaction not only improved the yield but also enhanced the concentration of Ca, K, and Zn in the flag leaf of barley while decreasing the Na concentration.

6.5 CONCLUSIONS

The knowledge of interactions among essential plant nutrients is important in formulating a balanced supply of fertilizers to crop plants. Nutrient interactions may affect the uptake and utilization of essential plant nutrients depending on the type of interaction. The interaction may be positive (synergistic), negative (antagonistic), and absent or neutral, depending on the nutrients and concentrations involved. Nutrient interaction is influenced by environmental factors. Environmental factors influencing nutrient interactions are climate and soil. In addition, nutrient interaction is also influenced by plant species and genotypes within species. These factors affect plant growth and development and thereby nutrient interactions. Furthermore, crop management practices can modify nutrient uptake by plants and hence influence nutrient interaction. In addition, soil microbial properties have profound effects on plant nutrient uptake through their effect on nutrient accumulation, depletion, immobilization, mineralization, and pH of the rhizosphere. Hence, these factors have a significant effect on nutrient interaction in crop plants. The interaction among major or macronutrients is generally positive and among micronutrients negative. However, there may be some exception in both groups of nutrients.

Interaction of N with other nutrients is also affected by the form of N absorption by plants (NO_3-N and NH_4-N). For example, the interaction of K with N is more with NO_3-N compared to NH_4-N. The uptake of N is adversely affected by salt-affected soils. In the salt-affected environment, plants are required to absorb essential nutrients from a dilute source in the presence of highly concentrated nonessential nutrients. Further, among the nutrients, N is one of the most widely limiting elements for crop production, when plants are subjected to salt stress. Based on a review of the literature, N has a positive interaction with most of the essential plant nutrients, except under saline soil conditions. Hence, adequate management of N is supposed to improve the uptake of other essential nutrients and consequently higher yields. In addition, a balanced supply of other nutrients with N can also improve biological N fixation in legumes, reduce the infestation of diseases and insects in cereals and legumes, and have a favorable influence on the crop quality and biochemical constituents of the produce. In addition, a balanced supply of N with other essential nutrients improves NUE and reduces the risk of environmental pollution. Balanced nutrient supply also controls many diseases and improves insect resistance and consequently improves the yield and NUE.

REFERENCES

Abbasi, M. K., M. M. Tahir, W. Azam, Z. Abbas, and N. Rahim. 2012. Soybean yield and chemical composition in response to phosphorus-potassium nutrition in Kashmir. *Agron. J.* 104:1476–1484.

Abdin, M. Z., A. Ahmad, N. Khan, I. Khan, and M. Iqbal. 2003. Sulphur interaction with other nutrients. In: *Sulphur in Plants*, eds., Y. P. Abrol, and A. Ahmad, pp. 359–374. Dordrecht: Kluwer Academic Publishers.

Adams, W. M. and F. M. R. Hughes. 1990. Irrigation development in desert environments. In: *Techniques for Desert Reclamation*, ed., A. S. Goudie, pp. 135–160. New York: John Wiley.

Aggarwal, R. K., P. Raina, and P. Kumar. 1987. Ammonium volatilization losses from urea and their possible management for increasing nitrogen use efficiency in an arid region. *J. Arid Environ.* 13:163–168.

Ahmad, A. and M. Z. Abdin. 2000. Photosynthesis and its related physiological variables in the leaves of Brassica genotypes as influenced by sulphur fertilization. *Physiol. Plant.* 110:144–149.

Ahmad, A., G. Abraham, and M. Z. Abdin. 1999. Physiological investigation of the impact of nitrogen and sulphur application on seed and oil yield of rapeseed (*Brassica campestris* L.) and mustard (*Brassica juncea* L.) genotypes. *J. Agron. Crop Sci.* 183:1925.

Ali, A. A., M. Ikeda, and Y. Yamada. 1985. Absorption, translocation, and assimilation of ammonium and nitrate nitrogen in rice plants as affected by the supply of potassium, calcium and magnesium. *J. Fac. Agric. Kyushu Univ. Jpn.* 30:113–124.

Ali, A. A., M. Ikeda, and Y. Yamada. 1987. Effect of the supply of potassium, calcium, and magnesium on the absorption, translocation, and assimilation of ammonium and nitrate nitrogen in wheat plants. *Soil Sci. Plant Nutr.* 33:585–594.

Alloway, B. J. 2008. Micronutrients and crop production: An introduction. In: *Micronutrient Deficiencies in Global Crop Production*, ed., B. J. Alloway, pp. 1–39. New York: Springer.

Amanuel, G., R. F. Kuhne, D. G. Tanner, and P. L. G. Vlek. 2000. Biological nitrogen fixation in faba bean (*Vicia faba* L.) in the Ethiopian highlands as affected by P fertilization and inoculation. *Biol. Fertil. Soils* 32:353–359.

Arnon, D. I. and P. R. Stout. 1939. Molybdenum as an essential element for higher plants. *Plant Physiol.* 14:599–602.

Aulakh, M. S. and I. M. Chhibba. 1992. Sulphur in soils and responses of crops to its application in Punjab. *Fertil. News* 37:33–45.

Aulakh, M. S. and S. S. Malhi. 2005. Interactions of nitrogen with other nutrients and water: Effect on crop yield and quality, nutrient use efficiency, carbon sequestration, and environmental pollution. *Adv. Agron.* 86:341–409.

Aulakh, M. S., N. S. Pasricha, and A. S. Azad. 1990. Phosphorus-sulphur interrelationships for soybean on P and S deficient soil. *Soil Sci.* 150:705–709.

Austin, R. B., J. Bingham, R. D. Blackwell, L. T. Evans, M. A. Ford, C. L. Morgan, and M. Taylo. 1980. Genetic improvements in winter wheat yields since 1900 and associated physiological changes. *J. Agric. Sci.* 94:675–689.

Bajwa, M. and J. Paul. 1978. Effect of continuous application of N, P, K and Zn on yield and nutrient uptake of irrigated wheat and maize and on available nutrients in a tropical arid brown soil. *J. Indian Soc. Soil Sci.* 26:160–165.

Baligar, V. C., N. K. Fageria, and Z. L. He. 2001. Nutrient use efficiency in plants. *Commun. Soil Sci. Plant Anal.* 32:921–950.

Barber, S. A. 1995. *Soil Nutrient Bioavailability: A Mechanistic Approach*, 2nd edition. New York: Wiley.

Bishnoi, S. K., B. R. Tripathi, and B. S. Kanwar. 1984. Availability of native N and its uptake by soybean in acid soils of Himachal Pradesh as affected by liming. *Indian J. Soil Sci. Bulletin* 13:255–262.

Black, C. A. 1993. *Soil Fertility Evaluation and Control*. Boca Raton, Florida: CRC Press.

Blevins, D. G. 1994. Uptake, translocation, and function of essential mineral elements in crop plants. In: *Physiology and Determination of Crop Yield*, ed., G. A. Paterson, pp. 259–275. Madison, Wisconsin: ASA, CSA, and SSSA.

Blevins, D. G. and K. M. Lukaszewski. 1998. Boron in plant structure and function. *Annu. Rev. Plant Physiol. Plant Mol. Biol.* 49:481–500.

Brennan, R. F. and M. D. A. Bolland. 2009. Comparing the nitrogen and phosphorus requirements of canola and wheat for grain yield and quality. *Crop Pasture Sci.* 60:566–577.

Brown, J. C. 1978. Mechanism of iron uptake by plants. *Plant Cell Environ.* 1:249–257.

Brown, P. H. 2008. Micronutrient use in agriculture in the United States of America: Current practices, trends and constraints. In: *Micronutrient Deficiencies in Global Crop Production*, ed., B. J. Alloway, pp. 267–286. New York: Springer.

Brown, L., D. Scholefield, E. C. Jewkes, N. Preedy, K. Wadge, and M. Butler. 2000. The effect of sulphur application on the efficiency of nitrogen use in two contrasting grassland soils. *J. Agric. Sci.* 135:131–138.

Brown, P. H., R. M. Welch, and E. E. Cary. 1987a. Nickel: A micronutrient essential for higher plants. *Plant Physiol.* 85:801–803.

Brown, P. H., R. M. Welch, E. E. Cary, and R. T. Checkai. 1987b. Beneficial effects of nickel on plant growth. *J. Plant Nutr.* 10:2125–2135.

Brown, P. H., R. M. Welch, and J. T. Madison. 1990. Effect of nickel deficiency on soluble anion, amino acid, and nitrogen levels. *Plant Soil* 125:19–27.

Broyer, T. C., A. B. Carlton, A. B. Johnson, and P. R. Stout. 1954. Chlorine: A micronutrient element for higher plants. *Plant Physiol.* 29:526–532.

Buah, S., T. A. Polito, and R. Killorn. 2000. No-tillage corn response to placement of fertilizer nitrogen, phosphorus, and potassium. *Commun. Soil Sci. Plant Anal.* 31:3121–3133.

Cakmak, I. 2008. Zinc deficiency in wheat in Turkey. In: *Micronutrient Deficiencies in Global Crop Production*, ed., B. J. Alloway, pp. 181–200. New York: Springer.

Cakmak, I., W. H. Pfeiffer, and B. McClafferty. 2010. Biofortification of durum wheat with zinc and ion. *Cereal Chem.* 87:10–20.

Camp, A. F. and B. R. Fudge. 1939. *Some Symptoms of Citrus Malnutrition in Florida*. Florida Experimental Station, Bulletin 335, Florida State University, Gainesville, Florida.

Christensen, N. W. and M. Brett. 1985. Chloride and liming effects on soil nitrogen form and take-all of wheat. *Agron. J.* 77:157–163.

Christensen, N. W., R. G. Taylor, T. L. Jackson, and B. L. Mitchell. 1981. Chloride effects on water potentials and yield of winter wheat infected with take-all root rot. *Agron. J.* 73:1053–1058.

Clarkson, D. T. and J. B. Hanson. 1980. The mineral nutrition of higher plants. *Annu. Rev. Plant Physiol.* 31:239–298.

Clarkson, D. T., L. R. Sarkar, and J. V. Purves. 1989. Depression of nitrate and ammonium transport in barley plants with diminished sulphate status: Evidence of co-regulation of nitrogen and sulphate intake. *J. Exp. Bot.* 40:953–963.

Daroub, S. M. and G. H. Snyder. 2007. The chemistry of plant nutrients in soil. In: *Mineral Nutrition and Plant Disease*, eds., L. E. Datnoff, W. H. Elmer, and D. M. Huber, pp. 1–7. St Paul, Minnesota: The American Phytopathological Society.

Daulay, H. S. and K. C. Singh. 1982. Effects of N and P rates and plant densities on the yield of rainfed sesamum. *Indian J. Agric. Sci.* 52:166–169.

Dev, G., R. C. Jaggi, and M. S. Aulakh. 1979. Study of nitrate-sulphate interaction on the growth and nutrient uptake by maize using ^{35}S. *J. Indian Soc. Soil Sci.* 27:302–307.

Dev, S. and U. C. Shukla. 1980. Nitrogen-zinc content in maize as affected by their different sources. *J. Indian Soc. Soil Sci.* 28:203–205.

Dibb, D. W. and L. F. Welch. 1976. Corn growth as affected by ammonium vs. nitrate absorbed from soil. *Agron. J.* 68:89–94.

Dibb, D. W. and W. R. Thomson, Jr. 1985. Interaction of potassium with other nutrients. In: *Potassium in Agriculture*, ed., D. Munson, pp. 515–533. Madison, Wisconsin: ASA, CSSA, and SSSA.

Distelfeld, A., I. Cakmak, Z. Peleg, L. Ozturk, A. M. Yazici, H. Budak, Y. Saranga, and T. Fahima. 2007. Multiple QTL effects of wheat Gpc-B1 locus on grain protein and micronutrient concentrations. *Physiol. Plant.* 29:635–643.

Dixon, N. E., C. Gazzola, R. L. Blakeley, and B. Zerner. 1975. Jack bean urease (EC 3.5.1.5). A metalloenzyme. A simple biological role for nickel? *J. Am. Chem. Soc.* 97:4131–4133.

Dwivedi, R. S. and N. S. Randhawa. 1973. Zinc nutrition and the energy value of cereal crops. *Curr. Sci.* 42:61–62.

Dwivedi, B. S., A. K. Shukla, V. K. Singh, and R. L. Yadav. 2003. Improving nitrogen and phosphorus use efficiencies through inclusion of forage cowpea in the rice-wheat systems in the Indo-Gangetic plains of India. *Field Crops Res.* 80:167–193.

Engel, R. E., P. L. Bruckner, and J. Eckhoff. 1998. Critical tissue concentration and chloride requirements for wheat. *Soil Sci. Soc. Am. J.* 62:401–405.

Epstein, E. and A. J. Bloom. 2005. *Mineral Nutrition of Plants: Principles and Perspectives*, 2nd edition. Sunderland, Massachusetts: Sinauer Associates, Inc.

Erenoglu, B., U. B. Kutman, Y. Ceylan, B. Yildiz, and I. Cakmak. 2011. Improved nitrogen nutrition enhances root uptake, root-to shoot translocation and remobilization of zinc (65Zn) in wheat. *New Phytol.* 189:438–448.

Eskew, D. L., R. M. Welch, and E. E. Cary. 1983. Nickel: An essential micronutrient for legumes and possibly all higher plants. *Science* 222:621–623.

Fageria, N. K. 1983. Ionic interactions in rice plants from dilute solutions. *Plant Soil* 70:309–316.

Fageria, N. K. 1984. Fertilization and mineral nutrition of rice. Editora Campus, Rio de Janeiro/EMBRAPA-CNPAF, Goiania, Brazil (in Portuguese).

Fageria, N. K. 1989. Tropical soils and physiological aspects of crops. EMBRAPA, Brasilia.

Fageria, N. K. 1992. *Maximizing Crop Yields*. New York: Marcel Dekker.

Fageria, N. K. and V. C. Baligar. 1999. Yield and yield components of lowland rice as influenced by timing of nitrogen fertilization. *J. Plant Nutr.* 22:23–32.

Fageria, N. K. 2000. Adequate and toxic levels of zinc for rice, common bean, corn, soybean and wheat production in cerrado soil. *Revista Brasileira Eng. Agri. Amb.* 4:390–395.

Fageria, N. K. 2009. *The Use of Nutrients in Crop Plants*. Boca Raton, Florida: CRC Press.

Fageria, N. K. 2010. Root growth of upland rice genotypes as influenced by nitrogen fertilization. Paper presented at the 19th World Soil Science Congress, Brisbane, Australia. 1–6 August, 2010.

Fageria, N. K. 2013. *The Role of Plant Roots in Crop Production*. Boca Raton, Florida: CRC Press.

Fageria, N. K. 2014. *Mineral Nutrition of Rice*. Boca Raton, Florida: CRC Press.

Fageria, N. K. and V. C. Baligar. 1997. Response of common bean, upland rice, corn, wheat, and soybean to soil fertility of an Oxisol. *J. Plant Nutr.* 20:1279–1289.

Fageria, N. K. and V. C. Baligar. 1999. Growth and nutrient uptake by common bean, lowland rice, corn, soybean, and wheat at different soil pH on an Inceptisol. *J. Plant Nutr.* 22:1495–1507.

Fageria, N. K. and V. C. Baligar. 2001. Improving nutrient use efficiency of annual crops in Brazilian acid soils for sustainable crop production. *Commun. Soil Sci. Plant Anal.* 32:1303–1319.

Fageria, N. K. and V. C. Baligar. 2005. Enhancing nitrogen use efficiency in crop plants. *Adv. Agron.* 88:97–185.

Fageria, N. K. and V. C. Baligar. 2008. Ameliorating soil acidity of tropical Oxisols by liming for sustainable crop production. *Adv. Agron.* 99:345–399.

Fageria, N. K., V. C. Baligar, and R. B. Clark. 2002. Micronutrients in crop production. *Adv. Agron.* 77:85–268.

Fageria, N. K., V. C. Baligra, and R. B. Clark. 2006. *Physiology of Crop Production*. New York: The Haworth Press.

Fageria, N. K., V. C. Baligar, and C. A. Jones. 2011a. *Growth and Mineral Nutrition of Field Crops*, 3rd edition. Boca Raton, Florida: CRC Press.

Fageria, N. K., V. C. Baligar, and R. J. Wright. 1990. Iron nutrition of plants: An overview on the chemistry and physiology of its deficiency and toxicity. *Pesq. Agropec. Bras.* 25:553–570.

Fageria, N. K. and H. R. Gheyi. 1999. *Efficient Crop Production.* Campina Grande, Brazil: Federal University of Paraiba.

Fageria, N. K., H. R. Gheyi, and A. Moreira. 2011b. Nutrient bioavailability in salt affected soils. *J. Plant Nutr.* 34:945–962.

Fageria, N. K., M. F. Moraes, E. P. B. Ferreira, and A. M. Knupp. 2012. Biofortification of trace elements in food crops for human health. *Commun. Soil Sci. Plant Anal.* 43:556–570.

Fageria, N. K., O. P. Morais, and A. B. Santos. 2010. Nitrogen use efficiency in upland rice. *J. Plant Nutr.* 33:1696–1711.

Fageria, N. K. and C. M. R. Souza. 1991. Upland rice, common bean, and cowpea response to magnesium application on an Oxisol. *Commun. Soil Sci. Plant Anal.* 22:1805–1816.

Fageria, N. K. and L. F. Stone. 2008. Micronutrient deficiency problems in South America. In: *Micronutrient Deficiencies in Global Crop Production*, ed., B. J. Alloway, pp. 245–266. New York: Springer.

Fageria, N. K. and F. J. P. Zimmermann. 1998. Influence of pH on growth and nutrient uptake by crop species in an Oxisol. *Commun. Soil Sci. Plant Anal.* 29:2675–2682.

Fismes, J., P. C. Vong, A. Guckert, and E. Frossard. 2000. Influence of sulphur on apparent N-use efficiency, yield and quality of oilseed rape (*Brassica napus* L.) grown on a calcerous soil. *Eur. J. Agron.* 12:127–141.

Fixen, P. E. 1993. Crop responses to chloride. *Adv. Agron.* 50:107–150.

Fixen, P. E., R. H. Gelderman, J. R. Gerwing, and B. G. Farber. 1987. Calibration and implementation of a soil Cl test. *J. Fert. Issues* 4:91–97.

Flowers, T. J. 1988. Chloride as a nutrient and as an osmoticum. *Adv. Plant Nutr.* 3:55–78.

Frota, J. N. E. and T. C. Tucker. 1978. Absorption rates of ammonium and nitrate by red kidney beans under salt and water stress. *Soil Sci. Soc. Am. J.* 42:753–756.

Garg, O. K. 1987. Physiological significance of zinc and iron: Retrospect and prospect. *Indian J. Plant Physiol.* 30:321–331.

Gerendas, J. and B. Sattelmacher. 1997. Significance of Ni supply for growth, urease activity and the concentration of urea, amino acids, and mineral nutrients of urea-grown plants. *Plant Soil* 190:153–162.

Gerendas, J., M. Sailer, M. L. Fendrich, T. Stahl, V. Mersch-Sundermann, and K. H. Muhling. 2008. Influence of sulfur and nitrogen supply on growth, nutrient status, and concentration of benzyl-isothiocyanate in cress (*Lepidium sativum* L.). *Sci. Food Agric.* 88:2576–2580.

Glass, A. D. and M. Y. Siddiqi. 1985. Nitrate inhibition of chloride influx in barley: Implications for a proposed chloride homeostat. *J. Exp. Bot.* 36:556–566.

Goedert, W. J. 1983. Management of the Cerrado soils of Brazil: A review. *J. Soil Sci.* 34:405–428.

Goos, R. J., B. E. Johnson, and B. M. Holmes. 1987. Effect of potassium chloride fertilization on two barley cultivars differing in common root rot reaction. *Can. J. Plant Sci.* 67:395–401.

Graham, R. D. 2008. Micronutrient deficiencies in crops and their global significance. In: *Micronutrient Deficiencies in Global Crop Production*, ed., B. J. Alloway, pp. 41–61. New York: Springer.

Grattan, S. R. and C. M. Grieve. 1999. Salinity-mineral nutrient relations in horticulture crops. *Scientia Horticulture* 78:127–157.

Gupta, U. C. 1979. Boron nutrition of crop plants. *Adv. Agron.* 31:273–307.

Hale, M. G. and D. M. Orcutt. 1987. *The Physiology of Plants under Stress.* New York: John Wiley & Sons.

Haneklaus, S., E. Bloem, and E. Schnug. 2007. Sulfur and plant disease. In: *Mineral Nutrition and Plant Disease*, eds., L. E. Datnoff, W. H. Elmer, and D. M. Huber, pp. 101–118. St. Paul, Minnesota: The American Phytopathological Society.

Hesse, H., V. Nikiforova, B. Gakiere, and R. Hoefgen. 2004. Molecular analysis and control of cysteine biosynthesis: Integration of nitrogen and sulfur metabolism. *J. Exp. Bot.* 55:1283–1292.

Hill, W. E. and L. G. Morrill. 1975. Boron, calcium and potassium interactions in Spanish peanuts. *Soil Sci. Soc. Am. Proc.* 39:80–83.

Hinsinger, P. 1998. How do plant roots acquire mineral nutrients? Chemical processes involved in the rhizosphere. *Adv. Agron.* 64:225–265.

Hitsuda, K., K. Toriyama, G. V. Subbarao, and O. Ito. 2010. Percent relative cumulative frequency approach to determine micronutrient deficiencies in soybean. *Soil Sci. Soc. Am. J.* 74:2196–2210.

Hulagur, B. F. and R. T. Dangarwla. 1983. Effect of zinc, copper and phosphorus fertilization on the content and uptake of nitrogen and secondary nutrients by hybrid maize. *Madras Agric. J.* 70:88–91.

Inal, A. and C. Tarakcioglu. 2001. Effects of nitrogen form on growth, nitrate accumulation, membrane permeability and nitrogen use efficiency of hydroponically grown bunch onion under boron deficiency and toxicity. *J. Plant Nutr.* 24:1521–1534.

Ishizuka, Y. 1978. *Nutrient Deficiencies of Crops*. ASPAC Food and Fertilizer Technology Center, Taipei, Taiwan.

Jackson, G. D. 2000. Effects of nitrogen and sulfur on canola yield and nutrient uptake. *Agron. J.* 92:644–649.

Janzen, H. H. and J. R. Bettany. 1984. Sulfur nutrition of rapeseed. I. Influence of fertilizer nitrogen and sulfur rates. S*oil Sci. Soc. Am. J.* 48:100–107.

Johnson, J. W. and H. F. Reetz. 1995. Adequate soil potassium increases nitrogen use efficiency by corn. *Better Crops Plant Food* 79:4.

Jones, J. B. and D. M. Huber. 2007. Magnesium and plant disease. In: *Mineral Nutrition and Plant Disease*, eds., L. E. Datnoff, W. H. Elmer, and D. M. Huber, 95–100. St. Paul, Minnesota: The American Phytopathological Society.

Kafkafi, U., M. Y. Siddiqi, R. J. Ritchie, A. D. M. Glass, and T. J. Ruth, 1992. Reduction of nitrate (13NO$_3$) influx and nitrogen (13N) translocation by tomato and melon varieties after short exposure to calcium and potassium chloride salts. *J. Plant Nutr.* 15:959–975.

Kaiser, D. E. and K. I. Kim. 2013. Soybean response to sulfur fertilizer applied as a broadcast or starter using replicated strip trials. *Agron. J.* 105:1189–1198.

Kaya, C. and D. Higgs. 2003. Relationship between water use and urea application in salt-stressed pepper plants. *J. Plant Nutr.* 26:19–30.

Keck, T. J., R. J. Wagenet, W. F. Campbell, and R. E. Knighton. 1984. Effects of water and salt stress on growth and acetylene reduction in alfalfa. *Soil Sci. Soc. Am. J.* 48:1310–1316.

Kemp, A. 1983. The effect of fertilizer treatments of grassland on the biological availability of magnesium to ruminants. In: *Role of Magnesium in Animal Nutrition*, eds., J. P. Fontenot, G. E. Bunce, K. E. Webb, Jr., and V. G. Allen, pp. 143–157. Blacksburg, Virginia: Virginia Polytechnic Institute.

Kene, D. R. and T. L. Deshpande. 1980. Effect of application of zinc, manganese and iron on the biological quality of grain of hybrid sorghum grown in black soils. *J. Indian Soc. Soil Sci.* 28:199–202.

Khan, M. G. and H. S. Srivastava. 1998. Changes in growth and nitrogen assimilation in maize plants induced by NaCl and growth regulators. *Bio. Plant.* 41:93–99.

Khattak, S. U., A. Khan, S. M. Shah, A, Zeb, and M. Iqbal. 1996. Effect of nitrogen and phosphorus fertilization on applied infestation and crop yield of three rapessed cultivars. *Pak. J. Zool.* 28:335–338.

Kim, H., M. Y. Hirai, H. Hayashi, M. Chinon, S. Naito, and T. Fujiwara. 1999. Role of *O*-acetyl-L-serine in the coordinated regulation of the expression of a soybean seed storage protein gene by sulfur and nitrogen nutrition. *Planta* 209:282–289.

Kim, S. J., T. Matsuo, M. Watanabe, and Y. Watanabe. 2002. Effect of nitrogen and sulphur application on the glucosinolate content in vegetable turnip rape (*Brassica rapa* L.). *Soil Sci. Plant Nutr.* 48:43–49.

Kiniry, J. R., G. McCauley, Y. Xie, and J. G. Arnold. 2001. Rice parameters describing crop performance of four U.S. cultivars. *Agron. J.* 93:1354–1361.

Klikocka, H., S. Haneklaus, E. Bloem, and E. Schnug. 2005. Influence of sulfur fertilization on infection of potato tubers with *Rhizoctonia solani* and *Streptomyces scabies*. *J. Plant Nutr.* 28:819–833.

Koprivova, A., M. Suter, R. O. Camp, C. Brunold, and S. Kopriva. 2000. Regulation of sulfate assimilation by nitrogen in *Arabidopsis*. *Plant Physiol.* 122:737–746.

Kumar, V., V. S. Ahlawat, and R. S. Antil. 1985. Interactions of nitrogen and zinc in pearl millet. I. Effect of nitrogen and zinc levels on dry matter yield and concentration and uptake of nitrogen and zinc in pearl millet. *Soil Sci.* 139:351–356.

Kutman. U. B., B. Yildiz, L. Ozturk, and I. Cakmak. 2010. Biofortification of durum wheat with zinc through soil and foliar applications of nitrogen. *Cereal Chem.* 87:1–9.

Kutman, U. B., B. Yildiz, and I. Cakmak. 2011. Effect of nitrogen on uptake, remobilization and partitioning of zinc and iron throughout the development of durum wheat. *Plant Soil* 342:149–164.

Kutman, U. B., B. Y. Kutman, Y. Ceylan, E. A. Ova, and I. Cakmak. 2012. Contributions of root uptake and remobilization to grain zinc accumulation in wheat depending on post-anthesis zinc availability and nitrogen nutrition. *Plant Soil* 361:177–187.

Loomis, W. D. and R. W. Durst. 1992. Chemistry and biology of boron. *Bio. Factors* 3:229–239.

Lopes, A. S. and F. R. Cox. 1977. A survey of the fertility status of surface soils under Cerrado vegetation in Brazil. *Soil Sci. Soc. Am. J.* 41:742–747.

Lopez-Lefebre, L. R., R. M. Rivero, P. C. Garcia, E. Sanchez, J. M. Ruiz, and L. Romero. 2002. Boron effect on mineral nutrients of tobacco. *J. Plant Nutr.* 25:509–522.

Loue, A. 1979. *The Interaction of Potassium with Other Growth Factors Particularly with Other Nutrients*, ed., International Potash Institute, pp. 407–433. Berne, Switzerland: International Potash Institute.

Ludlow, M. M. and R. C. Muchow. 1990. A critical evaluation of traits for improving crop yields in water-limited environments. *Adv. Agron.* 43:107–153.

Macleod, L. B. 1969. Effects of N, P, and K and their interactions on the yield and kernel weight of barley in hydrophonic culture. *Agron. J.* 61:26–29.

Mahajan, T. S. and K. R. Sonar. 1980. Effect of NaCl and Na_2SO_4 on dry matter accumulation and uptake of N, P, and K by wheat. *J. Maharastra Agric. Univ.* 5:110–112.

Malhi, S. S., S. Brandt, and K. S. Gill. 2003. Light fraction and total C and N in cultivated land versus grassland in a dark brown chernozemic soil. *Can. J. Soil Sci.* 83:145–153.

Malhi, S. S. and K. S. Gill. 2002. Effectiveness of sulphate-S fertilization at different growth stages for yield, seed quality and S uptake of canola. *Can. J. Plant Sci.* 82:665–674.

Malhi, S. S., D. W. McAndrew, and M. R. Carter. 1992. Effect of surface-applied Ca amendments and N on solonetzic soil properties and composition of barley. *Arid Soil Res. Rehab.* 6:71–81.

Malhi, S. S., G. Mumey, M. Nyborg, H. Ukrainetz, and D. C. Penney. 1995. Longevity of liming in western Canada: Soil pH crop yield and economics. In: *Plant–Soil Interactions at Low pH*, eds., R. A. Date, N. J. Grundon, G. E. Rayment, and M. E. Probert, pp. 703–710. Dordrecht: Kluwer Academic Publishers.

Malhi, S. S., M. Nyborg, H. G. Jahn, and D. C. Penney. 1988. Yield and nitrogen uptake of rapeseed (*Brassica campestris* L.) with ammonium and nitrate. *Plant Soil* 105:231–240.

Malhi, S. S., L. J. Piening, and D. J. Macpherson. 1989. Effect of copper on stem melanosis and yield of wheat: Sources, rates and methods of application. *Plant Soil* 119:199–204.

Marschner, H. 1995. *Mineral Nutrition of Higher Plants*, 2nd edition. New York: Academic Press.

Marschner, H., M. Treeby, and V. Romheld. 1989. Role of root-induced changes in the rhizosphere for iron acquisition in higher plants. *Z. Pflanzenern. Bodenk.* 152:197–204.

Mateo, P. I. Bonilla, E. Fernandez-valiente, and E. Sanchez-Maseo. 1986. Essentiality of boron for dinitrogen fixation in *Anabaena* sp. PCC 7119. *Plant Physiol.* 81:430–433.

Mcgrath, S. P. and F. J. Zhao. 1996. Sulphur uptake yield response and the interactions between N and S in winter oilseed rape (*Brassica napus* L.). *J. Agric. Sci. Cambr.* 126:53–62.

Mengel, K., E. A. Kirkby, H. Kosegarten, and T. Appel. 2001. *Principles of Plant Nutrition*, 5th edition. Dordrecht: Kluwer Academic Publishers.

Mengel, K., M. Voro, and G. Hehl. 1976. Effect of potassium on uptake and incorporation of ammonium nitrogen. *Plant Soil* 44:547–558.

Miller, M. H. 1974. Effects of nitrogen on phosphorus absorption by plants. In: *The Plant Root and Its Environment*, ed., E. W. Caron, pp. 643–668. Charlottesville, Virginia: University Press of Virginia.

Miley, W. N., G. W. Hardy, and M. B. Sturgis. 1969. Influence of boron, nitrogen and potassium on yield, nutrient uptake and abnormalities of cotton. *Agron. J.* 6:9–13.

Monge, E., E. Vale, and J. Abadia. 1987. Photosynthetic pigment composition of higher plants grown under iron stress. *Prog. Photosyn. Res.* 4:201–204.

Moraghan, J. T. and H. J. Mascagni, Jr. 1991. Environmental and soil factors affecting micronutrient deficiencies and toxicities. P. 371–425. In: *Micronutrient in Agriculture*, 2nd edition, eds., J. J. Mortvedt, F. R. Cox, L. M. Shuman, and R. M. Welch, pp. 371–425. Madison, Wisconsin: SSSA.

Muller, S., L. Heinrich, and I. Weigert. 1993. Influence of differentiated phosphorus and nitrogen fertilizer application on nutrient uptake and seed yield of faba beans (*Vicia faba* L.). *Bodenkultur* 44:127–133.

Murarka, I. P., T. L. Jackson, and D. P. Moore. 1973. Effects of N, K, and Cl on nitrogen components of Russet Burbank potato plants (*Solanum tuberosum* L.). *Agron. J.* 65:868–870.

Norton, R., R. Mikkelsen, and T. Jensen. 2013. Sulfur for plant nutrition. *Better Crops* 97:10–12.

Nuttall, W., F. H. Ukraintz, J. W. B. Stewart, and D. T. Spurr. 1987. The effect of nitrogen, sulphur and boron on yield and quality of rapeseed (*Brassica napus* L. and *Brassica campestris* L.). *Can. J. Soil Sci.* 67:545–559.

Okajima, H., I. Uritani, and H. K. Huang. 1975. *The Significance of Minor Elements on Plant Physiology*. Taipei, Taiwan: ASPAC Food and Fertilizer Technology Center.

Palfi, G. 1965. The effect of sodium salt on the nitrogen, phosphorus, potassium, sodium and amino acid content of rice shoots. *Plant Soil* 22:127–135.

Pan, W. L. 2012. Nutrient interactions in soil fertility and plant nutrition. In: *Handbook of Soil Sciences: Resource Management and Environmental Impacts*, eds., P. M. Huang, Y. Li, and M. E. Sumner, pp.16-1–16-13. Boca Raton, Florida: CRC Press.

Pasricha, N. S., M. S. Aulakh, G. S. Bahi, and H. S. Baddesha. 1987. Nutritional requirements of oilseed and pulse crops in Punjab (1976–1986). In: *Research Bulletin* 15, p. 92. Department of Soils, Ludhiana, India: Punjab Agricultural University.

Pasricha, N. S., G. S. Bahl, M. S. Aulakh, and K. S. Dhillon. 1991. *Fertilizer Use Research in Oilseed and Pulse Crops in India*. New Delhi: Indian Council of Agricultural Research.

Patel, M. S. and B. A. Golakia. 1986. Effect of calcium carbonate and boron application on yield and nutrient uptake by groundnut. *J. Indian Soc. Soil Sci.* 34:815–820.

Pederson, G. A., G. F. Brink, and T. E. Fairbrother. 2002. Nutrient uptake in plant parts of sixteen forages fertilized with poultry litter: Nitrogen, phosphorus, potassium, copper, and zinc. *Agron. J.* 94:895–904.

Peleg, Z., Y. Saranga, M. A. Yazici, T. Fahima, L. Ozturk, and I. Cakmak. 2008. Grain zinc, iron, and protein concentrations and zinc efficiency in wild emmer wheat under contrasting irrigation regimes. *Plant Soil* 306:57–67.

PPI. 1988. Effects of N and K fertilization in rice. Better Crops International, December 1988. p. 9.

Prasad, R. and J. F. Power. 1997. *Soil Fertility Management for Sustainable Agriculture*. Boca Raton, Florida: CRC Press.

Prasad, B., K. D. N. Singh, and B. P. Singh. 1986. Effect of long term use of fertilizer, lime and manures on farm and availability of N in an acid soil under multiple cropping system. *J. Indian Soc. Soil Sci.* 34:271–274.

Prosser, I. M., J. V. Purves, L. R. Saker, and D. T. Clarkson. 2001. Rapid disruption of nitrogen metabolism and nitrate transport in spinach plants deprived of sulphate. *J. Exp. Bot.* 52:113–121.

Raghuvanshi, R. K. S., R. K. Gupta, V. K. Paradkar, and D. D. Dubey. 1989. Response of cotton to N and P grown in sodic clay soils. *Indian J. Agron.* 34:18–20.

Randall, P. J., J. R. Freney, C. J. Smith, H. J. Moss, C. W. Wrigley, and I. E. Galbally. 1990. Effects of additions of nitrogen and sulfur to irrigated wheat at heading on grain yield, composition and milling and baking quality. *Aust. J. Exp. Agric.* 30:95–101.

Randall, P. J., K. Spencer, and J. R. Freney. 1981. Sulfur and nitrogen fertilizer effects on wheat. I. Concentrations of sulfur and nitrogen and the nitrogen to sulfur ratio in grain, in relation to the yield response. *Aust. J. Agric. Res.* 32:203–212.

Reneau, R. B., D. E. Bran Jr., and S. J. Donohue. 1986. Effect of sulphur on winter wheat grown in the coastal plain of Virginia. *Commun. Soil Sci. Plant Anal.* 17:149–158.

Reuveny, Z., D. Doughall, and P. Trinity. 1980. Regulatory coupling of nitrate and sulfate assimilation pathways in cultured tobacco cells. *Proc. Nat. Acad. Sci.* 77:6670–6672.

Roach, W. A. and C. Barclay. 1946. Nickel and multiple trace-element deficiencies in agricultural crops. *Nature* 157:696.

Robson, A.D. and J. B. Pitman. 1983. Interactions between nutrients in higher plants. In: *Inorganic Plant Nutrition. Encyclopedia of Plant Physiology, Vol 1*, eds., A. Lauchli, and R. L. Bieleski, pp. 147–180. New York: Springer-Verlag.

Romheld, V. 1987. Different strategies for iron acquisition in higher plants. *Physiol. Plant* 70:231–234.

Romheld, V. 1991. The role of phytosiderophores in acquisition of iron and other micronutrients in graminaceous species: Na ecological approach. *Plant Soil* 130:127–134.

Romheld, V. and H. Marschner. 1986. Mobilization of iron in the rhizosphere of different plant species. *Adv. Plant Nutr.* 2:155–204.

Sakal, R., A. P. Singh, and R. B. Sinha. 1988. Effect of different soil fertility levels on response of wheat to zinc application on caliorthent. *J. Indian Soc. Soil Sci.* 36:125–127.

Salt, D. E., M. Blaylock, P. B. A. Kumar Nanda, V. Dushenkov, B. O. Ensley, L. Chet, and I. Raskin. 1995. I. Phytoremediation: A novel strategy for removal of toxic metals from the environment using plants. *Biotechnology* 13:468–478.

Salvagiotti, F. and D. J. Miralles. 2008. Radiation interception, biomass production and grain yield as affected by the interaction of nitrogen and sulfur fertilization in wheat. *Eur. J. Agron.* 28:282–290.

Salvagiotti, F., J. H. M. Castellarin, D. J. Miralles, and H. M. Pedrol. 2009. Sulfur fertilization improves nitrogen use efficiency in wheat by increasing nitrogen uptake. *Field Crops Res.* 113:170–177.

Satyanarayana, T., V. P. Badanur, and G. V. Havanagi. 1978. Response of maize to N, P, and K on acid sandy loam soils of Bangalore. *Indian J. Agron.* 23:49–51.

Savithri, P. and K. M. Ramanathan. 1990. N use efficiency of rice as influenced by modified forms of urea and $ZnSO_4$ application in vertisol. *Madras Agric. J.* 77:216–220.

Scherer, N. W. 2001. Sulphur in crop production. *Eur. J. Agron.* 14:81–111.

Schonhof, I., D. Blankenburg, S. Muller, and A. Krumbein. 2007. Sulphur and nitrogen supply influence growth, product appearance and glucosinolate concentration of broccoli. *J. Plant Nutr. Soil Sci.* 170:65–72.

Schulthess, U., B. Feil, and S. C. Jutzi. 1997. Yield-independent variation in grain nitrogen and phosphorus concentration among Ethiopian wheats. *Agron. J.* 89:497–506.

Sexton, P. J., N. C. Paek, and R. Shibles. 1998. Soybean sulfur and nitrogen balance under varying levels of available sulfur. *Crop Sci.* 38:975–982.

Sharma, R. C. and E. L. Smith. 1986. Selection for high and low harvest index in three winter wheat populations. *Crop Sci.* 26:1147–1150.

Shi, R., Y. Zhang, X. Chen, Q. Sun, F. Zhang, V. Romheld, and C. Zou. 2010. Influence of long-term nitrogen fertilization on micronutrient density in grain of winter wheat (*Triticum aestivum* L.). *J. Cereal Sci.* 51:165–170.

Shuman, L. M. 1994. Mineral nutrition. In: *Plant–Environment Interactions*, ed., R. E. Wilkinson, pp. 149–182. New York: Marcel Dekker.

Sillanpaa, M. 1990. *Micronutrient Assessment at the Country Level: An International Study*. FAO Soils Bulletin No. 63, FAO, Rome.

Sinclair, A. H. and A. C. Edwards. 2008. Micronutrient deficiency problems in agricultural crops in Europe. In: *Micronutrient Deficiencies in Global Crop Production*, ed., B. J. Alloway, pp. 225–266. New York: Springer.

Singh, A. K. 1991. Response of pre-flood, early season maize to graded levels of nitrogen and phosphorus in *Gang diara* tract of Bihar. *Indian J. Agron.* 36:508–510.

Singh, M. 1992. The nitrogen-potassium interaction and its management. In: *Management of Nutrient Interactions in Agriculture*, ed., H. L. S. Tandon, pp. 21–37. New Delhi: Fertilizer Development and Consultation Organization.

Singh, M. V. 2008. Micronutrient deficiencies in crop and soils in India. In: *Micronutrient Deficiencies in Global Crop Production*, ed., B. J. Alloway, pp. 93–123. New York: Springer.

Singh, B. K. and R. P. Singh. 1985. Zinc deficiency symptoms in lowland rice as induced by modified urea materials applied at different rates of nitrogen in calcareous soils. *Plant Soil* 87:439–440.

Singh, V. and S. P. Singh. 1983. Effect of applied boron on the chemical composition of lentil plants. *J. Indian Soc. Soil Sci.* 31:169–170.

Singh, S., A. M. Kayastha, R. K. Asthana, and S. P. Singh. 2004. Response of garden pea to nickel toxicity. *J. Plant Nutr.* 27:1543–1560.

Singh, D. V. and B. R. Tripathi. 1974. Effect of N, P and K fertilization on the uptake of indigenous and applied zinc by wheat. *J. Indian Soc. Soil Sci.* 22:244–248.

Smith, J. B. and R. G. Heathcote. 1976. A new recommendation for the application of boronated superphosphate to cotton in northeastern Beune Plateau states. *Samarau Agric. Newsletter* 18:59–63.

Sommer, A L. and C. B. Lipman. 1926. Evidence on the indispensable nature of zinc and boron for higher green plants. *Plant Physiol.* 1:231–249.

Srinivas, K. and J. V. Rao. 1984. Response of French bean to N and P fertilization. *Indian J. Agron.* 29:146–149.

Takagi, S., K. Nomoto, and T. Takemoto. 1984. Physiological aspects of mugineiacid, a possible phytosiderophore of graminaceous plants. *J. Plant Nutr.* 7:469–477.

Tandon, H. L. S. 1992. *Management of Nutrient Interaction in Agriculture*. Fertilizer Development and Consultation Organization, New Delhi, India.

Taylor, R. G., T. I. Jackson, R. I. Powelson, and N. W. Christensen. 1981. Chloride, nitrogen form, lime, and planting date effects on take-all root rot of winter wheat. *Plant Dis.* 67:1116–1120.

Terry, D. 2004. AAPFCO Official Publication 57. Association of American Plant Food Control Officials, West Lafayette, Indiana.

Terry, N. 1980. Limiting factors in photosynthesis. I. Use of iron stress to control photochemical capacity *in vivo*. *Plant Physiol.* 65:114–120.

Terry, N. and J. Abadia. 1986. Function of iron in chloroplasts. *J. Plant Nutr.* 9:609–646.

Terman, G. l., J. C. Noggle, and C. M. Hunt. 1977. Growth rate-nutrient concentration relationship during early growth of corn as affected by applied N, P, and K. *Soil Sci. Soc. Am. J.* 41:363–368.

Thomas, S. G., P. E. Bilsborrow, T. J. Hocking, and J. Bennett. 2000. Effect of sulphur deficiency on the growth and metabolism of sugar beet (*Beta vulgaris* L.). *J. Sci. Food Agric.* 80:2057–2062.

Tisdale, S. L., W. L. Nelson, and J. D. Beaton. 1985. *Soil Fertility and Fertilizers*, 4th edition. New York: Macmillan.

Uauy, C., J. C. Brevis, and J. Dubcovsky. 2006a. The high grain protein content gene GpcB1 accelerates senescence and has pleiotropic effects on protein content in wheat. *J. Exp. Bot.* 57:2785–2794.

Uauy, C., A. Distelfeld, T. Fahima, A. Blechl, and J. Dubcovsky. 2006b. A NAC gene regulating senescence improves grain protein, zinc, and iron content in wheat. *Science* 314:1298–1301.

Verma, T. S. and R. M. Bhagat. 1990. Zinc and nitrogen interaction in wheat grown in limed and unlimed acid alfisol. *Fert. Res.* 2:29–35.

Voss, R. D. 1993. Corn. In: *Nutrient Deficiencies and Toxicities in Crop Plants*, ed., W. F. Bennett, pp. 11–14. St. Paul, Minnesota: The American Phytopathological Society.

Waals, J. H. V. and M. C. Laker. 2008. Micronutrient deficiencies in crops in Africa with emphasis on southern Africa. In: *Micronutrient Deficiencies in Global Crop Production*, ed., B. J. Alloway, pp. 201–224. New York: Springer.

Walker, K. C. and E. J. Booth. 2003. Sulphur nutrition and oilseed quality. In: *Sulphur in Plants*, eds., Y. P. Abrol, and A. Ahmad, pp. 323–339. Dordrecht: Kluwer Academic Publishers.

Wallace, A. 1990. Crop improvement through multi-disciplinary approaches to different types of stresses: Law of the maximum. *J. Plant Nutr.* 13:313–325.

Wang, C. H., T. H. Liem, and D. S. Mikkelsen. 1976. Sulphur deficiency: A limiting factor in rice production in the lower Amazon basin. II. Sulphur requirements for rice production. *IRI Res. Inst.* 48:9–30.

Warington, K. 1923. The effect of boric acid and borax on the broad bean and certain other plants. *Ann. Bot.* 37:629–672.

Waters, B. M., C. Uauy, J. Dubcovsky, and M. A. Grusak. 2009. Wheat (*Triticum aestivum* L.) NAM proteins regulate the translocation of iron zinc and nitrogen compounds from vegetative tissues to grain. *J. Exp. Bot.* 60:4263–4274.

Welch, R. M. 1981. The biological significance of nickel. *J. Plant Nutr.* 3:345–356.

Wilkinson, S. R., D. L. Grunes, and M. E. Sumner. 2000. Nutrient interactions in soil and plant nutrition. In: *Handbook of Soil Science*, ed., M. E. Sumner, pp. 89–111, Boca Raton, Florida: CRC Press.

Willett, I. R., P. Jackson, and B. A. Zarcinas. 1985. Nitrogen induced boron deficiency in Lucerne. *Plant Soil* 86:443–446.

Wilson, J. B. 1993. Macronutrient (NPK) toxicity and interactions in the grass *Festuca ovina*. *J. Plant Nutr.* 16:1151–1159.

Wood, B. W. and C. C. Reilly. 2007. Nickel and plant disease. In: *Mineral Nutritional and Plant Disease*, eds., L. E. Datnoff, W. H. Elmer, and D. M. Huber, pp. 215–231. St. Paul, Minnesota: The American Phytopathological Society.

Xue, Y. F., S. C. Yue, Y. Q. Zhang, Z. L. Cui, X. P. Chen, F. C. Yang, I. Cakmak, S. P. McGrath, F. S. Zhang, and C. Q. Zou. 2012. Grain shoot zinc accumulation in winter wheat affected by nitrogen management. *Plant Soil* 361:153–163.

Yadav, O. P. and H. R. Manchanda. 1979. Boron tolerance studies in gram and wheat grown on a sierozem sandy soil. *J. Indian Soc. Soil Sci.* 27:174–180.

Zhao, F. J., P. F. Bilsborrow, E. J. Evans, and S. P. McGrath. 1997. Nitrogen to sulphur ratio in rapeseed and in rapeseed protein and its use in diagnosing sulphur deficiency. *J. Plant Nutr.* 20:549–558.

Zhao, F. J., M. J. Hawkesford, and S. P. McGarth. 1999. Sulphur assimilation and effects on yield and quality of wheat. *J. Cereal Sci.* 30:1–17.

Zhao, F. J., Y. H. Su, S. J. Dunham, M. Rakszegi, Z. Bedo, S. P. McGrath, and P. R. Shewry. 2009. Variation in mineral micronutrient concentrations in grain of wheat lines of diverse origin. *J. Cereal Sci.* 49:290–295.

Zornoza, P., S. Robles, and N. Martin. 1999. Alleviation of nickel toxicity by ammonium supply to sunflower plants. *Plant Soil* 208:221–226.

Zou, C., X. Gao, R. Shi, X. Fan, and F. Zhang. 2008. Micronutrient deficiencies in crop production in China. In: *Micronutrient Deficiencies in Global Crop Production*, ed., B. J. Alloway, pp. 127–148. New York: Springer.

Zubillaga, M. M., J. P. Aristi, and R. S. Lavado. 2002. Effect of phosphorus and nitrogen fertilization on sunflower (*Helianthus annuus* L.) nitrogen uptake and yield. *J. Agron. Crop Sci.* 188:267–274.

7 Biological Nitrogen Fixation by Legumes

7.1 INTRODUCTION

The importance of legumes in agriculture is as old as the history of mankind. The family of Leguminosae is as diverse as it is large, comprising more than 19,000 species of plants, ranging from tiny herbs to huge trees (Ruiz-Diez et al., 2012). Despite this diversity (Lewis et al., 2005), only about 100 legumes have any substantial agricultural importance, although they are undoubtedly as important as grasses in global terms (Howieson et al., 2008). The legumes of agricultural importance grow on 12–15% of the earth's arable surface and account for 27% of the world's primary crop production, with grain legumes alone contributing 33% to the dietary protein nitrogen (N) needs for humans (Graham and Vance, 2003; Ruiz-Diez et al., 2012).

Furthermore, legumes have played a crucial role in agricultural production throughout history. Their success in N-deficient soils results from root nodules containing symbiotic *Rhizobium* bacteria that reduce N_2 to NH_3 (Phillips, 1980). Biological N fixation by legumes in association with rhizobia is known as dinitrogen (N_2) fixation. Dinitrogen fixation is defined as the conversion of molecular nitrogen (N_2) into ammonia (NH_3) and subsequently into organic N utilizable in biological processes (Soil Science Society of America, 2008). Rhizobium cells that reduce N_2 in root nodules are termed bacteroids. Biological N fixation resulting from symbiosis between legume plants and rhizobia provides a significant amount of N_2 to agricultural systems worldwide. It is one of the most spectacular natural phenomena (next to photosynthesis) and does not cause any hazard to the environment. In addition, the high costs of chemical fertilizers and developing agricultural systems that are ecologically sustainable have renewed interest in the process of dinitrogen fixation in recent years. Dinitrogen fixation is carried out by a variety of prokaryotic organisms that reduce atmospheric N_2 gas (which plant cannot assimilate) to NH_3, a form of N that plants readily incorporate into organic N (Layzell and Moloney, 1994). Lopez-Garcia et al. (2009) stated that inoculation of legume seeds with *Bradyrhizobium japonicum* strains selected to maximize atmospheric N_2 fixation has the potential of contributing to agricultural sustainability and conservation of this nutrient soil resource.

The quantity of N_2 reduction by prokaryotes at the global level is immense. The most important N_2-fixing agents in agricultural systems are the symbiotic associations between crop and forage/fodder legumes and rhizobia (Herridge et al., 2008). According to Delwiche (1983), the total world biological N fixation amounts to about 118×10^6 Mg year^{-1}. However, Brady and Weil (2002) calculated the total N fixation at about 139×10^6 Mg year^{-1} by legumes, nonlegumes, meadows and grassland, forest and woodland, and other vegetated land. Similarly, Brockwell and Bottomley (1995) and Sessitsch et al. (2002) reported that, globally, symbiotic N fixation has been estimated to amount to about 70 million metric tons per year. Herridge et al. (2008) reported that annual inputs of fixed N are calculated to be 2.95 Tg for pulses and 18.5 Tg for oilseed legumes. Soybean is the dominant crop legume, representing 50% of the global crop legume area and 68% of global production. Herridge et al. (2008) calculated that the soybean to fix 16.4 Tg N annually represents 77% of the N fixed by the crop legumes. Three of the largest soybean-producing countries are Brazil, the United States, and Argentina.

Rhizobia are well known for their capacity to establish a symbiosis with legumes. They inhabit root nodules, where they reduce atmospheric N and make it available to the plant (Sessitsch et al.,

2002). Furthermore, rhizobia are frequent rhizosphere colonizers of a wide range of plants and may also inhabit nonleguminous plants endophytically. In these rhizospheric and endophytic habitats, they may exhibit several plant growth-promoting effects, such as hormone production, phosphate solubilization, and the suppression of pathogens (Sessitsch et al., 2002).

Biological N fixation by legumes offers more flexible management than fertilizer N because the pool of organic N becomes slowly available to nonlegume species (Peoples et al., 1995a,b). Concomitant with N_2 fixation, the use of legumes in rotations offers control of crop diseases and pests (Graham and Vance, 2000; Sessitsch et al., 2002). Microorganisms capable of N_2 fixation may be divided into two groups, that is, those living free and those living symbiotically with higher plants. Microorganisms that form a symbiotic association have the greatest agricultural significance for N_2 fixation. *Rhizobium* and *Bradyrhizobium*, living symbiotically with legumes, are the most important N_2 fixers. According to Pohlhil (1981), there are about 19,700 known legume species of which about 80% grow symbiotically with *Rhizobium* or *Bradyrhizobium*. About 200 of these legume species are used as crop plants (Mengel et al., 2001).

Costs associated with commercial fertilizers continue to increase in response to energy prices and the large demand by domestic and international agricultural enterprises (Silveira et al., 2013). Alternative fertilizer options that maintain optimum forage production with minimum environmental impacts are the use of legume–grass mixtures in pastures for cattle raising. Dinitrogen-fixing legumes provide levels of N for pasture growth (Evers, 1985; Ocumpaugh, 1990), increase forage quality (Franzluebbers et al., 2004), and also enhance animal performance compared with grass alone (Hoveland et al., 1978).

Biological N fixation has a special importance in the twenty-first century due to the high cost of N fertilizers and low N use efficiency in the cropping systems, which may cause environmental pollution. In addition, biological N fixation is an important and integral component of sustainable agricultural systems (Sessitsch et al., 2002). Hence, the objective of this chapter is to review the latest information on biological N fixation by legumes. This information can be used by research scientists in planning their experiments and discussing experimental results related to N_2 fixation by legume crops. In addition, agronomists can also use this information in planning their cropping systems to improve N use efficiency.

7.2 MECHANISM OF NODULE FORMATION

One of the unique features of legumes is the formation of a symbiosis between the plant and soil bacteria of the genus *Rhizobium*, which results in the fixation of atmospheric N (Frame and Newbould, 1986). Gibson and Jordan (1983) reported that although nodule formation is regarded as a general characteristic of legumes, not all legumes have been reported to nodulate. The subfamily Papilionoideae contains a high proportion of nodulating genera (95%) (Allen and Allen, 1981). When nodulating legume plant growth is normal, the bacteria strain infects the roots and stimulates nodulation. There are several strains that are responsible for nodulation in legumes. The bacteria are free living in the soil, but fix N only in symbiosis (a mutually beneficial relationship) with the host legume. The most common type of symbiosis is between members of the plant family Leguminosae (also called Fabaceae) and soil bacteria, collectively called rhizobia, which include representatives of the genera *Rhizobium*, *Bradyrhizobium*, *Azorhizobium*, *Sinorhizobium*, and *Photorhizobium* (Epstein and Bloom, 2005). The attraction or recognition of a suitable host plant by rhizobia is through host plant lectins (carbohydrate-binding proteins) which interact selectively with microbial cell surface carbohydrates and serve as determinants of recognition or host specificity (Bauer, 1981).

Once the N_2 fixation begins, the nodule and its bacterial inhabitants maintain their dependency upon the host plant for both C and N. The reduction potential and ATP generated through photosynthesis are used to drive the carbon reduction cycle. Photoassimilates produced are subsequently allocated to the various sinks within the plant. The nodule and its subtending root system represent one of the strongest sinks, receiving an estimated 15–30% or more of the net photosynthate of the plant (Schubert, 1986). This supply of photosynthate transport via the phloem is used for the energy-yielding substrate and carbon skeletons to support the growth and maintenance of the nodule tissue. In addition, it is also used for the energy-consuming reactions associated with the reduction of N_2 in the endophyte and the assimilation of the NH_4^+ produced in the host cytosol and the synthesis of N-containing organic compounds for export from the nodule (Schubert, 1986).

Establishment of the legume–*Rhizobium* symbiosis involves infection of the host root and the subsequent formation of nodular growth containing approximately equal weights of root and bacterial cells (Bauer, 1981). Dart (1977) reported that *Rhizobium* cells can become attached to the host root surface within seconds or minutes after inoculation. Hence, attachment of *Rhizobium* cells to legume roots is likely to be one of the first steps in the required sequence of interactions leading to infection and nodulation (Bauer, 1981). Legume roots infected with rhizobia induce curling and branching of root hairs and the subsequent formation of a tubular structure called the infection thread. The infection thread develops inward from its point of origin near the most acutely curled region of the hair. Rhizobia are carried within the thread, usually single file, as the tip of the thread follows the movement of the nucleus toward the base of the hair cell. The infection thread passes through the wall of the hair cell and the adjacent cortical cell and branches into many newly divided cortical cells (Bauer, 1981). Rhizobia are released from the tips of the infection threads into the cytoplasm of the host cells, where they are surrounded by envelopes of the host plasma membrane. The enzyme nitrogenase is synthesized in the bacteria and converts dinitrogen into ammonia at the expense of the host plant photosynthate (Bauer, 1981).

The legume roots may be infected by N-fixing bacteria any time after root hairs are present on them. Infection normally occurs through young root hairs, although infection through the epidermal cell wall (Nutman, 1959) and, in peanut (*Arachis hypogaea*), through cells at the junction of the root hair cells and the epidermal and cortical cells (Chandler, 1978) is reported. The bacteria in the soil are attracted by secretions from the root, especially of tryptophan, which they convert into indole acetic acid (IAA) (Cobley, 1976). The nodules increase in size for about 3 weeks after infection, and vascular tissues derived from undifferentiated cells of the root cortex develop in it until they become continuous with the stele of the root. After 8–10 weeks, the nodule begins to break down and eventually falls from the root and disintegrates (Cobley, 1976).

Table 7.1 shows examples of host preference among bacteria species. Data in Table 7.1 show that bacteria of the genera *Rhizobium* and *Bradyrhizobium* provide the major biological source of fixed N in agricultural soils. The genus *Rhizobium* contains fast-growing, acid-producing bacteria, while *Bradyrhizobia* are slow growers that do not produce acid (Brady and Weil, 2002). The host plant provides carbohydrates for the N-fixing bacteria, and bacteria in exchange provide atmospheric

---

TABLE 7.1
Bacteria Species and Their Host Legumes

Bacteria Species	Preferred Host Legume
Rhizobium leguminosarum	Vetch, peas, lentils, and sweet pea
Rhizobium trifolii	Clovers
Rhizobium phaseoli	Dry bean and runner bean
Rhizobium meliloti	Sweet clover, alfalfa, and fenugreek
Rhizobium loti	Trefoils, lupinas, and chickpea
Bradyrhizobium japonicum	Soybean
Bradyrhizobium species	Cowpea, peanut pigeon pea, kudzu, crotolaria, and many other tropical legumes

fixed N for growth and development. If a given legume is grown on a specific area, N-fixing bacteria may be present. Under this situation, inoculation with N-fixing bacteria is not required. Otherwise, seeds of a legume should be inoculated with an appropriate strain of bacteria. Nowadays, commercial inoculants are available for legume species.

Both *Rhizobium* and *Bradyrhizobium* belong to the Rhizobiaceae family but are distinguished by their genetic properties and host specificity. In this respect, *Bradyrhizobium* species are not so selective as *Rhizobium* species (Werner, 1987). Some crop species can be infected by both types of N-fixing bacteria. For example, pegenon pea (*Cajanus cajan*) may be infected by *Rhizobium* as well as by *Bradyrhizobium*. Mengel et al. (2001) reported that *Bradyrhizobium* yielded a much higher N_2 fixation rate than *Rhizobium*, although *Rhizobium* grew faster than *Bradyrhizobium*. This difference is explained by a higher activity of the tricarboxylic acid cycle enzymes in *Bradyrhizobium* as compared with *Rhizobium* (Mengel et al., 2001). *Bradyrhizobium* is able to store appreciable amounts of a heteropolysaccharide in the nodules (Streeter and Salminen, 1993).

There are several legumes that fix biological N. However, most of the biological association studied between rhizobia and legumes is between soybean (*Glycine max* L. Merr.), pea (*Pisum sativum* L.), and clover (*Trifolium repens* L.) and the bacteria of the genus *Rhizobium* or *Bradyrhizobium*. Appunu et al. (2009) reported that soybean, an important N_2-fixing legume, forms N-fixing root nodules with diverse bacteria belonging to different genera and species. The slow-growing rhizobia that effectively nodule soybean are distributed among five different *Bradyrhizobium* species. These species are *Bradyrhizobium japonica*, *Bradyrhizobium elkanii*, *Bradyrhizobium liaoningense*, *Bradyrhizobium canariense*, and *Bradyrhizobium yuanmingense*.

Rhizobia infect cells within the roots of the legume, causing the formation of root outgrowths, called nodules, in which the bacteria proliferate and fix N_2 for export to the plant (Layzell and Moloney, 1994). The dinitrogen fixation requires or consumes a lot of energy. Hence, the yield of legumes is lower than that of cereals and N_2 is one of the reasons to divert production energy to nodule energy or N fixation energy.

Sprent (1989) has grouped legume nodules into two groups. In the determinate nodules group fall the nodules of soybean and dry bean (*Phaseolus vulgaris* L.). Nodules of these two species are usually spherical and all the infected cells within each individual nodule are of approximately the same age. In contrast, indeterminate nodules like pea, clover, and alfalfa are generally elongated due to the presence of a distinct meristem that continues to divide and produce new infected tissue throughout the life of the nodule. In both nodule types, the bacteria-infected cells are located in the central region of the nodule and are surrounded by layers of uninfected cortical cells (Layzell and Moloney, 1994).

Bauer (1981) reported that a broad range of evidence indicates that nodulation is self-regulated and optimized by the host. Ineffective nodules (i.e., nodules in which dinitrogen fixation does not take place) are usually formed in greater numbers than effective nodules. Hence, it can be concluded that nodule development is subject to negative feedback regulation by substances produced in effective nodules. Similarly, nodules are often found to occur in clusters, especially after a delayed inoculation, indicating that the frequency of subsequent nodulation below the cluster is diminished by self-regulation (Bauer, 1981).

Perennial legumes fix N during any time of active growth. In annual legumes, N fixation peaks at flowering. With seed formation, it ceases and the nodules fall off the roots. Rhizobia return to the soil environment to await their next encounter with legume roots. These bacteria remain viable in the soil for 3–5 years, but often at too low a level to provide significant optimal N-fixation capacity when legumes are replanted. Several articles or books are available on nodule formation in legumes. Readers may consult these articles, that is, nodule development (Robertson and Farden, 1980), characteristics of free-living and bacteroid rhizobia (Tsien et al., 1977), soil ecology and genetics (Abe and Higashi, 1979), *Rhizobium*–legume symbiosis (Dart, 1977), and infection of legumes by rhizobia (Bauer, 1981).

7.3 BIOCHEMISTRY OF NITROGEN FIXATION

Biological N by legumes is fixed mainly by two types of bacteria, *Rhizobium* and *Bradyrhizobium*. However, the key element in the process of N fixation is the enzyme *nitrogenase*. Biological N fixation is similar to industrial N fixation in that it produces ammonia from molecular N and hydrogen. N-fixing prokaryotes, in contrast to industrial processes, conduct this reaction at ambient temperatures and pressures. The reduction of N_2 to $2NH_3$, a six-electron transfer, is coupled with the reduction of two protons to evolve H_2 (Epstein and Bloom, 2005). The enzyme nitrogenase catalyzes the reduction of dinitrogen gas to ammonia according to the following equation (Brady and Weil, 2002):

$$N_2 + 8H^+ + 6e^- \text{ (nitrogenase)} \leftrightarrow 2NH_3 + H_2$$

The ammonia formed in the above reaction combines with organic acids to form amino acids and finally proteins as shown below:

$$NH_3 + \text{organic acids} \rightarrow \text{amino acids} \rightarrow \text{proteins}$$

The nitrogenase enzyme is a complex of two proteins, the smaller of which contains iron while the larger contains iron and molybdenum. To convert N_2 into NH_3 by nitrogenase requires energy, and this energy is supplied by plants through photosynthesis. The amount of energy required is about 355 kJ mol^{-1} NH_3 produced (Marschner, 1995). Calculations based on the carbohydrate metabolism of legumes show that a plant consumes 12 g of organic carbon per g of N_2 fixed (Heytler et al., 1984).

Nitrogenase is unique to N_2-fixation microorganisms and has been found, for example, in aerobic and anaerobic bacteria, blue-green algae, and root nodules of legumes. Nitrogenase enzyme is destroyed by free O_2, and it must be protected from exposure to O_2. When N fixation takes place in root nodules, one mechanism of protecting the enzyme from free oxygen is the formation of *leghemoglobin*. Leghemoglobin is virtually the same molecule as the *hemoglobin* that gives human blood its red color when oxygenated. This compound, which gives active nodules a red interior color, binds oxygen in such a way as to protect the nitrogenase while making oxygen available for respiration in other parts of the nodule tissue (Brady and Weil, 2002). The concentration of leghemoglobin is closely but not linearly correlated with the N_2-fixing capacity of root nodules (Werner et al., 1981).

For the nitrogenase reaction, energy in the form of a reductant and as ATP is essential. Energy from ATP and electrons from the electron carrier (usually ferredoxin) induce a conformational change in the iron protein and convert it into a powerful reductant capable of transferring electrons to the molybdenum-ion protein, which in turn reduces N_2 (Marschner, 1995). Reduction of N_2 to NH_3 requires 15–30 ATP molecules per N_2 molecule reduced (Shanmugam et al., 1978). The ATP energy is mainly produced during respiration (Evans and Barber, 1977). In *Rhizobiua*, between 30% and 60% of the energy supplied to the nitrogenase is released as H_2 (Schubert et al., 1978). Rhizobium strains are also capable of splitting the H_2 by hydrogenase, however, thus recycling the electrons for subsequent N_2 reduction (Marschner, 1995). Selection of *Rhizobium* strains with a generally higher recycling of electrons from H_2 may be important for higher efficiency of the N_2 fixation (Schubert et al., 1978; Marschner, 1995).

7.4 QUANTITY OF NITROGEN FIXATION BY LEGUMES

Introducing legumes into cereal-dominated cropping systems can provide many advantages, such as fixing great amounts of atmospheric N, which will then be partly available to the subsequent crop, reducing the occurrence of pests and weeds, and improving the quality of the soil (Peoples et al., 1995b;

Stevenson and Van Kessel, 1996; Unkovich et al., 1997; Van Kessel and Hartley, 2000; Ruisi et al., 2012). There is a wide range of values of dinitrogen fixation by crop and pasture legumes and by other diazotrophic systems. For the legumes, the values recorded depend, inter alia, on the crop grown, the effectiveness of the strains of rhizobia in fixing N with the host, and the environmental conditions (Gibson and Jordan, 1983). However, legumes have historically been used to maintain soil N fertility (Fauci and Dick, 1994). The legume–*Rhizobium* symbiosis is estimated to account for 40% of the world's fixed N (Ladha et al., 1992). Symbiotic N_2 fixation in legumes is determined by the formation of effective nodules on the roots. Formation of effective nodules depends on plant, soil, and climatic factors and their interactions. Hence, legumes have different N_2-fixation capabilities depending on the environmental conditions, management practices adopted, and type of legume species (Stute and Posner, 1993; Fageria et al., 2005).

There has to be an upper limit of biological N fixation. Herridge and Bergersen (1988) postulated a theoretical upper limit of 635 kg ha^{-1} for soybean and more than 300 kg ha^{-1} for pigeon pea (*Cajanus cajan* L. Huth) and peanut (*Arachis hypogaea* L.). However, under field conditions or farming systems, theoretical potential values are never achieved because of several limiting factors. N supplied by hairy vetch (*Vicia villosa* Roth) and crimson clover (*Trifolium incarnatum* L.) in cover crop experiments ranged from 72 to 149 kg N ha^{-1} (Hargrove, 1986; Ladha et al., 1988; Holderbaun et al., 1990). Henzell (1968) summarized many field experiments with tropical pasture legumes by concluding that they fixed 22–178 kg N_2 ha^{-1} year^{-1} under average conditions in northern Australia, but that, under better conditions, this could rise to 290 kg N_2 ha^{-1} year^{-1}. Soybean growing in the United States may have up to 400 kg N ha^{-1} of which as little as 25% or as much as 84% (Bezdicek et al., 1978) may have been fixed from the atmosphere, depending largely on the available N status of the soil (Gibson and Jordan, 1983). Smith and Hume (1987) reported that when soybean in symbiotic association with *Bradyrhizobium japonicum* can fix up to 200 kg N ha year^{-1}. In dry bean, the rate of N_2 fixation has been reported to vary from 25 to 71 kg ha^{-1} in a crop cycle of about 90–100 days (Graham, 1981). Bliss (1993) reported that it has been common to observe locally adapted bean cultivars able to fix at least 50 kg N ha^{-1} and about 40–50% of the plant N from fixation.

The quantity of N fixed by legumes depends on the legume species and environmental factors (Table 7.2). For good results, optimum environmental conditions are essential. These conditions are discussed in the next section. Brazil is the largest soybean-producing country in the world. Brazilian farmers do not apply N to soybean and its total N demand is met by inoculation by appropriate N-fixing bacteria. N fixed by legumes may be utilized by the host plant. When the host plant residues are incorporated into the soil, some fixed N may be added to the soil and can be used by the succeeding crop after decomposition of the residues. Some of the fixed N may be available to the nonfixing plants growing in association with N-fixing plants. In addition, the root remains in the soil may also add N to succeeding crops.

7.5 METHODS OF ASSESSING NITROGEN FIXATION

Various methods of N_2 fixation measurement have been proposed. These methods are acetylene reduction assay, nodule evaluation, determining N balance, N difference method, N fertilizer equivalence (NFE), and N isotopic technique. There are several techniques that have been used to measure the N_2 fixed by legumes in the cropping systems. These techniques are adopted to measure dinitrogen fixation both under controlled and field conditions. Measurements under controlled as well as field conditions are complementary to each other. However, more emphasis should be given to field experimentation to have practical applicability of the results. According to Ladha et al. (1988), the methods used have provided reasonable estimates of N_2, none of which are entirely satisfactory. However, N accumulation in the dry tissue of legume crops and ^{15}N dilution are widely reported references for N_2 fixation. Techniques of N_2 measurements have been summarized in the following section.

TABLE 7.2

Quantity of N Fixed by Principal Legumes

Crop Species	N$_2$ Fixed (kg ha^{-1} Crop^{-1})	Reference
Peanut (*Arachis hypogaea* L.)	40–80	Brady and Weil (2002)
Cowpea (*Vigna unguiculata* L. Walp.)	30–50	Brady and Weil (2002)
Alfalfa (*Medicago sativa* L.)	78–222	Heichel (1987)
Soybean (*Glycine max* L.)	50–150	Brady and Weil (2002)
Fava bean (*Vicia faba* L.)	177–250	Heichel (1987)
Hairy vetch (*Vicia villosa* Roth.)	50–100	Brady and Weil (2002)
Ladino clover (*Trifolium repens* L.)	164–187	Heichel (1987)
Red clover (*Trifolium pratense* L.)	68–113	Heichel (1987)
White lupine (*Lupinus albus* L.)	50–100	Brady and Weil (2002)
Field peas (*Pisum sativum* L.)	174–195	Heichel (1987)
Chickpea (*Cicer arietinum* L.)	24–84	Heichel (1987)
Pigeon pea (*Cajnus cajan* L. Huth.)	150–280	Brady and Weil (2002)
Kudzu (*Pueraria phaseoloides* Roxb. Benth)	100–140	Brady and Weil (2002)
Chick pea (*Cicer arietinum* L.)	24–84	Heichel (1987)
Greengram (*Vigna radiata* L. Wilczek.)	71–112	Chapman and Myers (1987)
Lentil (*Lens culinaris* L.)	57–111	Smith et al. (1987)

Source: Adapted from Brady, N. C. and R. R. Weil. 2002. *The Nature and Properties of Soils*, 13th edition. Upper Saddle River, New Jersey: Prentice Hall; Chapman, A. L. and R. J. K. Myers. 1987. Western Australia. *Aust. J. Exp. Agric.*, 27:155–163; Heichel, G. H. 1987. *Energy in Plant Nutrition and Pest Control*, pp. 63–80. Amsterdam: Elsevier Scientific Publishers; Smith, M. S., W. W. Frye, and J. J. Varco. 1987. *Adv. Soil Sci.* 7:95–139.

7.5.1 ACETYLENE REDUCTION ASSAY

The acetylene reduction assay depends on the ability of nitrogenase to reduce acetylene to ethylene, which is then released from the enzyme and measured by gas chromatographic procedures (Gibson and Jordan, 1983). In simple words, it is a technique for demonstrating or estimating nitrogenase activity by measuring the rate of acetylene (C_2H_2) reduction to ethylene (C_2H_4). The standard acetylene reduction assay method involves enclosing detached nodules or the nodulated root system in airtight containers and exposing them to an atmosphere containing C_2H_2. Gas samples are taken after a certain period of incubation and analyzed for ethylene by using gas chromatography. This technique is widely used in the research of dinitrogen fixation due to its high sensitivity and simplicity (Peoples and Herridge, 1990).

It is simple and relatively cheap to use, very sensitive, and many assays can be done in a short period of time. However, Peoples and Herridge (1990) reported that when results of dinitrogen fixation by this technique are compared with other techniques under field conditions, the acetylene reduction assay technique greatly underestimated N$_2$ fixation activity. Several factors may affect the C_2H_2 reduction to C_2H_4, that is, plant and nodule age, plant stress, soil fertility level, especially NO$_3$ and NH$_4$, and changes in O$_2$ partial pressure, making quantitative comparisons of N$_2$ fixation based on long-term C_2H_2 accumulation misleading (Peoples and Herridge, 1990).

In addition, Gibson and Jordan (1983) reported that the amount of dinitrogen fixation by this technique may lead to under- or overestimates. This may happen due to the overestimation of nitrogenase activity by this method, including the suppression of endogenous ethylene oxidation by acetylene in some soil system (Witty, 1979), and the apparent stimulation of nitrogenase activity by acetylene during longer-term assays (David and Fay, 1977), while a change in pO$_2$ conditions during the course of the assay may give overestimates or under-estimates of activity depending on the system. In addition, Gibson and Jordan (1983) reported that this technique has been of great benefit to N fixation research, especially under laboratory conditions. However, it is essential that great care be exercised in applying and interpreting field assays.

7.5.2 NODULE EVALUATION

Dinitrogen fixation depends on the formation of nodules on root system. Hence, their number, size, and distribution on the root system have been used to measure the symbiotic activity in legumes. The counting of the nodule number and size of the root system is simple, quick, and inexpensive. The degree of nodulation can be related to the degree of N_2 fixation. Poor nodulation indicates that the N_2 fixation process is low in a determined legume or legumes. Peoples and Herridge (1990) reported that the nodule evaluation technique can at best provide an indirect indication of a legume's potential to fix N_2 and cannot be used to quantify the amount of N_2 fixed.

7.5.3 DETERMINING THE NITROGEN BALANCE

The N_2 measured by this technique requires the determination of the quantity of N change in the soil under legume growth. It requires that all the N inputs in the soil such as rain water, dust, animal droppings, and weathering be measured. Similarly, all the N output sources such as denitrification, volatilization, leaching, erosion, and removal of crop or animal products within a given soil–plant system should be measured. The net increase in N under a legume is attributed to N_2 fixation. This technique requires many N input and output measurements and the inputs should be large enough to detect any difference in N accumulation or fixation. Experimental data should be repeated over several years to get meaningful results.

7.5.4 NITROGEN DIFFERENCE METHOD

In this method, N_2 fixation by a legume is compared with a nonfixing control. The difference in total N accumulated by the legume (N_2) and nonfixing control (N_1) legume is regarded as the contribution of N_2 fixation to legume growth. The fixation of N_2 can be calculated by using the following equation:

$$\text{Quantity of } N_2 \text{ fixed} = N_2 - N_1$$

This method is simple and can be used when total N analysis facilities are available. In this method, the choice of control crop is very important. Ideally, the two plant types should explore the same rooting volume, have the same ability to extract and utilize soil mineral N, and accumulate soil N over the same period of time (Peoples and Herridge, 1990). A non-N_2-fixing control may be a nonlegume, an uninoculated legume, or a nonnodulation legume, preferably an isoline of the test legume (Peoples and Herridge, 1990).

7.5.5 NITROGEN EQUIVALENCE

In the N equivalence technique, the N_2 fixation is assessed by growing N-fertilized non-N_2-fixing plants in plots alongside the unfertilized N-fixing test legume. The N fertilizer levels at which the yields of the nonfixing plants match those of the legume are equivalent to the amount of N_2 fixed (Peoples and Herridge, 1990). The value obtained is usually expressed as fertilizer N equivalence. Table 7.3 provides data of NFE of legume cover crops to succeeding nonlegume crops. The NFE values varied from 12 to 182 kg ha^{-1}. Smith et al. (1987) reported that NFE values range from 40 to 200 kg ha^{-1}, but more typically are between 75 and 100 kg ha^{-1}. Interseeding red clover

TABLE 7.3
N Fertilizer Equivalence (NFE) of Legume Cover Crops to Succeeding Nonlegume Crops

Legume/Nonlegume Crop	NFE (kg ha^{-1})
Hairy vetch/cotton	67–101
Hairy vetch + rye/corn	56–112
Hairy vetch/corn	78
Hairy vetch/sorghum	89
Hairy vetch/corn	78
Hairy vetch + wheat/corn	56
Crimson clover/cotton	34–67
Crimson clover/corn	50
Crimson clover/sorghum	19–128
Common vetch/sorghum	30–83
Bigflower vetch/corn	50
Subterranean clover/sorghum	12–103
Sesbania/lowland rice	50
Alfalfa/corn	62
Alfalfa/wheat	20–70
Arachis spp./wheat	28
Subterranean clover/wheat	66
White lupin/wheat	22–182
Arachis spp./corn	60
Pigeon pea/corn	38–49
Sesbania/potato	48
Mungbean/potato	34–148
Chickpea/wheat	15–65

Source: Compiled from Smith, M. S., W. W. Frye, and J. J. Varco. 1987. *Adv. Soil Sci.* 7:95–139; Kumar, K. and K. M. Goh. 2000. *Adv. Agron.* 68:197–319; Fageria, N. K., V. C. Baligar, and A. Bailey. 2005. *Commun. Soil Sci. Plant Anal.* 36:2733–2757.

(*Trifolium pratense* L.) into small grains is a common practice in the northeastern United States (Singer and Cox, 1998), and such practice can provide up to 85 kg N ha^{-1} to the subsequent corn crop (Vyn et al., 1999). Researchers in the southeastern United States have estimated that legumes such as hairy vetch can supply well over 100 kg N ha^{-1} to the following corn or grain sorghum crops (Hargrove, 1986; Blevins et al., 1990; Oyer and Touchton, 1990). On prairie soils in Kansas, Sweeney and Moyer (2004) found that grain sorghum following an initial kill-down of red clover and hairy vetch yielded as much as 131% more than continuous sorghum with estimated fertilizer N equivalencies exceeding 135 kg ha^{-1}.

7.5.6 NITROGEN ISOTOPIC TECHNIQUE

N has two stable isotopes that are designated as ^{14}N and ^{15}N. The ^{15}N isotope occurs in atmospheric N_2 at a constant abundance of 0.3663 atom% (Peoples and Herridge, 1990). If the isotopic concentrations are different in two sources of N (soil N and atmospheric N_2), the proportion P of plant N arising from one of them (in this case N_2) can be calculated by using the following equation (Peoples and Herridge, 1990):

$$P = \frac{x - y}{x - b}$$

where x, y, and b are, respectively, the isotopic compositions of the plant-available soil N of a N_2-fixing plant growing in the soil, and of a legume fully dependent on N_2 fixation for growth. In practice, x is obtained from the isotopic composition of a non-N_2-fixing reference plant (a nonlegume, uninoculated legume, or nonnodulating legume genotype) that is totally dependent on soil N (Peoples and Herridge, 1990).

7.5.7 ^{15}N ENRICHMENT

The ^{15}N enrichment method is generally regarded as the standard method for estimating legume N_2 fixation (Herridge and Danso, 1995). Its main advantage is that it provides a time-averaged estimate of Pfix, integrated for the period of plant growth to the time of harvest. The N_2 fixation is calculated by using the following equation:

$$Pfix = 1 - \frac{(atom\%^{15}N\ excess\ legume\ N)}{(atom\%\ ^{15}N\ excess\ soil - derived\ N)}$$

where atom% ^{15}N excess = (atom% ^{15}N sample) – (atom% ^{15}N air N_2), and atom% ^{15}N of air $N_2 = 0.3663$. The atom% ^{15}N of soil-derived N is generally estimated from the ^{15}N enrichment of the non-N_2-fixing reference plant grown in the same soil over the same period as the legume. The choice of an appropriate non-N_2-fixing reference plant is the single most important factor affecting the accuracy of estimate P (Peoples and Herridge, 1990).

The major assumption of both the ^{15}N-enriched and natural ^{15}N abundance methods is that the legume and nonlegume reference plants utilize soil N with the same isotopic composition (Herridge and Danso, 1995). With the enrichment system, this translates into the legume and nonfixing reference plants utilizing the same relative amounts of N from added ^{15}N and endogenous soil N (Herridge and Danso, 1995). This may not always occur and is the major drawback of this method (Witty et al., 1988). In addition, the high cost of instrumentation to measure ^{15}N plus the expense of the ^{15}N-labeled materials are real constraints to even greater use of the method.

7.5.8 ^{15}N ISOTOPE DILUTION TECHNIQUE

Data on the ^{15}N enrichment of biomass can be used to calculate %Ndfa according to Fried and Middleboe (1977) and Ruisi et al. (2012):

$$\%Ndfa = \left(1 - \frac{\delta\ ^{15}N_{legume}}{\delta\ ^{15}N_{cereal}}\right) \times 100$$

where $\delta\ ^{15}N_{legume}$ is the atom% ^{15}N excess of legume tissue and $\delta\ ^{15}N_{cereal}$ is the atom% ^{15}N excess of cereal tissue. The ^{15}N natural abundance of the atmosphere (0.3663% ^{15}N) was used to calculate the atom% ^{15}N excess of each species. The amount of N fixed by legume species can be estimated as follows (Ruisi et al., 2012):

$$N_{fixed} = Total\ legume\ N \times \frac{\%Ndfa}{100}$$

The amount of N derived from the soil (Ndfa) can be calculated by subtracting the amount of N fixed and the amount of N derived from the fertilizer (the latter can be calculated according to Giambalvo et al., 2010) from the total N in the aboveground biomass. The N balance, which represents the net potential contribution of N_2 fixation to the system, can be calculated by subtracting the amount of N removed in grains from the amount of N fixed, according to Evans et al. (2001).

Ruisi et al. (2012) calculated the N fixed by chickpea (*Cicer arietinum* L.), faba bean (*Vicia faba* L.), lentil (*Lens culinaris* Medik), and pea (*Pisum sativum* L.), and wheat was used as a cereal in the experiment. The percentage of N fixed differed by species in the order faba bean > chickpea > pea > lentil. On an average, the faba bean accumulated more N from the atmosphere and left more residual N in the soil than did the other three species. Ruisi et al. (2012) concluded that the faba bean makes the greatest contribution to the N balance, a result that is in agreement with Lopez-Belliodo et al. (2006), Walley et al. (2007), and Hauggard-Nielsen et al. (2009) who compared several grain legumes (chickpea, dry bean, faba bean, lentil, lupin, and pea). Ruisi et al. (2012) reported that it is important to note that when computing the amount of Ndfa on an aboveground biomass basis, one ignores fixed N in roots and nodules and from rhizodeposition, which can lead to a substantial underestimation of N_2 fixation. Research has shown that belowground contributions of fixed N may represent between 30% and 60% of the total N accumulated by legume crops (Peoples et al., 2009).

7.5.9 NATURAL ^{15}N ABUNDANCE

Almost all N transformations in the soil result in isotopic fractionation. The net effect is often a small increase in the ^{15}N abundance of soil N compared with atmospheric N_2 (Shearer and Kohl, 1986). In looking at such small differences in ^{15}N concentration, data are commonly expressed in terms of parts per thousand (δ ^{15}N or 0/00). The equation used to calculate δ ^{15}N is written as follows (Peoples and Herridge, 1990; Herridge and Danso, 1995):

$$\delta^{15}N = 100 \times \frac{(\text{atom}\%\ ^{15}N\ \text{sample}) - (\text{atom}\%\ ^{15}N\ \text{standard})}{(\text{atom}\%\ ^{15}N\ \text{standard})}$$

where the standard is usually atmospheric N_2 (0.3663 atom%). By definition, the δ ^{15}N of air N_2 is zero. The natural abundance method gives an integrated estimate of P over time as with ^{15}N enrichment studies, but it can be applied to established experiments (provided nonfixing reference material is available) because no pretreatment, that is, ^{15}N application, is necessary. An estimate of P is obtained by using the following equation (Peoples and Herridge, 1990; Herridge and Danso, 1995):

$$P = \frac{(\delta^{15}N\ \text{soil N}) - (\delta^{15}N\ \text{legume N})}{(\delta^{15}N\ \text{soil N}) - B}$$

The δ ^{15}N value of B is a measure of isotopic fractionation during N_2 fixation and is determined by analysis of the δ ^{15}N of total plant N of the nodulated legume grown in N-free media. Isotopic fractionation during N_2 fixation is minimal but not zero and should be taken into account when calculating Pfix (Peoples and Herridge, 1990). Although the principles of the natural abundance technique are the same as those underlying ^{15}N enrichment, the main limitations are quite different and have been reviewed by Shearer and Kohl (1986), Ledgard and Peoples (1988), and Peoples and Herridge (1990).

7.6 FACTORS AFFECTING DINITROGEN FIXATION

The environment of a plant may be defined as the sum of all external forces and substances affecting the growth, structure, and reproduction of that plant (Fageria et al., 2011). Crop environment

is composed of climatic and soil factors, which exert a great influence on plant growth and, consequently, yield. Climatic factors such as temperature and moisture supply play an important role in
crop production. Similarly, soil physical, chemical, and biological properties are directly related to
crop productivity (Fageria, 2013). Dinitrogen fixation is significantly affected by environmental factors. The most important environmental factors that affect dinitrogen fixation are soil water content
(Serraj et al., 1999; Pimratch et al., 2008), soil mineral N content (Jensen, 1987; Voisin et al., 2002),
and soil temperature (Halliday, 1975; Lie, 1971; Lira et al., 2003; Liu et al., 2013). When these
conditions are constant, however, the response of N_2 fixation can vary among species and cultivars
(Liu et al., 2013). Aside from cropping history, environmental factors such as soil moisture, pH,
and nutrient content can influence the persistence of rhizobia in the soil (Keyser and Munns, 1979).

People and Herridge (1990) reported that a large number of legumes can be produced in
N-impoverished soils without the addition of fertilizer N, and in many soils without depleting soil
N reserves. These authors further reported that these desirable characteristics of legumes can be
realized only when large amounts of atmospheric N_2 are fixed. The successful formation of a functional symbiosis is dependent on many physical, environmental, nutritional, and biological factors
and cannot be assumed to occur as a matter of course. If these factors are in the favorable range for
legume growth, dinitrogen fixation will also be at an optimum level. Some of these factors can be
manipulated in favor of N_2 fixation. These factors are discussed in the following sections.

7.6.1 Temperature

Temperature is one of the most important climatic factors affecting the growth and development
of plants. It affects various growth and metabolic processes in plants. In general, the rates in plant
processes are restricted when temperatures are too low to reach their maximum at somewhat higher
temperatures and decrease again when temperatures are too high. Crop species react differently to
temperature throughout their life cycles. Each species has a defined range of maximum and minimum temperatures within which growth occurs and an optimum temperature at which plant growth
progresses at its fastest rate (Hatefield et al., 2011). Vegetative development usually has a higher
optimum temperature than reproductive development. The progression of a crop through phono-
logical phases is accelerated by increasing temperatures up to the species-dependent optimum temperature. Exposure to higher temperatures causes faster development in food crops, which does not
translate into an optimum for maximum production because the shorter life cycle means a shorter
reproductive period and a shorter radiation interception period (Hatefield et al., 2011).

As with most plant tissues, the nodule metabolism increases with temperature, resulting in Q_{10}
values of approximately 2.0 (Layzell et al., 1983). Frame and Newbould (1986) reported that for
white clover (*Trifolium repens* L.) the temperature requirement to grow and fix N ranged from 9°C
to 27°C with an optimum of 26°C. Gibson and Jordan (1983) reported that legumes from temperate
regions nodulate rapidly at 28–30°C root temperature, but the optimum temperature for N_2 fixation
is 20–24°C and tends to decline as the plant ages. These authors also reported that legumes of a
tropical or subtropical origin exhibit a temperature optimum for nodulation and N fixation in the
range of 25–30°C. The maximum temperature for nodulation is about 36°C and the minimum is
about 15°C (Lindermann and Ham, 1979). However, Mengel et al. (2001) reported that maximum
N-fixing rates have been obtained at a high soil temperature (33°C). These authors further reported
that the potential N_2-fixing capacity of free-living bacteria is thus highest in subtropical and tropical
regions.

The minimum or maximum temperatures that can support N_2-fixing nodules vary greatly with
the legume species, and even within a single species. Variation has been reported in the tolerance
of symbioses for N_2 fixation at extremes in temperature (Thomas and Sprent, 1984). Consequently,
it seems that genetic potential may exist for realizing significant increases in N_2 fixation (Layzell
and Moloney, 1994). Weisz and Sinclair (1988) reported that the changes in nodule activity with
temperature are inversely correlated with changes in the nodule's resistance to O_2 diffusion.

Logically, much of the research has concentrated on root temperature as it is directed at the site of nodule development, but shoot temperatures are also important with respect to assimilating flow to and from the nodules (Gibson and Jordan, 1983). Studies with soybean (Schweitzer and Harper, 1980) and *Trifolium subterraneum* (Eckart and Ragues, 1980) concluded that shoot temperature is a major factor influencing nitrogenase activity. For example, soybean held at 27°C in the dark maintained the same level of nitrogenase activity as those kept under light for 72 h, but cooling the shoots to 18°C led to a marked drop in activity. The effect was attributed to a greater mobilization of carbohydrate.

Optimum temperatures for nodulation in dry bean, obtained from isolated root and whole plant studies, vary from 25°C to 30°C (Barrios et al., 1963; Small et al., 1968), and nodulation is markedly reduced at 12°C or 33°C (Graham, 1981). Pankhurst and Sprent (1976) reported the highest nitrogenase activity in isolated nodules at 20°C. Graham and Rosas (1979) reported that peak rates of N_2 (C_2H_2) fixation increases from 33.8 µmol C_2H_2 produced per plant per hour at 35/25°C day/night temperature regime to 73.08 µmol C_2H_2 produced per plant per hour at 25/15°C day/night temperature. Liu et al. (2013) reported that N_2 fixation was very sensitive to low temperature and photosynthetic rates. Potentially, cultivar breeding aiming at cold resistance and a higher photosynthetic rate could increase N_2 fixation in practice (Liu et al., 2013).

7.6.2 Soil Moisture

Moisture availability is one of the most important factors determining crop production. Seasonal water supply affects growth potential and N_2 fixation (Giller and Wilson, 1991; Peoples et al., 1992). Crop yield can be reduced both at very low and very high levels of moisture. Many studies have shown that biological N fixation can be adversely affected by both waterlogging and soil dehydration at critical times during the development and growth of legumes (Sall and Sinclair, 1991; Pena-Cabriales and Castellanos, 1993). Excess moisture reduces soil aeration and, thus, the supply of O_2 available to roots. With poor aeration, the activities of beneficial microorganisms and water and nutrient uptake by plants can be seriously inhibited, though aquatic plants and rice are adapted to and function well even when soils are saturated. Soil moisture deficits can cause the stomata in the leaf to close, reducing transpiration and helping maintain hydration of protoplasms, but also reducing photosynthesis. Moisture stress also causes reductions in both cell division and cell elongation and, hence, in growth (Fageria et al., 2011; Fageria, 2013).

The storage terms relate only to the portion of soil moisture available to the plant. Field capacity and a wilting coefficient or a permanent wilting point are the practical upper and lower limits of water availability for crops. The upper limit of soil water availability to plants is often considered the water contents after the saturated soil has freely drained for 2–3 days or the wetted soils have been subjected to pressures in the range from 5 to 30 kPa (kilopascals) or 0.05 to 0.3 bars (Unger et al., 1981). The lower value is generally applicable to light-textured soils and the higher value to heavy-textured soils. Root growth is better when the water content in the soil is around field capacity. Hence, dinitrogen fixation will also be maximum when soil moisture is at field capacity.

Worrall and Roughley (1976) reported that nodulation is affected by a reduction in root hair infection and suppression of nodule development at low soil water potential, that is -3.6×10^5 Pa (Pascal). Rewatering stressed plants to a rapid recommencement of root hair growth, followed by infection and nodulation initiation. In soils with low water potential, there is poor movement of rhizobia (Hamdi, 1971), and nodule development may be restricted to sites close to the crown. Gallacher and Sprent (1978) reported that the development of nodules initiated prior to the imposition of water stress is retarded by low water potential, although development is renewed when the stress is alleviated.

7.6.3 Soil Acidity

Soil acidity is a serious problem worldwide for agricultural production. According to Edwards et al. (1991), acid soils constrain agricultural production in more than 1.5 Gha worldwide, with the scope of the problem likely to increase as a result of acid rain, long-term N fertilization, and legume N_2 fixation. Theoretically, soil acidity is expressed by the concentration of hydrogen and aluminum ions. However, for crop production, soil acidity is much broader, including deficiency of macro- and micronutrients and low microbial activity (Graham, 1992; Graham and Vance, 2000). Legumes are particularly affected, acidity limiting both the survival and persistence of nodule bacteria in the soil and the process of nodulation itself (Graham and Vance, 2000).

Frame and Newbould (1986) reported that the growth of white clover under symbiotic conditions was almost nil at pH 4.0 and increased linearly from pH 4.0 to 6.0 and nodulation was reduced below pH 5.0. For most legumes to grow well and fix atmospheric N, pH values should be around 6.0. When the pH is lower than 5.5, Al toxicity may be a problem for legume growth and dinitrogen fixation (Frame and Newbould, 1986). In addition, at a lower pH (<5.0), P availability may be a problem due to the fixation of this element by Al and Fe oxides in acid soils, which may also limit dinitrogen fixation. Immobilization of Mo can not only cause apparent deficiency of this element in acid soils, but also can reduce nodulation in legumes. Adverse effects of Mo deficiency in dry beans grown in South America have been reported by Graham (1981).

7.6.4 Soil Fertility

Supply of an adequate rate of essential nutrients and in proper proportion is an important factor in determining the growth and yield of crops. In addition, some nutrients are essential for the development of nodules and, consequently, N_2 fixation. N-fixing bacteria have a relatively high requirement for P, S, Mo, and Fe because these nutrients are either part of the nitrogenase molecule or they are needed for its synthesis and use (Brady and Weil, 2002). However, higher amounts of NO_3–N and NH_4^+–N in the soil may limit the biological N fixation (Brady and Weil, 2002). In higher amounts, both forms of N inhibit nodule formation as well as nitrogenase activity (Layzell and Moloney, 1994). Bauer (1981) reported that host plants are able to sever their association with rhizobia quite dramatically when external N is plentiful. Thornton (1936) observed that exogenous nitrate caused fewer root hairs and fewer curled root hairs to be produced on alfalfa plants. He suggested that the reduced numbers of curled root hairs were responsible for the lower number of nodule forms in the presence of nitrate. Munns (1968a,b,c,d) confirmed and extended Thorn's observations in a classical series of papers. Munns concluded that the reduction in nodule number could not be attributed solely or even chiefly to interference by nitrate with any one of the nodulation processes. However, as indicated previously, nitrate caused both a reduction in the number of infection threads and an increase in the proportion of abortive infection threads (Munns, 1968c). Moreover, nitrate was able to delay nodulation by causing a delay in the formation of infected threads (Nutman, 1959).

Nitrate N has long been recognized as an inhibitor of nodulation since it is a highly efficient metabolic process for the legume (Bhangoo and Albritton, 1976). Peoples and Herridge (1990) reported that, in a soil with higher concentrations of mineral N, the legume can compensate for poor N_2 fixation by scavenging N from the soil. Although production in this situation may not be impaired, the net result of cropping with a legume with deficient nodulation is an exploitation of N reserves. Soil N fertility is lost and the potential benefit of the legume in a cropping sequence will not be realized (Peoples and Herridge, 1990).

Nitrate N is reported to be responsible for the decrease in phloem sap supply to nodules. It is well known that nitrate N alters the pattern of photosynthate partitioning in legume plants (Streeter, 1988), resulting in a decrease in phloem sap supply to nodules within 24–36 h of NH_3^- exposure (Vessey et al., 1988a,b). Minchin et al. (1989) suggested that NH_3^- may have an osmotic effect and inhibit nodule metabolism by increasing the diffusion barrier resistance within the nodule cortex. Kanayama et al. (1990) have reported that in soybean nodules, NH_3^- is converted first into nitrite and then into nitric oxide, which binds to leghemoglobin to form nitrosylleghemoglobin, a form of O_2-binding protein that is unable to facilitate the diffusion of O_2 to the bacteroids in the infected cells. They have proposed that nitrosylleghemoglobin formation may increase the resistance to O_2

diffusion in support of bacteroid respiration and thereby account for the greater O_2 limitation of nodule metabolism in NH_3^--inhibited nodules.

Excess N fertilizer may inhibit biological N fixation in legumes. However, a small amount of mineral N is needed by clover until nodules are formed and N fixation commences (Frame and Newbould, 1986). Hence, the use of starter fertilizer N is essential in soils of low fertility to improve N_2 fixation. Bethlenfalvay et al. (1978), Streeter (1988), and Waterer and Vessey (1993) also reported that both nodule growth and N_2 fixation of plants grown with a low N concentration in the root zone were stimulated compared with those without mineral N. Frame and Newbould (1986) also reported that the element needs of white clover and/or *Rhizobium,* especially for the N fixation process, include Co, Cu, Mg, Mn, Mo, Ni, B, Zn, S, and possibly Se. The addition of lime to raise the pH to 5.5 and above can markedly affect the availability of some essential nutrients (Mengel et al., 2001). In Australia and New Zealand, deficiencies of Co, Mo, S, and Zn are widespread, and the response of legumes to applications of one or all of these elements can be dramatic (Frame and Newbould, 1986).

Phosphorus deficiency is widespread in the bean-producing regions of South America and is perhaps the factor that most limits N_2 fixation on small farms (Graham, 1981). Graham and Rosas (1979) and Graham (1981) reported that among essential plant nutrients, the availability of P in adequate amounts is most important in biological N fixation. Plants engaged in symbiotic N_2 fixation generally have a higher requirement for P than those grown with N fertilization (Robson, 1983; Jungk, 1998). Higher ATP requirements for nitrogenase function (Ribet and Drevon, 1996; Al Niemi et al., 1997), plus P needs for signal transduction, membrane biosynthesis, and nodule development and function, contribute to this requirement (Graham and Vance, 2000). In consequence, the P concentration of pea and soybean nodules can be as high as 6 mg g^{-1} dry weight, while that of shoot P is only 2–3 mg g^{-1} dry weight (Israel, 1987). In low-P environments, P fertilization of N_2-dependent species leads to an increase in nitrogenase activity, nodule number, and mass and plant N accumulation (Robson et al., 1981; Israel, 1987). In clover and pea, the stimulation of N_2 fixation following P addition is through rapid enhancement of shoot growth, with a resultant influence on nodule parameters (Robson et al., 1981), whereas P deficiency in soybean and Stylosanthes appears to impact nodule function more directly (Gates, 1974; Israel, 1993).

7.6.5 CONCENTRATION OF OXYGEN

The two main reasons for lower oxygen levels in soils are waterlogging and poor structure, and under both conditions, poor nodulation has been reported (Gibson, 1977). Gibson (1977) reviewed the literature and reported poor nodulation of subterranean clover growing in soils with O_2 diffusion rates of 8×10^{-1} g cm^2 min^{-1} compared to soils having O_2 diffusion rates of 20×10^{-1} g cm^2 min^{-1}.

7.6.6 DROUGHT

Drought is a serious problem worldwide in crop production. Drought influences many physiological and biochemical processes in the plant, including legume nodulation. It is reported in the literature that nitrogenase activity is much more sensitive than photosynthesis (Durand et al., 1987; Sinclair et al., 1987; Layzell and Moloney, 1994). Water stress has been reported to be responsible for the collapse of lenticels (Pankhurst and Sprent, 1975), decrease in the respiratory capacity of bacteroids (Guerin et al., 1990), or decline in the leghemoglobin content of nodules (Guerin et al., 1990, 1991). However, the report available in the literature suggests that there is a significant variation in N_2 fixation among soybean genotypes under water stress (Sall and Sinclair, 1991). Hence, genotype selection is an important strategy in improving biological N fixation in legumes under drought stress. Sprent (1976) reported that both the nodule number and size were reduced by water deficit, with the specific nodule activity reduced by almost 90% in stressed plants.

Rhizobia populations have been shown to decrease during periods of drought in the absence of host plants (Furseth et al., 2011). Even though certain strains may react slightly differently, the correlation between a decreasing population and decreasing soil moisture is strong (Pena-Cabriales and Alexander, 1979). Rhizobium populations decline rapidly at first, then slowly after a period of time. Interestingly, when the soil is moistened, the population will stabilize, but not increase, then continue to decrease on further drying (Pena-Cabriales and Alexander, 1979). These results suggest that periods of drought may be more hazardous to rhizobia populations than periodic flooding, particularly in situations with no host plants (Furseth et al., 2011). Furseth et al. (2011) reported that the use of seed-applied rhizobia inoculants following severe flooding is not likely to produce a positive seed yield or quality under soil and environmental conditions.

7.6.7 CARBON DIOXIDE

Increasing CO_2 levels in the legume plant ambient improve the growth of legumes and, consequently, nodulation and N fixation (Gibson and Jordan, 1983). Improvements in N fixation with an increasing CO_2 concentration in the atmosphere by red clover (*Trifolium pretense* L.) (Wilson et al., 1933), white clover (*Trifolium repens* L.) (Masterson and Sherwood, 1978), soybean (*Glycine max* L. Merr.) (Gibson and Jordan, 1983), and pea (*Pisum sativum*) (Phillips et al., 1976) have been reported. The positive response can be attributed to higher levels of photosynthesis in the presence of increased ambient CO_2 and presumably to increased photosynthate supply to the roots (Gibson and Jordan, 1983).

7.6.8 INFLUENCE OF OTHER MICROORGANISM ON RHIZOBIA

The production of substances toxic to rhizobia by various soil organisms has been attributed to the poor colonization of *Rhizobium trifolii* in some soils (Gibson and Jordan, 1983). Actinomycetes may affect colonization and nodulation by *Rhizobium japonicum*, but not all actinomycetes are antagonistic to rhizobia (Antoun et al., 1978).

7.6.9 PATHOGENS AND PREDATORS

Various pathogens and predators such as fungi, viruses, nematodes, and insects can adversely affect the legume–*Rhizobium* symbiosis. Soil fungi may have a direct effect on nodulation through their inhibition of rhizobia (Gibson, 1974). Tu (1979) reported that nodule formation in soybean by *Rhizobium japonicum* was reduced by the pathogen *Phytophthora megasperma*. Tu et al. (1970) also reported that soybean plants infected with mosaic and bean pod mottle viruses showed reduced nodulation, and lower levels of N_2 fixation, with some evidence of host variety differences in the magnitude of the effect. Similarly, reductions in N_2 fixation by white clover (*Trifolium repens* L.) and dry bean (*Phaseolus vulgaris* L.) infected with a bean yellow mosaic virus have been reported by Smith and Gibson (1960) and Orellana and Fan (1978). Gibson and Jordan (1983) reported that although virus particles have been found in nodules, it is likely that the effect of viral infection on nodulation and N fixation is a consequence of the reduced physiological vigor of the host plants. Barker et al. (1972) reported that roots infected with nematodes can reduce the nodulation of various legumes because nematodes prefer nodule tissue rather than root tissue of host plants. Soybean N_2 fixation has been reduced when roots were infected by nematodes (Barker et al., 1972). Gibson and Jordan (1983) reviewed the literature on the effect of insects on nodulation of many legumes, including the faba bean (*Vicia faba* L.), pea (*Pisum sativum* L.), and cowpea (*Vigna unguiculata* L.), and reported that nodule formation and, consequently, N_2 fixation was adversely affected. He further reported that insects prefer nodule tissue over root tissue.

7.6.10 AGRICULTURAL CHEMICALS

In modern agriculture, the use of herbicides, fungicides, and insecticides is fundamental to control weeds, diseases, and insects. Graham (1981) reported that herbicides and insecticides applied at the recommended rates do little damage to nodulation or to *Rhizobium* in soil. Similarly, Gibson (1977) reported that herbicides at normal application rates have little, if any, effect on free-living rhizobia, but a number of reports indicate that they may interfere with nodulation. Gibson and Jordan (1983) reported that herbicides and insecticides generally have little effect on the viability of rhizobia. However, Gibson (1977) reported that the effect of herbicide on nodulation depends on the herbicide and also on the legume species. Many herbicides affect root development or chlorophyll synthesis and it is not surprising that they may influence nodule formation and function (Gibson, 1977). Strains of rhizobia vary in their response to different fungicides and it has been suggested that the selection of a fungicide-resistant mutant may provide inoculant strains with a competitive advantage in field situations (Gibson and Jordan, 1983).

7.6.11 CROP SPECIES AND GENOTYPES WITHIN SPECIES

There is a significant difference in dinitrogen fixation among crop species and genotypes within species. For example, soybean in the Brazilian Cerrado region (central part) produced a good yield (>4000 kg ha^{-1}) when seeds were inoculated with appropriate rhizobium. However, dry bean could not produce maximum grain yield without the application of the chemical form of N. The author and collaborators conducted greenhouse and field experiments testing promising dry bean genotypes using rhizobia and N treatments. Under greenhouse conditions, shoot and grain yield was significantly influenced by N and genotype treatments and N × genotype interactions were also significant for these two traits (Tables 7.4 and 7.5).

The N × G interaction indicates that responses of dry bean genotypes differ in the shoot dry weight and grain yield with the variation in N treatments (Table 7.4). The shoot dry weight at N_0 treatment varied from 3.26 to 10.47 g plant^{-1}, with an average value of 6.48 g plant^{-1}. At N_1 treatment, the variation in the shoot dry weight was from 4.45 to 10.92 g plant^{-1}, with an average value of 7.55 g plant^{-1}. Similarly, at N_2 treatment, the variation in the shoot dry weight was twofold between the lowest and highest shoot dry weight-producing genotypes. At N_3 treatment and across the four N levels, the minimum shoot dry weight was produced by the genotype BRSMG Talisma and the maximum shoot dry weight was produced by the genotype Diamante Negro. Overall, the maximum shoot dry weight was produced at the N_3 rate (200 mg N kg^{-1}) and the minimum shoot dry weight at the N_2 rate (inoculation with rhizobial strains + 50 mg N kg^{-1}). Overall, an increase in the shoot dry weight with the inoculation of dry bean genotypes with rhizobium strains was 16% and at 200 mg N kg^{-1} the increase in the shoot dry weight was 25% compared to control treatment. Variations in the shoot dry weight among dry bean genotypes have been reported by Fageria et al. (2011). These authors also reported the response of dry bean genotypes to N fertilization.

The grain yield of dry bean genotypes varied significantly by N and genotype treatments and the N × G interaction was also significant (Table 7.5). Across four N treatments, the minimum grain yield was produced by the genotype BRSMG Talisma and the maximum grain yield was produced by the genotype Diamante Negro. When the average values of N treatments were compared, the maximum grain yield was obtained with the addition of 200 mg N kg^{-1} soil and minimum grain yield with the treatment inoculation + 50 mg N kg^{-1} soil. The increase in grain yield with the inoculation of dry bean seeds with rhizobial strains was 7% compared to control treatment. The average increase in grain yield with the addition of 200 mg N kg^{-1} was 27% compared to 0 mg N kg^{-1} or control treatment. Overall, there was a significant decrease with the application of 50 mg N kg^{-1} + inoculation. Hence, there was a negative interaction between rhizobium and N application in small amounts. The grain yield results clearly show that dry bean genotypes did not

TABLE 7.4

Shoot Dry Weight (g Plant^{-1}) of 15 Dry Bean Genotypes at Different N and Rhizobium Inoculation Treatments

Genotype	N_0	N_1	N_2	N_3	Average
Aporé	4.45hi	5.59cd	4.18gh	5.44 g	4.91h
Pérola	6.21defg	6.36c	7.57ab	9.27bcd	7.25de
BRSMG Talisma	4.83gh	6.49c	3.79h	3.47h	4.64h
BRS Requinte	6.74cde	6.83c	4.85fgh	10.85b	7.32de
BRS Pontal	3.26i	6.56c	5.72defg	7.98de	5.88g
BRS 9435 Cometa	6.62cdef	6.46c	4.71fgh	9.89bc	6.91ef
BRS Estilo	5.11fgh	6.48c	4.42fgh	6.97efg	5.74g
CNFC 10408	7.47bcd	10.89a	5.53efg	7.14ef	7.75cd
CNFC 10470	6.66cdef	6.38c	3.59h	9.10cd	6.43fg
Diamante Negro	10.47a	10.78a	7.23abcd	12.96a	10.36a
Corrente	6.23defg	4.45d	7.17abcd	6.19fg	6.01g
BRS Valente	5.26efgh	9.13b	5.93cdef	9.54bcd	7.46de
BRS Grafite	8.08bc	10.28ab	8.63a	10.19bc	9.29b
BRS Marfim	8.94ab	10.92a	7.48abc	5.99fg	8.33c
BRS Agreste	6.89cd	5.66cd	6.82bcde	6.28efg	6.41fg
Average	6.48c	7.55b	5.84d	8.08a	
F-test					
N levels (N)	*				
Genotypes (G)	*				
N × G	*				
CVN (%)	8.54				
CVG (%)	7.58				

Source: Adapted from Fageria, N. K. et al. 2014. *Commun. Soil Sci. Plant Anal.* 45:111–125.

*Significant at the 1% probability level. Means in the same column followed by the same letter are not significantly different at the 5% probability level by Tukey's test. Average values were compared in the same line for significant differences among N rates. N_0 = 0 mg N kg^{-1} (control); N_1 = 0 mg N kg^{-1} + inoculation with rhizobial strains; N_2 = inoculation with rhizobial strains + 50 mg N kg^{-1}; and N_3 = 200 mg N kg^{-1}.

produce maximum grain yield either with inoculation with rhizobial strains or with the inoculation plus addition of a small amount of N. Hence, the addition of N in the inorganic form is essential to obtain the maximum yield of dry bean.

The N × genotype significant interactions for the maximum root length and root dry weight were observed, indicating that some genotypes were highly responsive to N treatments while others were not (Tables 7.6 and 7.7). The maximum root length varied from 11 to 28.33 cm at 0 mg N kg^{-1}, 10.33 to 22.67 cm at 0 mg kg^{-1} + inoculation with rhizobia, 11.33 to 24.33 cm at inoculation + 50 mg N kg^{-1}, and 12 to 26.67 cm at 200 mg N kg^{-1} soil treatment. The average values for these treatments were 18.2 cm for 0 mg N kg^{-1}, 15.31 cm for inoculation treatment, 16.04 cm for inoculation + 50 mg N kg^{-1}, and 18.71 cm for 200 mg N kg$^{-1.}$ Across four N levels, the values of root length varied from 13.33 to 19.66 cm. Variations in the root length of dry bean genotypes have been reported by Fageria and Moreira (2011) and Fageria (2009).

The root dry weight varied significantly among bean genotypes at different N treatments (Table 7.7). It was significantly higher at 200 mg N kg^{-1} soil treatment compared to other N treatments. The increase in the root dry weight was about 20% at 200 mg N kg^{-1} treatment compared to control treatment. Figures 7.3 and 7.4 show root growth at low and high N rates of two dry bean genotypes.

TABLE 7.5

Grain Yield (g Plant⁻¹) of 15 Dry Bean Genotypes at Different N and Rhizobium Inoculation Treatments

Genotype	N_0	N_1	N_2	N_3	Average
Aporé	7.19de	5.71g	8.54d	11.10de	8.13f
Pérola	11.30a	11.05bc	8.41d	13.16bc	10.98bc
BRSMG Talisma	8.90bcd	5.28g	5.42e	4.57g	6.04g
BRS Requinte	11.71a	10.37cd	8.13d	16.56a	11.69ab
BRS Pontal	5.57e	11.00bc	8.31d	14.04b	9.73de
BRS 9435 Cometa	10.50abc	9.20cde	8.62d	9.54ef	9.46e
BRS Estilo	7.28de	11.25bc	9.09cd	12.67bcd	10.07cde
CNFC 10408	8.82bcd	10.03cde	10.84bc	7.71f	9.35e
CNFC 10470	8.55cd	13.68a	9.60cd	14.29b	11.53ab
Diamante Negro	8.55cd	13.58a	13.33a	14.25b	12.43a
Corrente	11.38a	8.36de	8.90cd	11.48cde	10.03cde
BRS Valente	7.59de	12.90ab	5.76e	13.49b	9.93de
BRS Grafite	11.22a	6.10fg	9.67cd	10.83de	9.45e
BRS Marfim	10.88ab	8.12ef	8.40d	10.57e	9.49de
BRS Agreste	8.71cd	10.79c	11.71ab	10.70de	10.47cd
Average	9.21c	9.83b	8.98c	11.66a	
F-test					
N levels (N)	**				
Genotypes (G)	**				
N × G	**				
CVN (%)	3.39				
CVG (%)	7.03				

Source: Adapted from Fageria, N. K. et al. 2014. *Commun. Soil Sci. Plant Anal.* 45:111–125.

**Significant at the 1% probability level. Means in the same column followed by the same letter are not significantly different at the 5% probability level by Tukey's test. Average values were compared in the same line for significant differences among N rates. N_0 = 0 mg N kg⁻¹ (control); N_1 = 0 mg N kg⁻¹ + inoculation with rhizobial strains; N_2 = inoculation with rhizobial strains + 50 mg N kg⁻¹; and N_3 = 200 mg N kg⁻¹.

Variability in root growth among crop species and among genotypes of the same species is widely reported in the literature (O'Toole and Bland, 1987; Gregory, 1994; Fageria, 2009). This variability can be used in improving the yield of annual crops by incorporating vigorous root growth into desirable cultivars. Vigorous root growth is especially important when nutrient and water stresses are significant (Gregory, 1994). Ludlow and Muchow (1990), in their review of traits likely to improve yields in water-limited environments, place a vigorous rooting system high in their list of properties to be sought.

The genotypic variability in root growth of annual crops has been used to identify superior genotypes for drought-prone environments (Hurd et al., 1972; Gregory, 1994). Gregory and Brown (1989) reviewed the role of root characters in moderating the effects of drought and concluded that roots may have a direct effect, by increasing the supply of water available to the crop, or an indirect effect by changing the rate at which the supply becomes available. Where crops are grown on deep soils and water is stored throughout the whole soil profile, the depth of rooting has a major influence on the potential supply of water (Gregory, 1994). Rain may replenish the upper soil during the season, but later growth and grain filling in many crops are accomplished during periods of low rainfall when soil moisture stored deep in the profile must be utilized. Sponchiado et al. (1989)

TABLE 7.6

Maximum Root Length (cm) of 15 Dry Bean Genotypes at Different N and Rhizobium Inoculation Treatments

Genotype	N_0	N_1	N_2	N_3	Average
Aporé	21.33bcd	13.00cd	17.00bcd	22.33abc	18.41abcd
Pérola	25.33ab	11.67cd	11.33d	23.33ab	17.bcd
BRSMG Talisma	19.67cd	14.00cd	20.33abc	12.00d	16.50cde
BRS Requinte	24.33abc	13.33cd	21.67ab	26.67a	21.50a
BRS Pontal	18.33de	11.67cd	24.33a	14.00vd	17.08bcd
BRS 9435 Cometa	18.33de	17.33abc	12.33d	21.00abcd	17.25bcd
BRS Estilo	21.33bcd	22.67a	17.00bcd	17.67abcd	19.66ab
CNFC 10408	18.67d	21.00ab	11.67d	16.33bcd	16.91bcd
CNFC 10470	13.67ef	17.33abc	11.67d	19.00abcd	15.41def
Diamante Negro	13.33f	21.00ab	15.67bcd	16.67bcd	16.67bcde
Corrente	12.00f	14.00cd	17.33bcd	21.33abc	16.17def
BRS Valente	11.00f	12.33cd	14.33cd	15.67bcd	13.33f
BRS Grafite	28.33a	15.33bcd	14.00cd	19.67abcd	19.33abc
BRS Marfim	13.67ef	10.33d	14.33cd	16.33bcd	13.67ef
BRS Agreste	13.67ef	14.67cd	17.67abcd	18.67abcd	16.16def
Average	18.20a	15.31b	16.04b	18.71a	
F-test					
N levels (N)	**				
Genotypes (G)	**				
$N \times G$	**				
CVN (%)	17.94				
CVG (%)	13.07				

Source: Adapted from Fageria, N. K. et al. 2014. *Commun. Soil Sci. Plant Anal.* 45:111–125.
**Significant at the 1% probability level. Means in the same column followed by the same letter are not significantly different at the 5% probability level by Tukey's test. Average values were compared in the same line for significant differences among N rates. $N_0 = 0$ mg N kg^{-1} (control); $N_1 = 0$ mg N kg^{-1} + inoculation with rhizobial strains; $N_2 =$ inoculation with rhizobial strains + 50 mg N kg^{-1}; and $N_3 = 200$ mg N kg^{-1}.

reported that in dry bean, drought avoidance results from root growth and soil water extraction deep in the profile. N fertilization may increase crop root growth by increasing soil N availability (Fageria and Moreira, 2011). N also improves the production of lateral roots and root hairs, as well as increasing the rooting depth and root length density deep in the profile (Hansson and Andren, 1987). Hoad et al. (2001) reported that the surface application of N fertilizer increases root densities in the surface layers of the soil. Figure 7.1 shows the influence of N and rhizobia treatments on root growth of a dry bean genotype.

Fageria et al. (2012) studied the influence of N and rhizobia treatments on grain and straw yield of 15 dry bean genotypes under N and rhizobia treatments under field conditions (Tables 7.8 and 7.9). Grain and straw yields were significantly influenced by N and rhizobium and genotype treatments. The G × N interaction was also significant for these two traits. Hence, genotype response varied with the variation in N treatments. In the first year, overall, the grain yield varied from 1011.53 kg ha^{-1} produced at control treatment to 3463.53 kg ha^{-1} produced at 120 kg N ha^{-1}. The increase in grain yield was 38% with inoculation compared to control treatment. The overall increase in grain yield was 59% with the addition of 50 kg N ha^{-1} + inoculant. Similarly, the overall increase in grain yield was 81% with the addition of 120 kg N ha^{-1} compared to control treatments. In the second year,

TABLE 7.7

Root Dry Weight (g Plant^{-1}) of 15 Dry Bean Genotypes at Different N and Rhizobium Inoculation Treatments

Genotype	N_0	N_1	N_2	N_3	Average
Aporé	0.44bcdef	0.23def	0.26ef	0.58bcde	0.37cde
Pérola	0.33def	0.37bcd	0.26ef	0.56bcde	0.38cde
BRSMG Talisma	0.56bc	0.19f	0.29de	0.54bcde	0.39bcde
BRS Requinte	0.52bcde	0.25cdef	0.42abc	1.13a	0.58a
BRS Pontal	0.49bcdef	0.32cdef	0.47ab	0.51bcde	0.45bc
BRS 9435 Cometa	0.55bcd	0.39bcde	0.25ef	0.83ab	0.50ab
BRS Estilo	0.52bcde	0.38abc	0.31cde	0.57bcde	0.44bcd
CNFC 10408	0.28f	0.49ab	0.29de	0.24e	0.33e
CNFC 10470	0.27f	0.21ef	0.17f	0.21e	0.21f
Diamante Negro	0.66ab	0.54a	0.42abc	0.69bcd	0.57a
Corrente	0.41cdef	0.29cdef	0.36bcde	0.42cde	0.37cde
BRS Valente	0.31ef	0.32cdef	0.29de	0.41cde	0.33de
BRS Grafite	0.80a	0.39abc	0.38bcd	0.78abc	0.58a
BRS Marfim	0.40cdef	0.40abc	0.50a	0.47bcde	0.44bcd
BRS Agreste	0.41cdef	0.28cdef	0.27def	0.35de	0.33e
Average	0.46b	0.33c	0.33c	0.55a	
F-test					
N levels (N)	**				
Genotypes (G)	**				
N × G	**				
CVN (%)	19.89				
CVG (%)	18.72				

Source: Adapted from Fageria, N. K. et al. 2014. *Commun. Soil Sci. Plant Anal.* 45:111–125.

**Significant at the 1% probability level. Means in the same column followed by the same letter are not significantly different at the 5% probability level by Tukey's test. Average values were compared in the same line for significant differences among N rates. N_0 = 0 mg N kg^{-1} (control); N_1 = 0 mg N kg^{-1} + inoculation with rhizobial strains; N_2 = inoculation with rhizobial strains + 50 mg N kg^{-1}; and N_3 = 200 mg N kg^{-1}.

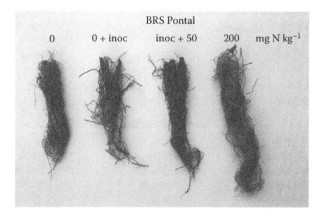

FIGURE 7.1 Root growth of dry bean genotype BRS Pontal at different N treatments. Left to right, 0 mg N kg^{-1}, 0 mg N kg^{-1} + inoculant, inoculant + 50 mg N kg^{-1}, and 200 mg N kg^{-1}. (Adapted From Fageria, N. K. et al. 2014. *Commun. Soil Sci. Plant Anal.* 45:111–125.)

TABLE 7.8

Grain Yield (kg ha⁻¹) of 15 Dry Bean Genotypes as Influenced by N + Rhizobium and Genotype Treatments

Genotype	0 kg N ha⁻¹	0 kg N ha + Inoculant	Inoculant + 50 kg N ha⁻¹	120 kg N ha⁻¹
		1st Year		
Aporé	1912.71b	3216.91ab	4086.29a	2688.52fg
Pérola	1729.63bc	3172.47ab	3660.37abc	3229.01cdef
BRSMG Talisma	1883.82bc	2642.34abcd	2953.70bcd	3103.33defg
BRS Requinte	2007.78ab	2846.66abcd	2872.71bcd	2652.34fg
BRS Pontal	2019.75ab	1870.37cd	2848.14bcd	3889.50bcd
BRS 9435 Cometa	1150.37c	2583.45abcd	2488.89d	4135.43ab
BRS Estilo	1954.69ab	1800.74cd	2673.45cd	4854.93a
BRS Notável	1759.75b	2250.49bcd	2472.96d	3434.81bcdef
BRS Ametista	2090.47ab	2823.45abcd	2939.50bcd	3731.60bcde
Diamante Negro	2517.77a	2390.49bcd	3861.97ab	2869.25efg
Corrente	1609.75bc	3674.56a	2700.99cd	4083.95abc
BRS Valente	2068.52ab	1921.23cd	2578.51d	2720.12fg
BRS Grafite	2043.21ab	1737.53d	2841.11bcd	2323.08 g
BRS Marfim	1937.04ab	2903.21abc	3288.52abcd	4192.71ab
BRS Agreste	1659.25bc	3227.40ab	3471.48abcd	4044.44abc
Average	1911.53d	2639.39c	3030.07b	3463.53a
		2nd Year		
Aporé	2856.66b	3375.60a	3373.45a	3752.75a
Pérola	2677.03bc	3062.21abc	3132.33abc	3196.97abc
BRSMG Talisma	2220.66de	2474.94def	2517.27abcd	2532.34bcd
BRS Requinte	2651.11bcd	2848.49bcd	2386.67bcd	3424.20ab
BRS Pontal	2655.22bc	2707.40bcd	2599.50abcd	2504.81bcd
BRS 9435 Cometa	2278.39cde	1770.49gh	2702.85abc	3639.13a
BRS Estilo	3375.18a	2976.66abc	3256.79abc	3227.11abc
BRS Notável	2358.64cde	3141.63ab	2714.07abc	3191.85abc
BRS Ametista	2613.23bcde	2114.26fg	2845.80abc	3355.80ab
Diamante Negro	1638.02f	1488.17 h	1687.53d	2096.54d
Corrente	2291.58cde	2621.23cde	2935.18abc	3066.02d
BRS Valente	2499.26bcde	2367.90def	2965.06abc	2985.92abcd
BRS Grafite	2216.91e	2586.99cdef	2367.90cd	2990.00abcd
BRS Marfim	2700.00bc	2802.26bcd	3343.18ab	3452.43ab
BRS Agreste	1769.51f	2127.24efg	2435.30abcd	2260.49cd
Average	2453.42d	2564.36c	2750.86b	3045.09a

Average of first year = 2751.62a

Average of second year = 2703.44b

		Average of Two Years		
Aporé	2384.69ab	3296.25a	3729.87a	3220.63bcde
Pérola	2203.33bc	3117.34abc	3396.35ab	3212.99bcde
BRSMG Talisma	2052.24bcd	2558.64bcdef	2735.53bc	2817.84cde
BRS Requinte	2329.44abc	2847.58abcd	2629.69c	3038.27cde
BRS Pontal	2337.49abc	2288.89def	2723.82bc	3197.16bcde
BRS 9435 Cometa	1714.38d	2176.97ef	2595.87c	3887.28ab
BRS Estilo	2664.94a	2388.71def	2965.12bc	4041.02a
BRS Notável	2059.19bcd	2696.06abcde	2593.52c	3313.33abcd

TABLE 7.8 (continued)
Grain Yield (kg ha⁻¹) of 15 Dry Bean Genotypes as Influenced by N + Rhizobium and Genotype Treatments

Genotype	0 kg N ha⁻¹	0 kg N ha + Inoculant	Inoculant + 50 kg N ha⁻¹	120 kg N ha⁻¹
BRS Ametista	2351.86abc	2468.85cdef	2892.65bc	3543.70abc
Diamante Negro	2077.90bcd	1939.33f	2774.75bc	2482.90e
Corrente	1950.66cd	3147.90ab	2818.08bc	3574.98abc
BRS Valente	2283.88abc	2144.57ef	2771.79bc	2853.02cde
BRS Grafite	2130.06bc	2162.26ef	2604.50c	2656.54de
BRS Marfim	2318.51abc	2852.73abcd	3315.85ab	3822.57ab
BRS Agreste	1714.38d	2677.33abcde	2953.39bc	3152.47bcde
Average	2171.53d	2584.22c	2900.05b	3254.31a
F-Test				
Year (Y)	*			
N + inoculant (N)	**			
Y × N	**			
Genotype (G)	**			
Y × G	**			
N × G	**			
Y × N × G	**			
CVY (%)	5.88			
CVN (%)	8.86			
CVG (%)	10.77			

Source: Adapted from Fageria, N. K. et al. 2012. Response of dry bean genotypes to nitrogen and rhizobia under field conditions. Paper presented at the Soil Fertility Meeting, 17–21 September 2012, Maceio, Brazil, Brazilian Soil Science Society/Federal University of Alagoas, Brazil.

*,**Significant at the 5% and 1% probability levels, respectively. Means followed by the same letter in the same column do not differ significantly at the 5% probability level by Tukey's test. Average values are compared in the same line.

the overall increase in grain yield was 5% at 0 kg N ha⁻¹ + inoculant treatment, 12% at the inoculant + 50 kg N ha⁻¹, and 24% at 120 kg N ha⁻¹ treatment compared to control treatment.

The average values of 2 years showed that the increase in grain yield across 15 dry bean genotypes compared to control treatment was 19% at 0 kg N ha⁻¹ + inoculant treatment, 34% at inoculant + 50 kg N ha⁻¹, and 50% at 120 kg N ha⁻¹. From the grain yield data, it can be concluded that inoculation improved the grain yield of dry bean genotypes but the increase was much higher when N was added. Hence, inoculation alone is not sufficient to achieve maximum yield of dry beans grown in Brazilian Oxisol and chemical N fertilization is essential. An improvement in dry bean yield with the addition of N has been reported by Soratto et al. (2004) and Pelegrin et al. (2009). These authors reported that dry bean produced maximum grain yield with the addition of 130 kg N ha⁻¹ in conventional cultivation and 182 kg N ha⁻¹ in the conservation or no-tillage system. Silveira and Damasceno (1993) recommended 72 kg N ha⁻¹ to achieve the maximum grain yield of dry bean. The improvement in grain yield of dry bean with the inoculation of seeds with *Rhizobium tropici* was also reported by Pelegrin et al. (2009) in a Brazilian Oxisol. However, the yield was 20% higher when 160 kg N ha⁻¹ was added compared to inoculation treatment. These authors also reported that the maximum yield of dry bean was obtained with the addition of 119 kg N ha⁻¹. The grain yield varied from year to year, and this variation may be related to the variation in environmental conditions (Fageria, 1992).

TABLE 7.9
Straw Yield (kg ha⁻¹) of 15 Dry Bean Genotypes as Influenced by N + Rhizobium and Genotype Treatments

Genotype	0 kg N ha⁻¹	0 kg N ha + Inoculant	Inoculant + 50 kg N ha⁻¹	120 kg N ha⁻¹
		1st Year		
Aporé	3490.37abc	5647.40a	5151.11a	5000.74abc
Pérola	2868.14abcd	4258.52abc	5068.88ab	4399.26abc
BRSMG Talisma	1977.81abcd	3805.18abc	3863.70abcd	3429.63bc
BRS Requinte	2905.92abcd	3485.92abc	3697.03abcd	3935.55abc
BRS Pontal	3564.44ab	1868.14c	3185.92cd	3597.77bc
BRS 9435 Cometa	3462.22abc	2615.55bc	3687.41abcd	4879.25abc
BRS Estilo	2537.03bcd	2554.07bc	4175.55abc	3222.22c
BRS Notável	3025.18abcd	3489.63abc	4332.59abc	3054.07c
BRS Ametista	1980.74d	3712.59abc	3648.89bcd	4465.92abc
Diamante Negro	2106.66cd	3129.63abc	3347.40cd	3343.70c
Corrente	3857.77ab	3520.74abc	3452.59cd	5000.00abc
BRS Valente	2136.29cd	2065.18bc	2508.88d	3547.40bc
BRS Grafite	3174.81abcd	3115.55abc	3151.11cd	6146.67a
BRS Marfim	2772.59bcd	2500.74bc	3902.96abcd	4341.48abc
BRS Agreste	4210.37a	4674.81ab	4231.11abc	5850.37ab
Average	2938.02c	3362.91bc	3827.01ab	4280.93a
		2nd Year		
Aporé	1521.48bcd	1502.96ab	1517.03bc	1745.18bcd
Pérola	1946.66abcd	1982.22ab	2268.14abc	2231.11abcd
BRSMG Talisma	1180.00cd	2505.11ab	2377.77abc	1582.22d
BRS Requinte	1800.00bcd	1703.70ab	2727.77a	2087.40abcd
BRS Pontal	1541.48bcd	1665.92ab	2224.14abc	1891.11bcd
BRS 9435 Cometa	1686.67bcd	1652.59ab	1486.66bc	1842.96bcd
BRS Estilo	1860.74bcd	2064.44ab	2465.93abc	2454.07abcd
BRS Notável	1140.74d	1177.77b	1511.85bc	1908.89bcd
BRS Ametista	1881.48bcd	1470.37ab	1464.44c	1679.25cd
Diamante Negro	2424.44ab	2126.66ab	2218.52abc	2578.44ab
Corrente	1607.40bcd	2005.92ab	1904.44abc	2530.11abc
BRS Valente	2195.55abc	1783.70ab	2308.15abc	2508.52abc
BRS Grafite	2956.29a	2659.63ab	2671.11ab	2931.11a
BRS Marfim	2058.52abcd	1482.96ab	1385.14c	1831.85bcd
BRS Agreste	1917.03abcd	2837.03a	2033.00abc	2328.88abcd
Average	1847.90	1908.06	2037	2142.07

Average of first year = 3607.17a
Average of second year = 1983.41b

		Average of Two Years		
Aporé	2505.92ab	3575.18ab	3334.07ab	3372.96abc
Pérola	2407.40abc	3120.37abc	3668.51a	3315.18abc
BRSMG Talisma	1577.40c	3155.15abc	3120.74abc	2505.92c
BRS Requinte	2352.96abc	2594.81abc	3212.40abc	3011.48bc
BRS Pontal	2552.96ab	1767.04c	2705.03bc	2744.44bc
BRS 9435 Cometa	2574.44ab	2134.07bc	2587.03bc	3361.11abc
BRS Estilo	2198.89bc	2309.26abc	3320.74ab	2838.15bc
BRS Notável	2082.96bc	2333.70abc	2922.22abc	2481.48c

TABLE 7.9 (continued)
Straw Yield (kg ha⁻¹) of 15 Dry Bean Genotypes as Influenced by N + Rhizobium and Genotype Treatments

Genotype	0 kg N ha⁻¹	0 kg N ha + Inoculant	Inoculant + 50 kg N ha⁻¹	120 kg N ha⁻¹
BRS Ametista	1931.11bc	2591.48abc	2556.66bc	3072.59bc
Diamante Negro	2265.55abc	2628.15abc	2782.96abc	2961.07bc
Corrente	2732.59ab	2763.33abc	2678.51bc	3765.05abc
BRS Valente	2165.92bc	1924.44c	2408.51c	3027.96bc
BRS Grafite	3065.55a	2872.59abc	2911.11abc	4538.89a
BRS Marfim	2415.55ab	1991.85bc	2644.05bc	3086.67bc
BRS Agreste	3063.70a	3755.93a	3132.05abc	4089.62ab
Average	2392.86c	2644.49c	2932.31b	3211.50a
F-test				
Year (Y)	**			
N + inoculant (N)	**			
Y × N	**			
Genotype (G)	**			
Y × G	**			
N × G	**			
Y × N × G	**			
CVY (%)	18.58			
CVN (%)	22.49			
CVG (%)	19.52			

Source: Adapted from Fageria, N. K. et al. 2012. Response of dry bean genotypes to nitrogen and rhizobia under field conditions. Paper presented at the Soil Fertility Meeting, 17–21 September 2012, Maceio, Brazil, Brazilian Soil Science Society/Federal University of Alagoas, Brazil.

**Significant at the 1% probability level. Means followed by the same letter in the same column do not differ significantly at the 5% probability level by Tukey's test. Average values are compared in the same line.

Overall, the straw yield varied from 2938 to 4281 kg ha⁻¹ under different N + inoculation treatments in the first year. In the second year, the variation in straw yield was 1848–2142 kg ha⁻¹. When the straw yield values were averaged across 2 years, it varied from 2393 to 3212 kg ha⁻¹. Both the N and inoculation treatments improved the straw yield significantly compared to control treatment. The improvement in straw yield with the addition of N was related to the increase in leaf area as well as shoot dry weight (Fageria and Santos, 2008). Straw yield is genetically controlled as well as influenced by environmental factors, including mineral nutrition (Nelson and Larson, 1984; Rasmusson and Gengenbach, 1994; Fageria and Santos, 2008). Figures 7.2 through 7.8 show the growth of dry bean genotypes at different N levels. It is very clear from these figures that the growth of all the genotypes was better at inoculate + 50 kg N ha⁻¹ and 120 kg N ha⁻¹ compared to control and control + inoculate treatments.

7.7 STRATEGIES TO ENHANCE BIOLOGICAL NITROGEN FIXATION

The greatest success in terms of modified agricultural practices arising from scientific research on biological N fixation has undoubtly been the development of rhizobial inoculants (Giller and Cadisch, 1995; Seneviratne et al. 2000). Soybean has been the only example where there has been widespread adoption, primarily due to the relative specificity of soybean for rhizobia (Seneviratne et al., 2000). Inoculation of soybean is a significant agency for the manipulation of rhizobia for

FIGURE 7.2 Dry bean genotype Perola growth at different N treatments in the second year of experimentation. Left to right 0 kg N ha^{-1}, 0 kg N ha^{-1} + inoculation with rhizobium, inoculation with rhizobium + 50 kg N ha^{-1}, and 120 kg N ha^{-1}.

FIGURE 7.3 Growth of dry bean genotype BRSMG Talisma at different N treatments in the second year of experimentation. Left to right 0 kg N ha^{-1}, 0 kg N ha^{-1} + inoculation with rhizobium, inoculation with rhizobium + 50 kg N ha^{-1}, and 120 kg N ha^{-1}.

improving crop productivity and fertility (Keyser and Li, 1992). It can lead to the establishment of a large rhizobial population in the plant rhizosphere and to improvement in nodulation and N$_2$ fixation, even under adverse soil N conditions (Peoples et al., 1995a,b).

Biological N fixation by legumes is an important biological process that deserves special attention due to its economic and environmental impacts. Attempts to increase N$_2$ fixation will require optimization of this process under the diverse environmental conditions to which crop plants are exposed. Some of the strategies that can be adopted to optimize the N$_2$ fixation process are selection of an appropriate strain for inoculate, appropriate rhizosphere conditions surrounding the plants, and also planting efficient crop species or genotypes within species.

7.7.1 SELECTION OF APPROPRIATE STRAINS FOR INOCULATION

Thies et al. (1991) quantified the indigenous soil rhizobia population, finding greater responses to rhizobia inoculation when populations were <10 cells g^{-1} of soil, and little to no responses existed

FIGURE 7.4 Growth of dry bean genotype BRS 9435 Cometa at different N treatments in the second year of experimentation. Left to right 0 kg N ha^{-1}, 0 kg N ha^{-1} + inoculation with rhizobium, inoculation with rhizobium + 50 kg N ha^{-1}, and 120 kg N ha^{-1}.

with rhizobia populations above 100 cells g^{-1} soil. If the indigenous soil rhizobia population is not sufficient, the selection of a suitable strain of rhizobia for the inoculation of legume seeds is the first step in the production of legume inoculants. The basic criteria were to be (i) the ability to form effective N$_2$-fixing nodules with all the hosts for which the culture is recommended, (ii) the capacity to do this under a wide range of agroecological conditions, and (iii) the rhizobia introduced into the soil on inoculated seeds must multiply rapidly in the soil solution and rhizosphere in numbers sufficient to allow prompt nodulation of the seedling plant. Effective nodulation is still the main criterion used in strain selection and forms that basis for host grouping, but various ecological characteristics, such as the ability to form nodules in competition with native rhizobial populations and the ability to survive and multiply in the soil, now receive greater emphasis (Date and Roughley, 1977).

The success of inoculation in the field depends upon the inoculant quality (Brockwell and Bottomley, 1995), the procedure used (Brockwell et al., 1988), operator competence, compatibility

FIGURE 7.5 Growth of dry bean genotype BRS Estilo at different N treatments in the second year of experimentation. Left to right 0 kg N ha^{-1}, 0 kg N ha^{-1} + inoculation with rhizobium, inoculation with rhizobium + 50 kg N ha^{-1}, and 120 kg N ha^{-1}.

FIGURE 7.6 Growth of dry bean genotype Corrente at different N treatments in the second year of experimentation. Left to right 0 kg N ha⁻¹, 0 kg N ha⁻¹ + inoculation with rhizobium, inoculation with rhizobium + 50 kg N ha⁻¹, and 120 kg N ha⁻¹.

with fertilizers, and the presence of toxic agrichemicals (People et al., 1995b). Even with a high-quality inoculant and good inoculation practice, failure can occur because of environmental factors influencing the survival of rhizobia (Brockwell and Bottomley, 1995). For example, 4–5% of soybean inoculum was recovered from the soil 24 h after sowing at 28°C, but there was a <0.2% survival sowing at 38°C (Brockwell et al., 1987).

Selection of appropriate rhizobium for adverse environmental conditions is an important step in improving N₂ fixation in legumes. It is also a notable example of the selection of an acid-tolerant strain of *Rhizobium tropici* for dry bean in Brazilian acid soils (Hungaria et al., 1997). This change occurred within a relatively short time span following the introduction of beans to the region and has also been reported in some acidic soil areas of Africa (Anyango et al., 1995). There is significant

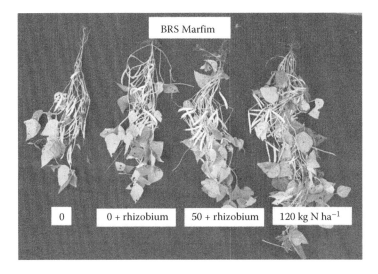

FIGURE 7.7 Growth of dry bean genotype BRS Marfim at different N treatments in the second year of experimentation. Left to right 0 kg N ha⁻¹, 0 kg N ha⁻¹ + inoculation with rhizobium, inoculation with rhizobium + 50 kg N ha⁻¹, and 120 kg N ha⁻¹.

FIGURE 7.8 Growth of dry bean genotype BRS Agreste at different N treatments in the second year of experimentation. Left to right 0 kg N ha⁻¹, 0 kg N ha⁻¹ + inoculation with rhizobium, inoculation with rhizobium + 50 kg N ha⁻¹, and 120 kg N ha⁻¹.

work to be done in identifying the additional acid-tolerant legume and rhizobial germplasm, as well as the deployment of acid-tolerant genes such as those that occur in *Rhizobium tropici* UMR1899 (Graham et al., 1994).

7.7.2 Adopting Appropriate Soil and Crop Management Practices

Adopting appropriate soil and crop management practices can improve biological N fixation in legumes. These practices include conservation tillage, crop rotation, planting higher N_2-fixing crop species or genotypes within species, pest, and disease control, plant nutrition and soil amelioration, and plant breeding and selection. A brief discussion of these practices in relation to biological N fixation is given in this section.

7.7.2.1 Conservation Tillage

Conservation tillage is widely adopted in many countries to save energy and conserve soil resources. Convention tillage may increase the oxidation of organic matter nitrification, which may increase nitrate N in the soil profile. The increase in nitrate level in the soil due to cultivation may be detrimental to biological N fixation as discussed earlier. Conservation tillage may create a favorable soil structure and improvement in the soil moisture and a favorable soil temperature. All these changes may bring favorable changes in the rhizosphere of legume crops for biological N fixation (Peoples and Herridge, 1990). Peoples and Herridge (1990) reported that soybean cultivated in subtropical Australia under conservation tillage improved nodulation and N_2 fixation compared with the cultivated system.

7.7.2.2 Crop Rotation

Peoples and Herridge (1990) reported that the quantity of soil nitrate available to a legume can be influenced by crop rotation. In addition to the effects on soil nitrate, different crop rotations can also influence biological N fixation through effects on legume growth (Peoples et al., 1992). Peoples and Herridge (1990) gave an example that soybean grown immediately after an oat crop fixed 244 kg N ha⁻¹ compared with 143 kg N ha⁻¹ fixed by soybean grown in previously fallowed soil. Similarly, Bergersen et al. (1989) reported that legumes can fix higher amounts of N and contributed more fixed N to the soil if sown immediately after a cereal crop than if grown in previously fallowed soil or immediately after a legume-based pasture.

7.7.2.3   Planting High N_2-Fixing Crop Species or Genotypes within Species

Planting high N_2-fixing crop species or genotypes within species is an important strategy in improving biological N fixation by legumes. There are evidences in the literature that the choice of species or cultivar significantly influences N_2 fixation in cropping systems (Wani et al., 1995; Peoples et al., 1995b). For example, in the Brazilian Cerrado region (central part), soybean in rotation with upland rice and corn can fix a significant amount of N that is not only sufficient for a good yield of soybean but that can also contribute to succeeding cereal crops. However, for planting dry bean in rotation with upland rice and corn, chemical N is required even for a maximum economic yield of dry bean because dry bean is considered as a poor N-fixing legume (Herridge and Danso, 1995).

Peoples et al. (1995a) reported that examples of variation in the genetic capacity of different species to grow and fix N under the same environmental conditions are presented for crop and pasture legumes, tree legumes, and green manures. In a comparison of upland food legumes in Thailand, for instance, black gram, green gram, and soybean all had similar growth patterns (64–73 days to maturity), but black gram appeared to be better adapted to the environment and fixed more N than the other species (Peoples et al. 1995b). However, in a related investigation in Australia, soybean had a much longer duration of growth than black gram (140 days vs. 95 days, respectively), and consequently had a greater growth potential and opportunity for N_2 fixation (Peoples et al., 1995b).

7.7.2.4   Pest and Disease Control

Pests and diseases influence legume crop growth, including leaf area and root growth, which influence the N_2 fixation capacity. Crop and pasture losses from 10% to >90% resulting from diseases have been reported (Johnston and Barbetti, 1987; Peoples et al., 1995a). Diseases or insects can be controlled by adopting appropriate crop rotation, by applying fungicides or insecticides at adequate rates and at the appropriate time, and by planting disease or insect-resistant crop species or genotypes and high-quality seeds (Peoples et al., 1995a).

7.7.2.5   Plant Nutrition and Soil Amelioration

Supply of adequate amounts and proportions enhances the growth and development of legume crops and, consequently, higher capacity of dinitrogen fixation. The root growth of most legumes has been significantly increased with the addition of essential macro- and micronutrients (Fageria and Moeira, 2011; Fageria, 2013). Similarly, the root growth of legumes such as dry bean, soybean, and tropical legume cover crops has been reported to improve with the addition of essential nutrients in low-fertility soils (Fageia, 2013). The root growth of most legumes improved with the addition of lime and gypsum in acid soils of the cerrado region of Brazil (Fageria and Moreira, 2011). Peoples et al. (1995) reported that applying starter N fertilizers to legumes may depress the N_2 fixation.

7.7.2.6   Plant Breeding and Selection

Enhancing crop legume dinitrogen fixation through plant breeding and selection is an important strategy. The potential for enhancing N_2 fixation through breeding and selection was recognized long ago (Phillips et al., 1971). However, progress has been slow (Herridge and Danso, 1995). Mytton (1983) and Graham and Temple (1984) noted that little attention had been given to enhancing N_2 fixation through breeding, and basic information necessary for understanding the expression of desired characters was largely absent. Herridge and Danso (1995) cited two main reasons for this apparent lack of success. First, it is a difficult task to combine a single, desirable trait such as N_2 fixation with other agronomic and yield traits. Second, still techniques for the accurate determination of N_2 fixed by legumes under field conditions are not available. To some extent, these observations remain relevant today. Herridge and Danso (1995) reported that for plant breeding and selection for enhanced N_2 fixation, the following criteria should be taken into consideration: (i) choice of traits as selection criteria that can be measured precisely and economically and allowing separation between efficient and inefficient genotypes, (ii) variability in legume germplasm in N_2 fixation, (iii) identification of genetically diverse parents, incorporating both agronomic and N_2 fixation, choice of selection units (individual plants or families) that facilitate the precise quantification of traits on interest and allow the production of progeny from selected plants, and (iv) use of a breeding procedure (i.e., mass selection, family selection) that provides maximum genetic gain for N_2 fixation and recombination with other desired agronomic traits.

Genetic variability among legume crop species in N fixation has been widely reported (Bliss, 1993; Herridge and Danso, 1995). Nutman (1984) concluded that N_2-fixing lines of the temperate forage species, red clover (*Trifolium pretense*), were superior because of an enlarged N_2 fixation system, rather than because of the increased efficiency of N_2 fixation. Superior plants nodulate earlier, leading to more and larger nodules. Piha and Munns (1987), Graham (1981), and Graham and Temple (1984) reported significant differences among dry bean genotypes in N_2 fixation. Bliss et al. (1989) and Bliss (1993) reported significant differences among bean genotypes in N_2 fixation. When considering future directions for breeding of dry bean, Bliss (1993) suggested that breeding plants with the capacity to nodulate and fix N_2 in the presence of soil nitrate, that is, nitrate tolerance, should be a priority. There is also evidence that variation exists for nitrate tolerance both in natural populations and in mutant lines (Park and Buttery, 1988, 1989). Other selection traits that may have merit in breeding programs for dry bean are early and late nodulation (Chaverra and Graham, 1992; Kipe-Nolt et al., 1993; Kipe-Nolt and Giller, 1993).

It may be possible to enhance N_2 fixation in a *Rhizobium*–legume symbiosis by selecting host-plant phenotypes as well as *Rhizobium* mutants. The number of root nodules is under some genetic control of the host plant. Two genes controlling the nodule number in peas have been identified, and the nodule number was correlated with the seed yield (Gelin and Blixt, 1964). Seetin and Barnes (1977) selected alfalfa genotypes with enhanced acetylene reduction activity in the presence of a commercial mixture of *Rhizobium* strains and transmitted this character to the progeny. That success in the greenhouse was extended to the field, whereas a ^{15}N technique showed that the elite alfalfa populations averaged 41% more annual N_2 fixation than a standard variety. Such results are quite promising from an applied point of view and definitely show that selection pressure on the symbiotic associations as a whole can be used to enhance N_2 fixation (Phillips, 1980).

Numerous studies have been reported that show the variation in N_2 fixation in soybean genotypes (Kucey et al., 1988; Herridge and Holland, 1992; Herridge and Danso, 1995). Variation in nitrate tolerance within a large and diverse germplasm collection of soybean was reported by Betts and Herridge (1987). Similarly, Serraj et al. (1992) also reported a large population of soybean genotypes for tolerance to nitrate. Plant mutagenesis was first used to generate peas with greatly enhanced nodulation and with a degree of nitrate tolerance (Jacobsen and Feenstra, 1984).

Fageria et al. (2014) studied the influence of N fertilization with or without rhizobial inoculation on growth of tops of 10 principal tropical cover crops. The N × cover crop interaction for shoot dry weight was significant, indicating that different responses of cover crops exist for varying levels of applied N and Bradyrhizobial inoculants (Table 7.10). At 0 mg N kg^{-1} (N_0) treatment, the shoot dry weight varied from 0.98 g $plant^{-1}$ produced by crotalaria to 10.89 g $plant^{-1}$ produced by lablab, with an average value of 5.53 g $plant^{-1}$. When seeds were treated with Bradyrhizobial inoculants (0 mg N kg^{-1} + inoculant or N_1), pueraria produced the minimal shoot dry weight and jack bean produced the maximal shoot dry weight. At N_2 treatment (100 mg N kg^{-1} + Bradyrhizobial inoculants or N_2), the differences in responses of these two cover crops were similar. However, when the N rate was increased to 200 mg kg^{-1} (N_3) treatment, pueraria produced the minimal shoot dry weight but the maximal shoot dry weight was produced by lablab. Across four N treatments, the maximal shoot dry weight was produced by the jack bean and the minimal shoot dry weight was produced by pueraria.

TABLE 7.10

Shoot Dry Weight of 10 Tropical Cover Crops as Influenced by N and Bradyrhizobial Inoculants

Cover Crops	Shoot Dry Weight (g Plant⁻¹)				
	N_0	N_1	N_2	N_3	Average
1. Crotalaria	0.98e	1.36de	1.95g	1.66e	1.48f
2. Smooth crotalaria	2.89bcd	3.05cd	3.84ef	3.56cd	3.33de
3. Showy Crotalaria	3.59bc	2.67cde	5.83d	4.00c	4.02cd
4. Calopo	2.27cde	3.01cd	2.78fg	2.13de	2.55e
5. Pueraria	1.25de	1.17e	1.21g	0.73e	1.09f
6. Pigeon pea	4.64b	4.46c	4.69de	3.56cd	4.33c
7. Lablab	10.89a	9.82b	11.89b	13.73a	11.58a
8. Black velvet bean	9.42a	9.16b	9.33c	13.37a	10.32b
9. Gray velvet bean	9.52a	11.90a	12.90b	12.66a	11.74a
10. Jack bean	9.85a	13.23a	16.73a	8.16b	11.99a
Average	5.53c	5.98bc	7.11a	6.36b	
F-test					
N rate (N)	**				
Cover crops (C)	**				
N × C	**				
CVN (%)	11.64				
CVC (%)	10.89				

Source: Adapted from Fageria, N. K. et al. (In press). *Commun. Soil Sci. Plant Anal.*

**Significant at the 1% probability level. Means in the same column followed by the same letter are not significantly different at the 5% probability level by Tukey's test. Average values were compared in the same line for significant differences among N rates. N_0 = 0 mg N kg⁻¹; N_1 = 0 mg N kg⁻¹ + Bradyrhizobial inoculants; N_2 = 100 mg N kg⁻¹ + Bradyrhizobial inoculants; and N_3 = 200 mg N kg⁻¹.

Overall, the maximal shoot dry weight was produced at N_2 treatment (100 mg N kg⁻¹ + inoculant). The increase in the shoot dry weight with the N_2 treatment was 29% compared to control treatment (N_0). Variation in the shoot dry weight among tropical legume cover crops has also been reported by Fageria et al. (2011). Similarly, improvement in the shoot dry weight of annual crops (cereals and legumes) with the addition of N was associated with an increase in the leaf area with the addition of N and improvement in the photosynthetic efficiency of plants (Marschner, 1995; Fageria et al. 2011). Engles and Marschner (1995) and Fageria et al. (2006) reported that N greatly influences the leaf growth, leaf area duration, and photosynthetic rate per unit leaf area to control the production of carbohydrates and other photosynthetic products (source activity), as well as the number and sizes of vegetative and reproductive storage organs (sink capacity).

Fageria et al. (2014) also studied the influence of N and rhizobia treatments on the maximum root length and root dry weight of 10 tropical legumes. The N × cover crops root length interaction was significant for root length (Table 7.11), indicating that some crop species were highly responsive to the applied N and Bradyrhizobial inoculants while others were not. In control treatment (N_0), maximal root length varied from 22 cm produced by showy crotalaria to 33 cm produced by smooth crotalaria, with an average value of 26.17 cm. At N_1 (0 mg N kg⁻¹ + Bradyrhizobial inoculants) and N_2 treatment (100 mg N kg⁻¹ + Bradyrhizobial inoculants), minimal root length was produced by the black velvet bean and maximum root length was produced by the pigeon pea. At N_3 (200 mg N kg⁻¹) treatment, the situation changed and the minimal root length of 20 cm was produced by the jack bean and maximum root length of 33.33 cm was produced by the gray velvet bean. Across four N

TABLE 7.11

Maximal Root Length of 10 Tropical Cover Crops as Influenced by N and Bradyrhizobial Inoculants

Cover Crops	Maximal Root Length (cm)				
	N_0	N_1	N_2	N_3	Average
1. Crotalaria	29.00ab	24.00cde	24.67cd	20.67e	24.58b
2. Smooth crotalaria	33.00a	25.33bcd	30.67ab	28.67abc	29.42a
3. Showy crotalaria	22.00d	21.67de	29.33bc	26.33cd	24.83b
4. Calopo	27.67bc	29.33ab	34.00ab	32.33ab	30.83a
5. Pueraria	23.33cd	24.33cde	24.00d	28.00bc	24.92b
6. Pigeon pea	30.00ab	30.67a	34.33a	23.00de	29.50a
7. Lablab	26.67bcd	29.33ab	31.67ab	26.33cd	28.50a
8. Black velvet bean	25.33bcd	20.00e	18.33e	32.33ab	24.00b
9. Gray velvet bean	27.67bc	29.33ab	30.67ab	33.33a	30.25a
10. Jack bean	23.67cd	27.67abc	21.00de	20.00e	23.08b
Average	26.17a	26.83a	27.87a	27.87a	26.99
F-test					
N rate (N)	NS				
Cover crops (C)	**				
N × C	**				
CVN (%)	9.27				
CVC (%)	6.75				

Source: Adapted from Fageria, N. K. (In press). *Commun. Soil Sci. Plant Anal.*

**, NS: Significant at the 1% probability level and not significant, respectively. Means in the same column followed by the same letter are not significantly different at the 5% probability level by Tukey's test. Average values were compared in the same line for significant differences among N rates. N_0 = 0 mg N kg^{-1}; N_1 = 0 mg N kg^{-1} + Bradyrhizobial inoculants; N_2 = 100 mg N kg^{-1} + Bradyrhizobial inoculants; and N_3 = 200 mg N kg^{-1}.

treatments, the minimal root length of 23.08 cm was produced by the jack bean and maximal root length of 30.83 cm was produced by calopo, with an average value of 26.99 cm. Variation in the root length is genetically controlled and varied among plant species and it is also influenced by environmental factors (Eghball et al., 1993; Costa et al., 2002; Fageria et al., 2006).

The root dry weight had a significant N × cover crop species interaction (Table 7.12), indicating a variation in the root dry weight with the variation in N and Bradyrhizobial inoculants. In control treatment (N_0), the shoot dry weight varied from 0.16 g plant^{-1} produced by pueraria to 2.01 g plant^{-1} produced by the gray velvet bean, with an average value of 0.72 g plant^{-1}. These two cover crops also produced minimal and maximal root dry weights at the N_1 (0 mg N kg^{-1} + inoculants) and N_2 (100 mg N kg^{-1} + inoculant) treatments. However, at the N_3 (200 mg N kg^{-1}) treatment, the minimum root dry weight was produced by crotalaria and the maximum root dry weight was produced by the black velvet bean. Across four N levels, the minimum root dry weight was produced by crotalaria and pueraria and the maximum root dry weight was produced by the gray velvet bean. Overall, the gray velvet bean produced about a 12-fold more root dry weight compared with the minimum root dry weight-producing cover crops such as crotalaria and pueraria. The root dry weight is an important trait in improving the organic matter content of the soil as well as in the absorption of water and nutrient (Sainju et al., 1998; Fageria et al., 2006). Vigorous root system also assimilates large amount of leachable nutrients from soil profile such as N and such nutrients are absorbed by the succeeding crops (Kristensen and Thorup-Kristensen, 2004; Fageria et al., 2011). The root dry weight was having a significant positive association with the shoot dry weight (Figure 7.9). Figures

TABLE 7.12

Root Dry Weight of 10 Tropical Cover Crops as Influenced by N and Bradyrhizobial Inoculants

Cover Crops	Root Dry Weight (g Plant⁻¹)				
	N_0	N_1	N_2	N_3	Average
1. Crotalaria	0.18d	0.13f	0.19d	0.14d	0.16e
2. Smooth crotalaria	0.23d	0.48ef	0.81bc	0.38d	0.47d
3. Showy Crotalaria	0.48bcd	0.51ef	0.77bc	0.43d	0.55d
4. Calopo	0.30cd	0.42ef	0.46cd	0.31d	0.37d
5. Pueraria	0.16d	0.15f	0.15d	0.16d	0.16e
6. Pigeon pea	0.73b	0.68de	0.48cd	0.21d	0.53d
7. Lablab	0.74b	1.42bc	1.72a	1.64b	1.38b
8. Black velvet bean	1.66a	1.74ab	0.50cd	2.08a	1.49b
9. Gray velvet bean	2.01a	1.91a	1.84a	1.71ab	1.87a
10. Jack bean	0.65bc	1.10cd	1.12b	0.94c	0.95c
Average	0.72a	0.85a	0.80a	0.80a	0.79
F-test					
N rate (N)	NS				
Cover crops (C)	**				
N × C	**				
CVN (%)	25.59				
CVC (%)	19.71				

Source: Adapted from Fageria, N. K. et al. (In press). *Commun. Soil Sci. Plant Anal.*

**, NS: Significant at the 1% probability level and not significant, respectively. Means in the same column followed by the same letter are not significantly different at the 5% probability level by Tukey's test. Average values were compared in the same line for significant differences among N rates. N_0 = 0 mg N kg⁻¹; N_1 = 0 mg N kg⁻¹ + Bradyrhizobial inoculants; N_2 = 100 mg N kg⁻¹ + Bradyrhizobial inoculants; and N_3 = 200 mg N kg⁻¹.

7.10 through 7.12 show the root growths of showy crotalaria, calopo, and lablab, respectively. The root growth varied with N treatments and overall, a more vigorous root system was produced at the N_2 (100 mg N kg⁻¹ + inoculant) treatment compared with the other three N treatments in three cover crops. Baligar et al. (1998) reported that the root dry weight of legume crops was higher with the addition of N compared with without N application treatment.

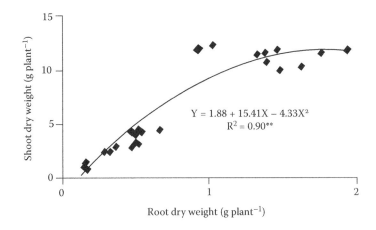

$$Y = 1.88 + 15.41X - 4.33X^2$$
$$R^2 = 0.90**$$

FIGURE 7.9 Relationship between the root dry weight and shoot dry weight of tropical legume cover crops.

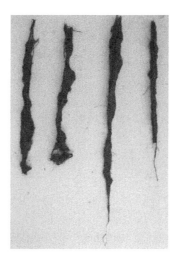

FIGURE 7.10 Showy crotalaria root growth at different N treatments. Left to right 0 mg N kg^{-1}, 0 mg N kg^{-1} + Bradyrhizobial inoculants, 100 mg N kg^{-1} + Bradyrhizobial inoculants, and 200 mg N kg^{-1}. (Adapted from Fageria, N. K. et al. (In press). *Commun. Soil Sci. Plant Anal.*)

FIGURE 7.11 Calopo root growth at different N treatments. Left to right 0 mg N kg^{-1}, 0 mg N kg^{-1} + Bradyrhizobial inoculants, 100 mg N kg^{-1} + Bradyrhizobial inoculants, and 200 mg N kg^{-1}. (Adapted from Fageria, N. K. et al. (In press). *Commun. Soil Sci. Plant Anal.*)

7.8 CONCLUSIONS

The majority of biological fixed N available for agriculture is formed by rhizobia in symbiosis with legumes. It is known as dinitrogen fixation (N$_2$). Biological N fixation or dinitrogen (N$_2$) fixation is one of the most important biochemical reactions occurring in growing legumes for life on the earth. It ranks second only to photosynthesis in importance as a biological fixation reaction and, as evidenced by this treatise, has been the subject of intensive research at all levels of interest. N is also fixed by some nonlegume crops, but it was not discussed in this chapter. Globally, a significant amount of N is fixed biologically each year. Globally, biological N fixation has been estimated to amount to at least 139 million metric tons of N per year. This amount is much more than the N produced by the fertilizer industry, which is estimated to be about 90 million metric tons per year.

FIGURE 7.12 Lablab root growth at different N treatments. Left to right 0 mg N kg⁻¹, 0 mg N kg⁻¹ + Bradyrhizobial inoculants, 100 mg N kg⁻¹ + Bradyrhizobial inoculants, and 200 mg N kg⁻¹. (Adapted from Fageria, N. K. et al. 2014. *Commun. Soil Sci. Plant Anal.* 45:111–125.)

The fixation is carried out by certain species of bacteria (mainly rhizobia and bradyrhizobia) in association with legume roots.

The key to biological N fixation is the enzyme *nitrogenase*, which catalyzes the reduction of dinitrogen gas to ammonia. The ammonia, in turn, is combined with organic acids to form amino acids and, ultimately, proteins. There are several techniques available for measuring N_2 fixation in legumes. However, the ^{15}N technique can measure N_2 fixation in legumes with reasonable precision in less time. The growth of N_2-fixing bacteria is influenced by many environmental factors. These factors are related to climate, soil, and plants. The important climatic factors that affect biological N fixation are temperature and soil moisture. Soil factors that influence dinitrogen fixation are soil pH and soil fertility. The deficiency of P, S, Mo, and Fe in the growth medium or soil strongly inhibits biological N fixation by legumes. Similarly, high levels of both NH_4^+ and NH_3^- can also limit N_2 fixation. Plant factors such as legume species and/or genotypes within species also influence biological N fixation. Without doubt, biological N fixation improves the N economy of soils. Hence, strategies to improve environmental factors in favor of the growth of legume crops will ultimately improve biological N fixation. Selection of crop genotypes with higher N fixation capacity is the most important strategy from an economic and environmental point of view. In addition, oil and crop management practices such as conservation tillage and appropriate crop rotation can improve the biological N fixation of legumes. Molecular biology can be used to improve biological N fixation in legumes, and this area needs to be explored in the future. In addition, it is hoped that microbial genetics can transfer N-fixing genes to rhizobium strains, to make elite strains for use in practical agronomy.

REFERENCES

Abe, M. and S. Higashi. 1979. The infectivity of *Rhizobium trifolli* into a minute excised root of white clove. *Plant Soil* 53:81–88.

Allen, O. N. and E. K. Allen. 1981. *The Leguminosae*. Madison, Wisconsin: University of Wisconsin Press.

Al Niemi, T. S., M. L. Kahn, and T. R. McDermott, 1997. P metabolism in the bean *Rhizobium tropici* symbiosis. *Plant Physiol.* 113:1233–1242.

Antoun, H., L. M. Bordeleau, C. Gagnon, and R. A. Lachance. 1978. Identification of actinomycetes antagonistic to infection of *Rhizobium meliloti. Can. J. Microbiol.* 24:1073–1975.

Anyango, B., J. K. Wilson, J. L. Beynon, and K. E. Giller. 1995. Diversity of rhizobia nodulation *Phaseolus vulgaris* in two Kenyan soils with constrasting pHs. *Appl. Environ. Microbial.* 61:416–421.

Appunu, C., N. Sasirekha, V. R. Prabavathy, and S. Nair. 2009. A significant proportion of indigenous rhizobia from India associated with soybean (*Glycine max* L.) distinctly belong to *Bradyrhizobium* and *Ensifer* genera. *Biol. Fertil. Soils* 46:57–63.

Baligar, V. C., N. K. Fageria, and M. Elrashidi. 1998. Toxicity and nutrient constraints on root growth. *HortScience* 33:960–965.

Barker, K. R., D. Husingh, and S. A. Johnston. 1972. A antagonistic interaction between *Heterodera glycine* and *Rhizobium japonicum* on soybean. *Phytopathology* 62:1201–1205.

Barrios, S., N. Raggio, and M. Raggio. 1963. Effect of temperature on infection of isolated bean roots by rhizobia. *Plant Physiol.* 38:171–174.

Bauer, W. D. 1981. Infection of legume4s by rhizobia. *Annu. Rev. Plant Physiol.* 32:407–449.

Bergersen, F. J., J. Brockwell, R. R. Gault, L. Morthorpe, M. B. Peoples, and G. L. Turner. 1989. Effects of available soil nitrogen and rates of inoculation on nitrogen fixation by irrigated soybeans and evaluation of $\delta^{15}N$ methods for measurement. *Aust. J. Agric. Res.* 40:763–780.

Bethlenfalvay, G. J., S. S. Abu-Shakra, and D. A. Phillips. 1978. Interdependence of nitrogen nutrition and photosynthesis in *Pisum sativum* L: I. Effect of combined nitrogen on symbiotic nitrogen fixation and photosynthesis. *Plant Physiol.* 62:127–130.

Betts, J. H. and D. F. Herridge. 1987. Isolation of soybean lines capable of nodulation and nitrogen fixation under high levels of nitrate supply. *Crop Sci.* 27:1156–1161.

Bezdicek, D. F., D. W. Evans, B. Abede, and R. E. Witters. 1978. Evaluation of plant and granular inoculum for soybean yield and N fixation under irrigation. *Agron. J.* 70:865–868.

Bhangoo, M. S. and D. J. Albritton. 1976. Nodulating and non-nodulating Lee soybean isolines response to applied nitrogen. *Agron. J.* 68:642–645.

Blevins, R. L., J. H. Herbek, and W. W. Frye. 1990. Legume cover crops as a nitrogen source for no-till corn and grain sorghum. *Agron. J.* 82:769–772.

Bliss, F. A. 1993. Breeding common bean for improves biological nitrogen fixation. *Plant Soil* 152:71–79.

Bliss, F. A., P. A. A. Perriera, R. S. Araujo, R. A. Henson, K. A. McFerson, M. G. Teixera, and C. C. Silva. 1989. Registration of five high nitrogen fixing common bean germplasm lines. *Crop Sci.* 29:240–241.

Brady, N. C. and R. R. Weil. 2002. *The Nature and Properties of Soils*, 13th edition. Upper Saddle River, New Jersey: Prentice Hall.

Brockwell, J. and P. J. Bottomley. 1995. Recent advances in inoculant technology and prospects for the future. *Soil Biol. Biochem.* 27:683–697.

Brockwell, J., R. J. Roughley, and D. F. Herridge. 1987. Population dynamics of *Rhizobium japonicum* strains used to inoculate three successive crops of soybean. *Aust. J. Agric. Res.* 38:61–74.

Brockwell, J., R. R. Gault, D. F. Herridge, L. J. Morthorpe, and R. J. Roughley. 1988. Studies on alternative means of legume inoculation: Microbial and agronomic appraisals of commercial procedures for inoculating soybeans with *Bradyrhizobium japonicum. Aust. J. Agric. Res.* 39:965–972.

Chandler, M. R. 1978. Some observations on infection of *Arachis hypogaea* L. by *Rhizobium. J. Exp. Bot.* 29:749–755.

Chapman, A. L. and R. J. K. Myers. 1987. Nitrogen contributions by grain legumes to rice grown in rotation on the cumunurra soils of the ord irrigation area, Western Australia. *Aust. J. Exp. Agric.* 27:155–163.

Chaverra, M. H. and P. H. Graham. 1992. Cultivar variation in traits affecting early nodulation of common bean. *Crop Sci.* 32:1432–1436.

Cobley, L. S. 1976. *An Introduction to the Botany of Tropical Crops.* New York: Longman.

Costa, C., L. M. Dwyer, X. Zhou, P. Dutilleul, L. M. Reid, and D. L. Smith. 2002. Root morphology of contrasting maize genotypes. *Agron. J.* 94:96–101.

Dart, P. J. 1977. Infection and development of leguminous nodules. In: *Treatise on Dinitrogen Fixation*, eds., R. W. F. Hardy, and W. S. Silver, pp. 367–372. New York: Wiley.

Date, R. A. and R. J. Roughley. 1977. Preparation of legume seed inoculants. In: *A Treatise on Dinitrogen Fixation*, eds., R. W. F. Hardy, and A. H. Gibson, pp. 243–275. New York: John Wiley & Sons.

David, K. A. V. and P. Fay. 1977. Effects of long-term treatment with acetylene on nitrogen fixing microorganisms. *Appl. Environ. Microbiol.* 34:640–646.

Delwiche, C. C. 1983. Cycling of elements in the biosphere. In: *Inorganic Plant Nutrition, Encyclopedia Plant Physiology*, New Series Vol. 15, eds., A. Lauchli, and R. L. Bieleski, pp. 212–238. New York: Springer.

Durand, J. L., J. E. Sheehy, and F. R. Minchin. 1987. Nitrogenase activity, photosynthesis and nodule water potential in soybean plants experiencing water deprivation. *J. Exp. Bot.* 38:311–321.

Eckart, J. F. and C. A. Raguse. 1980. Effects of diurnal variation in light and temperature on the acetylene reduction activity (nitrogen fixation) of subterranean clover. *Agron. J.* 72:519–523.

Edwards, D. G., H. A. H. Sharifuddin, M. N. M. Yusoff, N. J. Grundon, J. Shamshuddin, and M. Norhayati. 1991. The management of soil acidity for sustainable crop production. In: *Plant-Soil Interaction at Low pH*, eds., R. J. Wright, V. C. Baliagr, and R. P. Murrman, pp. 383–396. Dordrecht: Kluwer Academic Publishers.

Eghball, B., J. R. Settimi, J. W. Maranville, and A. M. Parkhurst. 1993. Fractal analysis for morphological description of corn roots under nitrogen stress. *Agron. J.* 85:147–152.

Engles, C. and H Marschner. 1995. Plant uptake and utilization of nitrogen. In: *Nitrogen Fertilization in the Environment*, ed., P. E. Bacon, pp. 41–81. New York: Marcel Dekker.

Epstein, E. and A. J. Bloom. 2005. *Mineral Nutrition of Plants; Principal and Perspectives*, 2nd edition. Sunderland, Massachusetts: Sinauer Associates.

Evans, H. J. and L. E. Barber. 1977. Biological nitrogen fixation for food and fiber production. *Science* 197:332–339.

Evans, J., A. M. McNeill, M. J. Unkovich, N. A. Fettell, and D. P. Heenan. 2001. Net nitrogen balances for cool-season grain legume crops and contribution to wheat nitrogen uptake: A review. *Aust. J. Exp. Agric.* 41:347–359.

Evers, G. W. 1985. Forage and nitrogen contribution of arrowleaf and subterranean clovers overseeded on bermudagrass and bahiagrass. *Agron. J.* 77:960–963.

Fageria, N. K. 1992. *Maximizing Crop Yields*. New York: Marcel Dekker.

Fageria, N. K. 2009. *The Use of Nutrients in Crop Plants*. Boca Raton, Florida: CRC Press.

Fageria, N. K. 2013. *The Role of Plant Roots in Crop Production*. Boca Raton, Florida: CRC Press.

Fageria, N. K. and A. Moreira. 2011. The role of mineral nutrition on root growth of crop plants. *Adv. Agron.* 110:251–332.

Fageria, N. K. and A. B. Santos. 2008. Yield physiology of dry bean. *J. Plant Nutr.* 31:983–1004.

Fageria, N. K., V. C. Baligar, and. A. Bailey. 2005. Role of cover crops in improving soil and row crop productivity. *Commun. Soil Sci. Plant Anal.* 36:2733–2757.

Fageria, N. K., V. C. Baligar, and R. B. Clark. 2006. *Physiology of Crop Production*. New York: The Haworth Press.

Fageria, N. K., V. C. Baligar, and C. A. Jones. 2011. *Growth and Mineral Nutrition of Field Crops*, 3rd edition. Boca Raton, Florida: CRC Press.

Fageria, N. K., E. P. Ferreira, V. C. Baligar, and A. M. Knupp. Growth of tropical legume cover crops as influenced by nitrogen fertilization and rhizobia. *Commun. Soil Sci. Plant Anal.* (In press).

Fageria, N. K., L. C. Melo, E. P. B. Ferreira, J. P. Oliveira, and A. M. Knupp. 2012. Response of dry bean genotypes to nitrogen and rhizobia under field conditions. Paper presented at the Soil Fertility Meeting, 17–21 September 2012, Maceio, Brazil, Brazilian Soil Science Society/Federal University of Alagoas, Brazil.

Fageria, N. K., L. C. Melo, E. P. B. Ferreira, J. P. Oliveira, and A. M. Knupp. 2014. Dry matter, grain yield, and yield components of dry bean as influenced by nitrogen and fertilization and rhizobia. *Commun. Soil Sci. Plant Anal.* 45:111–125.

Fauci, M. F. and R. P. Dick. 1994. Plant response to organic amendments and decreasing inorganic nitrogen rates in soil from a long term experiment. *Soil Sci. Soc. Am. J.* 58:134–138.

Frame, J. and P. Newbould. 1986. Agronomy of white clover. *Adv. Agron.* 40:1–88.

Franzluebbers, A. J., S. R. Wilkinson, and J. A. Stuedemann. 2004. Bermudagrass management in response to fertilization and defoliation regimens. *Agron. J.* 96:1400–1411.

Fried, M. and V. Middleboe. 1977. Measurements of amount of nitrogen fixed by a legume crop. *Plant Soil* 47:713–715.

Furseth, B. J., S. P. Conley, and J. M. Ane. 2011. Soybean response to rhizobia on previously flooded sites in southern Wisconsin. *Agron. J.* 103:573–576.

Gallacher, A. E. and J. I. Sprent. 1978. The effect of different water regimes on growth and nodule development of greenhouse-grown *Vicia faba*. *J. Exp. Bot.* 29:413–423.

Gates, C. T. 1974. Nodule and plant development in *Stylosanthes humilis*: Symbiosis response to phosphorus and sulphur. *Aust. J. Bot.* 22:45–56.

Gelin, O. and S. Blixt. 1964. Root nodulation in peas. *Agric. Hort. Genet.* 22:149–159.

Giambalvo, D., P. Ruisi, G. D. Miceli, A. S. Frenda, and G. Amato. 2010. Nitrogen use efficiency and nitrogen fertilizer recovery of durum wheat genotypes as affected by interspecific competitions. *Agron. J.* 102:707–715.

Gibson, A. H. 1974. Consideration of the growing legume as a symbiotic association. *Proc. Indian Natl. Sci. Acad.* 40B:741–767.

Gibson, A. H. 1977. The influence of the environment and managerial practices on the legume-rhizobium symbiosis. In: *A Treatise on Dinitrogen Fixation*, eds., R. W. F. Hardy, and A. H. Gibson, pp. 393–450. New York: John Wiley & Sons.

Gibson, A. H. and D. C. Jordan. 1983. Ecophysiology of nitrogen fixation systems. In: *Encyclopedia of Plant Physiology*, New Series, Vol. 12C, eds., O. L. Lange, P. S. Nobel, C. B. Osmond, and H. Ziegler, pp. 301–390. New York: Springer-Verlag.

Giller, K. E. and G. Cadisch. 1995. Future benefits from biological nitrogen fixation: An ecological approach to agriculture. *Plant Soil* 174:255–277.

Giller, K. E. and K. J. Wilson. 1991. *Nitrogen Fixation in Tropical Cropping Systems*. Wallingford, UK: CAB International.

Graham, P. H. 1981. Some problems of nodulation and symbiotic nitrogen fixation in *Phaseolus vulgaris* L: A review. *Field Crops Res*. 4:93–112.

Graham, P. H. 1992. Stress tolerance in *Rhizobium* and *Bradyrhizobium*, and nodulation under adverse soil conditions. *Can. J. Microbial*. 38:475–484.

Graham, P. H. and J. C. Rosas. 1979. Phosphorus fertilization and symbiotic nitrogen fixation in common beans (*Phaseolus vulgaris* L.). *Agron. J*. 71:925–927.

Graham, P. H. and S. R. Temple. 1984. Selection for improved nitrogen fixation in *Glycine max* (L.) Merr. and *Phaseolus vulgaris* L. *Plant Soil* 82:315–327.

Graham, P. H. and C. P. Vance. 2000. Nitrogen fixation in perspective: An overview of research and extension needs. *Field Crops Res*. 65:93–106.

Graham, P. H. and C. P. Vance. 2003. Legumes: Importance and constraints to greater use. *Plant Physiol*. 131:872–877.

Graham, P. H., K. J. Draeger, M. L. Ferrey, M. J. Conroy, B. E. Hammer, E. Martinez, S. R. Aarons, and C. Quinto. 1994. Acid pH tolerance in strains of *Rhizobium* and *Bradyrhizobium*, and initial studies on the basis for acid tolerance of *Rhizobium tropici* UMR1899. *Can. J. Microbial*. 40:198–207.

Gregory, P. J. 1994. Root growth and activity. In: *Physiology and Determination of Crop Yield*, ed., G. A. Peterson, pp. 65–93. Madison, Wisconsin: ASA, CSSA, and SSSA.

Gregory, P. J. and S. C. Brown. 1989. Root growth, water use and yield of crops in dry environments: What characteristics are desirable? *Aspects Appl. Biol*. 22:235–243.

Guerin, V., D. Pladys, J. C. Trinchant, and J. Rigaud. 1991. Proteolysis and nitrogen fixation in faba bean (*Vicia faba* L.) nodules under water stress. *Physiol. Plant*. 82:1–7.

Guerin, V., J. C. Trinchant, and J. Rigaud. 1990. Nitrogen fixation (C_2H_2 reduction) by broad bean (*Vicia faba* L.) nodules and bacteroids under water-restricted conditions. *Plant Physiol*. 92:595–601.

Halliday, J. 1975. An interpretation of seasonal and short term fluctuation in nitrogen fixation. Ph.D. dissertation. University of Western Australia, Perth.

Hamdi, Y. A. 1971. Soil-water tension and movement of rhizobia. *Soil Biol. Biochem*. 3:121–126.

Hansson, A. C. and O. Andren. 1987. Root dynamics in barley, Lucerne, and meadow fescue investigated with a minirhizotron technique. *Plant Soil* 103:33–38.

Hargrove, W. L. 1986. Winter legumes as nitrogen sources for no-till grain sorghum. *Agron. J*. 78:70–74.

Hatfield, J. L., K. J. Boote, B. A. Kimball, L. H. Zisks, R. C. Izaurralde, D. Ort, A. M. Thompson, and D. Wolfe. 2011. Climate impacts on agriculture: Implications for crop production. *Agron. J*. 103:351–370.

Hauggaard-Nielsen, H., S. Mundus, and E. S. Jensen. 2009. Nitrogen dynamics following grain legumes and subsequent catch crops and the effects on succeeding cereal crops. *Nutr. Cycling Agroecosyst*. 84:281–291.

Heichel, G. H. 1987. Legume nitrogen: Symbiotic fixation and recovery by subsequent crops. In: *Energy in Plant Nutrition and Pest Control*, ed., Z. R. Helsel, pp. 63–80. Amsterdam: Elsevier Scientific Publishers.

Henzell, E. F. 1968. Sources of nitrogen for Queensland pastures. *Trop. Grassl*. 2:1–17.

Herridge, D. F. and F. J. Bergersen. 1988. Symbiotic nitrogen fixation. In: *Advances in Nitrogen Cycling in Agricultural Ecosystems*, ed., J. R. Wilson, pp. 46–65. Wallingford, UK: CAB International.

Herridge, D. F. and S. K. A. Danso. 1995. Enhancing crop legume N_2 fixation through selection and breeding. *Plant Soil* 174:51–82.

Herridge, D. F. and J. F. Holland. 1992. Production of summer crops in northern New Wales. I. Effects of tillage and double cropping on growth, grain and N yield of six crops. *Aust. J. Agric. Res*. 43:105–122.

Herridge, D. F., M. B. Peoples, and R. M. Boddey. 2008. Global inputs of biological nitrogen fixation in agricultural systems. *Plant Soil* 311:1018.

Heytler, P. G., G. S. Reddy, and R. W. F. Hardy. 1984. *In vivo* energetics of symbiotic nitrogen fixation in soybeans. In: *Nitrogen Fixation and CO_2 Metabolism*, eds., P. W. Ludden, and I. E. Burris, pp. 283–292. New York: Elsevier.

Hoad, S. P., G., Russell, M. E., Lucas, and I. J. Bingham. 2001. The management of wheat, barley, and oat root systems. *Adv. Agron.* 74:193–246.

Holderbaun, J. F., A. M. Decker, J. J. Meisinger, F. R. Mulford, and L. R. Vough. 1990. Fall seeded legume cover crops for no-till corn in the humid east. *Agron. J.* 82:117–127.

Hoveland, C. S., W. B. Anthony, J. A. McGuire, and J. G. Starling. 1978. Beef cow-calf performance on coastal bermudagrass overeesed with winter annual clovers and grasses. *Agron. J.* 70:418–420.

Howieson, J. G., R. J. Yates, K. J. Foster, D. Real, and R. B. Besier. 2008. Prospects for the future use of legumes. In*: Nitrogen Fixing Leguminous Symbioses.* Vol. 7, eds., M. J. Dilworth, E. K. James, and J. I. Sprent, pp. 363–387. Dordrecht: Springer.

Hungaria, M., D. D. Andrade, A. Colozzi, and E. L. Balota. 1997. Interactions among soil organisms and bean and maize grown in monoculture or intercropped. *Pesq. Agropec. Bras.* 32:807–818.

Hurd, E. A., T. F. Townley-Smith, L. A. Patterson, and C. H. Owen. 1972. Techniques used in producing Wascana wheat. *Can. J. Plant Sci.* 52:689–691.

Israel, D. W. 1987. Investigation of the role of phosphorus in symbiotic dinitrogen fixation. *Plant Physiol.* 84:835–840.

Israel, D. W. 1993. Symbiotic dinitrogen fixation and host plant growth during development and recovery from phosphorus deficiency. *Physiol. Plant.* 88:29–30.

Jacobsen, E. and W. J. Feenstra. 1984. A new pea mutant with efficient nodulation in the presence of nitrate. *Plant Sci. Lett.* 33:337–344.

Jensen, E. S. 1987. Variation in intrate tolerance of nitrogen fixation in the pea/*Rhizobium* symbiosis. *Plant Breed.* 98:130:130–135.

Johnston, G. R. and M. J. Barbetti. 1987. Impact of fungal and virus diseases on pasture. In: *Temperate Pastures: Their Production, Use and Management*, eds., J. L. Wheeler, C. J. Pearson, and G. E. Robards, pp. 235–248. Melbourne: Australian Wool Corporation/CSIRO.

Jungk, A. O. 1998. Dynamics of nutrient movement at the soil-root interface. In: *Plant Roots: The Hidden Half*, eds., Y. Waisel et al., pp. 529–556. New York: Marcel Dekker.

Kanayama, Y., I. Watanabe, and Y. Yamamoto. 1990. Inhibition of nitrogen fixation in soybean plants supplied with nitrate. I. Nitrite accumulation and formation of nitrosylleghemoglobin in nodules. *Plant Cell Physiol.* 31:341–346.

Keyser, H. H. and D. N. Munns. 1979. Tolerance of rhizobia to acidity, aluminum, and phosphate. *Soil Sci Soc. Am. J.* 43:519–523.

Keyser, H. H. and F. Li. 1992. Potential for increasing biological nitrogen fixation in soybean. *Plant Soil* 141:119–135.

Kipe-Nolt, J. A. and K. E. Giller. 1993. A field evaluation using the ^{15}N isotope dilution method of lines of *Phaseolus vulgaris* L. bred for increased nitrogen fixation. *Plant Soil* 152:107–114.

Kipe-Nolt, J. A., H. Vargas, and K. E. Giller. 1993. Nitrogen fixation in breeding lines of *Phaseolus vulgaris* L. *Plant Soil* 152:103–106.

Kristensen, H. L. and K. Thorup-Kristensen. 2004. Root growth and nitrate uptake of three different catch crops in deep soil layers. *Soil Sci Soc Am J* 68:529–537.

Kucey, R. M. N., P. Snitwongse, P. Chaiwanakupt, P. Wadisirisuk, C. Siripaibool, T. Arayangkool, N. Boonkerd, and R. J. Rennie. 1988. Nitrogen fixation (^{15}N dilution) with soybean under Thai field conditions. *Plant Soil* 108:33–41.

Kumar, K. and K. M. Goh. 2000. Crop residues and management practices: Effects on soil quality, soil nitrogen dynamics, crop yield, and nitrogen recovery. *Adv. Agron.* 68:197–319.

Ladha, J. K., R. P. Pareek, and. M. Becker. 1992. Stem nodulation legume-rhizobium symbiosis and its agronomic use in lowland rice. *Adv. Soil Sci.* 20:147–192.

Ladha, J. K. I. Watanabe, and S. Saono. 1988. Nitrogen fixation by leguminous green manure and practices for its enhancement in tropical lowland rice. In: *Sustainable Agriculture: Green Manure in Rice Farming*, ed., The International Rice research Institute, pp. 165–183. Los Banos, the Philippines: IRRI.

Layzell, D. B. and A. H. M. Moloney. 1994. Dinitrogen fixation. In: *Physiology and Determination of Crop Yield*, eds., K. J. Boote, J. M. Bennett, T. R. Sinclair, and G. M. Paulsen, pp. 311–335. Madison, Wisconsin: ASA, CSSA, and SSSA.

Layzell, D. B., P. Rochman, and D. T. Canvin. 1983. Low root temperature and nitrogenase activity in soybean. *Can. J. Bot.* 62:965–971.

Ledgard, S. F. and M. N. Peoples. 1988. Measurement of nitrogen fixation in the field. In: *Advances in Nitrogen Cycling in Agricultural Ecosystems*, ed., J. R. Wilson, pp. 351–367. Wallingford, UK: CAB International.

Lewis, G., B. Schrire, B. Mackinder, and M. Lock. 2005. *Legumes of the World.* Kew: Royal Botanical Gardens.

Lie, T. A. 1971. Symbiotic nitrogen fixation under stress conditions. *Plant Soil* 35:117–118.

Lindermann, W. C. and G. E. Ham. 1979. Soybean plant growth, nodulation, and nitrogen fixation as affected by root temperature. *Soil Sci. Soc. Am. J.* 43:1134–1137.

Lira, M. A. Jr., C. Costa, and D. L. Smith. 2003. Effects of addition of flavonoid signals and environmental factors on nodulation and nodule development in the pea. *Aust. J. Soil Res.* 41:267–276.

Liu, Y., L. Wu, C. A. Watson, J. A. Baddeley, X. Pan, and L. Zhang. 2013. Modeling biological dinitrogen fixation of field pea with a process based simulation model. *Agron. J.* 105:670–678.

Lopez-Belliodo, L., R. J. Lopez-Bellido, R. Redondo, and J. Benitez. 2006. Faba bean nitrogen fixation in a wheat-based rotation under rainfed Mediterranean conditions: Effects of tillage system. *Field Crops Res.* 98:253–260.

Lopez-Garcia, S. L., A. Perticari, C. Piccinetti, L. Ventimiglia, N. Arias, J. J. Battista, M. J. Althabegoiti et al. 2009. In-furrow inoculation and selection for higher motility enhances the efficacy of *Bradyrhizobium japonicum* nodulation. *Agron. J.* 101:357–363.

Ludlow, M. M. and R. C. Muchow. 1990. A critical evaluation of traits for improving crop yields in water-limited environments. *Adv. Agron.* 43:107–153.

Masterson, C. L. and M. T. Sherwood. 1978. Some effects of increased atmospheric carbon dioxide on white clover (*Trifolium repens*) and pea (*Pisum sativum*). *Plant Soil* 49:421–426.

Mengel, K., E. A. Kirkby, H. Kosegarten, and T. Appel. 2001. *Principles of Plant Nutrition*, 5th edition. Dordrecht: Kluwer Academic Publishers.

Minchin, F. R., M. Becana, and J. I. Sprent. 1989. Short-term inhibition of legume N_2 fixation by nitrate. II. Nitrate effects on nodule oxygen diffusion. *Planta* 180:46–52.

Munns, D. N. 1968a. Nodulation of *Medicago sativa* in solution culture. I. Acid sensitive steps. *Plant Soil* 28:129–146.

Munns, D. N. 1968b. Nodulation of *Medicago sativa* in solution culture. II. Compensating effects of nitrate and prior nodulation. *Plant Soil* 28:246–257.

Munns, D. N. 1968c. Nodulation of *Medicago sativa* in solution culture. III. Effects of nitrate on root hairs and infection. *Plant Soil* 29:33–47.

Munns, D. N. 1968d. Nodulation of *Medicago sativa* in solution culture. IV. Effects of indole-3-acetate in relation to acidity and nitrate. *Plant Soil* 29:257–261.

Mytton, L. R. 1983. Host plant selection and breeding for improved symbiotic efficiency. In: *The Physiology, Genetics and Nodulation of Temperate Legumes*, eds., D. G. Jones, and D. R. Davies, pp. 373–393. London: Pitman.

Nelson, C. J. and K. L. Larson. 1984. Seedling growth. In: *Physiological Basis of Crop Growth and Development*, ed., M. B. Tesar, pp. 93–129. Madison, Wisconsin: American Society of Agronomy and Crop Science Society of America.

Nutman, P. S. 1959. Some observations on root hair infection by nodule bacteria. *J. Exp. Bot.* 10:250–262.

Nutman, P. S. 1984. Improving nitrogen fixation in legumes by plant breeding: The relevance of host selection experiments in red clover (*Trifolium pratense* L.) and subterranean clover (*Trifolium Subterraneum* L.). *Plant Soil* 82:285–301.

Ocumpaugh, W. R. 1990. Coastal bermudagrass-legume mixtures vs. nitrogen fertilizer for grazing in a semi-arid environment. *J. Prod. Qual.* 26:371–376.

Orellana, R. G. and F. F. Fan. 1978. Nodule infection by bean yellow mosaic virus in *Phaseolus vulgaris*. *Appl. Environ. Microbiol.* 36:814–818.

O'Toole, J. C. and W. L. Bland. 1987. Genotypic variation in crop plant root systems. *Adv. Agron.* 41:91–145.

Oyer, L. J. and J. T. Touchton. 1990. Utilization legume cropping systems to reduce nitrogen fertilizer requirements for conservation-tilled corn. *Agron. J.* 82:1123–1127.

Pankhurst, C. E. and J. I. Sprent. 1975. Effects of water stress on the respiratory and nitrogen fixing activity of soybean root nodules. *J. Exp. Bot.* 91:287–304.

Pankhurst, C. E. and J. I. Sprent. 1976. Effects of temperature and oxygen tension on the nitrogenase and respiratory activities of turgid and water stressed soybean and french bean root nodules. *J. Exp. Bot.* 27:1–9.

Park, S. J. and B. R. Buttery. 1988. Nodulation mutants of white bean (*Phaseolus vulgaris* L.) induced by ethyl-methane sulphonate. *Can. J. Plant Sci.* 68:199–202.

Park, S. J. and B. R. Buttery. 1989. Identification and characterization of common bean (*Phaseolus vulgaris* L.) lines well nodulated in the presence of high nitrate. *Plant Soil* 119:237–244.

Pelegrin, R., F. M. Mercante, I. M. N. Otsubo, and A. K. Otsubo. 2009. Response of common bean crop to nitrogen fertilization and inoculation in Mato Grosso do Sul state of Brazil. *Rev. Bras. Ci. Solo* 33:219–226.

Pena-Cabriales, J. J. and M. Alexander. 1979. Survival of rhizobium in soils undergoing drying. *Soil Sci. Soc. Am. J.* 43:962–966.

Pena-Cabriales, J. J. and J. Z. Castellanos. 1993. Effects of water stress on N_2 fixation and grain yield of *Phaseolus vulgaris* L. *Plant Soil* 152:151–155.

Peoples, M. B. and D. F. Herridge. 1990. Nitrogen fixation by legumes in tropical and subtropical agriculture. *Adv. Agron.* 44:155–223.

Peoples, M. B., M. J. Bell, and H. V. A. Bushby. 1992. Effect of rotation and inoculation with *Bradyrhizobium* on nitrogen fixation and yield of peanut (*Arachis hypogaea* L., cv Virginia Bunch). *Aust. J. Agric. Res.* 43:595–607.

Peoples, M. B., J. Brockwell, D. F. Herridge, I. J. Rochester, B. J. R. Alves, S. Urquiaga, R. M. Boddey et al. 2009. The contributions of nitrogen fixing crop legumes to the productivity of agricultural systems. *Symbiosis* 48:1–17.

Peoples, M. B., D. F. Herridge, and J. K. Ladha. 1995a. Biological nitrogen fixation: An efficient source of nitrogen for sustainable agricultural production. *Plant Soil* 174:3–28.

Peoples, M. B., J. K. Ladha, and D. F. Herridge. 1995b. Enhancing legume N_2 fixation through plant and soil management. *Plant Soil* 174:83–101.

Phillips, D. A. 1980. Efficiency of symbiotic nitrogen fixation in legumes. *Annu. Rev. Plant Physiol.* 31:29–49.

Phillips, D. A., K. D. Newell, S. A. Hassell, and C. E. Felling. 1976. The effect of CO_2 enrichment on root nodule development and symbiotic N_2 reduction in *Pisum sativum* L. *Am. J. Bot.* 63:356–362.

Phillips, D. A., J. G. Torrey, and R. H. Burris. 1971. Extending symbiotic nitrogen fixation to increase mans food supply. *Science* 174:169–171.

Piha, M. I. and D. N. Munns. 1987. Nitrogen fixation capacity of field grown bean compared to other grain legumes. *Agron. J.* 79:690–696.

Pimratch, S., S. Jogloy, N. Vorasoot, B. Toomsan, A. Patanothai, and C. C. Holbrook. 2008. Relationship between biomass production and nitrogen fixation under drought-stress condition in peanut genotypes with different levels of drought resistance. *J. Agron. Crop Sci.* 194:15–25.

Pohlhil, R. M. 1981. The Papillionideae. In: *Advances in Legume Systematics*, eds., R. M. Pohlhil, and P. H. Raven, pp. 191–204. London: Royal Botanic Gardens Kew.

Rasmusson, D. C. and B. G. Gengenbach. 1984. Genetics and use of physiological variability in crop breeding. In: *Physiological Basis of Crop Growth and Development*, ed., M. B. Tesar, pp. 291–321. Madison, Wisconsin: American Society of Agronomy and Crop Science Society of America.

Ribet, J. and J. J. Drevon. 1996. The phosphorus requirement of N_2 fixing and urea-fed *Acacia mangium*. *New Phytol.* 132:383–390.

Robertson, J. G. and K. J. F. Farnden. 1980. The ultrastructure and metabolism of the developing legume root nodule. In: *The Biochemistry of Plants*, eds., P. K. Stumpf, and E. Conn, New York: Academic Press.

Robson, A. D. 1983. Mineral nutrition. In: *Nitrogen Fixation of Legumes*, Vol. 3, ed., W. J. Broughton, pp. 35–55. Oxford, UK: Clarendon Press.

Robson, A. D., G. W. O'Hara, and L. K. Abbott. 1981. Involvement of phosphorus in nitrogen fixation by subterranean clover (*Trifolium subterraneum* L.). *Aust. J. Plant Physiol.* 8:81–87.

Ruisi, P., D. Giambalvo, G. D. Miceli, A. S. Frenda, S. Saia, and G. Amato. 2012. Tillage effects on yield and nitrogen fixation of legumes in Mediterranean conditions. *Agron. J.* 104:1459–1466.

Ruiz-Diez, B., S. Fajardo, and M. Fernandez-Pascual. 2012. Selection of rhizobia from agronomic legumes grown in semiarid soils to be employed as bioinoculants. *Agron. J.* 104:550–559.

Sainju, U. M., B. P. Singh, and W. F. Whitehead. 1998. Cover crop root distribution and its effect on soil nitrogen cycling. *Agron. J.* 90:511–518.

Sall, K. and T. R. Sinclair. 1991. Soybean genotypic differences in sensitivity of symbiotic nitrogen fixation to soil dehydration. *Plant Soil* 133:31–37.

Schubert, K. R. 1986. Products of biological nitrogen fixation in higher plants: Synthesis, transport, and metabolism. *Annu. Rev. Plant Physiol.* 37:539–574.

Schubert, K. R., N. T. Jennings, and H. J. Evans. 1978. Hydrogen reactions of nodulated leguminous plants. *Plant Physiol.* 61:398–401.

Schweitzer, L. E. and J. E. Harper. 1980. Effect of light, dark and temperature on root nodule activity (acetylene reduction) of soybeans. *Plant Physiol.* 65:51–56.

Seetin, M. W. and D. K. Barnes. 1977. Variation among alfalfa genotypes for rate of acetylene reduction. *Crop Sci.* 17:783–787.

Seneviratne, G., L. H. J. Van Holm, and E. M. H. G. S. Ekanayake. 2000. Agronomic benefits of rhizobial inoculant use over nitrogen fertilizer application in tropical soybean. *Field Crops Res.* 68:199–203.

Serraj, R., J. J. Drevon, M. Obaton, and A. Vidal. 1992. Variation in nitrate tolerance of nitrogen fixation in soybean (*Glycine max*)-*Bradyrhizobium* symbiosis. *J. Plant Physiol.* 140:366–371.

Serraj, R., T. R. Sincalir, and L. C. Purcell. 1999. Symbiotic N_2 fixation response to drought. *J. Exp. Bot.* 50:143–155.

Sessitsch, A., J. G. Howieson, X. Perret, H. Antoun, and E. Martinez. 2002. Advances in rhizobium research. *Critical Rev. Plant Sci.* 21:323–378.

Shanmugam, K. T., F. O'Gara, K. Andersen, and R. C. Valentine. 1978. Biological nitrogen fixation. *Annu. Rev. Plant Physiol.* 29:29:263–276.

Shearer, G. and D. H. Kohl. 1986. N_2 fixation in field settings: Estimations based on natural ^{15}N abundance. *Aust. J. Plant Physiol.* 13:699–755.

Silveira, P. M. and M. A. Damasceno. 1993. Rate and fractional application of nitrogen and potassium in irrigated dry bean. *Pesq. Agropec. Bras.* 28:1269–1276.

Silveira, M. C., F. M. Rouquette, Jr., V. A. Haby, and G. R. Smith. 2013. Impacts of thirty-seven years of stocking on soil phosphorus distribution in bermudagrass pastures. *Agron. J.* 105:922–928.

Sinclair, T. R., R. C. Muchow, J. M. Bennett, and L. C. Hammond. 1987. Relative sensitivity of nitrogen and biomass accumulation to drought in field-grown soybean. *Agron. J.* 79:986–991.

Singer, J. W. and W. J. Cox. 1998. Agronomics of corn production under different crop rotations. *J. Prod. Agric.* 11:462–468.

Small, J. G. C., Hough, M. C., B. Clarke, and N. Grobbelaar. 1968. The effect of temperature on nodulation of whole plants and isolated roots of *Phaseolus vulgaris* L. S. *Afr. J. Sci.* 64:218–224.

Smith, J. H. and P. B. Gibson. 1960. The influence of temperature on growth and nodulation of white clover infected with bean yellow mosaic virus. *Agron. J.* 52:5–7.

Smith, D. L. and D. J. Hume. 1987. Comparison of assay methods for N_2 fixation utilizing white bean and soybean. *Can. J. Plant Sci.* 67:11–19.

Smith, M. S., W. W. Frye, and J. J. Varco. 1987. Legume winter cover crops. *Adv. Soil Sci.* 7:95–139.

Soil Science Society of America. 2008. *Glossary of Soil Science Terms*. Madison, Wisconsin: SSSA.

Soratto, R. P., M. A. C. Carvalho, and O. Arf. 2004. Chlorophyll content and grain yield of common bean as affected by nitrogen fertilization. *Pesq. Agropec. Bras.* 39:895–901.

Sponchiado, B. N., J. W. White, J. A. Castillo, and P. G. Jones. 1989. Root growth of four common bean cultivars in relation to drought tolerance in environments with contrasting soil types. *Exp. Agric.* 25:249–257.

Sprent, J. I. 1976. Nitrogen fixation by legumes subjected to water and light stresses. In: *Symbiotic Nitrogen Fixation in Plants*, ed., P. S. Nutman, pp. 405–420. London: Cambridge University Press.

Sprent, J. I. 1989. Tansley review no. 15. Which steps are essential for the formation of functional legume nodules? *New Phytol.* 111:129–153.

Stevenson, F. C. and C. Van Kessel. 1996. A landscape scale assessment of the nitrogen and non-nitrogen benefits of pea in a crop rotation. *Soil Sci. Soc. Am. J.* 60:1797–1805.

Streeter, J. 1988. Inhibition of legume nodule formation and N_2 fixation by nitrate. *Crit. Rev. Plant Sci.* 7:1–23.

Streeter, J. G. and S. O. Salminen. 1993. Distribution of two types of polysaccharide formed by *Bradyrhizobium japonicum* bacteroids in nodules of field grown soybean plants (*Glycine max* L. Merr.). *Soil Biol. Biochem.* 25:1027–1032.

Stute, J. K. and J. L. Posner. 1993. Legume cover option for grain rotations in Wisconsin. *Agron. J.* 85:1128–1132.

Sweeney, D. W. and J. L. Moyer. 2004. In-season nitrogen uptake by grain sorghum following legume green manures in conservation tillage systems. *Agron. J.* 96:510–515.

Thies, J. E., P. W. Singleton, and B. B. Bohlool. 1991. Modeling symbiotic performance of introduced rhizobia in the field by use of indices of indigenous population size and nitrogen status of the soil. *Appl. Environ. Microbiol.* 57:29–37.

Thomas, R. J. and J. I. Sprent. 1984. The effects of temperature on vegetative and early reproductive growth of a cold-tolerant and a cold sensitive line of *Phaseolus vulgaris* L. I. Nodulation, growth and partitioning of dry matter, carbon and nitrogen. *Ann. Bot.* 53:579–588.

Thornton, H. G. 1936. Action of sodium nitrate on infection of lucerne root hairs by nodule bacteria. *Proc. R. Soc. Lond. Ser. B.* 119:47–92.

Tsien, H. C., P. S. Cain, and E. L. Schmidt. 1977. Viability of *Rhizobium* bacteroids. *Appl. Environ. Microbial.* 34:845–856.

Tu, J. C. 1979. Evidence of differential tolerance among some root fungi to rhizobial parasitism in vitro. *Physiol. Plant Path.* 14:171–177.

Tu, J. C., R. E. Ford, and S. S. Quiniones. 1970. Effects of soybean mosaic virus and/or bean pod mottle virus infection on soybean nodulation. *Phytopathology* 60:518–523.

Unger, P. W., H. V. Eck, and J. T. Musick. 1981. Alleviating plant water stress, In: *Modifying the Root Environment to Reduce Crop Stress*, eds., G. F. Arkin, and H. Taylor, pp. 61–96. St. Joseph, Michigan: Am. Soc. Agric. Eng., St. Joseph, Michigan.

Unkovich, M. J., J. S. Pate, and P. Sanford. 1997. Nitrogen fixation by annual legumes in Australian Mediterranean agriculture. *Aust. J. Agric. Res.* 48:267–293.

Van Kessel, C. and C. Hartley. 2000. Agricultural management of grain legumes: Has it led to an increase in nitrogen fixation? *Field Crops Res.* 65:165–181.

Vessey, J. K., K. B. Walsh, and D. B. Layzell. 1988a. Oxygen limitation of N_2 fixation in stem-girdled and nitrate-treated soybean. *Physiol. Plant.* 73:113–121.

Vessey, J. K., K. B. Walsh, and D. B. Layzell. 1988b. Can a limitation in phloem supply to nodules account for the inhibitory effect of nitrate on nitrogenase activity in soybean: *Physiol. Plant.* 74:137–146.

Voisin, A., S., C. Salon, N. G. Munier-Jolain, and B. Ney. 2002. Effect of mineral nitrogen on nitrogen nutrition and biomass partitioning between the shoot and roots of pea. *Plant Soil* 242:251–262.

Vyn, T. J., K. L. Janovicek, M. H. Miller, and E. G. Beauchamp. 1999. Soil nitrate accumulation and corn response to preceding small-grain fertilization and cover crops. *Agron. J.* 91:17–24.

Walley, F. L., G. W. Clayton, P. R. Miller, P. M. Carr, and G. P. Lafond. 2007. Nitrogen economy of pulse crop production in the northern great plains. *Agron. J.* 99:1710–1718.

Wani, S. P., O. P. Rupela, and K. K. Lee. 1995. Sustainable agriculture in the semi-arid tropics through biological nitrogen fixation in grain legumes. *Plant Soil* 174:29–50.

Waterer, J. G. and J. K. Vessey. 1993. Effect of low static nitrate concentrations on mineral nitrogen uptake, nodulation, and nitrogen fixation in field pea. *J. Pant Nutr.* 16:1775–1789.

Weisz, P. R. and T. R. Sinclair. 1988. Soybean nodule gas permeability, nitrogen fixation and diurnal cycles in soil temperature. *Plant Soil* 109:227–234.

Werner, D. 1987. *Plant and Microbial Symbiosis*. Stuttgart: Verlag.

Werner, D., J. Wilcockson, R. Tripf, E. Morschel, and H. Papen. 1981. Limitation of symbiotic and associated nitrogen fixation by developmental stages in the system *Rhizobium japonicum* with *Glycine max* and *Azospirillum brasilense* with grasses, e.g. *Triticum aestivum*. In: *Biology of Inorganic Nitrogen and Sulfur*, eds., H. Bothe, and A. Trebst, pp. 299–308. New York: Springer-Verlag.

Wilson, P. W., E. B. Fred, and M. R. Salmon. 1933. Relation between carbon dioxide and elemental nitrogen assimilation in leguminous plants. *Soil Sci.* 35:145–163.

Witty, J. F. 1979. Acetylene reduction assay can overestimate nitrogen fixation in soil. *Soil Biol. Biochem.* 11:209–210.

Witty, J. F., R. J. Rennie, and C. A. Atkins. 1988. ^{15}N addition methods for assessing N_2 fixation under field conditions. In: *World Crops: Cool Season Food Legumes*, ed., R. J. Summerfield, pp. 715–730. London: Kluwer Academic Publishers.

Worrall, V. S. and R. J. Roughley. 1976. The effect of moisture stress on infection of *Trifolium subterraneum* L. by *Rhizobium trifolii*. *J. Exp. Bot.* 27:1233–1241.

8 Management Practices to Improve Nitrogen Use Efficiency in Crop Plants

8.1 INTRODUCTION

The use of nitrogen (N) is one of the most important single factors in increasing crop productivity in the last half century. The past 60 years have brought marked advances in the capacity to manufacture and apply plant-available N as commercial fertilizers. These advances, however, have not diminished the importance of problems related to N management but have created a greater appreciation of the importance of avoiding yield-limiting N deficiencies (Blackmer, 2000). Close correlations are commonly found between historical crop yields and annual N application rates (Sinclair and Horie, 1989). Similarly, Fageria and Baligar (2005) reported that N is the most limiting nutrient for crop production in many of the world's agricultural areas and its efficient use is important for the economic sustainability of cropping systems. Adopting appropriate nitrogen management practices is essential to improve N use efficiency in crop plants. Further, it also promotes the sustainability of cropping systems. In addition, the dynamic nature of N and its propensity for loss from soil–plant systems creates a unique and challenging environment for its efficient management. Adequate N management practices to improve its efficiency involve soil management factors, nitrogen fertilizer management factors, crop management factors, and plant management factors. Hence, the nitrogen economy of crop production involves considerable complexity due to the involvement of several factors and their interactions. Some of these practices are already mentioned in other chapters, and hence there is duplication. However, some duplication is unavoidable due to the importance of some practices in some of the other chapters.

Appropriate N management practices are essential not only to improve crop yields and reduce the cost of production, but its effective management is also related to the reduction of environmental pollution. Culman et al. (2013) reported that a major goal of current agronomic research is to make annual grain crop production more sustainable, with management strategies that supply adequate fertility to meet crop demand while conserving soil and water quality. Environmental pollution by N involved the contamination of groundwater by NO_3^- N and also greenhouse gases such as NO and N_2O. Nitric oxide, which is the dominant component of the so-called NO_x, has no direct effect on the earth's radiation balance, but it is very active chemically and plays a critical role in its interaction with oxidants such as ozone (Campbell et al., 1995). Because NO_x is eventually oxidized to nitric acid, it will also contribute to acid precipitation.

Modern agricultural technologies are often responsible for increased emissions of gases causing atmospheric pollution (especially N_2O and CH_4), and agriculture accounts for an estimated 84% of global anthropogenic N_2O emissions (Smith et al., 2008). Pang et al. (2009) also reported that agricultural soils are known to be responsible for a large proportion (70–81%) of the increase in N_2O emissions to the atmosphere, mainly due to the use of nitrogen fertilizer. Nitrous oxide is mainly produced in the soil by nitrification and denitrification, which are particularly controlled by the soil water-filled pore space, soil temperature, availability of labile organic carbon, soil pH, and the proportion and amount of ammonium and nitrate present (Bouwman, 1998; Dobbie and Smith, 2003; Martins et al., 2003; Pang et al., 2009).

The use of chemical fertilizers is fundamental to improve crop yields and maintain sustainability of the cropping systems. From 1960 to 1990, the use of N fertilizer in modern agricultural systems has increased in parts of Asia, North America, and West Europe. This was accompanied by a steady increase in the yield of most annual crops (Food and Agricultural Organization of the United Nations, 2000). On the other hand, negative impacts of N compounds on the atmosphere, groundwater, and other components of the ecosystems have also been reported (Socolow, 1999; Presterl et al., 2003). For example, excessive N fertilization in intensive agricultural areas of China results in serious environmental problems such as eutrophication (Li et al., 2010), greenhouse gas emissions (Xue et al., 2012), and soil acidification (Guo et al., 2010). Galloway et al. (2004) estimated that out of 268 Mt per year of reactive N (all N compounds except N_2) added every year to continents and inland water, of which 100 Mt was coming from manufactured fertilizers, 48 Mt was lost as nitrate, 53 Mt as ammonia, 46 Mt as nitrogen oxides, and 11 Mt as nitrous oxide. Improving N recovery by crops is one of the options identified by Galloway et al. (2004) to reduce the amount of reactive N in the environment (Heffer and Prudhomme, 2013). Currently, improving N management and development-related policies are major issues in crop production and environmental protection in China (Xue et al., 2012) and other countries (Fageria, 2013). Looking into this scenario, appropriate management of this nutrient is necessary not only to improve the yield but also to reduce environmental pollution.

Overall, nitrogen use efficiency is a function of the capacity of the soil to supply adequate levels of N, and ability of plants to absorb, transport in roots and shoots, and to remobilize to other parts of the plant (Baligar et al., 2001). Plant interaction with environmental factors such as solar radiation, rainfall, and temperature, and their response to diseases, insects and allelopathy, and root microbes have a greater influence on nitrogen use efficiency in plants. Optimizing N use efficiency is an imperative from agronomic, economic, and environmental perspectives. The nutrient use efficiency in crops can be increased by adopting various agronomic practices as well as by understanding the molecular nature of relevant traits and using this knowledge in breeding programs (Malik and Rengel, 2013). Significant progress has been made both agronomically and at the molecular level toward characterizing nutrient use efficiency. Some examples of this success are supplying nutrients to match the crop demand (Wiesler, 1998), crop rotation (Raun and Johnson, 1999), and precision agriculture (Wong et al., 2005), which can be adopted for improving nitrogen use efficiency. Nitrogen-efficient genotypes of several crop species have been identified (Fageria et al., 2008, 2011; Rengel and Damon, 2008). Genetic variability in root growth among crop species and genotypes within species has been identified (Fageria and Moreira, 2011; Fageria, 2013, 2014). There is evidence in the literature that some plant species and genotypes within species have a capacity to grow and yield well on soils low in available nutrients (Rengel and Marschner, 2005; Brennan and Bolland, 2007; Damon and Rengel, 2007; Malik and Rengel, 2013). The objective of this chapter is to discuss the latest development in nitrogen management in crop production and suggest measures to improve N use efficiency in crop plants.

8.2 SOIL MANAGEMENT PRACTICES

Efficient agricultural production and management practices are required to meet the growing demand for food without compromising the environment and agricultural resources (Heumesser et al., 2013). In this context, adopting adequate soil management practices is essential to improve the nitrogen use efficiency in crop plants. These management practices may include improvement in the physical, chemical, and biological properties of the soil. Improvement in the physical, chemical, and biological properties of the soil creates favorable conditions for the plant growth, including root system, which can absorb more nutrients and water and utilize N more efficiently. Some physical properties can be improved by proper land preparation during the sowing of a particular crop. If a soil is compacted, it is better to have deep ploughing to break the hard pan and incorporate plant residues into the soil.

The impact of cultivation practices or soil tillage practices on physical properties of the soil such as structure, water retention, aggregate stability, aeration, bulk density, and biochemical processes (mineralization, nitrification) is of considerable importance in the effective sustainable management of soils and nutrient use efficiency including N. Adequate preparation of land may improve soil aeration and drainage, improve root penetration, and influence the water retention properties (Fageria, 2013). Lal (1991) reported that appropriate tillage operations are a powerful tool to overcome problems associated with infiltration, surface crusting, poor drainage, soil compaction, burial of weeds and surface debris, and pest management. All these changes in physical properties of the soil may improve N use efficiency.

8.2.1 Physical Properties

Physical properties of the soil that influence plant growth and development and consequently nitrogen use efficiency are soil temperature, soil water content, soil texture, soil structure, and soil bulk density. These soil properties are responsible for the germination of seeds and consequently the emergence of seedlings. In addition, root penetration, water-holding capacity, soil water infiltration rate, soil aeration, soil microbial activities, and availability of nutrients are also determined by these soil properties. Hence, if these soil properties are favorable or can be modified and/or improved in favor of higher crop growth, nitrogen use efficiency will improve. However, some properties such as soil texture are difficult to modify by normal soil management practices. Hence, the discussion in this section will be on the remaining soil physical properties.

8.2.1.1 Favorable Soil Temperature

Soil temperature plays a role in many important functions and processes by regulating the biological and chemical reaction rates (Paul and Clark, 1966; Soil Survey Staff, 1999). Oxidation–reduction reactions, C storage, CO_2 efflux, nutrient availability, and bacterial mineral reduction rates have all been correlated to soil temperature (Abdollahi and Nedwell, 1979; Koerselman et al., 1993; Schimel et al., 1994; Fang and Moncrieff, 2001; Davidson and Janssens, 2006; Vaughan et al., 2009; Salisbury and Stolt, 2011). As such, soil temperature plays an important role in soil formation, part of the climate state factors described in Jenny's (Jenny, 1941) factors of soil formation. Soil temperature has been used in definitions of the biological zero and growing season in soil taxonomy and wetland science (Soil Survey Staff, 1975; Salisbury and Stolt, 2011).

Favorable soil temperature influences the growth and development of crop plants and consequently higher nitrogen use efficiency. On the other hand, a too high or too low temperature influences plant growth adversely and nitrogen use efficiency decreases. High and/or low temperatures influence the photosynthetic rate, plant water relations, flowering, and fruit set in temperate and tropical crops (Abrol and Ingram, 1996). Ali et al. (2009) reported that, in the winter season, in crops such as wheat, an increase in the mean minimum and maximum temperatures during the reproductive stage adversely affected grain yields in northern India. Baker and Allen (1993) reported water requirements and decreased yields in rice, soybean, and citrus under higher maximum and minimum temperatures. For every rise in the day/night temperature above 28/21°C, the rice yield declined by 10%. Ali et al. (2009) in a review of the literature reported that due to the unprecedented heat wave in northern India in 2004, 4.4 million tons of loss in productivity was reported in the wheat crop. These authors further reported that breeding for suitable varieties, improved crop management, and changes in planting dates to some extent would help in overcoming the effects of climatic change. However, evolving low-cost methods, which can be easily adopted by small farmers, is a major challenge (Ali et al., 2009).

Available reports indicated that the adverse effect in plants by abiotic stresses such as drought, salinity, and high and low temperature is alleviated by microbial inoculation (Timmusk and Wagner, 1999; Redman et al., 2002; Han and Lee, 2005; Ait et al., 2006; Marquez et al., 2007; Ali et al., 2009). Ali et al. (2009) reported that the *Pseudomonas* sp. strain AKM-P6 can enhance the tolerance of sorghum seedlings to elevated temperatures by inducing physiological and biochemical changes in the plant. Tolerance of plants to abiotic stresses depends on the type of plant and environmental factors. A number of rhizobacteria, including *Pseudomonas* (Hong et al., 1991), are known to promote plant growth. Timmusk and Wagner (1999) reported that inoculation with *Paenibacillus polymyxa* protected *Arabidopsis thaliana* from drought stress by increasing the expression of stress-induced gene *Erd15*.

Ali et al. (2009) reported that inoculation did not affect plant growth and biomass at ambient temperatures, whereas at elevated temperatures, significant differences in plant growth and survival were observed due to inoculation, probably because *Pseudomonas* inoculation triggers some stress responsive mechanisms that enable the plants to tolerate high temperature. The SDS-PAGE analysis of leaf proteins revealed the presence of three high-molecular-weight polypeptides in the inoculated plants exposed to elevated temperatures (Ali et al., 2009). Induction of heat shock proteins has been reported in several plants (Howarth, 1991). McLellan et al. (2007) observed that the rhizosphere fungus *Paraphaeosphaeria quadriseptata* enhanced the thermotolerance of *Arabidopsis thaliana* through the induction of HSP101 and HSP70 proteins. Similarly, Ali et al. (2009) suggested that the inoculation with the *Pseudomonas* sp. strain AKM-P6 enhanced the tolerance of sorghum seedlings to high temperature due to the synthesis of high-molecular-weight proteins and also to the improvement of cellular metabolite levels.

Tropical plant species are greatly affected in growth and development by low temperature (Allen and Ort, 2001). Photosynthesis is often the first physiological process to be inhibited by chilling. Because the Calvin–Benson cycle enzymes are more sensitive to low temperature than photochemical reactions, inhibition of the enzyme-catalyzed reactions of the Calvin–Benson cycle by chilling reduces the utilization of absorbed light energy for CO_2 assimilation and results in an increased photosynthetic electron flux to O_2 (Allen and Ort, 2001; Alam and Jacob, 2002; Lu et al., 2013). All these changes ultimately lead to lower N use efficiency by crop plants. Optimum soil temperature for the growth and development of important field crops is given in Table 8.1.

8.2.1.2 Supply of Adequate Soil Moisture

Supply of adequate moisture during the crop growth cycle is essential to achieve maximum economic yield and consequently higher nitrogen use efficiency. Excess as well as low soil moisture adversely affects plant growth and consequently lower nitrogen use efficiency. However, there may be some exceptions, such as flooded rice culture. Fernandez and Laird (1959) showed that, under optimum soil moisture, wheat yielded 24 kg grain kg^{-1} applied N but only 11 kg grain kg^{-1} when water was liming. Hence, the agronomic efficiency of wheat was less than half when the soil moisture was limiting.

The recovered N in the grain decreased from around 20% to 4% when the water supply was restricted in the experiment of Spratt and Gasser (1970), and from around 30% with water kept at field capacity in the experiments of Thompson et al. (1975). In such experiments, water supply would affect the growth rate as well as the availability of soil nitrogen (Novoa and Loomis, 1981). Spratt and Gasser (1970) found that the recovery of N was greater from nitrate than from ammonium under adequate water supply, but with limited water, the difference was smaller, probably reflecting the much greater mobility of nitrate in soil. The use of irrigation is the best practice to improve soil moisture content for crop production wherever possible. In addition, the use of conservation tillage and mulching may also help in improving the soil moisture content.

8.2.1.3 Improving Bulk Density

Bulk density is a physical property of the soil that can be used as a simple index to the general structural condition of the soil. Although it cannot be interpreted in a specific manner as with the degree of aggregation, aggregate stability, or pore size distribution, bulk density does provide a general index to air–water relations and impedance to root growth. Bulk density is defined as the mass of dry soil per unit bulk volume. The value is expressed as megagram per cubic meter (Mg m^{-3}) or

TABLE 8.1

Optimum Soil Temperature for the Growth and Development of Important Crop Species

Crop Species	Temperature (°C)
Alfalfa	21–27
Barley	18–20
Dry bean	28–30
Cassava	25–29
Corn	25–30
Coffee	20–26
Cotton	28–30
Cowpea	20–35
Mung bean	20–30
Oats	15–20
Peas	18–22
Potato	20–23
Peanut	24–33
Pigeon pea	18–29
Rice	25–30
Soybean	22–27
Sesame	25–27
Sorghum	27–30
Sugarbeet	18–24
Sugarcane	25–30
Sunflower	23–25
Sweet potato	24–26
Tobacco	22–26
Tomatoes	25–30
Wheat	20–30

Source: Adapted from Voorhees, W. B., R. R. Allmares, and C. E. Johnson. 1981. *Modifying the Root Environment to Reduce Crop Stress*, pp. 217–266. St. Joseph, Michigan: Am. Soc. Agric. Eng.; Fageria, N. K. 1989. *Tropical Soils and Physiology of Field Crops*. Brasilia: EMBRAPA-CNPAF; Fageria, N. K. 1992. *Maximizing Crop Yields*. New York: Marcel Dekker.

gram per cubic cm (g cm^{-3}). If 1 cm^3 of the soil weighs 1.3 g, the soil has a bulk density equal to 1.3 g cm^{-3}. Dry weight refers to the weight of the soil dried at 105–110°C and the volume of the soil refers to the combined volume of the soil solids and pore space.

The bulk density of most surface soils usually ranges from 1.0 to 1.6 g cm^{-3} or from 1.0 to 1.6 Mg m^{-3}. Compacted subsoils have bulk density values reaching 1.8 g cm^{-3} or 1.8 Mg m^{-3} or more. High bulk density values (greater than 1.6 Mg m^{-3}) indicate limited soil aeration, very slow water movement, poor drainage, and impedance to root growth. High bulk density for surface soils indicates the presence of soil crust that critically retards seed germination. Bulk density can be improved with the addition of organic manures and incorporation of crop residues into the soil. In addition, the use of appropriate crop rotation can also improve soil bulk density. The adequate value of bulk density in the soils permits better root growth and distribution in the soil profile. Plant roots in a noncompacted soil have a more extensive feeding zone than those in a similar soil with a compacted layer. In the presence of a hard pan, the development of the root system is greatly impaired. Roots tend to spread along the surface of the hard pan layer. Better root growth with wider or uniform distribution can improve nutrient uptake and use efficiency by crop plants.

8.2.1.4 Improving Soil Structure

The development of well-structured soils is one of the major objectives for achieving sustainable agricultural systems (Diaz-Zorita et al., 2004; Bhattacharyya et al., 2013). The clustering of soil particles (sand, silt, and clay) into aggregates or peds, and their arrangement into various patterns, is termed soil structure. Soil structure is a physical manifestation of the processes involved in the development of soil bodies. It is one of the differentiating characteristics among soils that it is always included in the morphological description of soil profiles. From the agronomic standpoint, soil structure affects plant growth through its influence on infiltration, percolation, and retention of water, soil aeration, and mechanical impedance to root growth. Six et al. (2004) reported that important aspects for optimal functioning of soils that would deliver services for agriculture and reduce soil erosion are soil structure, a property largely defined by aggregation. It follows that well-aggregated soil is critical for nutrient and water use efficiency. Aggregates provide porosity, which not only improves water retention and the recovery of nutrients and water for crops but also enables better root growth.

Soil structure influences crop productivity because it is related to soil organic matter content (Bird et al., 2000), microbial biomass (Franzluebbers et al., 1996), N mineralization (Hassink, 1994), and crop N requirement (Oberle and Keeney, 1990). For example, Oberle and Keeney (1990) reported that corn N requirements were lower on irrigated loamy sandy soils than on silt loam soils in a study conducted in the northern U.S. corn belt. The contribution of soil texture to crop yield variability across the landscape has been reported by many workers (McConkey et al., 1996; Cox et al., 2003; Cambouris et al., 2006; Nyiraneza et al., 2012), indicating the need for variable-rate N fertilization (Nyiraneza et al., 2012). Other studies, such as those conducted in eastern Canada by Nolin et al. (1989) and Leclerc et al. (2001), have demonstrated that soil texture is the most important component in the soil fertility classification system.

Soil compaction is a serious soil structural problem on many fine-textured soils. Compaction of soil affects nearly all soil properties and functions, which in turn affect the growth and productivity of plants (Gregorich et al., 2011). Compacted soils are characterized by high strength, high bulk density, and low hydraulic conductivity and air-filled porosity (Lowery and Schuler, 1991; Blanco-Canqui and Lal, 2008). Soil with high strength impedes the growth (Montagu et al., 2001; Bengough et al., 2006), distribution (Kasper et al., 1991), and function (Tardieu, 1994) of roots. With these restrictions, root systems develop superficially and as a consequence, the roots explore a smaller volume of soil and hence intercept a limited amount of water and nutrients (Oussible et al., 1992). Lower matric hydraulic conductivity and porosity can restrict soil gas diffusion and water availability. Reduced O_2 content in compacted surface soils, resulting in part from reduced porosity and soil structure degradation (Topp et al., 2000), can in turn affect the transport, absorption, and transformation (e.g., mineralization) of nutrients (Lipies and Stepniewski, 1995). Gregorich et al. (2011) reported that compaction degrades the soil structure and consequently reduces N uptake by corn. As a consequence, relatively large amounts of postharvest soil N in compacted soils increase the risk of N loss to the environment by denitrification or leaching.

Soil structure can be improved with the addition of organic manures and appropriate crop rotation. The use of fertilizers in adequate amount for crop production has a positive effect on crop residue (Campbell et al., 1993a); in turn, this will have a direct effect on soil aggregation (Campbell et al., 1995). The latter will influence soil structure, erodibility, soil workability, and water infiltration. Hence, the adoption of proper fertilization practice is imperative for maintaining the soil in good physical condition for crop production. Legumes (green manure and particularly grass–alfalfa hay crops) are very effective in promoting good soil aggregation (Birch, 1959).

Tillage operations are major factors involved in soil structure degradation on arable lands (Kay, 1990). Tillage management can have a profound impact on soil structural properties. Conservation tillage can improve the structure of arable lands. The results of several studies showed that surface soil in zero-tilled plots had significantly greater aggregate mean weight diameter and available water capacity than soil that had been tilled (Mikha and Rice, 2004; Prakash et al., 2004; Gulde et al., 2008).

8.2.2 CHEMICAL PROPERTIES

Soil chemical properties play a significant role in determining nitrogen use efficiency in crop plants. Important chemical properties related to N availability, uptake, and use efficiency are soil acidity, soil fertility, soil salinity and alkalinity, use of balanced nutrient supply, and reducing allelopathy. These soil chemical properties should be favorable for plant growth to have higher nitrogen use efficiency.

8.2.2.1 Liming Acid Soils

Soil acidity is a major constraint to crop production for large areas worldwide. Soils become acidic during geological evolution, especially in areas of high rainfall, because bases are leached to lower profiles leaving surface layers acidic. Furthermore, acidity is associated with the release of protons (H^+) during the transformation and cycling of C, N, and S in soil–plant systems (Bolan and Hedley, 2003). Sumner and Noble (2003) reported that top soils affected by acidity account for 30% of the total ice-free land areas of the world, with the Americas, Africa, and Asia accounting for the largest portions. In tropical South America, 85% of the soils are acidic, and approximately 850 million ha of this area is underutilized (Fageria and Baligar, 2001a). Theoretically, soil acidity is measured in terms of H^+ and Al^{3+} concentrations in soil solutions. In Crop production, for crop production, soil acidity involves many factors that adversely affect plant growth and development. Plant growth on acidic soils can be limited by deficiencies of N, P, K, Ca, Mg, or Mo; toxicity of H, Al, or Mn; reduced organic matter breakdown and nutrient cycling by microflora; and reduced uptake of nutrients by plant roots and inhibition of root growth (Marschner, 1991). In Brazilian Oxisols, deficiencies of most essential macro- and micronutrients have been reported for the production of upland rice, corn, wheat, dry bean, and soybean (Fageria and Baligar, 1997a). The positive effects of liming on crop growth may be associated with the amelioration of one or more of the abovementioned factors (Haynes, 1984), and possibly from reduced weed growth (Haynes, 1984; Legere et al., 1994).

Soil acidity adversely affects the morphological, physiological, and biochemical processes in plants and consequently the N uptake and use efficiency (Raven, 1975; Feldman, 1980; Foy, 1984; Baligar et al., 1997; Grewal and Williams, 2003; Fageria and Baligar, 2005). External pH affects root growth by influencing apoplastic auxin translocation from the zone of synthesis (i.e., root tips) to the zone of root elongation (Raven 1975; Feldman, 1980). Excessive H^+ concentrations have been reported to cause severe reductions in the shoot and root growth of wheat (Johnson and Wilkinson, 1993) and sorghum (Wilkinson and Duncan, 1989). Reduction in root growth and nodulation of alfalfa in acidic soil has been reported by Grewal and Williams (2003). Apart from H^+ toxicity, excessive Al^{3+} and Mn^{2+} as well as deficiencies of nutrients have been found to be major factors contributing to the reduction in plant growth and nutrient uptake in acidic soils (Foy, 1984). Figure 8.1 shows that the grain yield of dry bean decreased significantly with increasing Al levels in the Brazilian Oxisol. Acidity also decreases N uptake and use efficiency by reducing N mineralization, nitrification, and nodulation as well as by reducing the root development of crop plants (Grewal and Williams, 2003; Menzies, 2003).

Liming is the most common and effective practice for reducing soil acidity-related problems. Lime significantly increased grain yields of annual crops such as common bean, corn, and soybean grown on Brazilian Oxisols (Fageria, 2001; Fageria and Baligar, 2001a, 2003b). Figures 8.2 through

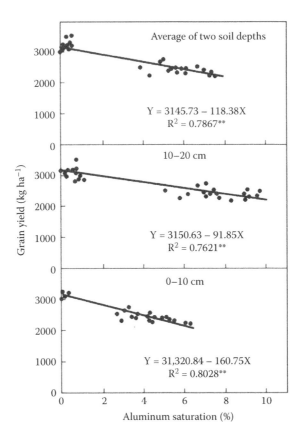

FIGURE 8.1 Relationship between aluminum saturation and grain yield of dry bean. (From Fageria, N. K. 2008. *Commun. Soil Sci. Plant Anal.* 39:845–857. With permission.)

8.6 show that dry bean yields increased significantly and in a quadratic fashion with increased soil pH, base saturation, calcium saturation, magnesium saturation, and potassium saturation in the range of 5.2–7.0 in a Brazilian Oxisol. Across two soil depths, maximum yields calculated on the basis of regression equation were obtained at a soil pH of 6.5, base saturation of 67%, Ca saturation of 48%, and Mg saturation of 19% (Fageria, 2008). It has been reported by many scientists that Ca added with NH_4–N increases plant N use efficiency because of more rapid absorption, greater rates of tillering in cereals, greater metabolite deposition in seeds, and possibly increases in photosynthesis (Fenn et al., 1991, 1993, 1995; Bailey, 1992). Alexander et al. (1991) and Sung and Lo (1990) reported that enhanced NH_4^+ absorption causes substantial increase in photosynthesis.

For correcting soil acidity, dolomitic lime [$CaMg(CO_3)_2$], which has both Ca and Mg, should be used. Dolomitic lime may supply both Ca and Mg and can maintain balances between these two elements. The equations below illustrate the kind of reactions that follow the addition of dolomitic lime to an acidic soil:

$$CaMg(CO_3)_2 + 2H^+ \rightarrow 2HCO_3^- + Ca^{2+} + Mg^{2+}$$
$$2HCO_3 + 2H^+ \rightarrow 2CO_2 + 2H_2O$$
$$CaMg(CO_3)_2 + 4H^+ \rightarrow Ca^{2+} + Mg^{2+} + 2CO_2 + 2H_2O$$

The above equations show that the acidity-neutralizing reactions of lime occur in two steps. In the first step, Ca and Mg react with H to replace these ions with Ca^{2+} and Mg^{2+} on the exchange sites

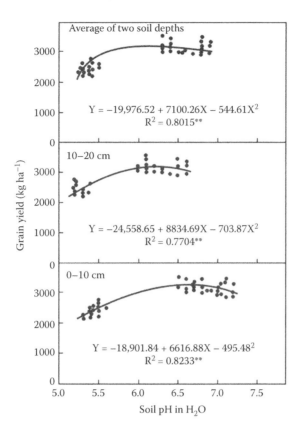

FIGURE 8.2 Relationship between soil pH and grain yield of dry bean. (From Fageria, N. K. 2008. *Commun. Soil Sci. Plant Anal.* 39:845–857. With permission.)

(negatively charged particles of clay or organic matter), forming HCO_3^-. In the second step, HCO_3^- reacts with H^+ to form CO_2 and H_2O to increase pH. The liming reaction rate is mainly determined by the soil moisture and temperature and quantity and quality of liming material. To get maximum benefits from liming or for improving crop yields, liming materials should be applied in advance of crop sowing and thoroughly mixed into the soil. Selected soil chemical property changes with lime applied to a Brazilian Oxisol are presented in Tables 8.2 and 8.3.

The quantity of lime required for specific crops can be determined by laboratory methods (Adams, 1984; Fageria and Baligar, 2003b). However, the best method for lime quantity determination for a given crop is crop yield versus lime rate curves (Fageria and Baligar, 2005). Some of the curves developed for the dry bean, soybean, and corn in a Brazilian Oxisol are presented and discussed by Fageria (2001) and Fageria and Baligar (2005). The author developed a lime response curve for soybean grown on a Brazilian Oxisol (Figure 8.7). The maximum grain yield of soybean was obtained with the addition of 12.6 Mg dolomitic lime per hectare. In tropical America, the quantity of lime required is calculated by taking into account Al, Ca, and Mg according to the following equation (Fageria et al., 1990):

$$Lime\ rate\ (Mg\ ha^{-1}) = (2 \times Al) + \{2 - (Ca + Mg)\}$$

where Al, Ca, and Mg are in $cmol_c\ kg^{-1}$ soil.

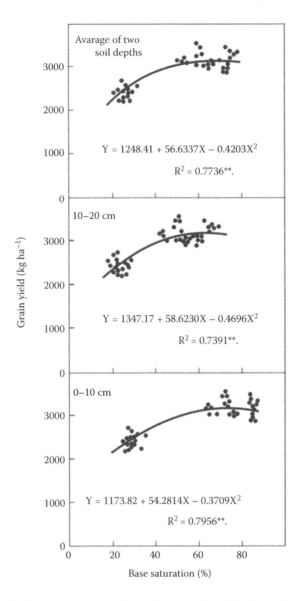

FIGURE 8.3 Relationship between base saturation and grain yield of dry bean. (From Fageria, N. K. 2008. *Commun. Soil Sci. Plant Anal.* 39:845–857. With permission.)

In addition, in Brazil the lime rate is also determined on the basis of base saturation by using the following formula (Fageria et al., 1990):

$$\text{Lime rate (Mg ha}^{-1}) = [\text{CEC } (B_2 - B_1)/\text{TRNP}] \times df$$

where CEC is the cation exchange capacity or total exchangeable cations (Ca^{2+}, Mg^{2+}, K^+, $H^+ + Al^{3+}$) in $cmol_c$ kg^{-1}, B_2 is the desired optimum base saturation, B_1 is the existing base saturation, TRNP is the total relative neutralizing power of the liming material, and df is the depth factor, 1 for 20 cm depth and 1.5 for 30 cm depth.

For Brazilian Oxisols, the desired optimum base saturation for most cereals is in the range of 50–60%, and for legumes it is in the range of 60–70% (Fageria et al., 1990). However, there may

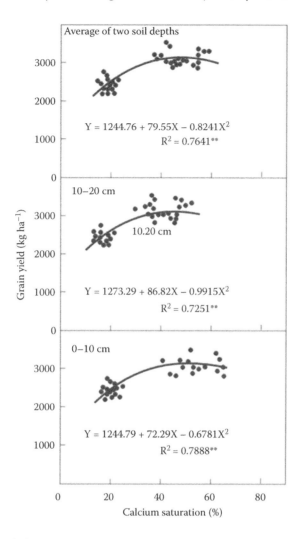

FIGURE 8.4 Relationship between calcium saturation and grain yield of dry bean. (From Fageria, N. K. 2008. *Commun. Soil Sci. Plant Anal.* 39:845–857. With permission.)

be exceptions, such as upland rice, which is very tolerant of soil acidity and can produce good yield at a base saturation lower than 50%. Specific optimal base saturation values for important annual crops grown on Brazilian Oxisols are given in Table 8.4. The quantity of lime required depends on the quality of liming material, crop species, soil pH, concentrations of Ca, Mg, and Al in the soil, soil type, and economic considerations.

8.2.2.2 Adequate Soil Fertility

Soil fertility refers to the ability of a soil to supply the nutrient elements in the amounts, forms, and proportions required for maximum plant growth. It is measured in terms of the amount of the available forms of the nutrient elements in the soil at any given time. The fundamental components of soil fertility are the essential plant nutrients. There are 17 essential plant nutrients for plant growth. The essential nutrients from the soil occur there largely as constituents of minerals and organic matter and in smaller amounts in the so-called exchangeable (adsorbed) ionic form. The adsorbed and ionic forms are extracted readily when contacted by plant roots and provide the active fertility

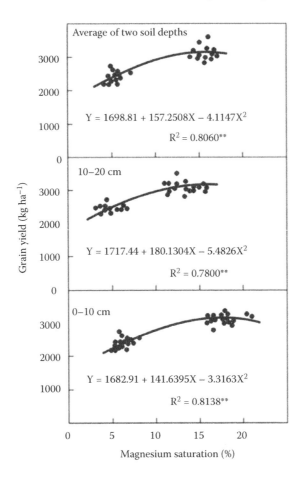

FIGURE 8.5 Relationship between magnesium saturation and grain yield of dry bean. (From Fageria, N. K. 2008. *Commun. Soil Sci. Plant Anal.* 39:845–857. With permission.)

of the soil. Nutrient elements that are not immediately available to plants, such as those in primary and secondary minerals and in semiresistant organic combinations, comprise the potential fertility of the soil. Often, crop production is highly dependent upon the rate of transfer from the potential to the active form.

Maintaining soil fertility at an adequate level is very important not only to improve crop yields but also maintain the sustainability of the cropping systems. Soils of low fertility used for crop production provide low yield as well as degrade more rapidly compared with soils of high fertility. Crops produced on high-fertility soils produced a higher yield of straw and root biomass, which is responsible for maintaining soil organic matter content at an adequate level (Fageria, 2013). The export of grain or other produce from the land takes with it large quantities of N, P, K, and S (Campbell et al., 1995). If these nutrients are not replaced (e.g., by addition of fertilizers), soil organic matter and the fertility of the soil will gradually decline. Because fertilizers and the use of legumes in the rotation increase the N supplying power of the soils, their regular use, year after year, can lead to a significant improvement in the fertility of the soil (Campbell et al., 1995). Fageria and Baligar (1997b) reported that an integrated nutrient management system should be adopted in improving the nutrient use efficiency of annual crops. A model of integrated nutrient management is presented by these authors (Figure 8.8).

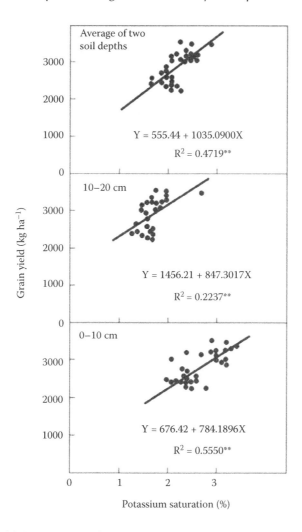

FIGURE 8.6 Relationship between potassium saturation and grain yield of dry bean. (From Fageria, N. K. 2008. *Commun. Soil Sci. Plant Anal.* 39:845–857. With permission.)

8.2.2.3 Controlling Soil Salinity and Alkalinity

Soil salinity and alkalinity problems exist in many regions of the world. There are more than 800 million hectares of land that suffer from salinization and alkalinization in the world (Lv et al., 2013). The control of soil salinity and alkalinity is an important strategy in efficient N management in crop production. A large part of world's agricultural land is affected by salts and this imposes serious limitations on crop growth and productivity (Lauchili and Luttge, 2002; Flexas et al., 2007; Guo et al., 2013) and consequently on N use efficiency (Fageria, 2013). In naturally saline soils, the predominant ions are Na^+, Ca^{2+}, Mg^{2+}, K^+, Cl^-, SO_4^{2-}, CO_3^{2-}, and NO_3^- and the main salts responsible are NaCl, Na_2SO_4, $NaHCO_3$, and Na_2CO_3 (Lauchli and Luttge, 2002). Many studies suggest that saline soils present two distinct forms of stress, that is, salt stress (principally NaCl and Na_2SO_4) and alkali or pH stress ($NaHCO_3$ and Na_2CO_3) (Guo et al., 2013). High pH conditions not only damage the plant directly but also cause deficiencies of micronutrients (Fageria, 2013).

In addition, salt stress induces an osmotic stress and direct ion injury by disrupting ion homeostasis and the ion balance within cells (Niu et al., 1995; Ghoulam et al., 2002; De-Lacerda et al., 2003). Salinity- and alkalinity-tolerant plant species have the capacity to absorb and utilize N more

TABLE 8.2

Values of Soil pH and Ca^{2+} and Mg^{2+} Contents after Harvest of Eight Crops (Upland Rice, Common Bean, Corn, and Soybean) Grown in Rotation for 4 Years on a Brazilian Oxisol at Two Soil Depths

Lime Rate (Mg ha^{-1})	pH in H$_2$O	Ca^{2+} (cmol$_c$ kg^{-1})	Mg^{2+} (cmol$_c$ kg^{-1})
0–20 cm Soil Depth			
0	5.6	1.9	1.0
4	6.0	2.3	1.1
8	6.2	3.0	1.2
12	6.4	3.1	1.2
16	6.5	3.3	1.3
20	6.8	3.8	1.4
20–40 cm Soil Depth			
0	5.5	1.7	0.9
4	5.9	1.9	1.0
8	6.1	2.3	1.1
12	6.2	2.4	1.1
16	6.3	2.6	1.2
20	6.7	3.3	1.3

Source: Adapted from Fageria, N. K. 2001. *Pesq. Agropec. Bras.* 36:1419–1424.

efficiently compared to sensitive plant species (Fageria, 2013). Plants respond to environmental stresses by reprogramming various physiological processes, including selective ionic absorption, which reduces ion toxicity, and the accumulation of osmoregulation substance to maintain continuous water absorption at a low soil water potential. It has been shown that plants respond differently to alkaline salts (NaHCO$_3$ and Na$_2$CO$_3$) and neutral salts (NaCl and Na$_2$SO$_4$) (Yang et al., 2007).

Guo et al. (2010) compared the response of wheat seedlings to saline and alkaline stress, and showed that alkaline stress inhibits fructan synthesis, while saline stress increases it. Yang et al. (2008) reported that *Suaeda glauca* had similar osmotic adjustment responses to saline and alkaline stresses in shoots, including the accumulation of proline, organic acids, and inorganic ions, but the mechanisms governing ionic balance under these stresses were different. Under saline stress, *Suaeda glauca* accumulated more inorganic ions than organic acids, while it was the reverse under alkaline stress. Islam et al. (2011) reported that Na concentration in leaves of foxtail millet (*Setaria italica* L.) and proso millet (*Panicum miliaceum* L.) was increased two times by alkaline than in saline stress. However, the tolerance to alkaline stress was different between the two millet species, with higher tolerance in proso millet than in foxtail millet. In rice, the inorganic anions were dominant in maintaining intracellular ionic equilibrium under salt stress; however, organic acids, especially malate and citrate, were the dominant components in maintaining intracellular ion balance under alkaline stress (Wang et al., 2011).

Lv et al. (2013) studied the effects of saline–alkaline stress in rice and found that seed germination is most strongly affected by osmotic stress, indicating that water availability is the determining factor for rice seed germination under saline–alkaline stress. Meanwhile, seedlings grew the least under alkaline stress, which may be ascribed to severe cellular injury in the root system in addition to ion toxicity from high salinity. The alkaline treatment significantly reduced the total

TABLE 8.3
Selected Soil Chemical Properties after Harvest of Three Dry Bean Crops at Two Soil Depths as Influenced by Liming Treatments

Soil Property	Lime Rate (Mg ha⁻¹)			F-Test	CV (%)
	0	12	24		
0–10 cm Soil Depth					
pH	5.5c	6.5b	7.0a	*	2
Base saturation (%)	32.6a	73.1b	85.6c	*	8
H + Al (cmol$_c$ kg⁻¹)	6.2a	2.1b	1.1c	*	12
Acidity saturation (%)	67.6a	26.9b	14.4c	*	14
Ca (cmol$_c$ kg⁻¹)	2.1c	4.2b	4.9a	*	12
Mg (cmol$_c$ kg⁻¹)	0.6b	1.4a	1.4a	*	12
CEC (cmol$_c$ kg⁻¹)	9.3a	8.1b	7.6b	*	7
Organic matter (g kg⁻¹)	19.9	19.8	19.9	NS	1
0–20 cm Soil Depth					
pH	5.3c	6.1b	6.5a	*	3
Base saturation (%)	24.2c	51.3b	66.0a	*	10
H + Al (cmol$_c$ kg⁻¹)	6.8a	4.1b	2.8c	*	8
Acidity saturation (%)	75.8a	48.7a	33.9c	*	9
Ca (cmol$_c$ kg⁻¹)	1.6c	3.1b	3.9a	*	15
Mg (cmol$_c$ kg⁻¹)	0.4c	1.1b	1.2a	*	15
CEC (cmol$_c$ kg⁻¹)	8.9a	8.4b	8.1b	*	7
Organic matter (g kg⁻¹)	17.9	17.7	18.0	NS	13

Source: Adapted from Fageria, N. K., V. C. Baligar, and R. W. Zobel. 2007. *Commun. Soil Sci. Plant Anal.* 38:1637–1653.

*, NS: Significant at the 1% probability level and not significant, respectively.

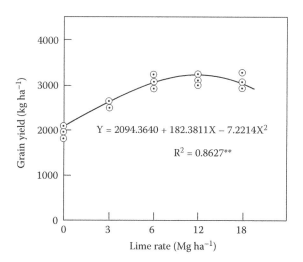

$$Y = 2094.3640 + 182.3811X - 7.2214X^2$$

$$R^2 = 0.8627**$$

FIGURE 8.7 Response of soybean to liming grown on a Brazilian Oxisol.

TABLE 8.4

Optimal Base Saturation for Important Annual Crops Grown on Brazilian Oxisols

Crop Species	Type of Experiment	Plant Part Measured	Base Saturation (%)	Reference
Common bean	Field	Grain yield	60	Fageria and Santos (2005)
Common bean	Field	Grain yield	69	Fageria and Stone (2004)
Upland rice	Field	Grain yield	40	Fageria and Baligar (2001)
Common bean	Field	Grain yield	70	Lopes et al. (1991)
Corn	Field	Grain yield	59	Fageria (2001)
Soybean	Field	Grain yield	63	Fageria (2001)
Upland rice	Field	Grain yield	50	Lopes et al. (1991)
Upland rice	Field	Grain yield	30	Sousa et al. (1996)
Common bean	Field	Grain yield	71	Fageria and Stone (2004)
Corn	Field	Grain yield	60	Raij et al. (1985)
Wheat	Field	Grain yield	60	Lopes et al. (1991)
Soybean	Field	Grain yield	60	Raij et al. (1985)
Cotton	Field	Grain yield	60	Raij et al. (1985)
Sugarcane	Field	Cane yield	50	Raij et al. (1985)
Soybean	Field	Grain yield	61	Gallo et al. (1986)

Source: Adapted from Fageria, N. K. 2001. *Pesq. Agropec. Bras.* 36:1419–1424; Fageria, N. K., and V. C. Baligar. 2001. *Commun. Soil Sci. Plant Anal.* 32:1303–1319; Fageria, N. K., and A. B. Santos. 2005. Influence of base saturation and micronutrient rates on their concentration in the soil and bean productivity in cerrado soil in a no-tillage system. Paper presented at the VIII National Bean Congress, Goiânia, Brazil, 18–20 October 2005; Fageria, N. K., and L. F. Stone. 1999. Acidity management of cerrado and varzea soils of Brazil. Santo Antônio de Goiás, Brazil: EMBRAPA Arroz e Feijão Document No. 92; Gallo, P. B. et al. 1986. *Rev. Bras. Ci. Solo* 10:253–258; Lopes, A. S., M. C. Silva, and L. R. G. Guilherme. 1991. Soil acidity and liming. Technical Bulletin 1, São Paulo, Brazil: National Association for Diffusion of Fertilizers and Agricultural Amendments; Raij, B. V. et al. 1985. Fertilizer and lime recommendations for the State of São Paulo, Brazil. Technical Bulletin 100, Campinas, Brazil: Agronomy Institute; Sousa, D. M. G., L. N. Miranda, and E. Lobato. 1996. Evaluation methods of lime requirements in cerrado soils. Cerrado Center of EMBRAPA Technical Circular 33, Planaltina, Brazil: Cerrado Center of EMBRAPA.

biomass, total root length, root surface area, root number, and root volume, while it increased the root diameter and Na^+/K^+ ratio. Further, these authors found that rice seedlings accumulate significant amounts of osmolytes, most abundantly in response to alkaline treatment, probably in an attempt to cope with cellular damages from the stress.

Soil salinity and alkalinity can be reduced by adopting many soil and plant management practices. Management practices that can improve crop yields and consequently nutrient use efficiency by crop plants grown on salt-affected soils are use of soil amendments to reduce the effect of salts, the application of farmyard manures to create favorable plant growth environments, the leaching of salts from the soil profile, and the planting of salt-tolerant crop species or genotypes within species (Fageria, 2014). The addition of fertilizers, especially potassium, may also help in reducing salinity effects and improving nutrient use efficiency. The maintenance of an internal positive turgor potential of plants exposed to saline conditions is an important factor for maintaining growth. This is accomplished by the uptake of ions, chiefly K^+, Na^+, and Cl^-, as well as by synthesizing organic metabolites. Potassium is the most abundant cation in the cytoplasm, and, in glycophytes, it plays an important role in osmotic adjustment (Marschner, 1995). Thus, the application of high K^+ fertilization might enhance the capacity for osmotic adjustment of plants growing in saline habitats. Planting salt-tolerant crop species or genotypes within species is an attractive, economical, and

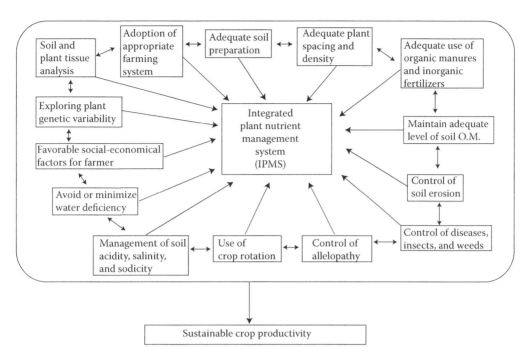

FIGURE 8.8 Integrated plant nutrient management for sustainable crop production. (From Fageria, N. K., and V. C. Baligar. 1997b. *Int. J. Trop. Agric.* 15:1–18. With permission.)

environmentally sound practice to improve crop growth, including rice on saline soils (Fageria, 2014). Salt-tolerant crop species are barley, cotton, oats, rye, triticale, sugar beet, guar, and canola or rapeseed (Fageria, 2014)

8.2.2.4 Use of Organic Manures

Organic manures are products from the processing of animals or vegetable substances that contain a reasonable amount of plant nutrients to be of value as fertilizers. Brosius et al. (1998) reported that plant- and animal-based organic by-products may substitute for commercial fertilizers and enhance the chemical and biological attributes of soil quality in agricultural production systems. Organic matter increases the soil's abilities to hold and make available essential plant nutrients and to resist the natural tendency of the soil to become acidic (Cole et al., 1987). Furthermore, the addition of organic manures to acid soils has been shown to increase soil pH, decreases Al saturation, and thereby improves conditions for plant growth (Alter and Mitchell, 1992; Reis and Rodella, 2002; Wong and Swift, 2003). All these processes improve N use efficiency. Miyazawa et al. (1993) reported that crop residues of wheat and corn and 20 plant species utilized as green manure increased soil pH and decreased the Al content of the soil. Several mechanisms have been proposed for reducing acidity by organic manures. These mechanisms include specific adsorption of organic anions on hydrous Fe and Al surfaces and the corresponding release of hydroxyl ions, which increase soil pH (Hue, 1992; Wong and Swift, 2003). Adsorption of Al by organic matter sites and the subsequent isolation of the inorganic phase to maintain the equilibrium Al activity in soil solution have been proposed to increase soil pH (Wong et al., 1998; Wong and Swift, 2003). Chelating agents released by decomposing organic matter may detoxify Al ions.

Plant roots decay in the soil and form soil organic matter. Active roots also release organic acids such as citrate, malate, and tartrate. These organic acids react strongly with Al and convert it into less toxic organically bound forms (Yang et al., 2000). The organic Al affinities or stability constants are in the order of citrate > tartrate > malate (Hue et al., 1986). The decrease in Al activity by the addition of organic matter has been reported by Kochain (1995). The functional groups involved in metal complexation by organic matter are COOH and OH (Wong and Swift, 2003). Surface application or surface incorporation of organic matter also decreased phytotoxic subsoil Al^{3+} activities because dissolved organic matter (DOM) that leached into the subsoil formed nontoxic Al–DOM complexes (Hue, 1992; Liu and Hue, 1996; Hue and Licudine, 1999; Willert and Stehouwer, 2003). The combined application of $CaCO_3$ and organic matter in lime-stabilized biosolids decreased subsoil acidity and increased subsoil Ca saturation, compared with $CaCO_3$ alone (Tan et al., 1985; Brown et al., 1997; Willert and Stehouwer, 2003). This effect was attributed to increases in Ca mobility caused by Ca–DOM complexes (Willert and Stehouwer, 2003).

Additional benefits of organic matter addition to acid soils are improving nutrient cycling and availability to plants through direct additions as well as through modification in soil physical and biological properties. A complementary use of organic manures and chemical fertilizers has proven to be the best soil fertility management strategy in the tropics (Makinde and Agboola, 2002; Fageria and Baligar, 2005). Enhanced soil organic matter increases soil aggregation and the water-holding capacity, provides the source of nutrients, and reduces P fixation, toxicities of Al and Mn, and leaching of nutrients (Baligar and Fageria, 1999). Build-up of organic matter through the addition of crop and animal residues increases the population and species diversity of microorganisms and their associated enzyme activities and respiration rates (Kirchner et al., 1993; Weil et al., 1993). The use of organic compost may result in a soil that has greater capacity to resist the spread of plant pathogenic organisms. The improvement in the overall soil quality may produce more vigorous-growing and high-yielding crops (Brosius et al., 1998). All these changes create a favorable soil environment for N uptake and use efficiency in crop plants.

8.2.2.5 Use of Balanced Nutrient Supply

The use of balanced nutrient supply improves N use efficiency in crop plants. Duan et al. (2011) reported that unbalanced inorganic fertilization resulted in a decreased yield of corn and nitrogen use efficiency. These authors reported that nitrogen use efficiency in corn was increased from 20% to 45% by the application of P in adequate amounts along with N in corn. In addition, these authors further reported that nitrogen use efficiency in corn was further improved by up to 70% with the continuous application of farmyard manure. The application of farmyard manure is known to improve soil physical and chemical properties, maintain soil fertility, and supply nutrients to crops on time that will minimize N losses (Duan et al., 2011). Improvement in nitrogen use efficiency with the addition of P in adequate amounts along with N has been reported by Almeida et al. (2000) and Sa and Israel (1998). Wang et al. (2010) also reported that the use of adequate amounts of N and P improved plant growth and N use efficiency in crop plants. Wang et al. (2010) reported that agronomic efficiency could be improved from 3.5 kg grain kg^{-1} N under the N treatment to 16.3 kg grain kg^{-1} N under the NP treatment in China in corn.

8.2.2.6 Crop Residue Management

Crop residues are defined as the parts of the plants left in the field after the crops have been harvested and thrashed or left after the pastures are grazed (Kumar and Goh, 2000). They are a tremendous natural resource and should not be considered as a waste. Crop residues are important for erosion control, soil water storage, filling gaps in various agroecosystems-based modeling, and a sink for atmospheric carbon (Aguilar et al., 2012). Data on nitrogen benefits and nitrogen recoveries from residues show that a considerable potential exists from residues, especially leguminous residues, not only in meeting the N demands of the succeeding crops but also in increasing the long-term fertility of the soils (Kumar and Goh, 2000). The amount of N that recycle into agricultural fields through residues may add 25–100 Tg of N $year^{-1}$ (Mosier and Kroeze, 1998). In addition, crop residues and their proper management affect the soil quality either directly or indirectly (Fageria, 2002).

A number of studies have shown that the management of straw residues has the potential to alter the magnitude of N immobilization. Managing the residue particle size may significantly alter the decomposition rates (Sims and Frederick, 1970; Ambes and Jensen, 1997; Angers and Recous, 1997). During the initial decomposition, microbial colonizers must penetrate the straw epidermis with its protecting cuticula and then the layers of strongly lignified sclerenchyma of the internode before colonizing less lignified parenchymatic cells (Ambus and Jensen, 2001). Breaking up plant material may therefore facilitate microbial attack (Amato et al., 1984). In addition, finely ground plant material mixes more intimately with the soil accelerating microbial breakdown (Angers and Recous, 1997). Furthermore, fine particles may also be physically protected against decomposition by adsorption to clay and other soil constituents (Jensen, 1997). Ambus and Jensen (2001) reported that adequate management of crop residues may regulate soil N mineralization–immobilization and match the N release synchronously with the crop requirement.

Incorporation of crop residue into the soil improves soil fertility, including N, and crop productivity by increasing C sequestration and reducing the emission of greenhouse gases among other parameters (Biau et al., 2013). Wilhelm et al. (2004) and Lal (2005) also reported that the incorporation of stover has many benefits, including the prevention of soil erosion, maintenance of soil organic matter and soil structure by humification, and as a source of energy for soil biota. Stover is also an important source of macronutrients (NPK) and micronutrients such as S, Cu, B, Zn, and Mo (Mubarak et al., 2002). Stover improves soil organic carbon (SOC), which is a key of the CO_2 sink, maintaining the productivity of agriculture while reducing greenhouse gas emissions and mitigating global climate change (Christopher et al., 2009).

The benefits of high SOC levels include the sequestration of atmospheric CO_2 as well as better soil quality (Blanco-Canqui and Lal, 2009; Benjamin et al., 2010). Plant residues influence N cycling in soils because they are the primary sources and sinks for C and N (Dinnes et al., 2002). Incorporation of crop residues into the soil provides substantial amounts of nutrients, including N for succeeding crops (Carranca et al., 1999; Ambus and Jensen, 2001). In the long term, straw incorporation has resulted in the increased N mineralization potential in rice and nonrice systems (Bacon, 1990). Sustained increases in the microbial biomass have been observed following many seasons of straw incorporation compared with burning (Powlson et al., 1987; Bird et al., 2000).

When plant residues having C/N ratios greater than 20/1 are incorporated into the soil, the available N is immobilized during the first few weeks by the decomposing microbial populations present (Doran and Smith, 1991; Somda et al., 1991; Green and Blackmer, 1995). However, some workers have reported that net immobilization is likely to occur following the addition of plant material with C/N ratios above ~25:1 (Brady and Weil, 2002; Burgess et al., 2002). Cereal straws (rice, corn, wheat, and barley) usually have high C/N ratios (Table 8.5) and may induce temporary N deficiency in crops due to N immobilization by soil microbial populations when straw is not incorporated or decomposes in advance. However, this temporary adverse effect of N immobilization can be alleviated by applications of about 15 kg N ha^{-1} under most cropping systems (Fageria and Baligar, 2005).

Legume crop residues are effective sources of N (Bremer and Van Kessel, 1992; Haynes et al., 1993). When released in synchrony with the crop N demand, crop residue N is a particularly desirable source of N as losses to the environment are minimized (Stute and Posner, 1995; Soon et al., 2001). Legume residues generally have high N contents and lower C/N ratios compared with cereals (Table 8.5). During the mineralization of leguminous materials, up to 50% of the amount of N can be released within 2 months of incorporation into the soil (Kirchmann and Bergqvist, 1989). Besides providing N, crop residues can provide effective weed control and consequently improve N use efficiency if managed properly. Winter weed residues reduced weed seedling emergence by 45% (Crutchfield et al., 1986) and biomass by 60% in corn (Wicks et al., 1994). Crop residues suppress weed emergence by reducing light penetration and soil temperature fluctuations (Teasdale and Mohler, 1993).

Optimizing the N fertility level and water availability can influence residue production and the SOC sequestration potential (Follett, 2001; Halvorson et al., 2006, 2009; Halvorson and Jantalia, 2011). Crop residue production under no-tillage and irrigation should be sufficient to increase SOC storage in the semiarid central Great Plains (Varvel and Wilhelm, 2008; Halvorson et al., 2009;

TABLE 8.5
C/N Ratio of Straw of Major Cereal and Legume Crops

Crop Species	Growth Stage/Age in Days	C/N Ratio	Reference
Corn residues (*Zea mays* L.)	Physiological maturity	67	Burgess et al. (2002)
Rice straw (*Oryza sativa* L.)	Physiological maturity	69	Eagle et al. (2001)
Rice straw (*Oryza sativa* L.)	Physiological maturity	56	Davelouis et al. (1991)
Sorghum (*Sorghum bicolor* L. Moench)	Vegetative	22.0	Clement et al. (1998)
Barley straw (*Hordeum vulgare* L.)	Physiological maturity	99.1	Larnez and Janzen (1996)
Ryegrass (*Lolium multiflorum* Lam)	Vegetative	30	Kuo and Jellum (2002)
Rye (*Secale cereale* L.)	Heading	40	Rannells and Wagger (1996)
Alfalfa hay (*Medicado sativa* L.)	Not given	15.9	Larney and Janzen (1996)
Pea straw (*Pisum sativum* L.)	Physiological maturity	21	Fauci and Dick (1994)
Pea hay (*Pisum sativum* L.)	Not given	15.4	Larnez and Janzen (1996)
Red clover (*Trifolium pratense* L.)	101 days	13.7	Kirchmann (1988)
White clover (*Trifolium repens* L.)	101 days	10.7	Kirchmann (1988)
Yellow trefoil (*Medicago lupulina* L.)	101 days	10.1	Kirchmann (1988)
Persian clover (*Trifolium resupinatum* L.)	101 days	15.8	Kirchmann (1988)
Egyptian clover (*Trifolium alexandrium* L.)	101 days	16.7	Kirchmann (1988)
Subterranean clover (*Trifolium subterraneum* L.)	101 days	11.4	Kirchmann (1988)
Cowpea (*Vigna unguiculata* L. Walp.)	Green pods	13.9	Clement et al. (1998)
Sunnhemp (*Crotalaria juncea* L.)	Mature pods	20.2	Clement et al. (1998)
Soybean (*Glycine max* L. Merr.)	Vegetative	17.9	Clement et al. (1998)
Pigeon pea (*Cajanus cajan* L. Millspaugh)	Not given	25.9	Clement et al. (1998)
Wild indigo (*Indigofera tinctoria* L.)	Flowering	15.8	Clement et al. (1998)
Sesbania (*Sesbania rostrata* Bremek & Oberm)	Vegetative	27.8	Clement et al. (1998)
Sesbania (*Sesbania emerus* Aubl. Urb.)	Vegetative	26.5	Clement et al. (1998)
Hairy vetch (*Vicia villosa* Roth)	Vegetative	12	Kuo and Jellum (2002)
Hairy vetch (*Vicia villosa* Roth)	Flowering	18	Sainju et al. (2002)
Hairy vetch (*Vicia villosa* Roth)	Early bloom	17	Rannelles and Wagger (1996)
Crimson clover (*Trifolium incarnatum* L.)	Midbloom	11	Rannelles and Wagger (1996)
Tropical kudzu (*Pueraria phaseoloides*)	Not given	19	Davelouis et al. (1991)

Source: Adapted from Burgess, M. S., G. R. Mehuys, and C. A. Madramootoo. 2002. *Soil Sci. Soc. Am. J.* 66:1350–1358; Clement, A., J. K. Ladha, and F. P. Chalifour. 1998. *Agron. J.* 90:149–154; Davelouis, J. R., P. A. Sanchez, and J. C. Alegre. 1991. *TropSoils Technical Report 1988–1989*, pp. 286–289. Raleigh, NC: North Carolina State University; Kirchmann, H. 1988. *Acta Agric. Cand.* 38:25–31; Kuo, S. and E. J. Jellum. 2002. *Agron. J.* 94:501–508; Larney, F. J. and H. H. Janzen. 1996. *Agron. J.* 88:921–927; Eagle, A. J. et al. 2001. *Agron. J.* 93:1346–1354; Fauci, M. F. and R. P. Dick. 1994. *Soil Sci. Soc. Am. J.* 58:134–138; Rannells, N. N. and M. J. Wagger. 1996. *Agron. J.* 88:777–782; Sainju, U., Singh, B. P., and S. Yaffa. 2002. *Agron. J.* 94:594–602.

Benjamin et al., 2010; Halvorson and Schlegel, 2012). Follett et al. (2005) reported an increase in SOC under an irrigated, no-till (NT) production system with high N fertility in Mexico. Varvel and Wilhelm (2008) reported significant increases in SOC levels in the 0–7.5 cm soil depth for irrigated continuous corn and corn–soybean rotations in Nebraska.

8.2.2.7   Use of Cover Crops

Cover crops have been shown to provide many beneficial effects on crop growth and development, if properly managed. The beneficial effects of cover crops include reduced soil erosion, increased biological diversity (i.e., microbes, insects, and birds), increased nutrient cycling and biological N_2

fixation, increased soil organic matter, improved weed control, and increased crop yield (Pimentel et al., 1992, 1995; Sainju and Singh, 1997; Williams et al., 1998; Altieri, 1999; Reddy et al., 2003; Wortman et al., 2012). While cover crops have traditionally been used as a soil conservation tool (Pimentel et al., 1995), there is increasing interest in using cover crops to enhance agronomic crop performance. However, maximizing agronomic benefits associated with cover crops will depend on appropriate species selection and residue management (Ashford and Reeves, 2003; Wortman et al., 2012).

The use of legumes as cover crops has been shown to reduce synthetic N input demands by 50–100% depending on the species, the duration of cover crop growth, and the subsequent crop N requirement (Biederbeck et al., 1996; Burket et al., 1997; Wortman et al., 2012). While legume species have a potential for biological nitrogen fixation, faster-growing cover crop species (i.e., grasses and mustard species) may be more useful in scavenging nitrates and nutrient cycling (Dabney et al., 2001). A mixture of legume and nonlegume species may maximize the benefits of biological N_2 fixation and nutrient cycling, as legumes can increase N availability to other crop species in mixture leading to increased productivity (Kuo and Sainju, 1998; Mulder et al., 2002). In addition, termination method and residue management can influence N mineralization, soil availability, and crop uptake (Sainju and Singh, 2001). Incorporation of a cover crop residue via a field disk or plow often results in rapid N mineralization and plant availability, but management of the residue on the soil surface has been shown to result in greater crop N uptake and yield (Sainju and Singh, 2001). Hence, residue management on the soil surface with conservation tillage methods may be effective in syncing N mineralization and availability with crop demand and uptake (Parr et al., 2011; Wortman et al., 2012).

8.2.3 Improving Soil Microbial Biomass

Biological soil properties are related to many changes in the soil environment, which determine the availability of nitrogen for plant growth and development. Microbial and biochemical soil properties have been suggested as early and sensitive indicators of soil quality as they manifest themselves over shorter timescales and are central to the ecological function of a soil (Karlen et al., 1994; Bandick and Dick, 1999; Fageria, 2002). Soil microbial and enzyme activities in particular are increasingly used as indicators of soil quality because of their relationship to decomposition and nutrient cycling, ease of measurement, and rapid response to change in soil management (Dick, 1984; Dilly et al., 2003; Geisseler and Horwath, 2009). In a long-term study, Kandeler et al. (1999) found that enzyme activities were significantly increased in the top 10 cm of the profile after 2 years of minimum and reduced tillage to conventional tillage.

One of the management practices that significantly affects microbial biomass in the soil is tillage. Minimum tillage favors higher microbial biomass compared to conventional tillage. In addition, seasonal fluctuations in soil moisture, temperature, and substrate availability can also have large effects on microbial biomass and activity. Franzluebbers et al. (1994) found that the soil microbial biomass changed significantly during the cropping season in all crop sequences and tillage regimes under investigation. Bausenwein et al. (2008) also reported significant effects of sampling dates on microbial biomass and enzyme activities under minimum tillage.

In general, conservation tillage practices leave a significant amount of plant residue on the soil surface. This results in increased soil organic matter content in the topsoil, which in turn leads to higher microbial biomass and activity (Geisseler and Horwath, 2009). Several studies have shown that reduced tillage increases the organic matter content of the soil and results in physical and chemical stratified soils, with more nutrients and organic matter localized near the surface (Logan et al., 1991; Cannell and Hawes, 1994; West and Post, 2002).

Including grain legumes in crop rotations has other benefits, including breaking disease and pest cycles in cereals (Stevenson and Van Kessel, 1996), improving soil structure (Chan and Heenan, 1991), and improving biological soil health through increased microbial biomass, diversity, and activity (Lupwayi and Kennedy, 2007). The decline in pathogen populations with crop rotation results from natural mortality (due to the absence of a susceptible host) and the antagonistic activities of microorganisms that coexist in the root zone (Peters et al., 2003). Lupwayi et al. (2012) reported that the economic and environmental benefits of including grain legumes in crop rotations may tempt farmers to grow them more frequently than recommended, resulting in potential changes to the soil chemical, physical, and biological properties.

Crop diversity in rotations affects soil microbial communities because different plant species produce different root exudates (Garbeva et al., 2008), and different crop residues returned to the soil after harvest offer heterogeneous substrates to soil microorganisms during residue decomposition (Lupwayi et al., 2004). Monocultures have been shown to result in less soil microbial biomass, activity, and diversity than diversified crop rotation (Acosta-Martinez et al., 2010; Li et al., 2010; Lupwayi et al., 2012). These effects are implicated as major causes of yield decline in monocultures because soil microorganisms are the main drives of nutrient cycling and biological pest control in the soil (Lupwayi et al., 2012).

Lupwayi et al. (2012) reported that a field pea grown in monoculture had less microbial biomass and bacterial functional diversity than a field pea rotated with wheat. Nayyar et al. (2009) also reported less microbial biomass, enzyme (dehydrogenase, phosphatase, and urease) activities, and arbuscular mycorrhizal colonization of the field pea in field pea monoculture than in the wheat–pea rotation. Reductions in the soil microbial biomass in crops grown in monoculture relative to the same crops grown in rotations have been reported in soybean in Argentina (Vargas Gil et al., 2011), sugarcane in Australia (Holt and Mayer, 1998; Pankhurst et al., 2005), and cotton in the southern United States (Acosta-Martinez et al., 2010). Low microbial biomass carbon in field pea monoculture relative to a rotation with wheat is probably related to the low crop biomass and organic C returned to the soil with field pea residues under pea–pea rotation.

Soil invertebrates (i.e., earthworm) and functional groups (bacteria) contribute to a wide range of soil services vital in agricultural soils, such as water and air movement and nutrient cycling (Dominati et al., 2010) through feeding, excretion, burrowing, casting, and litter incorporation (Lavelle et al., 2006). Even in intensive systems, it would appear that invertebrates still have the ability to contribute to soil quality, as Schon et al. (2011) found that invertebrates improved plant uptake of N even in an artificial high-fertility, compacted pasture soil (Schon et al., 2012). Similarly, several authors have reported that invertebrates contribute to a wide variety of soil services and as such can be regarded as indicators of these services and soil conditions (Bardgett, 2005; Lavelle et al., 2006). Improvement in soil physical conditions and the addition of food resources available to the soil food web can increase soil invertebrates (Schon et al., 2012).

8.3 NITROGEN FERTILIZER MANAGEMENT PRACTICES

Mineral nutrients in various forms have been applied to crops for thousands of years. However, during the twentieth century, science has addressed the task of identifying the essential nutrients and recommending the amounts, forms, and timing needed for optimal crop growth (Angus et al., 1993). In this context, improving the efficiency of fertilizer N use is vital to achieve and sustain high crop yields and reduce losses of N that can potentially deteriorate environmental quality. Fertilizer N is being increasingly recognized as an expensive input and also a source of nitrate contamination in the groundwater (Bijay-Singh and Yadvinder-Singh, 2004). Appropriate modification in fertilizer source or management practices can lead to a reduction in losses of N and increased fertilizer N use efficiency (Thind et al., 2010).

Culman et al. (2013) reported that the management choices that growers make can have dramatic impacts on the local ecosystem services. For example, the N fertilizer type, rate, and application in relationship to crop demand are important regulators of N cycling efficiency and loss pathways; this has direct and indirect consequences for water and soil quality (Robertson, 1997;

Snapp et al., 2005; Syswerda et al., 2012). Similarly, Fageria (2009) also reported that adopting adequate nitrogen fertilizer management practices can produce higher crop yields and consequently improves nitrogen use efficiency. These practices include the use of an effective source of nitrogen, the use of appropriate methods of nitrogen application, the use of adequate rates, the adoption of appropriate timing of nitrogen application, the use of nitrification inhibitors, and foliar fertilization.

8.3.1 USE OF AN EFFECTIVE SOURCE OF NITROGEN

The use of an effective source is fundamental in improving N use efficiency and consequently reaping a higher yield of crops. There are several sources of nitrogen. Major N fertilizers available in the world market along with their N contents are presented in Table 8.6. Urea and ammonium sulfate are the main nitrogen carriers worldwide in annual crop production. However, urea is generally favored by the growers over ammonium sulfate due to the lower application cost because urea has a higher N analysis than ammonium sulfate (46% vs. 21% N). Tisdale et al. (1993) reported that the most cost-effective granular form of N is urea $[CO(NH_2)_2]$ as it has a high N concentration and lower relative manufacturing, handling, storage, and transportation costs. Once applied to the soil, urea is hydrolyzed by the enzyme urease to ammonia N (NH_3^-), which temporarily creates a high concentration of NH_3^-, and then converts into ammonium N (NH_4^+). The conversion from NH_3^- into NH_4^+ can be delayed by dry soil conditions or coarse-textured soils, which increases the potential for volatilization in wet, windy conditions, or phytotoxicity to seeds and plants when seed-placed (Tisdale et al., 1993).

In developed countries like the United States, anhydrous NH_3 is an important N source for annual crop production. At normal pressures, NH_3 is a gas and is transported and handled as liquid under pressure. It is injected into the soil to prevent loss through volatilization. The NH_3 protonates to form NH_4^+ in the soil and becomes XNH_4^+, which is stable (Foth and Ellis, 1988). The major advantages of anhydrous NH_3 are its high N analysis (82% N) and low cost of transportation and handling. However, specific equipment is required for storage, handling,

---

TABLE 8.6
Major Nitrogen Fertilizers, Their Chemical Formulas, and N Contents

Common Name	Formula	N (%)
Ammonium sulfate	$(NH_4)_2SO_4$	21
Urea	$CO(NH_2)_2$	46
Anhydrous ammonia	NH_3	82
Ammonium chloride	NH_4Cl	26
Ammonium nitrate	NH_4NO_3	35
Potassium nitrate	KNO_3	14
Sodium nitrate	$NaNO_3$	16
Calcium nitrate	$Ca(NO_3)_2$	16
Calcium cyanamide	$CaCN_2$	21
Ammonium nitrate sulfate	$NH_4NO_3(NH_4)_2SO_4$	26
Nitrochalk	$NH_4NO_3 + CaCO_3$	21
Monoammonium phosphate	$NH_4H_2PO_4$	11
Urea ammonium nitrate	$CO(NH_2)_2 + NH_4NO_3$	32
Diammonium phosphate	$(NH_4)_2HPO_4$	18

Source: From Fageria, N. K. 2009. *The Use of Nutrients in Crop Plants*. Boca Raton, Florida: CRC Press. With permission.

and application. Hence, NH_3 is not a popular N carrier in developing countries. Stanford (1973) and Campbell et al. (1993b) reported that ammonium nitrate is generally superior to urea, which may volatilize easily.

Fageria et al. (2010, 2011a,b,c) conducted field and greenhouse experiments compared to ammonium sulfate and urea as sources of N in upland and lowland rice production. Grain yield expressed in relative yield was significantly increased by urea as well as ammonium sulfate fertilization and the increase was in quadratic fashion (Figure 8.9). In fertilizer experiments, 90% of the relative yield is considered as an economic index and this index was used to calculate an adequate N rate (Figure 8.9). Based on 90% of the relative yield (corresponding to 5750 kg grain ha⁻¹) which was obtained with the application of 84 kg N ha⁻¹ in case of ammonium sulfate. Similarly, in the case of urea, 90% of the relative grain yield (corresponding to 4811 kg grain ha⁻¹) was obtained with the application of 130 kg N ha⁻¹.

Fageria et al. (2011c) also conducted a greenhouse experiment comparing ammonium sulfate and urea as sources of N in lowland rice production (Figure 8.10). Grain yield significantly ($P < 0.01$) and quadratically increased with increasing N rate from 0 to 400 mg kg⁻¹ of soil by ammonium sulfate as well as urea source. Maximum grain yield was obtained with the application of 168 mg N kg⁻¹ of soil by ammonium sulfate and 152 mg N kg⁻¹ of soil by urea. The variation in grain yield was 5.5–22.8 g plant⁻¹, with an average yield of 15.9 g plant⁻¹ by ammonium sulfate and 8.0–19.7 g plant⁻¹, with an average value of 14.4 g plant⁻¹ by urea fertilization. Ammonium sulfate accounted for 90% variability in grain yield, whereas urea accounted for 78% variability in grain yield. Across six N rates, ammonium sulfate produced a 10% higher grain yield compared to urea. In addition, average across two N sources (160 mg N kg⁻¹), ammonium sulfate produced 22.5 g grain yield per plant and urea produced 18.5 g grain yield per plant. The application of ammonium sulfate at the rate of 160 mg N kg⁻¹ produced a 22% higher grain yield compared to urea at the same rate of N. This means that ammonium sulfate was a superior fertilizer for lowland rice grain yield compared to urea.

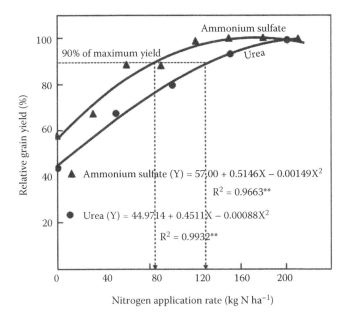

FIGURE 8.9 Relative grain yield of lowland rice as influenced by ammonium sulfate and urea fertilization. (From Fageria, N. K., A. B. Santos, and M. F. Moraes. 2010. *Commun. Soil Sci. Plant Anal.* 41:1565–1575. With permission.)

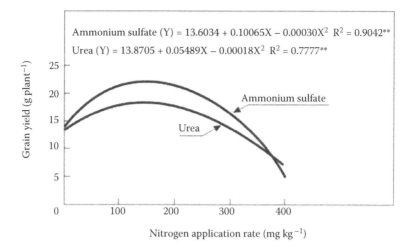

FIGURE 8.10 Relationship between nitrogen rate applied by ammonium sulfate and urea and grain yield of lowland rice. (From Fageria, N. K., A. B. Santos, and A. M. Coelho. 2011c. *J. Plant Nutr.* 34:371–386. With permission.)

Similarly, Fageria et al. (2011c) studied the influence of ammonium sulfate and urea fertilization on the panicle number of lowland rice (Figure 8.11). The number of panicles increased significantly and quadratically with increasing N rates from 0 to 400 mg kg^{-1} of soil by both ammonium sulfate and urea sources of N (Figure 8.11). Panicle response to N fertilization was similar for both the N sources; however, the magnitude of response was higher in the case of ammonium sulfate.

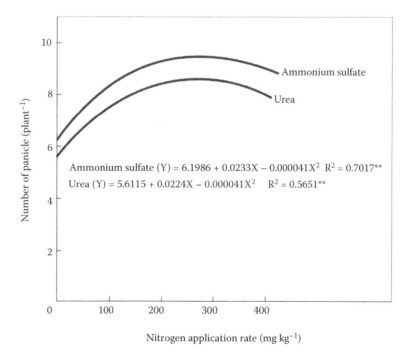

FIGURE 8.11 Relationship between nitrogen rate applied by ammonium sulfate and urea and panicle number of lowland rice. (From Fageria, N. K., A. B. Santos, and A. M. Coelho. 2011c. *J. Plant Nutr.* 34:371–386. With permission.)

Ammonium sulfate accounted for 70% variability in panicle number, whereas urea accounted for about 57% variability in panicle number. This means that ammonium sulfate was a superior fertilizer for panicle production in lowland rice compared to ammonium sulfate. Overall, ammonium sulfate produced 8% higher panicles compared to urea.

Fageria et al. (2011b) also studied the influence of ammonium sulfate and urea on growth, yield, and yield components of upland rice (Figures 8.12 through 8.16). Nitrogen source × N rate interactions for grain yield, shoot dry weight, number of panicle, plant height, and root dry weight were significant, indicating variability between two N sources for grain yield, yield components, and growth parameters of upland rice. Hence, values of these characteristics are presented at two N sources at different N rates. The grain yield increased significantly in a quadratic fashion, when the N rate was increased in the range of 0–400 mg kg^{-1} of soil, using ammonium sulfate and urea sources of N. Based on the regression equation, maximum grain yield was obtained with the application of 380 mg N kg^{-1} of soil with the ammonium sulfate. Similarly, maximum grain yield was obtained with the application of 271 mg N kg^{-1} of soil by urea. Fageria et al. (2006) reported that maximum grain yield of upland rice in a Brazilian Oxisol was obtained with the application of 400 mg N kg^{-1} of soil through ammonium sulfate. At the lower as well as higher N

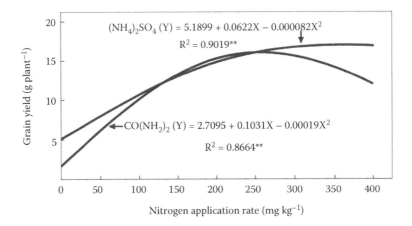

FIGURE 8.12 Grain yield of upland rice as influenced by two nitrogen sources. (From Fageria, N. K., A. Moreira, and A. M. Coelho. 2011b. *J. Plant Nutr.* 34:361–370. With permission.)

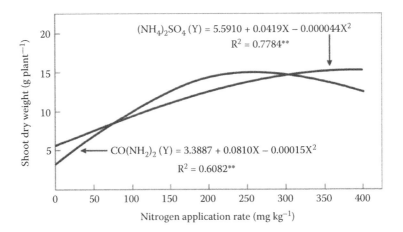

FIGURE 8.13 Shoot dry weight of upland rice as influenced by two nitrogen sources. (From Fageria, N. K., A. Moreira, and A. M. Coelho. 2011b. *J. Plant Nutr.* 34:361–370. With permission.)

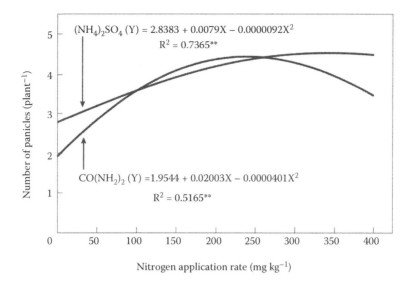

FIGURE 8.14 Number of panicles of upland rice as influenced by two nitrogen sources. (From Fageria, N. K., A. Moreira, and A. M. Coelho. 2011b. *J. Plant Nutr.* 34:361–370. With permission.)

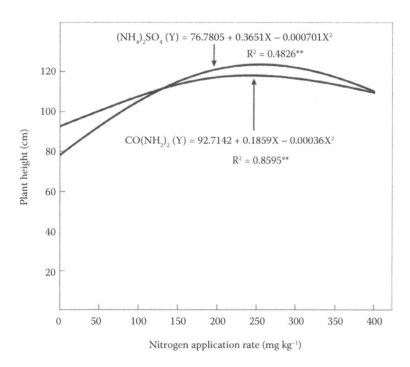

FIGURE 8.15 Plant height of upland rice as influenced by two nitrogen sources. (From Fageria, N. K., A. Moreira, and A. M. Coelho. 2011b. *J. Plant Nutr.* 34:361–370. With permission.)

rates, ammonium sulfate produced higher grain yield compared to urea. However, at the intermediate N rate (125–275 mg N kg^{-1}), urea was slightly superior in producing grain yield compared to ammonium sulfate. Across the six N rates, ammonium sulfate produced a 12% higher grain yield compared to urea. The superiority of ammonium sulfate at higher N rates compared to urea may be associated with the higher acidity-producing capacity of ammonium sulfate compared

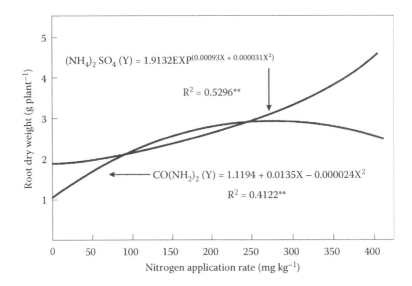

FIGURE 8.16 Root dry weight of upland rice as influenced by two nitrogen sources. (From Fageria, N. K., A. Moreira, and A. M. Coelho. 2011b. *J. Plant Nutr.* 34:361–370. With permission.)

to urea. Rice is an acid-tolerant plant and its growth was linearly increased when Al saturation in the Brazilian Oxisol soil was increased from 0% to 30% (Fageria et al., 2011a). Data in Figure 8.16 show that the response of root growth to N sources was different. Ammonium sulfate produced much higher root growth at the lower N rate (<75 mg kg^{-1}) and at the higher N rate (250 mg N kg^{-1}).

8.3.2 USE OF APPROPRIATE METHODS OF NITROGEN APPLICATION

Nitrogen is a mobile nutrient in the soil–plant system and it is different from P and K, which are immobile in the soil–plant system. Hence, it can be moved from a distance to plant roots and can be absorbed. However, if it is broadcast and there is no rainfall or irrigation, it may lead to loss due to evaporation or volatilization. Hence, application in the band or incorporating into the soil may reduce its loss and improve N use efficiency (Fageria, 2009). Campbell et al. (1995) also reported that the banding of N fertilizers is superior to broadcasting. Fertilizer banded with the planter, also known as starter fertilizer, is a common practice in Minnesota for corn (Kaiser and Kim, 2013). Research with N banded to the side and below the seed has shown positive yield benefits for soybean (Osborne and Riedell, 2006). Banding fertilizer with the planter can save on application costs of broadcast fertilizer, which is commonly spread before the crop sowing.

In furrow placement of fertilizer is attractive to growers because of the potential to increase nutrient availability (Mengel et al., 1988). The most commonly banded nutrients are N, P, but K and S are also included in some sources commonly used (Kaiser and Rubin, 2013a,b). In furrow placement is often justified for its potential to stimulate early growth, although responses in grain yield are not always seen (Kaiser et al., 2005; Kim et al., 2013). Additionally, using a starter fertilizer with a high N/P ratio can enhance P absorption by reducing soil P immobilization (Kaiser and Rubin, 2013a). Placing ammonium polyphosphate fertilizer in a band with the seed resulted in increased plant biomass and a higher concentration of P in plant tissue compared with deeper band placement (Mengel et al., 1988). Furrow application of N has been reported to be beneficial to increase the biomass of corn (Bermudez and Mallarino, 2004) and soybean (Kaiser and Rubin, 2013a).

Total N plus K applied is typically used to make suggestions on how much fertilizer can be safely applied with the seed (Kaiser and Rubin, 2013b). It is commonly recommended that in furrow applications should not exceed 11.2 kg ha^{-1} of N and K combined (Laboski, 2008). It is not clear if this information can be used for different soil types. This lack of consistency is probably due to differences in soil texture and soil moisture. Rehm (1999) reported that 10–15–0 (N–P–K) can be safely applied with the seed at rates up to 132 kg ha^{-1} as long as the soil moisture is adequate. Niehues et al. (2004) reported that applying 22 kg ha^{-1} as NH_4NO_3 (34–0–0, N–P–K) and K kg ha^{-1} as KCl (0–0–50, N–P–K) did not have any effect on plants standing on a silt loam soil. Kaiser and Rubin (2013b) reported that 10.6 kg N + K ha^{-1} could be applied for clay loam and silt loam soils while the rate was 5.7 kg for fine sand.

Soil moisture is important to take into account when applying fertilizer directly with the seed. Raun et al. (1986) found significant stand loss as a result of the relatively dry soil conditions. Salt damage from fertilizer is often more severe on dry soils because the ion concentration in the dry solution becomes greater (Kaiser and Rubin, 2013b). A large ion concentration around the root will move water out of the plant cells and induce injury (Laboski, 2008). Fertilizer sources that cause a higher amount of damage to seedlings are those that release free NH_3, including urea and diammonium phosphate. These sources cause severe damage when placed in direct contact with germinating seeds (Kaiser and Rubin, 2013b).

Nitrogen fertilizers are broadcast and mixed into soil before crop sowing. They may also be applied in rows below seeds at sowing and may be banded in rows beside seeds at planting or pre-emergence. During postemergence, fertilizers may be side-dressed, injected into the subsurface, and top-dressed. Fertilizers that are mixed into the soil or injected into the subsurface are more efficient methods of N application compared to broadcast and left on the soil surface. The side-dress application, N fertilization several weeks after corn emergence, has maximized the efficiency of fertilizer N in most situations (Fageria and Baligar, 2005). Placement of urea or ammonium sulfate in the anaerobic layer of flooded rice is an important strategy to avoid N losses by nitrate leaching and denitrification (Fageria and Baligar, 2005). Savant et al. (1982) reported that lowland rice recovered 50–61% of applied N from deep placed urea supergranules when the average N recovery efficiency of this crop is less than 40% (Fageria et al., 2003a,b).

The use of strip-till (ST) is another N management practice to improve its efficiency in crop plants. ST is a minimal tillage practice that allows for deep, banded placement of dry, liquid, or gas-based N fertilizers within tilled, planted rows without requiring tillage of the entire field (Nash et al., 2013). Because only the crop rows are tilled, ST allows for many of the soil conservation and fertility benefits associated with NT practices to be maintained while lowering the potential for significant N loss occurring with a fall fertilizer application through deep band placement. An additional benefit of tilling the soil in the seeded row is that it breaks up soil aggregates incorporating surface residues that lower bulk density, and increases internal drainage and drying within the seedbed. These effects of tillage ultimately allow for earlier warming of the soil in the spring, which has been reported to increase plant emergence and growth (Randall and Vetsch, 2005). Although studies are limited, ST has been found to produce yields similar to those of NT (Al-Kaisi and Licht, 2004; Vetsch and Randall, 2004; Al-Kaisi and Kwaw-Mensah, 2007; Nash et al., 2013).

8.3.3 USE OF ADEQUATE RATE

The use of an adequate rate is fundamental for meeting plant N demands, minimizing N losses, and improving N use efficiency in crop production (Lopez-Bellido et al., 2006). The application of the most optimal N fertilizer rate is a major factor in determining economic viability, crop productivity, and environmental quality. An adequate rate of mobile nutrients in a soil–plant system such as N is determined by experimental data generated under field conditions. Before making N recommendations for a given crop, it is essential that field experiments should be conducted under different agroecological conditions for several years. This is essential to get average values of different environmental conditions such as climate and soil types. An adequate time period to repeat field experiments for fertilizer recommendations is 3 years. However, if it is conducted at several locations, a 2 year period can also serve as a reasonably good time period to make recommendations.

When conducting field experiments for fertilizer recommendations, care should be taken to select appropriate rates or levels of a given nutrient. The levels selected should create a wide range of nutrients in the soil to satisfy the crop needs for a given nutrient from a deficiency level to a sufficiency level. Grain yield is the best parameter to determine nutrient requirements or fertilizer recommendations for grain crops. Hence, grain yield should be determined at physiological maturity and the relationship between grain yield and nutrient rates should be determined. Analysis of variance should be used in data analysis, and the quadratic regression model is used to describe the yield response to the applied fertilizer rate. The quadratic response function is the most common functional form to evaluate the yield response to fertilizer rates. The quadratic model is a second-order polynomial function written as

$$Y = a + bx + cx^2$$

where Y is the estimated yield, X is the application rate of the nutrient, and a, b, and c are coefficients estimated by fitting the model to the data. The quadratic function assumes that crop yield will increase at a decreasing rate as the nutrient application rate increases until the maximum yield is achieved at

$$N(Y_{max}) = b/2c$$

where $N(Y_{max})$ is the level of the applied nutrient that achieves the maximum yield. Yield decreases past this point.

The author conducted field experiments for upland and lowland rice grown under Brazilian conditions using promising genotypes. Results related to lowland rice response to N rates are presented in Figure 8.17. The response of lowland rice to N fertilization was highly significant and quadratic when N was applied in the range of 0–200 kg ha⁻¹. Based on a quadratic regression equation, a maximum grain yield of about 6250 kg ha⁻¹ was obtained with the application of 171 kg N ha⁻¹. Half of the N was applied at the time of sowing and half at 45 days after sowing corresponding to the active tillering growth stage. Similarly, the author also studied the response of upland rice to N

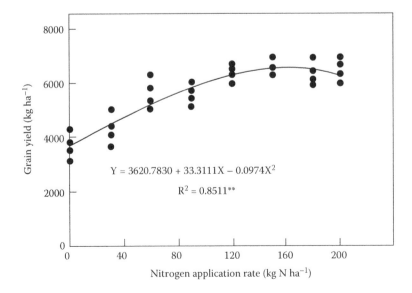

FIGURE 8.17   Response of lowland rice to nitrogen application.

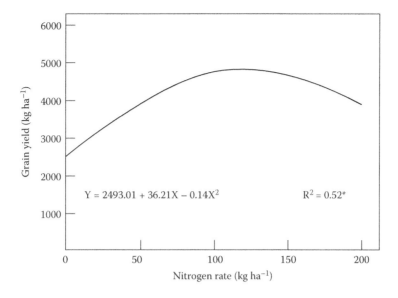

$$Y = 2493.01 + 36.21X - 0.14X^2 \qquad R^2 = 0.52^*$$

FIGURE 8.18 Response of upland rice to N fertilization.

fertilization grown on a Brazilian Oxisol (Figure 8.18). The response was significant and quadratic. A maximum grain yield of about 5000 kg ha^{-1} was obtained with the addition of 129 kg N ha^{-1} applied twice times. Half of the N was applied at sowing and the remaining half at 45 days after sowing corresponding to the active tillering growth stage.

The author also studied an adequate N rate for 12 lowland rice genotypes in a field experiment for two consecutive years (Table 8.7). The yield of 12 genotypes increased significantly in a quadratic fashion with the addition of the N fertilizer from 0 to 200 kg ha^{-1}. A maximum grain yield

TABLE 8.7
Relationship between Nitrogen Rate (X) and Grain Yield (Y) of 12 Lowland Rice Genotypes

Genotype	Regression Equation	R^2	NRMY
BRS Tropical	$Y = 4149.51 + 42.34X - 0.126X^2$	0.62**	168
BRS Jaçanã	$Y = 3500.10 + 39.75X - 0.152X^2$	0.54**	131
BRA 02654	$Y = 3596.41 + 32.46X - 0.126X^2$	0.57**	129
BRA 051077	$Y = 3367.42 + 32.59X - 0.134X^2$	0.58**	122
BRA 051083	$Y = 4142.04 + 25.47X - 0.059X^2$	0.79**	215
BRA 051108	$Y = 3375.72 + 28.58X - 0.083X^2$	0.78**	172
BRA 051126	$Y = 3961.13 + 36.66X - 0.106X^2$	0.82**	173
BRA 051129	$Y = 3598.17 + 37.82X - 0.138X^2$	0.58**	137
BRA 051130	$Y = 3991.41 + 42.91X - 0.143X^2$	0.86**	150
BRA 051134	$Y = 3843.46 + 35.04X - 0.103X^2$	0.75**	170
BRA 051135	$Y = 3566.95 + 32.53X - 0.092X^2$	0.88**	177
BRA 051250	$Y = 3306.92 + 40.79X - 0.151X^2$	0.75**	135

Note: NRMY = nitrogen rate for maximum grain yield calculated by quadratic regression equations. Values are averages of 2 years' data.
**Significant at the 1% probability level.

of genotypes was obtained with the application of 122–215 kg N ha^{-1} depending on the genotypes. Half of this N should be applied at sowing in the furrow and the remaining half as top-dressing at the active tillering growth stage. Fageria and Baligar (2001b) reported that in a field experiment for 3 years, maximum grain yield was obtained with the application of 209 kg N ha^{-1} in the first year, 163 kg N ha^{-1} in the second year, and 149 kg N ha^{-1} in the third year. The average across 3 years' maximum grain yield (6465 kg ha^{-1}) was achieved with the application of 171 kg N ha^{-1} in this experiment (Fageria and Baligar, 2001b). Singh et al. (1998) reported that the maximum average grain yield of 7700 kg ha^{-1} of 20 lowland rice genotypes was obtained at 150–200 kg N ha^{-1}.

8.3.4 ADOPTING APPROPRIATE TIMING OF APPLICATION

The timing of N application during crop growth is an important strategy in improving N use efficiency. The N application according to plant needs may improve its efficiency and avoid its loss from the soil–plant system. In other words, synchronizing of N application with N demand of plants is an important strategy in improving N use efficiency. Application timing is one of the factors that can influence the efficiency with which applied N is utilized by crops, and research on the optimum application timing of fertilizer N has been extensive (Randall et al., 2003; Randall and Vetsch, 2005; Fageria, 2009, 2013, 2014). It has been reported by Matson et al. (1997) and Tilman et al. (2002) that nutrient use efficiency is increased by appropriately applying fertilizers and by better matching the temporal and spatial nutrient supply with plant uptake. Applying fertilizers during periods of highest crop uptake, at or near the point of uptake (roots and leaves), as well as in smaller and more frequent applications has the potential to reduce losses while maintaining or improving the crop yield quantity and quality (Matson et al., 1996; 1997; Cassman et al., 2002). Rose and Bowden (2013) reported that split application of the N fertilizer after crop emergence improves N use efficiency, because plant roots had a chance to penetrate to depth and crop sink sizes are sufficient to take up significant quantities of the soil-mobile nitrate.

N is lost from the soil–plant system via volatilization, leaching, denitrification, or runoff (Fageria and Baligar, 2005; Fageria et al., 2006). This suggests that there is more N available for loss at any time during the crop growing season if N is applied only once during crop growth. Hence, splitting N fertilizer applications during crop growth can reduce nitrate leaching and improve N use efficiency. For lowland rice under Brazilian conditions, applying half of the N in a band at sowing and the remaining 6–7 weeks later should increase both N fertilizer use efficiency and N uptake by minimizing leaching opportunity time and better timing the N application to N uptake (Fageria and Baligar, 1999). Fageria and Baligar (1999) reported that the agronomic efficiency of N in lowland rice was higher when N was applied in a three-split application (one-third at sowing + one-third at active tillering + one-third at panicle initiation) compared with the entire N applied at sowing. Split application of N in sandy soils and high rainfall areas is most desirable. A study conducted by Fageria and Prabhu (2004) in the Brazilian Inceptisol showed that N fractionated into two or three equal doses produced a higher grain yield of lowland rice compared with the total applied at sowing (Figure 8.19).

Split applications of the N fertilizer are often recommended as a way to reduce N losses and improve NUE. Sainz et al. (2004) conducted a field experiment on corn to evaluate the effect of urea rate (0, 70, 140, and 210 kg N ha^{-1}) at planting or the six-leaf stage (V6) and reported increased grain yield (10.5 vs. 11.2 Mg ha^{-1}) and N uptake (168 vs. 192 kg ha^{-1}) when the N fertilizer was applied at the V6 stage. In another experiment, N recovery by a corn crop was 58% for application at planting and increased to 71% when N was applied at the V6 stage (Sainz et al., 1997). Same authors also reported that the N losses were 5.5% when N was applied at sowing and reduced to 1% when it was applied at the V6 stage.

The author also studied in another field experiment the influence of different rates and timing of N application on lowland rice yield for three consecutive years (Table 8.8). Overall, the grain yield was higher at 200 kg N ha^{-1} compared to 150 kg N ha^{-1} under all timing treatments. The maximum grain yield at 200 kg N ha^{-1} (one-third at sowing + one-third at 45 days after sowing + one-third

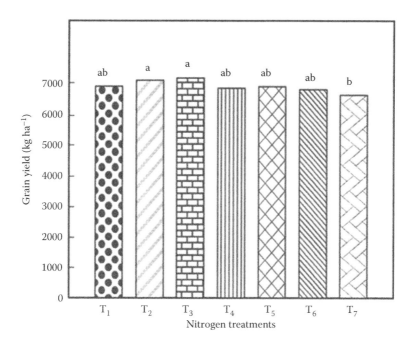

FIGURE 8.19 Grain yield of lowland rice as influenced by N timing treatments. T_1 = all the N applied at sowing, T_2 = 1/3 N applied at sowing + 1/3 N applied at active tillering + 1/3 N applied at the initiation of panicle primordia, T_3 = 1/2 N applied at sowing + 1/2 N applied at active tillering, T_4 = 1/2 N applied at sowing + 1/2 N applied at the initiation of panicle primordia, T_5 = 2/3 N applied at sowing + 1/3 N applied at active tillering, T_6 = 2/3 N applied at sowing + 1/3 N applied at initiation of primordia floral, and T_7 = 1/3 N applied at sowing + 2/3 N applied at 20 days after sowing. (Adapted from Fageria, N. K. and A. S. Prabhu. 2004. *Pesq. Agropec. Bras.* 39:123–129.)

at initiation of panicle primordial) was 49% higher compared to control treatment and 13% higher compared to the highest-yielding 150 kg N ha^{-1} (one-third at sowing + one-third at 45 days after sowing + one-third at initiation of panicle primordial). Fageria and Baligar (2005) and Fageria et al. (2011a) reported that the maximum grain yield of lowland rice under Brazilian conditions was obtained with the addition of 200 kg N ha^{-1}. Fageria and Baligar (2001b) also reported a significant quadratic response of lowland rice yields in a Brazilian Inceptisol when the N fertilizer was applied in the range of 0–210 kg ha^{-1}.

Meyers (1992) reviewed a series of flooded rice experiments in Indonesia and concluded that there are technical solutions to increase nitrogen use efficiency. One solution is to use various split applications and another is to place N deep into the soil to increase nitrogen use efficiency. Nyiraneza et al. (2010) reported that split application of the N fertilizer can help to synchronize N supply with wheat N demand. Because crops, including wheat, remove around 50% of the applied N fertilizer, unused N can be lost through leaching, denitrification, or volatilization.

8.3.5 NITRIFICATION INHIBITORS

The use of nitrification inhibitors may be one of the options to improve nitrogen use efficiency in crop plants. There is plenty of literature showing positive as well as no effects of nitrification inhibitors on N use efficiency in crop plants. For example, slow-release N fertilizer developed by coating urea granules with sulfur has been tested for rice and was found to be superior to common urea in almost all types of soils (Bijay-Singh and Katyal, 1987). Thind et al. (2010) reported that neem (*Azadirachta indica*)-coated urea applied to rice can result in high N use efficiency as it contains

TABLE 8.8
Grain Yield (kg ha⁻¹) of Lowland Rice as Affected by Nitrogen Timing Treatments

Treatments[a]	1st Year	2nd Year	3rd Year	Average
T_1	4677.66b	4069.44d	4486.66c	4411.26d
T_2	6934.00a	5069.44bc	4763.19bc	5588.87c
T_3	6605.67ab	5402.78b	5298.75abc	5769.06bc
T_4	8065.00a	4708.33bcd	4730.69bc	5834.67bc
T_5	6890.00a	5001.38bc	4884.72bc	5592.03c
T_6	6782.33a	4583.22cd	4627.22c	5330.96c
T_7	7256.66a	6250.00a	5790.97a	6432.54ab
T_8	7174.33a	6305.55a	5555.55ab	6345.14ab
T_9	8347.00a	6527.77a	4818.33bc	6564.37a
T_{10}	7086.33a	5416.67b	4953.05abc	5818.68bc
T_{11}	7776.33a	4583.33cd	4475.69c	5611.78c
Average	7054.12a	5265.28b	4944.07c	
F-test				
Year (Y)	**			
N timing (NT)	**			
Y × NT	*			
CV (%)	7.97			

[a] T_1 = control (0 kg N ha⁻¹), T_2 = 150 kg N ha⁻¹ (total at sowing), T_3 = 150 kg N ha⁻¹ (1/2 at sowing + 1/2 at 45 days after sowing), T_4 = 150 kg N ha⁻¹ (1/3 at sowing +1/3 at 45 days after sowing + 1/3 at initiation of panicle primordial), T_5 = 150 kg N ha⁻¹ (1/4 at sowing + 1/2 at initiation of tillering + 1/4 at 45 days after sowing), T_6 = 150 kg N ha⁻¹ (2/3 at sowing + 1/3 at 45 days after sowing), T_7 = 200 kg N ha⁻¹ (total at sowing), T_8 = 200 kg N ha⁻¹ (1/2 at sowing + 1/2 at 45 days after sowing), T_9 = 200 kg N ha⁻¹ (1/3 at sowing +1/3 at 45 days after sowing + 1/3 at initiation of panicle primordial), T_{10} = 200 kg N ha⁻¹ (1/4 at sowing + 1/2 at initiation of tillering + 1/4 at 45 days after sowing), and T_{11} = 200 kg N ha⁻¹ (2/3 at sowing + 1/3 at 45 days after sowing).

*,**Significant at the 5% and 1% probability level, respectively. Means followed by the same letter in the same column are not significantly different at the 5% probability level by Tukey's test.

nitrification inhibitor properties. Similar results were also reported by Agarwal et al. (1980) and Singh and Singh (1986).

Nash et al. (2013) reported that over a 3-year study, conducted at one location, NT/injected anhydrous ammonia with nitrapyrin at preplant produced at least 2 Mg ha⁻¹ higher corn grain yields than all other management systems with a fall in N application. De Datta and Buresh (1989) and Campbell et al. (1995) reported that generally less than 40% of N applied to lowland rice is recovered. Most of the losses are through ammonia volatilization, leaching, and denitrification. This loss can be significantly reduced by the use of urease inhibitors (Fillery and De Datta, 1989).

8.3.6 FOLIAR FERTILIZATION

Foliar application of nutrients is an important crop management strategy in maximizing crop yields. It can supplement soil fertilization. When nutrients are applied to soils, they are absorbed by plant roots and are translocated to aerial parts. In case of foliar application, the nutrients penetrate the cuticle of the leaf or the stomata and then enter the cells (Fageria et al., 2009). Hence, crop response occurs in short time in foliar application compared to soil application. The rate at which an ion passes through the cuticle, and generally the epidermal tissues of the leaves, depends on many factors, including the concentration and the physical and chemical properties of the sprayed ion. In foliar sprays, macronutrient concentrations of generally less than 2% are used to avoid leaf burning. Plant age should also be considered in selecting nutrient concentration. Old plants are more tolerant of higher concentration of salts compared to young plants. In foliar fertilization, droplet size and fertilizer solubility should be carefully controlled since it will affect crop response. Foliar fertilization in food crops may not increase yield but may increase the protein content of grains, if applied during anthesis or flowering (Fageria et al., 2009).

Urea is the most popular form of N used for foliar fertilization in crop plants. Yamada (1962) reported that the greater effectiveness of urea when applied to foliage resided in its nonpolar organic properties. Urea containing the ^{15}N label has been used to measure rates of absorption and translocation of foliar-applied N, because it permits the direct determination of the uptake and translocation of foliar-applied N (Gerik et al., 1998). Oosterhuis et al. (1989) and Baolong (1989) reported that the sympodial leaf rapidly took up foliar-applied N. They found that 30% and 47% of applied N were recovered within 1 and 24 h after application, respectively. Approximately 70% of the foliar-applied urea N was absorbed by 8 days after application. There have been similar reports for soybean when urea was sprayed on the foliage of this crop (Vasilas et al., 1980). Foliar-applied ^{15}N to cotton was rapidly translocated from the closest treated leaf to the bolls and was first detected 6 h after application (Baolong, 1989). Baolong (1989) found that about 70% of the total foliar-applied ^{15}N urea was found in the cotton bolls, with less than 5% remaining in the leaves, petioles, bracts, and branches.

Gerik et al. (1998) reviewed the literature on foliar N fertilization and reported that absorption was more rapid in young leaves than in old leaves on many crops. Bondada et al. (1997) reported the correlation between increasing leaf cuticle thickness as the leaf aged and decreased absorption of foliar-applied ^{15}N in cotton. Many factors can affect the uptake of foliar-applied urea, including the condition of the leaf and the prevailing environment. It has been reported that the leaf water status affects the physical structure of the cotton leaf cuticle (Oosterhuis et al., 1991) and consequently affects the absorption of the foliar-applied nutrients (Kannan, 1986). Baolong (1989) reported that water deficit stress impeded the absorption of foliar-applied urea N by sympodial leaves, as well as the subsequent translocation within the branch of cotton. Furthermore, application made either in the late afternoon or early morning was more effectively absorbed than those made at midday, and this was more pronounced for water-stressed plants (Gerik et al., 1998). These authors further reported that this was associated with the crystallization of urea on the leaf surface and also with changes in the cuticle caused by water stress.

The yield response of field crops to foliar fertilization of N is highly variable. Positive results with foliar fertilization of N in soybean have been associated with high-yielding environments. Yield responses to foliar fertilization are generally not positive when yield is low or nutrients are at an optimum level in the soil. Leaf damage due to higher concentration of foliar fertilization may be one of the reasons for either yield decreases or the lack of yield increases. Foliar spray of nutrients should be avoided at high temperature during the day to avoid leaf burning. Similarly, windy days may drift the applied nutrient solution, and rain immediately after application may wash out the sprayed material and reduce its efficiency. In rice, foliar spray of nutrients should not be done after flowering because this may cause spikelet discoloration. Foliar fertilization cannot substitute soil application. It is simply a nutrient corrective technique in crops during the growth cycle when soil application is ineffective due to immobilization of soil-applied nutrients or the cost or methods of application are prohibitive (Fageria et al., 2009).

8.4   CROP MANAGEMENT PRACTICES

Crop management practices are defined as the practices adopted during the crop growth cycle to produce maximum economic yield. There are several crop management practices that can be adopted during planting and/or the crop growth cycle to improve nitrogen use efficiency. These practices are adopting conservation tillage, appropriate crop rotation, and water management.

8.4.1 Use of Conservation Tillage

Conservation tillage is related to tillage operations. Hence, it is logical to first define what tillage is. Tillage is defined as the mechanical manipulation of the soil profile for any purpose, but in agriculture it is usually restricted to modifying soil conditions and/or managing crop residues and/or weeds and/or incorporating chemicals for crop production (Soil Science Society of America, 2008). Conservation tillage, on the other hand, can be defined as any tillage sequence, the object of which is to minimize or reduce the loss of soil and water operationally, a tillage or tillage and planting combination that leaves a 30% or greater cover of crop residue on the surface (Soil Science Society of America, 2008). In addition to conservation tillage, no-tillage or zero tillage terms are also used in the literature and these terms are related to conservation tillage and need to be defined. No-tillage or zero tillage is defined as a procedure whereby a crop is planted directly into the soil with no primary or secondary tillage since the harvest of the previous crop; usually a special planter is necessary to prepare a narrow, shallow seedbed immediately surrounding the seed being planted. NT is sometimes practiced in combination with subsoiling to facilitate seeding and early root growth, whereby the surface residue is left virtually undisturbed except for a small slot in the path of the subsoil shank (Soil Science Society of America, 2008).

Two other terms—minimum tillage and conventional tillage—are also frequently used in the soil preparation operations. Hence, these terms should also be defined to correlate with conservation tillage. Minimum tillage is defined as the minimum use of primary and/or secondary tillage necessary for meeting crop production requirements under the existing soil and climatic conditions, usually resulting in fewer tillage operations than for conventional tillage (Soil Science Society of America, 2008). Similarly, conventional tillage is defined as the primary and secondary tillage operations normally performed in preparing a seedbed and/or cultivating for a given crop grown in a given geographical area, usually resulting in <30% cover of crop residues remaining on the surface after completion of the tillage sequence (Soil Science Society of America, 2008). Conservation tillage is practiced on approximately 95 million ha worldwide, with the largest area in South America (approximately 47%), North America (approximately 40%), and Australia (approximately 9%); the remaining areas are located in the rest of the world, including Asia, Europe, and Africa (Mchunu et al., 2011).

Several studies have shown that the use of conservation tillage in cropping systems can improve soil organic matter, soil moisture, and total available soil N (Sharifi et al., 2008; Lafond et al., 2011; Zakeri et al., 2012). Zakeri et al. (2012) reported that the use of conservation tillage improves not only C and N but also the overall soil quality. Across the three major soil zones of Saskatchewan, continuous NT increased potentially mineralizable soil N by 16–40 kg ha^{-1} compared to conventional tillage (Liang et al., 2004). In the black soils of this province, Schoenau et al. (2008) measured greater soil-available N and P following 28 years of continuous no-tillage compared to 5 years of continuous NT history in the same field. As a result of the improved soil quality and soil moisture, the biological nitrogen fixation of legume crops has gradually increased in NT systems (Matus et al., 1997; Van Kessel and Hartley, 2000). Blanco-Canqui and Lal (2008) and Wuest and Schillinger (2011) reported that NT is an attractive method of farming because it can lower production costs, leave the soil in a less erodible condition, and improve the soil quality compared with tillage-based systems. Wust and Schillinger (2011) further reported that wheat farmers in many regions of the world have developed NT systems that are economically and environmentally superior to tillage-based systems.

Improved soil and crop management practices, such as reduced tillage and continuous cropping, can increase dryland N storage to a depth of 20 cm compared to a traditional farming system (Sherrod et al., 2003; Sainju et al., 2006). Besides reducing N mineralization, NT can conserve surface residues and soil water more than conventional tillage (Farahani et al., 1998). As a result, crops can use soil water more efficiently in conservation tillage (Deibert et al., 1986; Aase and Pikul, 1995), which can reduce or eliminate summer fallow by increasing cropping intensity (Farahani et al., 1998; Peterson et al., 1998).

8.4.2 Adopting Appropriate Crop Rotation

Crop rotation is defined as a planned sequence of crops growing in a regularly recurring succession on the same area of land, in contrast to a continuous culture of one crop or growth of a variable sequence of crops (Soil Science Society of America, 2008). Growing crops in rotation has many positive effects that are responsible for higher and/or sustainable yields. These effects are control of diseases, insects, and weeds, increase in nutrient use efficiency, improve in water use efficiency (WUE), improvement in soil fertility, and control of allelopathy. In an appropriate crop rotation, legumes are rotated with cereals. For example, in the Brazilian central part locally known as the "Cerrado" region, growing soybean, upland rice, dry bean, and corn rotation is a common practice. Legumes in rotation with cereals are beneficial in many ways (Yau et al., 2003; Krupinsky et al., 2006). In addition to enhancing soil N status and suppression of cereal diseases and pests, there is an N-sparing effect (Chalk, 1998), and growth-promoting substances released from decaying legume residues give healthier wheat roots (Stevenson and Van Kessel, 1996). Jones and Arous (1999) and Jones and Singh (2000) reported that shorter and earlier-maturing legumes enhance the yield of the subsequent cereal crops grown in rotation.

Including legumes in rotation with cereals is an appropriate rotation. However, many studies have shown that even growing cereals in rotation or including oil crops in rotation with cereals has positive effects on yields compared to growing cereals in monoculture (Yau and Ryan, 2012). Yau and Ryan (2012) reported that growing safflower (*Carthamus tinctorius* L.), an oil crop, before barley increased barley yields and was comparable to or better than after some legumes. In the northern Great Plains, growing wheat and barley following safflower increased wheat yield by 30% and barley yield by 23% (Krupinsky et al., 2004), while flax (*Linum usitatissimum* L.) yield was increased one-fold (Tanaka et al., 2005). Wheat in Western Australia is generally grown in rotation with pastures, lupins, and other cereals. The supply of N for wheat is mostly from biological fixed N from previous legume (Angus et al., 1993). Campbell et al. (1995) reported that sweet clover used as green manure in rotation with wheat increased wheat yield compared to fallow–wheat rotation, probably a result of N fixation, as evidenced by a 57% higher N supplying power in the soil under green manure.

Crops grown in rotation or with shorter fallow periods often have higher annualized biomass N than monocrop systems or those with longer fallow periods (Copeland and Crookston, 1992; Halvorson et al., 2002). Including pea in rotation with spring wheat and barley can not only sustain their yields by efficiently using soil water but also reduce N fertilization rates by supplying supplemental N from the pea residue due to its higher N concentration (Miller et al., 2002; Sainju et al., 2009, 2013). The increase in WUE is a result of less soil water use by pea than spring wheat and barley, thereby leaving more water available for succeeding crops and increasing their yields (Miller et al., 2002). Other benefits of crop rotation compared to monoculture include control of weeds, diseases, and pests (Vigil et al., 1997; Miller et al., 2002), reduction in farm inputs, and improvement in economic environmental sustainability (Gregory et al., 2002).

8.4.3 Improvement in Water Use Efficiency

Although one-third of the earth's surface is occupied by water, only about 2.5% of the total water on earth is freshwater (Shiklomanov, 1993). The water shortage on a global scale may be arguable (Heumesser et al., 2013). Some sources consider that there is no serious threat of water shortages in the future (Bruinsma, 2003), whereas others estimate that taking environmental water requirements into account, serious water shortages are likely to occur (Smakhtin et al., 2004; Heumesser et al., 2013).

Globally, agriculture is the most extensive and by far the largest area of human management ecosystems (Monfreda et al., 2008; Foley et al., 2011). The FAO estimates that about 1.53 billion ha of land, 11% of the world's total land area, is currently used for crop production, 80% of which is rainfed (Gebremedhin et al., 2012). Furthermore, in the last few decades, the main threats to agricultural development in many regions all over the world are increasing water deficits and decreasing soil retention capacity in the face of growing water demands. Furthermore, global food production depends on water not only in the form of precipitation but also, and critically so, in the form of available water resources for irrigation (Kedziora and Kundzewicz, 2013). Although irrigated land, representing about 18% of global agricultural land (more than 240 million ha), produces 1 billion metric tons of grain annually, or about half the world's total supply, this is because irrigated crops yield on average two to three times more than their rainfed counterparts (Somerville and Briscoe, 2001; Kedziora and Kundzewicz, 2013). In addition, up to 45% of the world agriculture lands are subject to continuous or frequent drought, wherein 38% of the world human population resides (Bot et al., 2000). Therefore, it is imperative to improve WUE in crop plants through agronomic practices as well as cultivar improvement (Hussain et al., 2012).

Agriculture is the principal user of all water resources, accounting for 70% of all withdrawals (e.g., rainfall, water from rivers, lakes, and aquifers) (Heumesser et al., 2013). In comparison, 10% is assigned to domestic uses and 20% to industrial uses (FAO, 2003). However, according to Nair et al. (2013), the agricultural sector uses >80% of the developed freshwater supply of the world. Water is recognized as an extremely important limiting factor in world food production second only to land area (Wittwer, 1975; Nair et al., 2013). Because both the arable area and availability of freshwater are limited and crop yield loss due to water availability exceeds that from all other causes, efficient use of water is a vital aspect of modern-day agriculture (Gleick, 2003; Fedoroff et al., 2010).

Crops typically give a large response to applied fertilizer in favorable environments and a small, zero, or negative response in an unfavorable environment (Angus et al., 1993). In this context, the availability of water in adequate amounts is one of the factors determining crop responses to applied fertilizers. Crops may not be able to use N efficiently if water is a limiting factor for growth and production. This may result in increased residual N accumulation in the soil after crop harvest, which can degrade environmental quality through increased N leaching into the groundwater and emissions of greenhouse gases, such as N_2O (Wang et al., 2013).

Water plays a significant role in crop production. It is essential for many physiological and biochemical processes in the plants, which determine yield and quality. WUE is defined as dry matter or the harvested portion of the crop produced per unit of water consumed (Soil Science Society of America, 2008). Crop WUE originates in the economic concept of crop productivity and therefore is now known as crop water productivity (CWP) (Jabro et al., 2012). CWP is defined as the amount of water required per unit of yield and is a vital parameter to assess the performance of irrigated and rainfed agriculture (www.fao.org/Landandwater/aglw/cropwater). CWP varies according to the soil type, crop species, climatic conditions, and crop management practices adopted. In western Australia, despite the relatively reliable winter rainfall in the region, the low water storage capacity of the soils allows marked variation in the between-season water balance and so causes large variation in the crop demand for N (Angus et al., 1993).

Land degradation can exacerbate drought because it affects water availability, quality, and storage (Diouf, 2001; Bossio et al., 2010; Sileshi et al., 2011). Therefore, measures that mitigate land degradation are important to increase water productivity and reduce the risk of crop failure under rainfed cropping systems (Sileshi et al., 2012). In addition, WUE can be increased by increasing biomass production with the same amount of water use, the same amount of biomass production with decreased water use, or a combination of both (Blum, 2005). Xin et al. (2009) evaluated 341 sorghum genotypes for transpiration efficiency (TE, the ratio of biomass produced to water transpired) based on biomass production in controlled environments and reported that TE had little correlation with the water transpired and a large correlation with the biomass produced. They concluded that increased biomass production rather than decreased transpiration accounted for increased TE. This result is in contrast with the selection of increased TE genotypes using the C isotope discrimination method that are often associated with decreased transpiration, growth, and biomass production (Condon et al., 2002; Impa et al., 2005; Blum, 2009). Tanner and Sinclair (1983) reported that WUE or TE within a species is relatively constant and cannot be manipulated. Xin et al. (2009), however, identified considerable genetic variability in TE for sorghum under controlled environments and reported that TE can be improved through improving the biomass production. They further concluded that identifying high TE genotypes based on biomass accumulation is a useful approach to select for high TE in sorghum. These results widen the scope for improving TE or WUE through exploiting traits or mechanisms that improve biomass production.

Important soil management practices that can improve WUE are conservation tillage, increased soil organic matter content, reduced length of fallow periods, contour farming, furrow dikes, control of plowpans, crop selection, and use of appropriate crop rotation (Nielsen et al., 2002; Fageria and Stone, 2013). It is possible to increase WUE by 25–40% through soil management practices that involve tillage and by 15–25% by modifying nutrient management practices (Hatfield et al., 2001). Also, precipitation use efficiency can be enhanced through the adoption of more intensive cropping systems in semiarid environments and increased plant populations in more temperate and humid environments (Hatfield et al., 2001).

Skip-row planting and tie-ridging can improve WUE in arid and semiarid environments. Typically, the skip-row planting arrangement is one or two rows planted with 0.75 m spacing alternating with one or two rows that are not planted (Mesfin et al., 2010). Similarly, tie-ridges typically consist of interrow furrows of 20–30 cm depth that are blocked with earthen ties spaced according to the slope of the land, water infiltration rate, and expected intensity of rainfall (Lal, 1977; Gusha, 2002; Brhane et al., 2006). Clark and Knight (1996) observed increased grain yield with skip-row planting of grain sorghum when the mean grain yield was <2 Mg ha^{-1}, but decreased grain yield when the mean grain yield was 3 Mg ha^{-1}. On the High Plains of the United States, skip-row planting of corn and grain sorghum was found to have a grain yield advantage compared with conventional planting when the mean grain yield was <4 Mg ha^{-1} and no disadvantage at 4–5 Mg ha^{-1} (Mesfin et al., 2010). Increased evaporative loss of soil water with skip-row planting is a concern, although saving more deep soil water for later in the season can compensate for these losses (Myers et al., 1986). With skip-row planting, more soil water is out of reach of the plant roots until later into the season when it gradually becomes available as the roots extend (Milroy et al., 2004). In northern Ethiopia, Brhane et al. (2006) found a sorghum grain yield increase of 62% with tie-ridging compared with flat planting. Highland pulse grain yield was increased with tie-ridging by 31–96% in northern Ethiopia (Brhane and Wortman, 2008). Sanders et al. (1996) estimated that the adoption of tie-ridging for small-scale sorghum production in Africa increased farm income by 12%.

In many arid and semiarid regions of the world, drought limits crop productivity (Habibzadeh et al., 2013). Management practices that can provide support to plants to withstand water deficits stress would improve crop production (Sylvia et al., 1993) and consequently N use efficiency. Considerable evidence or research data are available to suggest that arbuscular mycorrhizal fungi (AMF) have the potential to increase the tolerance of their host plants to water deficit stress (Al-Karaki and Al-Raddad, 1997; Al-Karaki and Clark, 1998; Davies et al., 2002; Auge, 2004; Habibzadeh et al., 2013). Arbuscular mycorrhizal symbiosis has been reported in 70–90% of studied land plants (Smith and Read, 2008). The pioneering studies of Allen and Boosalis (1983) indicated a possible role of AMF hyphae in water uptake and transfer to host plants. AMF have been shown to affect the water balance in plants under both well-watered and water-deficit stressed conditions (Auge, 2001).

In addition, the use of crop species that requires less water can be an important strategy in a water scarce environment. Table 8.9 shows the variation in WUE of principal food, vegetable, and fruit crops under Brazilian conditions. It is very clear from the data in Table 8.9 that there is a large variation in water use among crop species. In addition, WUE can also be improved by planting drought-resistant crop species and genotypes within species. Sorghum and pearl millets are good examples of drought-tolerant crop species. Sorghum is the fifth major cereal crop in the world in terms of production and acreage. In addition, sorghum is one of the most drought-tolerant cereal crops currently under cultivation (Blum, 2004) and radiation use efficiency (RUE) (Kiniry et al., 1989; Muchow and Sinclair, 1994).

TABLE 8.9
Water Use Efficiency (WUE) of Principal Food, Vegetable, and Fruit Crops in Brazil

Crop	Scientific Name	WUE (kg ha^{-1} mm^{-1})[a]	Reference
Rice	*Oryza sativa* L.	4.6	Coelho et al. (2009)
Dry bean	*Phaseolus vulgaris* L.	3.9	Coelho et al. (2009)
Corn	*Zea mays* L.	17.2	Coelho et al. (2009)
Wheat	*Triticum aestivum* L.	9.6	Coelho et al. (2009)
Soybean	*Glycine max* L. Merr.	5.7	Coelho et al. (2009)
Cotton	*Gossypium hirsutum* L.	6.8	Lima et al. (1999)
Garlic	*Allium sativum* L.	12.3	Lima et al. (1999)
Onion	*Allium cepa* L.	40.9	Lima et al. (1999)
Potato	*Solanum tuberosum* L.	38.2	Coelho et al. (2009)
Tomato	*Lycopersicon esculentum* Mill.	167.6	Coelho et al. (2009)
Banana	*Musa paradisiaca* L.	11.6	Coelho et al. (2009)
Water melon	*Citrullus lanatus* (Thunb.) Matsum. & Nakai	29.8	Lima et al. (1999)
Cantaloupe	*Cucumis melo* L.	23.1	Coelho et al. (2009)
Grape	*Vitis vinifera* L.	37.7	Lima et al. (1999)

Source: Adapted from M. A. Coelho, E. F., Coelho Filho, and A. J. P. Silva. 2009. In: Simpósio nacional sobre o uso da água na agricultura, 3, 2009, Passo Fundo. Disponível em: http://www.upf.br/coaju/download/Eugenio.pdf. Acesso em: 11 set. 2009; Lima, J. E. F. W., R. S. A. Ferreira, and D. Christofidis. 1999. In: *O estado das águas no Brasil*, pp. 73–101. Brasília: MME, MMA/SRH, OMM.

[a] Calculated on the basis of data of water consumed by the crops given by the authors and average yield of the irrigated crops.

Benjamin and Nielsen (2004) and Benjamin et al. (2013) reported that plant response to dry soil conditions can vary with the plant species and growth stage of a plant within a species. They found that the root system of soybean was relatively unaffected by water-deficit conditions, whereas field pea and chickpea responded to the water deficit by growing more roots deeper in the soil profile. Campbell et al. (1995) reported that in the semiarid climate of southwestern Saskatchewan, wheat yield, and therefore the nitrogen use efficiency response to N, was directly influenced by available water.

Matching phenology to the water supply is an important strategy in improving WUE and consequently N use efficiency. Genotypic variation in growth duration is one of the most obvious means of matching seasonal transpiration with the water supply and thus maximizing the water transpiring (Ludlow and Muchow, 1990). Early flowering tends to give a higher yield and greater yield stability than later flowering, if rain does not occur during the latter half of the growing season. Moreover, if it enables a cultivar to escape drought during the critical reproductive stages, harvest index is improved (Ludlow and Muchow, 1990). Data presented in Table 8.10 show clearly that short-duration cultivars of upland rice were having higher grain yields compared to longer-duration cultivars planted in the Oxisols of the central part of Brazil.

Selecting crop genotypes having a vigorous root system is also an important strategy in increasing the water and nutrient uptake and consequently higher N use efficiency. Fageria et al. (2012) studied the root system of dry bean genotypes under two P levels. The maximum root length and root dry weight were significantly influenced by the P level as well as genotype treatments (Table 8.11). The maximum root length varied from 8.00 to 29.67 cm, with an average value of 18.86 cm at a low P level. At a high P level, the maximum root length varied from 17.00 to 30.67 cm, with an average value of 22.65 cm. There was a 20% increase in maximum root length at a high P level

TABLE 8.10

Grain Yield of Upland Rice Cultivars Having Different Growth Cycles at Two P Levels

Flowering Days	Growth Cycle (Days)	Grain Yield (kg ha⁻¹)	
85 (25)	110	Low P (2.3 mg kg⁻¹)	High P (4.9 mg kg⁻¹)
95 (15)	120	1591	2093
106 (30)	130	1216	1320
111 (2)	135	1071	1164

Source: From Fageria, N. K. 1992. *Maximizing Crop Yields*. New York: Marcel Dekker. With permision.

Note: Values in parentheses represent the number of cultivars tested. P was extracted by the Mehlich 1 extracting solution.

compared to a low P level. The root dry weight varied from 0.21 to 0.54 g plant⁻¹ at a low P level. Similarly, at a high P level, the root dry weight varied from 0.60 to 1.97 g plant⁻¹, with an average value of 1.27 g plant⁻¹. The average increase in root weight with the addition of P was 234% compared with control treatment. Figures 8.20 through 8.22 show that the root growth of three dry bean genotypes was more vigorous at a high P level compared to a low P level. Improvement in the root dry weight with the addition of P in dry bean is reported by Fageria (2009).

The traditional view is that a large vigorous root system through avoidance of plant water deficit is a major feature of high yield in water-limited environments (Ludlow and Muchow, 1990) and consequently higher N use efficiency in crop species or genotypes of the same species. Some evidences suggest that vigorous and deep roots may have additional benefits for water extraction and root function because water uptake continues at night, resulting in an increase in the soil water content of upper soil layers and presumably of roots in these layers (Ludlow and Muchow, 1990).

In addition to several regular agricultural practices, WUE can be improved by growing crops in soils enhanced with water-holding amendments such as polymers (Johnson and Leah, 1990). These polymers are becoming more important in regions with insufficient water availability (Monnig, 2005). Applying superb-sorbent polymers can increase the water-holding capacity of soils and reduce the detrimental effects of short-term drought in drought-prone arable areas (Karmini and Naderi, 2007; Rostampour et al., 2013). Polymers absorb and store water and nutrients in gel form and undergo cycles of hydration and dehydration according to moisture demand, increasing both water and nutrient use efficiency in crop plants (Lentz and Sojka, 1994; Rostampour et al., 2013). A superabsorbent polymer can hold 400–1500 times as much water as its dry hydrogel (Boman and Evans, 1991). Polymers are safe and nontoxic, and decompose to CO_2, water, NH_4, and K^+ without any residue (Mikkelsen, 1994). They have also been reported to improve soil physical properties and reduce soil erosion, runoff, and nutrient loss (Shainberg et al., 1990, 1994; Rostampour et al., 2013).

8.5 PLANT MANAGEMENT PRACTICES

Plant management practices such as the use of adequate seed rate and spacing, the use of efficient crop species and genotypes within species, and the control of plant diseases, insects, and weeds are important management practices to improve nitrogen use efficiency in crop plants. If these practices are adopted properly at the right time, they have a significant impact on N use efficiency.

TABLE 8.11

Maximum Root Length and Root Dry Weight of 30 Dry Bean Genotypes at Two Phosphorus Levels

Genotype	Maximum Root Length (cm)		Root Dry Weight (g Plant^{-1})	
	0 mg P kg^{-1}	200 mg P kg^{-1}	0 mg P kg^{-1}	200 mg P kg^{-1}
1. Aporé	21.00c–f	22.00b–h	0.33ab	1.47a–d
2. Pérola	22.33b–d	25.00a–g	0.44ab	1.02a–d
3. BRSMG Talisma	20.33c–f	30.67a	0.32ab	1.29a–d
4. BRS Requinte	28.00ab	23.66a–h	0.51ab	1.53a0d
5. BRS Pontal	20.67c–f	19.00e–h	0.54a	1.57a–d
6. BRS 9435 Cometa	19.67c–g	26.33a–d	0.43ab	1.35a–d
7. BRS Estilo	22.00b–e	21.33c–h	0.43ab	1.00a0d
8. BRSMG Majestoso	23.33a–d	30.67a	0.26ab	1.33a–d
9. CNFC 10429	29.67a	29.00ab	0.36ab	1.12a–d
10. CNFC 10408	19.00c–h	20.00d–h	0.27ab	0.88b–d
11. CNFC 10467	19.33c–g	28.00a–c	0.39ab	1.49a–d
12. CNFC 10470	20.67c–f	26.00a–e	0.36ab	1.45a–d
13. Diamante Negro	18.67d–h	23.67a–h	0.42ab	1.87ab
14. Corrente	15.67e–i	17.00 h	0.30ab	0.74d
15. BRS Valente	25.33a–c	25.67a–f	0.47ab	1.97a
16. BRS Grafite	15.00f–i	20.33d–h	0.42ab	1.42a–d
17. BRS Campeiro	20.33c–f	20.33d–h	0.38ab	1.17a–d
18. BRS 7762 Supermo	13.33 g–j	18.67f–h	0.43ab	1.31a–d
19. BRS Esplendor	21.67b–e	24.00a–h	0.52ab	1.06a–d
20. CNFP 10104	23.67a–d	21.67c–h	0.49ab	1.49a–d
21. Bambuí	22.33b–d	19.33d–h	0.24ab	1.16a–d
22. BRS Marfim	13.67 g–j	18.33gh	0.40ab	0.86b–d
23. BRS Agreste	13.67 g–j	19.33d–h	0.35ab	1.81a–c
24. BRS Pitamda	15.00f–i	24.00a–h	0.21b	1.05a–d
25. BRS Verede	23.00b–d	25.67a–f	0.41ab	1.96a
26. EMGOPA Ouro	8.00j	22.00b–h	0.26ab	0.79cd
27. BRS Radiante	10.67ij	18.67f–h	0.23ab	0.60d
28. Jalo Precoce	13.33 g–j	21.00c–h	0.25ab	1.03a–d
29. BRS Executivo	12.67 h–j	21.00c–h	0.38ab	1.20a–d
30. BRS Embaixador	13.67 g–j	25.33a–g	0.30ab	1.13a–d
Average	18.86b	22.65a	0.38b	1.27a
F-test				
P levels (P)	**		**	
Genotype (G)	**		**	
P × G	**		**	
CVP (%)	8.07		72.39	
CVG (%)	12.78		26.28	

Source: From Fageria, N. K. et al. 2012. *Commun. Soil Sci. Plant Anal.* 43:1–15. With permission.

**Significant at the 1% probability level. Means followed by the same letter in the same column or same line (P levels) are not significantly different at the 5% probability level by Tukey's test.

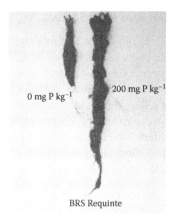

FIGURE 8.20 Root growth of dry bean genotype BRS Requinte at two P levels.

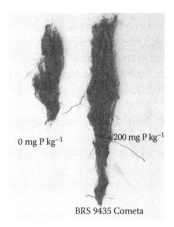

FIGURE 8.21 Root growth of dry bean genotype BRS 9435 Cometa at two P levels. (From Fageria, N. K. et al. 2012. *Commun. Soil Sci. Plant Anal.* 43:1–15. With permission.)

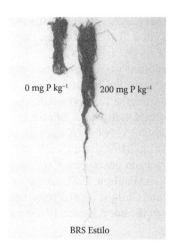

FIGURE 8.22 Root growth of dry bean genotype BRS Estilo at two P levels. (From Fageria, N. K. et al. 2012. *Commun. Soil Sci. Plant Anal.* 43:1–15. With permission.)

8.5.1 Use of Adequate Seed Rate and Row Spacing

Adequate seed rate and row spacing are important factors in determining the number of plants per unit area and consequently N uptake and use efficiency. Seeds used for sowing crops should have a germination of more than 85% to get good plant stands. Row spacing is determined experimentally for each crop under each agroclimatic condition. Adequate seed rate and row spacing do not involve extra cost to the farmers but reflect significantly in yield determination. For example, in the Brazilian Cerrado region or the central part of Brazil, the upland rice seed rate is about 80 kg ha^{-1} and row spacing about 30 cm. Similarly, the lowland rice seed rate is about 110 kg ha^{-1} and row spacing varied from 18 to 20 cm. Dry bean is planted about 12–15 seeds per meter row and row spacing is about 40 cm in the Cerrado region of Brazil.

8.5.2 Use of Efficient/Tolerant Crop Species and Genotypes within Species

It is estimated that a 1% increase in nitrogen use efficiency saves about 1.1 billion US$ annually worldwide (Kant et al., 2010). Hence, to minimize the loss of N, reduce environmental pollution, and decrease input cost, it is crucial to plant crop species or genotypes with higher nitrogen use efficiency. Fageria and Baligar (2005) reported that planting crop species and genotypes within species having a higher N uptake and use efficiency is an important strategy in modern agriculture. Differences in N uptake and utilization among crop species and cultivars within species for wheat, sorghum, corn, ryegrass, rice, and soybean have been reported (Moll and Kamprath, 1977; Pollmer et al., 1979; Reed et al., 1980; Traore and Maranville, 1999; Fageria, 2014). Similarly, many researchers have found significant variations of N use efficiency among lowland rice genotypes (Broadbent et al., 1987; Singh et al., 1998; Fageria and Barbosa Filho, 2001; Fageria and Baligar, 2003b). Pandey et al. (2001) reported that agronomic efficiency of N was higher in sorghum compared to pearl millet and corn over four N rates (45, 90, 235, and 180 kg N ha^{-1}). Fowler (2003) reported significant yield differences among wheat genotypes with increasing N rates from 0 to 240 kg ha^{-1}.

Plant species and genotypes within species differ in their ability for N uptake, utilization, and conversion into economic entity under low input (Singh, 2013). Lemaire et al. (1996) observed that N uptake by corn and sorghum were similar at high N input, but under N limitation, sorghum plants acquired significantly more N than corn. A more developed and branched root system in sorghum possibly permitted it to scrounge for N shortage (Hirel et al., 2007). The existence of genotypic variation for nitrogen use efficiency has been recognized for a long time (Smith, 1934; Springfield and Salter, 1934). Svecnjak and Rengel (2005) reported differences among spring canola genotypes in nitrogen use efficiency, which resulted in greater biomass with parallel performance at high and low N inputs.

Isfan (1993) reported highly significant variation among oat genotypes in both yield and physiological efficiency of absorbed N. According to this author, ideal genotypes could be those that absorb relatively high amounts of N from the soil and fertilizers, produce high grain yields per unit of absorbed N, and store relatively little N in the straw. Similarly, many workers found corn genotype differences for the absorption and utilization of N (Kamprath et al., 1982; Moll et al., 1982, 1987; Anderson et al., 1984). Lynch and White (1992) and Lynch and Rodriguez (1994) reported genetic variability in N use efficiency of dry bean genotypes. Experiments with the U.S. corn belt (Balko and Russell, 1980), tropical (Muruli and Paulsen, 1981; Lafitte and Edmeades, 1994; Banziger et al., 1997), and European maize (Bertin and Gallais, 2000; Presterl et al., 2003) indicated that genotypes can differ considerably in their N use efficiency. Hence, breeding for adaptation to low soil N seems feasible.

A direct relationship between the N fertilizer rate and corn plant growth and grain yields has been widely demonstrated (Jokela and Randall 1989; Zhang et al., 1993; McCullough et al., 1994; Costa et al., 2002). However, studies with conventional corn hybrids (Chevalier and Schrader, 1997) have shown that corn genotypes vary in their response to N availability, reflecting variations in their relative abilities to absorb native or fertilizer N from the soil. Similar observations have been reported by Moll et al. (1982) in the absorption and utilization of N by corn genotypes.

Figure 8.23 shows the responses of four lowland rice genotypes to N fertilization. These genotypes differ in yield response to applied N and can be grouped into three classes according to their responses to N fertilization. The first group was efficient and responsive to N. The genotype that produced above-average yields compared to all the genotypes tested at the low N level responded well to applied N—this was CNAi 9018. The second classification was efficient and nonresponsive. The genotype that produced well at low N rates did not respond well at higher N rates—this was CNAi 8569. The third group was genotypes that produced low at low N rates, but responded well to higher N rates. These have been designated as inefficient and responsive—these were Bigua and Jaburu. From a practical point of view, the genotypes that fell into the efficient and responsive group would be the most desirable, because they can produce well at low soil N levels and also respond well to applied N. Thus, this group could be utilized with low as well as high input technology with reasonably good yields. The second most desirable group would be efficient nonresponsive. Genotypes of this type can be planted under a low N level and still produce more than average yields. The inefficient responsive genotypes could be used in breeding programs for their N-responsive characteristics. Similarly, Figure 8.24 shows the linear and quadratic responses of five lowland rice genotypes grown on a Brazilian Inceptsol. Figure 8.25 shows the responses of upland rice genotypes to N fertilization. All the genotypes were having similar responses but the magnitude of response was different.

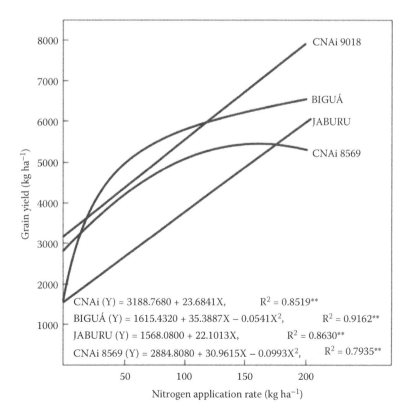

CNAi (Y) = 3188.7680 + 23.6841X, $R^2 = 0.8519**$
BIGUÁ (Y) = 1615.4320 + 35.3887X − 0.0541X², $R^2 = 0.9162**$
JABURU (Y) = 1568.0800 + 22.1013X, $R^2 = 0.8630**$
CNAi 8569 (Y) = 2884.8080 + 30.9615X − 0.0993X², $R^2 = 0.7935**$

FIGURE 8.23 Response of lowland rice genotypes to N fertilization. (From Fageria, N. K., Stone, L. F., and Santos, A. B. 2003b. *Soil Fertility Management of Irrigated Rice*. Santo Antônio de Goiás, Brazil: EMBRAPA—Rice and Bean Research Center. With permission.)

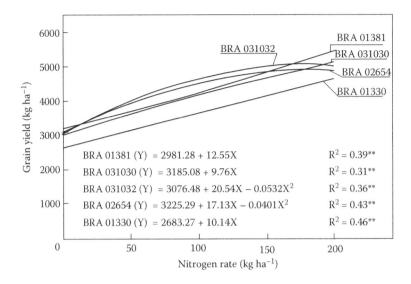

FIGURE 8.24 Response of lowland rice genotypes to nitrogen fertilization.

Data in Figures 8.26 and 8.27 show a grain yield and a straw yield of 10 upland rice genotypes at two soil pH. Genotypes were having significant differences in grain yield as well as in straw yield at two pH levels. Based on the grain yield efficiency index, genotypes were classified as tolerant, moderately tolerant, and susceptible to soil acidity (Figure 8.28). Figures 8.29 through 8.31 show the root growth of upland rice genotypes at two lime levels. It is very clear from these figures that upland rice genotypes were having different root growth at two lime levels, which might be responsible for the different uptake of nutrients and water and different responses to soil acidity.

Oilseed rape (*Brassica napus* L.) is an important oil crop that needs a large amount of nutrients (Zhang et al., 2010). Owing to the low efficiency of use of N fertilizers (Schjoerring et al., 1995), a large amount of the N fertilizer is applied to obtain maximum economic yield (Zhang et al., 2010). It

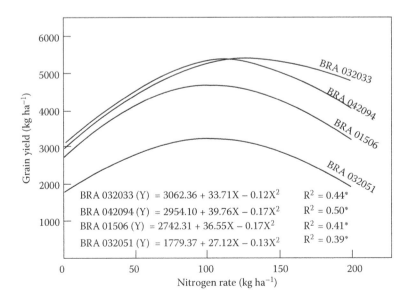

FIGURE 8.25 Response of upland rice genotypes to nitrogen fertilization.

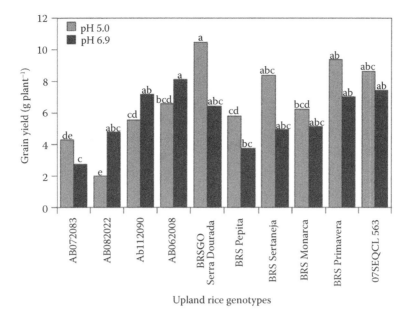

FIGURE 8.26 Grain yield of upland rice genotypes at two pH levels. Same letters on the bar at the same pH level do not differ significantly at the 5% probability level.

is estimated that the rate of supply of the N fertilizer to oilseed rape has reached 200–300 kg N ha^{-1} (Schjoerring et al., 1995; Wiesler et al., 2001a), and it is still increasing year by year (Zhang et al., 2010). Not much work has been conducted on the breeding of this crop. However, limited studies show that a large difference exists in the genotypes of oilseed rape for N efficiency (Wiesler et al., 2001a,b). It is also reported that cultivars with a high N harvest index has high N use efficiency

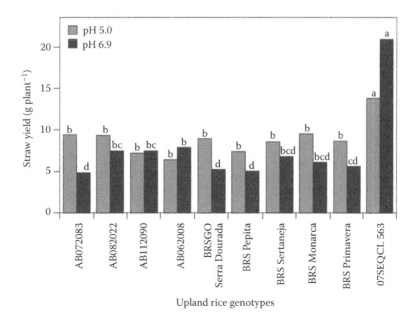

FIGURE 8.27 Straw yield of upland rice genotypes at two pH levels. Same letters on the bar at the same pH level do not differ significantly at the 5% probability level.

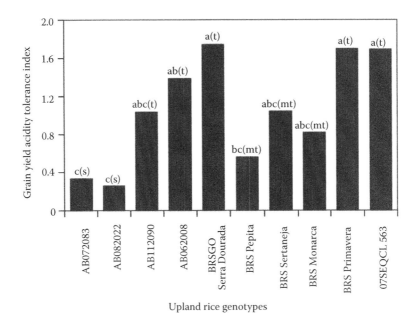

FIGURE 8.28 Classification of upland rice genotypes to acidity tolerance. Same letters on the bar at the same pH level do not differ significantly at the 5% probability level. Letters in parentheses indicate: t = tolerant, mt = moderately tolerant, and s = susceptible.

(Zhang et al., 2010). The increase of activities of asparaginate synthetase and glutamine synthetase of oilseed rape through gene transformation can raise the N use efficiency (Seifert et al., 2004). Zhang et al. (2010) also reported a variation in N use efficiency among rape cultivars. These authors also reported that under low N supply, high N use efficiency cultivars had longer roots, more lateral roots, higher amounts of reuse of nitrate from the stem and leaves, and higher nitrate reductase activities in leaves.

Several reasons have been cited why some genotypes are more efficient in N utilization compared to others (Thomason et al., 2002). Nutrient absorbing efficiency is one of the important causes of variation of nutrient efficiency in crop plants (Dhugga and Waines, 1989; Zhang et al., 2010).

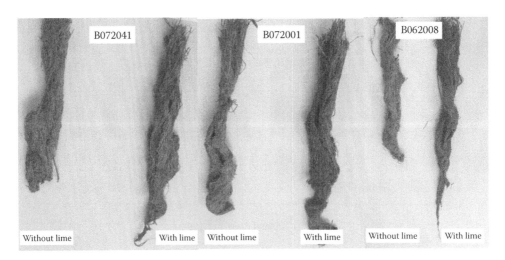

FIGURE 8.29 Root growth of upland rice with and without lime.

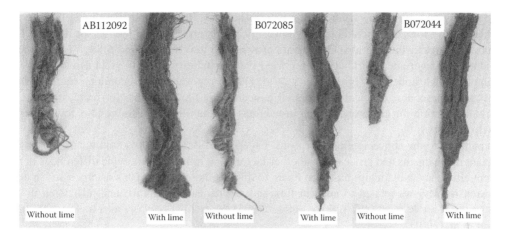

FIGURE 8.30   Root growth of upland rice with and without lime.

Moll et al. (1982) reported that N use efficiency differences among corn hybrids were due to the differing utilization of N already accumulated in the plant prior to anthesis, especially at low N levels. Eghball and Maranville (1991) reported that N use efficiency generally parallels water use efficiency in corn. Hence, both N use and water use efficiency traits might be selected simultaneously where such parallels exist. Kanampiu et al. (1997) reported that wheat cultivars with higher grain harvest indexes had higher N use efficiencies. Cox et al. (1985) reported that wheat cultivars that accumulate large amounts of N early in the growing season do not necessarily have high N use efficiency. Plants must convert this accumulated N into grain N and also assimilate N after anthesis to produce high N use efficiency. Forms of N uptake (NH_4^+ vs. NO_3^-) may also have effects on N use efficiency (Thomason et al., 2002). Plants with a preferential uptake of NH_4^+ during grain fill may provide increased N use efficiency over plants without this preference (Tsai et al., 1992). Ammonium N supplied to high-yielding corn genotypes increased the yield over plants supplied with NO_3^- during critical ear development (Pan et al., 1984). Salsac et al. (1987) reported that NH_4^+ assimilation processes require 5 ATP (adenosine triphosphate) mol^{-1} of NH_4^+, whereas NO_3^- assimilation processes require 20 ATP mol^{-1} NO_3^-. This energy-saving mechanism may be responsible for higher N use efficiency in NH_4^+–N.

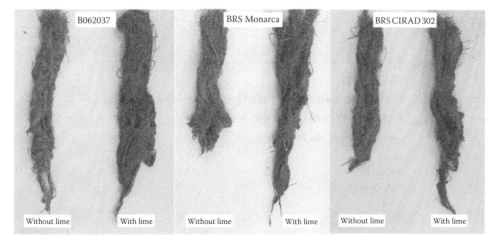

FIGURE 8.31   Root growth of upland rice with and without lime.

Regarding the genotypic variability for N use efficiency, Rosielle and Hamblin (1981) reported that heritability for grain yield is usually lower for plants grown under low versus high N. Thus, potential progress would be lower for plants grown with low N compared to high N target environments. Banziger and Lafitte (1997) reported that heritability of grain yield usually decreases for plants grown under low N. Banziger et al. (1997) reported that secondary traits (ears per plant, leaf senescence, and leaf chlorophyll concentration) are valuable for increasing the efficiency of selection for grain yield when broad-sense heritability of grain yield is low under low N environments.

In addition to the abovementioned reasons, Fageria and Baligar (2005) summarize various soil and plant mechanisms and processes and other factors that influence genotypic differences in plant nutrient efficiency. No attempt has been made to discuss these mechanisms or processes in detail. For extensive reviews related to nutrient flux and mechanisms of uptake and utilization in soil–plant systems, see Mengel et al. (2001), Barber (1995), Marschner (1995), Fageria et al. (2008), and Baligar et al. (2001).

In addition to planting N-efficient crop species and genotypes within species, the use of acid-tolerant crop species or genotypes within species is another important plant management practice to improve N use efficiency. *Stylosanthes guianensis* (Aublet) Sw., which has a tropical origin, is an important forage legume with high yield and quality, tolerant to soil acidity, and excellent adaption to infertile soils in tropical and subtropical regions.

8.5.3 CONTROL OF DISEASES, INSECTS, AND WEEDS

Diseases, insects, and weeds of agricultural crops are as old as agriculture itself (Fageria, 1992). The resultant losses in economic terms are impossible to estimate accurately because the severity of diseases, insects, and weeds varies greatly from place to place, crop to crop, season to season, and from year to year owing to changes in environmental factors (Fageria, 1992). Kramer (1967) estimated that the average worldwide losses for the main agricultural crops were 11.8% for diseases and 12.2% for insect pests. The average combined losses caused by diseases, insects, and weeds are put at 33.7%. Similarly, Oerke (2006) also reported that yield losses of corn, rice, and wheat by weeds and pests varied from 28.2% to 37.4% at the world level (Table 8.12).

The use of crop protection measures by the use of herbicides, insecticides, and fungicides, like nutrients, has played a huge part in increasing and sustaining the yield of arable crops around the world. Oerke and Dehne (1997) reviewed the literature and field experiments from around the world and calculated the percentage of potential losses prevented by control measures, that is, the efficacy of control. In 1991–1993, the efficacy reached only 34–38% in rice, wheat, and corn, but was 43% in soybean and potatoes. The efficacy was 55% for weeds, 31% for pests, and 23% for diseases. On

---

TABLE 8.12

Global Estimates of Actual Yield Losses (%) of Corn, Rice, and Wheat by Weeds and Pests

Losses by Weeds and Pests	Corn	Rice	Wheat
Weeds	209.5	10.2	7.7
Animal pests	9.6	15.1	7.9
Pathogens	8.5	10.8	10.2
Virus	2.7	1.4	2.4
Total	31.2	37.4	28.2

Source: Adapted from Oerke, E. C. 2006. *J. Agric. Sci.* 144:31–43.

a regional basis, the efficacy was 61% in west Europe, 56% in North America and Oceania, and 37% in the rest of the world. Both the potential and actual losses have increased in both actual and relative terms since the early 1960s, when the yields were lower and cropping less intensive. In 205 German wheat trials, the losses in 1985 and 1990 owing to diseases increased from 11% when the attainable yield was 4 Mg ha^{-1} to 20% when the yield was 11 Mg ha^{-1} (Jaggard et al., 2010). This indicates that as the yield rises in the future, more efficient crop protection measures will be needed to sustain productivity.

The control of diseases, insects, and weeds is an important factor in improving N use efficiency in crop production (Fageria and Gheyi, 1999). Crops infested with diseases, insects, and weeds have a lower photosynthetic efficiency and lower rate of absorption of water and nutrients, and the competition for light, water, and nutrients consequently reduces yields and results in low N use efficiency. In cereals or legumes, N is responsible for increasing yield components such as panicle or head numbers, grain numbers, grain weight, and pod numbers. When these components are adversely affected, N utilization by crop plants is decreased. In wheat crop, yield components such as spike weight, kernel weight, kernel number per spike, and number of spikes per plant can be negatively affected by stripe rust when infestation intensity is high at the heading and milk stages (Schultz and Line, 1992).

The presence of necrotic lesions on plant leaves is well known to decrease light interception and consequently leaf photosynthesis (Fageria and Gheyi, 1999). Boote et al. (1983) proposed a classification of diseases based on their effect on the physiology of the crop. For cercospora leaf spots, they hypothesized that disease effects on photosynthesis were mediated through loss of leaf area index (LAI), senescence acceleration, self-shading of healthy leaf area by leaf spots (light stealing), and a toxic effect of leaf spot disease on the photosynthetic mechanisms of the remaining leaves. Bourgeois and Boote (1992) reported that the late leaf spot in peanut induced by *Cercosporidium personatum* decreased the LAI, and that canopy photosynthesis was inversely proportional to total disease severity, which is an expression of both defoliation and necrotic area.

Planting disease- and insect-resistant crop cultivars is an important strategy in controlling the adverse effects of these biotic factors (Lynch and Mack, 1995; Sherwood et al., 1995) and improving yield and consequently N use efficiency. Soybean cultivars differ in sclerotinia stem rot tolerance and yield losses may not always occur under low levels of disease due to the yield compensation of nearby soybean plants (Hart, 1998). Yield reduction caused by sclerotinia stem rot or white mold may range from 147 to 370 kg ha^{-1} for every 10% increase in disease severity, depending on the environment and cultivar (Nelson et al., 2002). Planting soybean cultivars that are tolerant to selerotinia stem rot is strongly recommended to reduce yield loss (Grau and Radke, 1984; Boland and Hall, 1987; Kim et al., 1999; Yang et al., 1999) and improves nutrient use efficiency.

Disease, insects, and weeds can also be controlled by other crop management practices. One typical example is the control of irrigated rice diseases by an appropriate planting date in the state of Rio Grande do Sul of Brazil. Rio Grande do Sul is the largest lowland rice-producing state in Brazil. The average lowland rice yield in this state is about 7 Mg ha^{-1}, the highest in the country. Disease severity (brown spots, grain spots, *Rhizoctonia solani*, and rice blast fungus) was minimum when the rice crop was planted from October 1 to October 10 and maximum when it was planted from December 1 to December 12 (Table 8.13). Grohs et al. (2010) reported that brown spot disease infestation in lowland or irrigated rice was almost nil when planted between October 1 and October 15 and increased linearly when sowing was delayed from November 15 to December 1. The timing of crop management practices is increasingly based on the crop development stage as a means of improving efficacy (McMaster et al., 2012). For example, management practices for controlling *Fusarium graminearum* head blight in wheat fields focus on timing the fungicide application at flowering (Del Ponte et al., 2007). The efficacy of the fungicide application decreases as the variation of plant reaching flowering increases within the field (McMaster et al., 2012).

TABLE 8.13

Rating of Diseases Occurrence (Brown Spots, Grain Discoloration, Blast, and *Rhizoctonia solani*) in Irrigated Rice According to Date of Sowing in the State of Rio Grande do Sul of Brazil

Sowing Date	Probability of Disease Occurrence (%)	Disease Rating	Grain Yield (Mg ha^{-1})	Whole Grains (%)
1 October–10 October	35	Very low to low	8.30	64
15 October–30 October	45	Low	8.15	63
1 November–11 November	65	Low to medium	8.00	62
15 November–30 November	65	Medium to high	7.45	59
1 December–12 December	75	High to very high	6.65	56

Source: Adapted from Grohs, D. P. et al. 2010. Criterion for diseases management in irrigated rice. IRGA Technical Bulletin/Technical Report 7, 48p, Cachoeirinha, Rio Grande do Sul, Brazil: IRGA.

Note: The cultivars planted were BR IRGA 410 and IRGA 417.

8.6 QUANTITATIVE GENETICS AND MOLECULAR APPROACH TO IMPROVE NITROGEN USE EFFICIENCY

Quantitative genetics and molecular approach have been considered as important strategies in developing crop cultivars having a higher N use efficiency (Malik and Rengel, 2013). Molecular genetics can be used to identify key elements controlling the process of N remobilization. Nitrogen remobilization efficiency is subject to genetic variability (Malik and Rengel, 2013). Nitrogen remobilization from vegetative parts to grain is an important strategy in improving the yield of crops (Fageria et al., 2011a). Nitrogen remobilization in crop species can be manipulated by modifying physiological traits and the use of transgenic plants or mutants with modified capacity for N or carbon assimilation and recycling (Masclaux-Daubresse et al., 2001, 2010; Malik and Rengel, 2013). Significant variation in N remobilization from vegetative parts to grain has been reported for wheat and corn (Kichey et al., 2007).

Improvement in plant components that contribute to grain formation and filling can increase the total grain yield in cereals (Singh, 2013). A number of plant characteristics, including plant height, number of tillers, time of anthesis, number of panicles per plant, panicle length, number of grains per panicle, and size of grain, contribute to the final yield. Both classical and molecular genetics have provided information regarding the genes that control several important yield-controlling traits (Singh, 2013). In sorghum, six genes attributed to maturity (Rooney et al., 2000) and four genes responsible for dwarfism (Hadley, 1957) have been identified by classical genetics while controlling regimes of the genome for plant height (Rami et al., 1998), maturity (Lin et al., 1995), number of tillers (Paterson et al., 1995), seed weight (Pereira and Lee, 1995), panicle characteristics (Pereira and Lee, 1995), and stay green (Tao et al., 2000).

The identification of quantitative trait loci (QTL) for many morphological traits, grain yield, and its components under different N levels (low and high) has been reported for many crop species (An et al., 2006; Malik and Rengel, 2013). Grain yield, grain protein content, and leaf senescence are genetically controlled and that prompted mapping QTL for leaf senescence on various plant species (Malik and Rengel, 2013). In wheat (Joppa et al., 1997), barley (Mickelson et al., 2003) identified QTL related to N remobilization from senescing leaves for grain filling. Colonization of QTL for flag leaf remaining green and grain yield was confirmed by Verma et al. (2004) in wheat, and these traits are positively correlated in a phenotype. Similarly, in rice, Jiang et al. (2004) reported 46 QTL for stay-green-related traits and reported a positive correlation between stay-green and grain yield.

Nitrogen uptake from the soil–plant system and its translocation from root to shoot and grain are important physiological and biochemical processes in plants which are associated with N use efficiency. These processes can be better understood through a molecular approach or studies. Nitrogen translocation processes in plants are controlled by transport proteins (Malik and Rengel, 2013). The identification and characterization of transport systems that direct the flow of N metabolites between cellular compartments, in tissues and organs, and throughout the plants is therefore crucial for N distribution and improving N use efficiency in crop plants (Malik and Rengel, 2013). Key enzymes (two enzymes involved in the conversion of ammonium into amino acids in plants, that is, glutamine synthetase, GS, and glutamate synthase, GOGAT) and genes for N use efficiency have been identified (Hirel and Lea 2011; Malik and Rengel, 2013). Transgenic rice lines showed an increase in N use efficiency relative to control rice (Shrawat et al., 2008). A comparative study between the control plants and transgenic rice at the transcriptome level revealed that distinct functional classes of genes were differentially expressed in roots and shoots. However, none of the N uptake and assimilatory genes was upregulated or down-regulated (Beatty et al., 2009).

The rapid adoption of transgenic insect-protected corn hybrids has occurred during the past 15 years in North and South America (Traxler, 2006). For example, the benefits of hybrids with transgenic protection against Western corn rootworm (*Diabrotica virgifera virgifera*) include improved consistency of insect control, healthier root systems, advancements in environmental and farmer safety, and increased yields (Bender et al., 2013). These transgenic hybrids result in significantly less root damage and stunting (Vaughn et al., 2005), which in turn may allow them to accumulate more water and mineral nutrients compared to their nontransgenic isolines (Bender et al., 2013).

8.7 CONCLUSIONS

Nitrogen nutrition has been considered as one of the most important factors in increasing crop productivity in the last half of the twentieth century. It is one of the nutrients required by plants in large amounts and is also quite expensive. In addition, N recovery efficiency in crop plants is lower than 50%. Hence, a large amount of N is lost in the soil–plant system. This not only increases the cost of production but also creates environmental pollution. Under these situations, adopting appropriate nitrogen management practices in crop production is essential to increase crop yields and to reduce the cost of production. Further, adequate use of N also maintains sustainability of the cropping systems and also reduces environmental pollution. Improving the soil physical, chemical, and biological properties can improve the crop yields and N use efficiency. Crop responses to N and N use efficiency are very similar for similar types of crops in the tropics and temperate climates. Further, the same factors influence efficiency. Hence, greater efficiency of N use is obtained when N is placed in a band rather than broadcast, when N is applied to the plant in the right amount near the seeding time, and when N availability is in synchrony with N requirements by the crop. Ammonium sulfate has proven to be superior to upland and lowland rice compared to urea. This may be related to the tolerance of rice to soil acidity and soils that are deficient in S, because ammonium sulfate generates higher acidity compared to urea and has about 24% S which urea does not contain.

The use of organic manures and inclusion of legumes in crop rotation can increase the soil organic matter and N content of the soil, which may improve N use efficiency. In addition, the use of conservation tillage and improving WUE can also have higher N use efficiency by crop plants. The use of nitrogen-efficient/acidity-tolerant crop species and genotypes within species is an important strategy from an economic and environmental point of view, because a significant variation among crop species and genotypes of the same species has been found among most crop plants. Finally, the use of an integrated nutrient management model can improve the yield of crops and consequently higher N use efficiency. Quantitative genetics and molecular approach have a great potential in increasing N use efficiency in crop plants. However, so far, only limited success has been achieved through this approach in developing crop cultivars of higher N use efficiency. This may be due to the lack of interest and/or better cooperation between soil scientists, physiologists, and breeding groups. Based on the author's assumptions and calculations, it will be possible to increase the yield of field or grain crops by 60%, a target to feed the world population of 9.1 billion people by the year 2050. To achieve this goal, improving/maintaining soil fertility (especially the use of N in adequate amounts and proportions) will play a pivotal role in future crop production processes.

REFERENCES

Aase, J. K. and J. L. Pikul, Jr. 1995. Crop and soil response to long-term tillage practices in the northern Great Plains. *Agron. J.* 87:652–656.

Abdollahi, H. and D. B. Nedwell. 1979. Seasonal temperature as a factor influencing bacterial sulfate reduction in salt marsh sediment. *Microb. Ecol.* 5:73–79.

Abrol, Y. P. and K. T. Ingram. 1996. Effect of higher day and night temperatures on growth and yields of some crop plants. In: *Global Climate Change and Agricultural Production*, eds., F. Bazzaz, and W. Sombroek, pp. 123–140. New York: John Wiley & Sons.

Acosta-Martinez, V., G. Burow, T. M. Zobeck, and V. G. Allen. 2010. Soil microbial communities and function in alternative systems to continuous cotton. *Soil Sci. Soc. Am. J.* 74:1181–1192.

Adams, F. 1984. Crop response to lime in the southern United States. In: *Soil Acidity and Liming*, 2nd edition, ed., F. Adams, pp. 211–265. Madison, Wisconsin: ASA, CSSA, SSSA.

Agarwal, S. R., H. Shankar, and M. M. Agarwal. 1980. Effect of slow release nitrogen and nitrification inhibitors on rice-wheat sequence. *Indian J. Agron.* 35:337–340.

Aguilar, J., R. Evans, M. Vigil, and C. S. T. Daughtry. 2012. Spectral estimates of crop residue cover and density for standing and flat wheat stubble. *Agron. J.* 104:271–279.

Ait, B. E., J. Nowwak, and C. Clement. 2006. Enhancement of chilling resistance of inoculated grapevine plantlets with a plant growth promoting rhizobacterium; *Burkholderia phytofirmans* strain PsJn. *Appl. Environ. Microbiol.* 72:7246–7252.

Alam, B. and J. Jacob. 2002. Overproduction of photosynthetic electrons is associated with chilling injury in green leaves. *Photosynthetica* 40:91–95.

Alexander, K. G., M. H. Miller, and E. G. Beauchamp. 1991. The effect of an NH_4^+-enhanced nitrogen source on the growth and yield of hydroponically grown maize (*Zea mays* L.). *J. Plant Nutr.* 145:31–44.

Ali, S. K., V. Sandhya, M. Gover, N. Kishore, L. V. Rao, and B. Venkateswarlu. 2009. *Pseudomonas* sp. strain AKM-P6 enhances tolerance of sorghum seedlings to elevated temperatures. *Biol. Fertil. Soils* 46:45–55.

Al-Kaisi, M. and D. Kwaw-Mensah. 2007. Effect of tillage and nitrogen rate on corn yield and nitrogen and phosphorus uptake in a corn-soybean rotation. *Agron. J.* 99:1548–1558.

Al-Kaisi, M. and M. A. Licht. 2004. Effect of strip tillage on corn nitrogen uptake and residual soil nitrate accumulation compared with no tillage and chisel plow. *Agron. J.* 96:1164–1171.

Al-Karaki, G. N. and A. Al-Raddad. 1997. Effects of arbuscular mycorrhizal fungi and drought stress on growth and nutrient uptake of two wheat genotypes differing in drought resistance. *Mycorrhiza* 7:83–88.

Al-Karaki, G. N. and R. B. Clark. 1998. Growth, mineral acquisition and water use by mycorrhizal wheat grown under water stress. *J. Plant Nutr.* 21:263–276.

Allen, M. F. and M. G. Boosalis. 1983. Effects of two species of VA mycorrhizal fungi on drought tolerance of winter wheat. *New Phytol.* 93:67–76.

Allen, D. J. and D. R. Ort. 2001. Impact of chilling temperatures on photosynthesis in warm climate plants. *Trends Plant Sci.* 6:36–42.

Almeida, J. P. F., U. A. Hartwig, M. Frehner, J. Nosberger, and A. Luscher. 2000. Evidence that P deficiency induces N feedback regulation of symbiotic N_2 fixation in white clover (*Trifolium repens* L.). *J. Exp. Bot.* 51:1289–1297.

Alter, D. and A. Mitchell. 1992. Use of vermicompost extract as an aluminum inhibitor in aqueous solutions. *Commun. Soil Sci. Plant Anal.* 23:231–240.

Altieri, M. 1999. The ecological role of biodiversity in agroecosystems. *Agric. Ecosyst. Environ.* 74:19–31.

Amato, M., R. B. Jackson, J. H. A. Butler, and J. N. Ladd. 1984. Decomposition of plant material in Australian soils. II. Residual organic ^{14}C and ^{15}N from legume plant parts decomposition under field and laboratory conditions. *Aust. J. Soil Res.* 22:331–341.

Ambus, P. and E. S. Jensen. 1997. Nitrogen mineralization and denitrification as influenced by particle size. *Plant Soil* 197:261–270.

Ambus, P. and E. S. Jensen. 2001. Crop residue management strategies to reduce N-losses-interaction with crop N supply. *Commun. Soil Sci. Plant Anal.* 32:981–996.

An, D. G., Y. J. Su, and Q. Y. Liu. 2006. Mapping QTL for nitrogen uptake in relation to the early growth of wheat. *Plant Soil* 284:73–84.

Anderson, E. L., E. J. Kamprath, and R. H. Moll. 1984. Nitrogen fertility effects on accumulation, remobilization, and partitioning of N and dry matter in corn genotypes differing in prolificacy. *Agron. J.* 76:397–404.

Angers, A. and S. Recous. 1997. Decomposition of wheat and rye residue as affected by particle size. *Plant Soil* 189:197–203.

Angus, J. F., J. W. Bowden, and B. A. Keating. 1993. Modeling nutrient responses in the field. *Plant Soil* 155/156:57–66.

Ashford, D. L. and D. W. Reeves. 2003. Use of a mechanical roller-crimper as an alternative kills method for cover crops. *Am. J. Alternative Agric.* 18:37–45.

Auge, R. M. 2001. Water relations, drought and vesicular arbuscular mycorrhizal symbiosis. *Mycorrhiza* 11:373–381.

Auge, R. M. 2004. Arbuscular mycorrhizae and soil/plant water relations. *Can. J. Soil Sci.* 84:373–381.

Bacon, P. E. 1990. Effects of stubble and N fertilization management on N availability and uptake under successive rice (*Oryza sativa* L.) crops. *Plant Soil* 121:11–19.

Bailey, J. S. 1992. Effects of gypsum on the uptake, assimilation and cycling of ^{15}N-labeled ammonium and nitrate-n by perennial ryegrass. *Plant Soil* 143:19–31.

Baker, J. T. and L. H. Allen Jr. 1993. Contrasting crop species response to CO_2 and temperature: Rice, soybean, and citrus. *Vegetatio* 104:239–260.

Baligar, V. C. and N. K. Fageria. 1999. Plant nutrient efficiency: Towards the second paradigm In: *Soil Fertility, Soil Biology and Plant Nutrition Interrelationships*, eds., J. Q. Siqueira, F. M. S. Moreira, A. S. Lopes, L. R. G. Guilherme, V. Faquin, A. E. Furtini Neto, and J. G. Carvallo, pp. 183–204. Lavras, Brazil: Brazilian Society of Soil Science and University of Lavras.

Baligar, V. C., G. V. E. Pitta, E. E. G. Gama, R. E. Schaffert, A. F. D. Bahia Filho, and R. B. Clark. 1997. Soil acidity effects on nutrient use efficiency in exotic maize genotypes. *Plant Soil* 192:9–13.

Baligar, V. C, N. K. Fageria, and Z. L. He. 2001. Nutrient use efficiency in plants. *Commun. Soil Sci. Plant Anal.* 32:921–950.

Balko, L. G. and W. A. Russell. 1980. Effects of rate of nitrogen fertilizer on maize inbred lines and hybrid progeny. I. Prediction of yield responses. *Maydica* 25:65–79.

Bandick, A. K. and R. P. Dick. 1999. Field management effects on soil enzyme activities. *Soil Biol. Biochem.* 31:1471–1479.

Banziger, M. and H. R. Lafitte. 1997. Efficiency of secondary traits for improving maize for low nitrogen target environments. *Crop Sci.* 37:1110–1117.

Banziger, M., F. J. Betran, and H. R. Lafitte. 1997. Efficiency of high nitrogen selection environments for improving maize low-nitrogen target environment. *Crop Sci.* 37:1103–1109.

Baolong, Z. 1989. The absorption and translocation of foliar applied nitrogen in cotton. M.S. Thesis, Fayetteville: University of Arkansas.

Barber, S. A. 1995. *Soil Nutrient Bioavailability: A Mechanistic Approach*, 2nd edition. New York: John Wiley & Sons.

Bardgett, R. D. 2005. *The Biology of Soil; A Community and Ecosystem Approach*. Oxford: Oxford University Press.

Bausenwein, U., A. Gattinger, U. Langer, A. Embacher, H. P. Hartmann, M. Sommer, J. C. Munch, and M. Schloter. 2008. Exploring soil microbial communities and soil organic matter, variability and interactions in arable soils under minimum tillage practices. *Appl. Soil Ecol.* 40:67–77.

Beatty, P. H., A. K. Shrawt, and R. T. Carroll. 2009. Transcriptome analysis of nitrogen efficient over expressing alanine aminotransferase. *Plant Biotech. J.* 7:562–576.

Bender, R. R., J. W. Haegele, M. L. Ruffo, and F. E. Below. 2013. Nutrient uptake, partitioning and remobilization in modern transgenic insect-protected maize hybrid. *Agron. J.* 105:161–170.

Bengough, A. G., M. F. Bransby, J. Hans, S. J. McKenna, T. J. Roberts, and T. A. Valentine. 2006. Root responses to soil physical conditions: Growth dynamics from field to cell. *J. Exp. Bot.* 57:437–447.

Benjamin, J. G. and D. C. Nielsen. 2004. A method to separate plant roots from soil and analyze root surface area. *Plant Soil* 267:225–234.

Benjamin, J. G., A. D. Halvorson, D. C. Nielsen, and M. M. Mikha. 2010. Crop management effects on crop residue production and changes in soil organic carbon in the central Great Plains. *Agron. J.* 102: 990–997.

Benjamin, J. G., D. C. Nielsen, M. F. Vigil, M. M. Mikha, and F. J. Calderon. 2013. A comparison of two models to evaluate soil physical property effects on corn root growth. *Agron. J.* 105:713–720.

Bermudez, M. and A. P. Mallarino. 2004. Corn response to starter fertilizers and tillage across and within fields having no-till management histories. *Agron. J.* 96:776–785.

Bertin, P. and A. Gallais. 2000. Genetic variation for nitrogen use efficiency in a set of recombinant maize inbred lines. I. Agrophysiological results. *Maydica* 45:53–66.

Bhattacharyya, R., S. C. Pandey, J. K. Bist, J. C. Bhatt, H. S. Gupta, M. D. Tuti, D. Mahanta et al. 2013. Tillage and irrigation effects on soil aggregation and carbon pools in the Indian sub-Himalayas. *Agron. J.* 105:101–112.

Biau, A., F. Santiveri, and J. Lloveras. 2013. Stover management and nitrogen fertilization effects on corn production. *Agron. J.* 105:1264–1270.

Biederbeck, V. O., O. T. Bouman, C. A. Campbell, L. D. Bailey, and G. E. Winkelman. 1996. Nitrogen benefits from four green manure legumes in dryland cropping systems. *Can. J. Plant Sci.* 76:307–315.

Bijay-Singh and J. C. Katyal. 1987. Relative efficacy of some new urea-based nitrogen fertilizers for growing wetland rice on a permeable alluvial soil. *J. Agric. Sci.* 109:27–31.

Bijay-Singh and Yadvinder-Singh. 2004. Balanced fertilization for environmental quality—Punjab experience. *Fert. News* 49:107–113.

Birch, H. F. 1959. Further observations on humus decomposition and nitrification. *Plant Soil* 11:262–286.

Bird, J. A., W. R. Horwath, A. J. Eagle, and C. van Kessel. 2001. Immobilization of fertilizer nitrogen in rice: Effects of straw management practices. *Soil Sci. Soc. Am. J.* 65:1143–1152.

Bird, M. I., E. M. Veenendaal, C. Moyo, J. Lloy, and P. Frost. 2000. Effect of fire and soil texture on soil carbon in subhumid savanna. *Geoderma* 94:71–90.

Blackmer, A. M. 2000. Bioavailability of nitrogen. In: *Handbook of Soil Science*, ed., M. E. Sumner, pp. 3–18. Boca Raton, Florida: CRC Press.

Blanco-Canqui, H. and R. Lal. 2008. No-tillage and soil profile carbon sequestration: Na on farm assessment. *Soil Sci. Soc. Am. J.* 72:693–701.

Blanco-Canqui, H. and R. Lal. 2009. Crop residue removal impacts on soil productivity and environmental quality. *Crit. Rev. Plant Sci.* 28:139–163.

Blum, A. 2004. Sorghum physiology. In: *Physiology and Biology Integration for Plant Breeding*, ed., H. T. Nguyen and A. Blum, pp. 141–223. New York: Marcel Dekker.

Blum, A. 2005. Drought resistance, water use efficiency and yield potential: Are they compatible dissonant or mutually exclusive? *Aust. J. Agric. Res.* 56:1159–1168.

Blum, A. 2009. Effective use of water (EUW) and not water use efficiency (WUE) is the target of crop yield improvement under drought stress. *Field Crops Res.* 112:119–123.

Bolan, N. S. and M. J. Hedley. 2003. Role of carbon, nitrogen, and sulfur cycles in soil acidification. In: *Handbook of Soil Acidity,* ed., Z. Rengel, pp. 29–56. New York: Marcel Dekker.

Boland, G. J. and R. Hall. 1987. Evaluating soybean cultivars for resistance to *Sclerotinia sclerotiorum* under field conditions. *Plant Dis.* 71:934–936.

Boma, D. C. and R. Y. Evans. 1991. Calcium inhibition of polyacrylamide gel hydration is partially reversible by potassium. *HortScience* 26:1063–1065.

Bondada, B. R., D. M. Oosterhuis, and R. J. Norman. 1997. Cotton leaf age, epicuticular wax, and nitrogen-15 absorption. *Crop Sci.* 37:807–811.

Boote, K. J., J. W. Jones, J. W. Mishoe, and R. D. Berger. 1983. Coupling pests to crop growth simulators to predict yield reductions. *Phytopathology* 73:1581–1587.

Bossio, D., K. Geheb, and W. Critchley. 2010. Managing water by managing land: Addressing land degradation to improve water productivity and rural livelihoods. *Agric. Water Manage.* 97:536–542.

Bot, A. J., F. O. Nachtergaele, and A. Young. 2000. Land resource potential and constraints at regional and country levels. World soil resources report 90. Land and water development division. Food and Agricultural Organization, Rome.

Bourgeois, G. and K. J. Boote. 1992. Leaflet and canopy photosynthesis of peanut affected by late leaf spot. *Agron. J.* 84:359–366.

Bouwman, A. F. 1998. Nitrogen oxides and tropical agricultural. *Nature* 392:866–867.

Brady, N. C. and R. R. Weil. 2002. *The Nature and Properties of Soils*, 13th edition. Upper Saddle River, New Jersey: Prentice Hall. 960p.

Bremer, E. and C. Van Kessel. 1992. Plant-available nitrogen from lentil and wheat residues during a subsequent growing season. *Soil Sci. Soc. Am. J.* 56:1155–1160.

Brennan, R. F. and M. D. A. Bolland. 2007. Comparing the potassium requirements of canola and wheat. *Aust. J Agric. Res.* 58:359–366.

Brhane, G. and C. S. Wortman. 2008. Tie-ridge tillage for high altitude pulse production in northern Ethiopia. *Agron. J.* 100:447–453.

Brhane, G., C. S. Wortman, M. Mamo, H. Gebrekidan, and A. Belay. 2006. Micro-basin tillage for grain sorghum production in semi-arid areas of northern Ethiopia. *Agron. J.* 98:124–128.

Broadbent, F. E., S. K. De Datta, and E. V. Laureles. 1987. Measurement of nitrogen utilization efficiency in rice genotypes. *Agron. J.* 79:786–791.

Brosius, M. R., G. K. Evanylo, L. R. Bulluck, and J. B. Ristaino. 1998. Comparison of commercial fertilizer and organic by products on soil chemical and biological properties and vegetable yields. In: *Beneficial Co-Utilization of Agricultural, Municipal and Industrial By-Products*, eds., S. Brown, J. S. Angle, and L. Jacobs, pp. 195–202. Dordrecht: Kluwer Academic Publishers.

Brown, S. R., R. Chaney, and J. S. Angle. 1997. Subsurface liming and metal movement in soils amended with lime-stabilized biosolids. *J. Environ. Qual.* 26:724–732.

Bruinsma, J. 2003. *World Agriculture: Towards 2015/2030. An FAO Perspective.* London: Earth Scan Publications Ltd.

Burgess, M. S., G. R. Mehuys, and C. A. Madramootoo. 2002. Nitrogen dynamics of decomposing corn residue components under three tillage systems. *Soil Sci. Soc. Am. J.* 66:1350–1358.

Burket, J. Z., D. D. Hemphill, and R. P. Dick. 1997. Winter cover crops and nitrogen management in sweet corn and broccoli rotations. *HortScience* 32:664–668.

Cambouris, A. N., M. C. Nolin, B. J. Zebarth, and M. R. Laverdiere. 2006. Soil management zones delineated by electrical conductivity to characterize spatial and temporal variations in potato yield and in soil properties. *Am. J. Potato Res.* 83:381–395.

Campbell, C. A., R. J. K. Myers, and D. Curtin. 1995. Managing nitrogen for sustainable crop production. *Fertilizer Res.* 42:277–296.

Campbell, C. A., A. P. Moulin, D. Curtin, G. P. Lafound, and L. Townley-Smith. 1993a. Soil aggregation as influenced by cultural practices in Saskatchewan. I. Black chernozemic soils. *Can. J. Soil Sci.* 73:579–595.

Campbell, C. A., R. P. Zentner, F. Selles, B. G. McConkey, and F. B. Dyck. 1993b. Nitrogen management for spring wheat grown annually on zero-tillage: Yields and nitrogen use efficiency. *Agronomy J.* 85: 107–114.

Cannell, R. Q. and J. D. Hawes. 1994. Trends in tillage practices in relation to sustainable crop production with reference to temperate climates. *Soil Tillage Res.* 30:245–282.

Carranca, C., A. Varennes, and D. Rolston. 1999. Biological nitrogen fixation by faba bean, pea and chickpea, under field conditions, estimated by the ^{15}N isotope dilution technique. *Eur. J. Agron.* 9:109–116.

Cassman, K. G., A. Dobermann, and D. Walters. 2002. Agroecosystems, nitrogen use efficiency, and nitrogen management. *AMBIO* 31:132–140.

Chalk, P. M. 1998. Dynamics of biological fixed N in legume-based rotations: A review. *Aust. J. Agric. Res.* 49:303–316.

Chan, K. Y. and D. P. Heenan. 1991. Differences in surface soil aggregation under six different crops. *Aust. J. Exp. Agric.* 31:683–686.

Chevalier, P. and L. E. Schrader. 1997. Genotypic differences in nitrate absorption and portioning of N among plant parts in maize. *Crop Sci.* 17:897–901.

Christopher, S. F., R. Lal, and U. Mishra. 2009. Regional study of no-till effects on carbon sequestration in the Midwestern United States. *Soil Sci. Soc. Am. J.* 73:207–216.

Clark, L. E. and T. O. Knight. 1996. Grain production and economic returns from dryland sorghum in response to tillage systems and planting patterns in the semi-arid southwestern USA. *J. Prod. Agric.* 9:249–256.

Clement, A., J. K. Ladha, and F. P. Chalifour. 1998. Nitrogen dynamics of various green manure species and the relationship to lowland rice production. *Agron. J.* 90:149–154.

Coelho, E. F., M. A. Coelho Filho, and A. J. P. Silva. 2009. Agriculture irrigada: otimização da eficiência de irrigação e do uso da água. In: Simpósio nacional sobre o uso da água na agricultura, 3, 2009, Passo Fundo. Disponível em: http://www.upf.br/coaju/download/Eugenio.pdf. Acesso em: 11 set. 2009.

Cole, C. V., J. Williams, M. Shaffer, and J. Hanson. 1987. Nutrient and organic matter dynamics as components of agricultural production systems models. In: *Soil Fertility and Organic Matter as Critical Components of Production Systems*, ed., R. F. Follett, pp. 147–166. Madison, Wisconsin: ASA, CSSSA, and SSSA.

Condon, A. G., R. A. Richards, G. D. Rebetzke, and G. D. Farquhar. 2002. Improving intrinsic water-use efficiency and crop yield. *Crop Sci.* 42:122–131.

Copeland, P. J. and R. K. Crookston. 1992. Crop sequence affects nutrient composition of corn and soybean grown under high fertility. *Agron. J.* 84:503–509.

Costa, C., L. M. Dwyer, D. W. Stewart, and D. L. Smith. 2002. Nitrogen effects on grain yield and yield components of leafy and nonleafy maize genotypes. *Crop Sci.* 42:1556–1563.

Cox, M. C., C. O. Qualset, and D. W. Rains. 1985. Genetic variation for nitrogen assimilation and translocation in wheat. II. Nitrogen assimilation in relation to grain yield and protein. *Crop Sci.* 25:435–440.

Cox, M. S., P. D. Gerard, M. C. Wardlaw, and M. J. Abshire. 2003. Variability of selected soil properties and their relationships with soybean yield. *Soil Sci. Soc. Am. J.* 67:1296–1302.

Crutchfield, D. A., G. A. Wicks, and O. C. Burnside. 1986. Effect of winter wheat (*Triticum aestivum* L.) straw mulch level on weed control. *Weed Sci.* 34: 110–114.

Culman, S. W., S. S. Snapp, M. Ollenburger, B. Basso, and L. R. Dehaan. 2013. Soil and water quality rapidly respond to the perennial grain kernza wheatgrass. *Agron. J.* 105:735–744.

Dabney, S. M., J. A. Delgado, and D. W. Reeves. 2001. Using winter cover crops to improve soil and water quality. *Commun. Soil Sci. Plant Anal.* 32:1221–1250.

Damon, P. M. and Z. Rengel. 2007. Wheat genotypes differ in potassium efficiency under glass house and field conditions. *Aust. J. Agric. Res.* 58:816–825.

Davelouis, J. R., P. A. Sanchez, and J. C. Alegre. 1991. Green manure incorporation and soil acidity amelioration. In: *TropSoils Technical Report* 1988–1989, ed., T. P. McBridge, pp. 286–289. Raleigh, NC: North Carolina State University.

Davidson, E. A. and I. A. Janssens. 2006. Temperature sensitivity of soil carbon decomposition and feedback to climate change. *Nature* 440:165–173.

Davies, F. T., V. Olalde-Portugal, L. Aguilera-Gomez, M. J. Alvarado, R. C. Ferreira-Cerrato, and T. W. Boutton. 2002. Alleviation of drought stress of chile ancho pepper (*Capsicum annuum* L. cv. San Luis) with arbuscular mycorrhiza indigenous to Mexico. *Sci. Hortic.* 92:347–359.

De Datta, S. K. and R. J. Buresh. 1989. Integrated nitrogen management in irrigated rice. *Adv. Soil Sci.* 10:143–169.

Deibert, E. J., E. French, and B. Hoag. 1986. Water storage and use by spring wheat under conventional tillage and no-tillage in continuous and alternate crop-fallow systems in the northern Great Plains. *J. Soil Water Conserv.* 41:53–58.

De-Lacerda, C. F., J. Cambraia, M. A. Oliva, H. A. Ruiz, and J. T. Prisco. 2003. Solute accumulation and distribution during shoot and leaf development in two sorghum genotypes under salt stress. *Environ. Exp. Bot.* 49:107–120.

Del Ponte, E. M., J. M. C. Fernandes, and G. C. Bergstrom. 2007. Influence of growth stage on fusarium head blight and deoxynivalenol production in wheat. *J. Phytopathol.* 155:577–581.

Dhugga, K. S. and J. G. Waines. 1989. Analysis of nitrogen accumulation and use in bred durum wheat. *Crop Sci.* 29:1232–1239.

Diaz-Zorita, M., J. H. Grove, L. Murdock, J. Herbeck, and E. Perfect. 2004. Soil structural disturbance effects on crop yields and soil properties in a no-tillage production system. *Agron. J.* 96:1651–1659.

Dick, W. A. 1984. Influence of long-term tillage and crop rotation combinations on soil enzyme activities. *Soil Sci. Soc. Am. J.* 48:569–574.

Dilly, O., H. P. Blume, and Munch, J. C. 2003. Soil microbial activities in Luvisols and Anthrosols during 9 years of region typical tillage and fertilization practices in northern Germany. *Biochemistry* 65: 319–339.

Dinnes, D. L., D. L. Karlen, D. B. Jaynes, T. C. Kaspar, J. L. Hatfield, T. S. Colvin, and C. A. Cambardella. 2002. Nitrogen management strategies to reduce nitrate leaching in tile-drained midwestern soils. *Agron. J.* 94:153–171.

Diouf, A. 2001. Monitoring land-cover changes in semi-arid regions: Remote sensing data and field observations in the Ferol, Senegal. *J. Arid Environ.* 48:129–148.

Dobbie, K. E. and K. A. Smith. 2003. Nitrous oxide emission factors for agricultural soils in Great Britain: The impact of soil water-filled pore space and other controlling variables. *Glob. Chang. Biol.* 9:204–218.

Dominati, E., M. Patterson, and A. D. Mackay. 2010. A framework for classifying and quantifying the natural capital and ecosystem services of soils. *J. Ecol. Econ.* 69:1858–1868.

Doran, J. W. and M. S. Smith. 1991. Role of cover crops in nitrogen cycling. In: *Cover Crop for Clean Water*, ed., W. L. Hargroce, pp. 85–90. Proc. Int. Conf. Jackson, TN. 9–11 April 1991. Ankeny, Iowa: Soil and Water Conserve. Soc.

Duan, Y., M. Xu, B. Wang, X. Yang, S. Huang, and S. Gao. 2011. Long term evaluation of manure application on maize yield and nitrogen use efficiency. *Soil Sci. Soc. Am. J.* 75:1562–1573.

Eagle, A. J., J. A. Bird, J. E. Hill, W. R. Horwath, and C. V. Kessel. 2001. Nitrogen dynamics and fertilizer use efficiency in rice following straw incorporation and winter flooding. *Agron. J.* 93:1346–1354.

Eghball, B. and J. W. Maranville. 1991. Interactive effects of water and nitrogen stresses on nitrogen utilization efficiency, leaf water status and yield of corn genotypes. *Commun. Soil Sci. Plant Anal.* 22:1367–1382.

Fageria, N. K. 1989. *Tropical Soils and Physiological Aspects of Crops*. Brasilia, Brazil: EMBRAPA.

Fageria, N. K. 1992. *Maximizing Crop Yields*. New York: Marcel Dekker.

Fageria, N. K. 2001. Effect of liming on upland rice, common bean, corn, and soybean production in cerrado soil. *Pesq. Agropec. Bras.* 36:1419–1424.

Fageria, N. K. 2002. Soil quality vs. environmentally-based agriculture. *Commun. Soil Sci. Plant Anal.* 33:2301–2329.

Fageria, N. K. 2008. Optimum soil acidity indices for dry bean production on an oxisol in no-tillage system. *Commun. Soil Sci. Plant Anal.* 39:845–857.

Fageria, N. K. 2009. *The Use of Nutrients in Crop Plants.* Boca Raton, Florida: CRC Press.

Fageria, N. K. 2013. *The Role of Plant Roots in Crop Production.* Boca Raton, Florida: CRC Press.

Fageria, N. K. 2014. *Mineral Nutrition of Rice.* Boca Raton, Florida: CRC Press

Fageria, N. K. and V. C. Baligar. 1997a. Response of common bean, upland rice, corn, wheat, and soybean to soil fertility of an Oxisol. *J. Plant Nutr.* 20:1279–1289.

Fageria, N. K. and V. C. Baligar. 1997b. Integrated plant nutrient management for sustainable crop production. An overview. *Int. J. Trop. Agric.* 15:1–18.

Fageria, N. K. and V. C. Baligar. 1999. Yield and yield components of lowland rice as influenced by timing of nitrogen fertilization. *J. Plant Nutr.* 22:23–32.

Fageria, N. K. and V. C. Baligar. 2001a. Improving nutrient use efficiency of annual crops in Brazilian acid soils for sustainable crop production. *Commun. Soil Sci. Plant Anal.* 32:1303–1319.

Fageria, N. K. and V. C. Baligar. 2001b. Lowland rice response to nitrogen fertilization. *Commun. Soil Sci. Plant Anal.* 32:1405–1429.

Fageria, N. K. and V. C. Baligar. 2003b. Fertility management of tropical acid soils for sustainable crop production. In: *Handbook of Soil Acidity*, ed., Z. Rengel, pp. 359–385. New York: Marcel Dekker.

Fageria, N. K. and V. C. Baligar. 2005. Enhancing nitrogen use efficiency in crop plants. *Adv. Agron.* 88:97–185.

Fageria, N. K. and M. P. Barbosa Filho. 2001. Nitrogen use efficiency in lowland rice genotypes. *Commun. Soil Sci. Plant Anal.* 32:2079–2089.

Fageria N. K. and H. R. Gheyi. 1999. *Efficient Crop Production.* Campina Grande, Paraiba, Brazil: Federal University of Paraiba.

Fageria, N. K. and A. Moreira. 2011. The role of mineral nutrition on root growth of crop plants. *Adv. Agron.* 110:251–331.

Fageria, N. K. and A. S. Prabhu. 2004. Blast control and nitrogen management in lowland rice cultivation. *Pesq. Agropec. Bras.* 39:123–129.

Fageria, N. K. and A. B. Santos. 2005. Influence of base saturation and micronutrient rates on their concentration in the soil and bean productivity in cerrado soil in no-tillage system. Paper presented at the *VIII National Bean Congress*, Goiânia, Brazil, 18–20 October 2005.

Fageria, N. K. and L. F. Stone. 1999. *Acidity Management of Cerrado and Varzea Soils of Brazil.* Santo Antônio de Goiás, Brazil: EMBRAPA Arroz e Feijão Document No. 92.

Fageria, N. K. and L. F. Stone. 2013. Water and nutrient use efficiency in food production in South America. In: *Improving Water and Nutrient Use Efficiency in Food Production Systems*, ed., Z. Rengel, pp. 275–296. Ames, Iowa: John Wiley & Sons.

Fageria, N. K., V. C. Baligar, and R. B. Clark. 2006. *Physiology of Crop Production.* New York: The Howorth Press.

Fageria, N. K., V. C. Baligar, and D. G. Edwards. 1990. Soil-plant nutrient relationships at low pH stress. In: *Crops as Enhancers of Nutrient Use*, eds., V. C. Baligar and R. R. Duncan, pp. 475–507. San Diego, California: Academic Press.

Fageria, N. K., V. C. Baligar, and C. A. Jones. 2011a. *Growth and Mineral Nutrition of Field Crops,* 3rd edition. Boca Raton, Florida: CRC Press.

Fageria, N. K., V. C. Baligar, and Y. C. Li. 2008. The role of nutrient efficient plants in the twenty first century. *J. Plant Nutr.* 31:1121–1157.

Fageria, N. K., V. C. Baligar, and R. W. Zobel. 2007. Yield, nutrient uptake, and soil chemical properties as influenced by liming and boron application in common bean in a no-tillage system. *Commun. Soil Sci. Plant Anal.* 38:1637–1653.

Fageria, N. K., M. P. Barbosa Filho, A. Moreira, and C. M. Guimaraes. 2009. Foliar fertilization of crop plants. *J. Plant Nutr.* 32:1044–1064.

Fageria, N. K., L. C. Melo, J. P. Pereira, and A. M. Coelho. 2012. Yield and yield components of dry bean genotypes as influenced by phosphorus fertilization. *Commun. Soil Sci. Plant Anal.* 43:1–15.

Fageria, N. K., A. Moreira, and A. M. Coelho. 2011b. Yield and yield components of upland rice as influenced by nitrogen sources. *J. Plant Nutr.* 34:361–370.

Fageria, N. K., A. B. Santos, and A. M. Coelho. 2011c. Growth, yield and yield components of lowland rice as influenced by ammonium sulfate and urea fertilization. *J. Plant Nutr.* 34:371–386.

Fageria, N. K., A. B. Santos, and M. F. Moraes. 2010. Influence of urea and ammonium sulfate on soil acidity indices in lowland rice production. *Commun. Soil Sci. Plant Anal.* 41:1565–1575.

Fageria, N. K., N. A. Slaton, and V. C. Baligar. 2003a. Nutrient management for improving lowland rice productivity and sustainability. *Adv. Agron.* 80:63–152.

Fageria, N. K., Stone, L. F., and Santos, A. B. 2003b. *Soil Fertility Management of Irrigated Rice.* Santo Antônio de Goiás, Brazil: EMBRAPA—Rice and Bean Research Center.

Fang, C. and J. B. Moncrieff. 2001. The dependence of soil CO_2 efflux on temperature. *Soil Biol. Biochem.* 33:155–165.

Farahani, H. J., G. A. Peterson, and D. G. Westfall. 1998. Dryland cropping intensification: A fundamental solution to efficient use of precipitation. *Adv. Agron.* 64:197–223.

Fauci, M. F. and R. P. Dick. 1994. Plant response to organic amendments and decreasing inorganic nitrogen rates in soil from a long-term experiment. *Soil Sci. Soc. Am. J.* 58:134–138.

Fedoroff, N. V., D. S. Batisti, R. N. Beachy, P. J. M. Coper, D. A. Fishhoff, and C. N. Hodges. 2010. Radically rethinking agricultural for the 21st century. *Science* 327:833–834.

Feldman, L. J. 1980. Auxin biosynthesis and metabolism in isolated roots of *Zea mays*. *Planta* 49:145–150.

Fenn, L. B., B. Hasanein, and C. M. Burks. 1995. Calcium-ammonium effects on growth and yield of small grains. *Agron. J.* 87:1041–1046.

Fenn, L. B., R. M. Taylor, M. L. Binzel, and C. M. Burks. 1991. Calcium simulation of ammonium absorption by onion. *Agron. J.* 83:840–843.

Fenn, L. B., R. M. Taylor, and C. M. Burks. 1993. Influence of plant age on calcium stimulated ammonium absorption by radish and onion. *J. Plant Nutr.* 16:1161–1177.

Fernandez, R. and R. T. Laird. 1959. Yield and protein content of wheat in central Mexico as affected by available soil moisture and nitrogen fertilization. *Agron. J.* 51:33–36.

Fillery, I. R. P. and S. K. De Datta. 1989. Ammonium volatilization from nitrogen source applied to rice fields. I. Methodology, ammonia fluxes and nitrogen-15 loss. *Soil Sci. Soc. Am. J.* 50:80–86.

Flexas, J., A. Diaz-Espejo, J. Galmes, R. Kaldenhoff, H. Medrano, and M. Ribas-Carbo. 2007. Rapid variations of mesophyll conductance in response to changes in CO_2 concentration around leaves. *Plant Cell Environ.* 30:1284–1298.

Food and Agricultural Organization of the United Nations (FAO). 2000. FAO statistical databases [Online]. [1p.] Available at: http://apps.fao.org/ (verified 10 January 2003). FAO, Rome.

Food and Agricultural Organization of the United Nations (FAO). 2003. Unlocking the water potential of agriculture. (accessed 3 September 2012, at ftp://ftp.fao.org/agl/docs/kyotofactsheet_e.pdf).

Foley, J. A., N. Ramankutty, K. A. Brauman, E. S. Cassidy, J. S. Gerber, and M. Johnston. 2011. Solutions for a cultivated planet. *Nature* 478:337–342.

Follett, R. F. 2001. Soil management concepts and carbon sequestration in cropland soils. *Soil Tillage Res.* 61:77–92.

Follett, R. F., J. Z. Castellanos, and E. D. Buenger. 2005. Carbon dynamics and sequestration in an irrigated Vertisol in central Mexico. *Soil Tillage Res.* 83:148–158.

Foth, H. D. and B. G. Ellis. 1988. *Soil Fertility.* New York: John Wiley & Sons.

Fowler, D. B. 2003. Crop nitrogen demand and grain protein concentration of spring and winter wheat. *Agron. J.* 95:260–265.

Foy, C. D. 1984. Physiological effects of hydrogen, aluminum and manganese toxicities in acid soils. In: *Soil Acidity and Liming*, 2nd edition, ed., F. Adams, pp. 57–97. Madison, Wisconsin: ASA.

Franzluebbers, A. J., F. M. Hons, and D. A. Zuberer. 1994. Seasonal changes in soil microbial biomass and mineralizable C and N in wheat management systems. *Soil Biol. Biochem.* 26:1469–1475.

Franzluebbers, A. J., R. L. Haney, F. M. Hons, and D. A. Zuberer. 1996. Active fractions of organic matter in soils with different texture. *Soil Biol. Biochem.* 28:1367–1372.

Gallo, P. B., Mascarenhas, H. A. A., Quaggio, J. A., and Bataglia, O. C. 1986. Differential responses of soybean and sorghum to liming. *Rev. Bras. Ci. Solo* 10:253–258.

Galloway, J. N., F. J. Dentener, and D. G. Capone. 2004. Nitrogen cycles: Past, present, and future. *Biogeochemistry* 70:153–226.

Garbeva, P., J. D. V. Elsas, and J. A. V. Veen. 2008. Rhizosphere microbial community and its response to plant species and soil history. *Plant Soil* 302:19–32.

Gebremedhin, M. T., H. W. Loescher, and T. D. Tsegaye. 2012. Carbon balance of no-till soybean with winter wheat cover crop in the southeastern United States. *Agron. J.* 104:1321–1335.

Geisseler, D. and W. R. Horwath. 2009. Short term dynamics of soil carbon, microbial biomass, and soil enzyme activities as compared to longer term effects of tillage in irrigated row crops. *Biol. Fertil. Soils* 46:65–72.

Gerik, T. J., D. M. Oosterhuis, and H. A. Torbert. 1998. Managing cotton nitrogen supply. *Adv. Agron.* 64:115–147.

Ghoulam, C., A. Foursy, and K. Fares. 2002. Effects of salt stress on growth, inorganic ions and proline accumulation in relation to osmotic adjustment in five sugar beet cultivars. *Environ. Exp. Bot*. 47:39–50.

Gleick, P. H. 2003. Global freshwater resources: Soft-path solutions for the 21st century. *Science* 302:1524–1528.

Grau, C. R. and V. L. Radke. 1984. Effects of cultivars and cultural practices on *Sclerotinia* stem rot of soybean. *Plant Dis*. 68:56–58.

Green, C. J. and A. M. Blackmer. 1995. Residue decomposition effects on nitrogen availability to corn following corn or soybean. *Soil Sci. Soc. Am. J*. 59:1065–1070.

Gregorich, E. G., D. R. Lapen, B. L. Ma, N. B. McLaughlin, and A. J. Vandenbygaart. 2011. Soil and crop response to varying levels of compaction, nitrogen fertilization, and clay content. *Soil Sci. Soc. Am. J*. 75:1483–1492.

Gregory, P. J., J. S. I. Ingram, R. Anderson, R. A. Betts, V. Brovkin, and T. N. Chase. 2002. Environmental consequences of alternative practices for intensifying crop production. *Agric. Ecosyst. Environ*. 88:279–290.

Grewal, H. S. and R. Williams. 2003. Liming and cultivars affect root growth, nodulation, leaf to stem ratio, herbage yield, and elemental composition of alfalfa on an acid soil. *J. Plant Nutr*. 26:'1683–1696.

Grohs, D. P., V. G. Menezes, G. R. D. Funk, and C. M. Mundstock. 2010. Criterion for diseases management in irrigated rice. IRGA Technical Bulletin/Technical Report 7, 48p, Cachoeirinha, Rio Grande do Sul, Brazil: IRGA.

Gulde, S., H. Chung, W. Amelung, C. Chang, and J. Six. 2008. Soil carbon saturation controls labile stable carbon pool dynamics. *Soil Sci. Soc. Am. J*. 72:605–612.

Guo, J. H., X. J. Liu, Y. Zhang, J. L. Shen, W. X. Han, W. F. Zhang, P. Christie, K. W. T. Goulding, P. M. Vitousek, and F. S. Zhang. 2010. Significant acidification in major Chinese crop lands. *Science* 327:1008–1010.

Guo, R., J. Zhou, G. X. Ren, and W. Hao. 2013. Physiological responses of linseed seedlings to iso osmotic polyethylene glycol, salt, and alkali stress. *Agron. J*. 105:764–772.

Gusha, A. C. 2002. Effects of tillage on soil micro relief, surface depression storage and soil water storage. *Soil Tillage Res*. 76:105–114.

Habibzadeh, Y., A. Pirzad, M. R. Zardashti, J. Jalilian, and O. Eini. 2013. Effects of arbuscular mycorrhizal fungi on seed and protein yield under water deficit stress in mung bean. *Agron. J*. 105:79–84.

Hadley, H. H. 1957. An analysis of variation in height in sorghum. *Agron. J*. 49:144–147.

Halvorson, A. D. and C. P. Jantalia. 2011. Nitrogen fertilization effects on irrigated no-till corn production and soil carbon and nitrogen. *Agron. J*. 103:1423–1431.

Halvorson, A. D. and A. J. Schlegel. 2012. Crop rotation effect on soil carbon and nitrogen stocks under limited irrigation. *Agron. J*. 104:1265–1273.

Halvorson, A. D., B. J. Wienhold, and A. L. Black. 2002. Tillage nitrogen, nitrogen, and cropping system effects on soil carbon sequestration. *Soil Sci. Soc. Am. J*. 66:906–912.

Halvorson, A. D., R. F. Follett, C. A. Reule, and S. Del Grosso. 2009. Soil organic carbon and nitrogen sequestration in irrigated cropping systems of the central Great Plains. In: *Soil Carbon Sequestration and the Greenhouse Effect*, 2nd edition, eds., R. Lal and R. F. Follett, pp. 141–157. Madison, Wisconsin: ASA.

Halvorson, A. D., A. R. Mosier, C. A. Reulw, and W. C. Bausch. 2006. Nitrogen and tillage effects on irrigate continuous corn yields. *Agron. J*. 98:63–71.

Han, H. S. and K. D. Lee. 2005. Plant growth promoting rhizobacteria effect on antioxidant status, photosynthesis, mineral uptake and growth of lettuce under soil salinity. *Res. J. Agric. Biol. Sci*. 1:210–215.

Hart, P. 1998. *White Mold in Soybeans. Soybean Facts*. Frankenmuth, Michigan: Michigan Soybean Promotion Committee.

Hassink, J. 1994. Effect of soil texture and grassland management on soil organic C and N rates of C and N mineralization. *Soil Biol. Biochem*. 26:1221–1231.

Hatfield, J. L., T. J. Sauer, and J. H. Prueger. 2001. Managing soils to achieve greater water use efficiency: A review. *Agron. J*. 93:271–280.

Haynes, R. J. 1984. Lime and phosphate in the soil–plant system. *Adv. Agron*. 37:249–315.

Haynes, R. J., R. J. Martin, and K. M. Goh. 1993. Nitrogen fixation, accumulation of soil nitrogen and nitrogen balance for some field grown legumes crops. *Field Crops Res*. 35:85–92.

Heffer, P. and M. Prudhomme. 2013. Nutrients as limited resources: Global trends in fertilizer production and use. In: *Improving Water and Nutrient Use Efficiency in Food Production Systems,* ed., Z. Rengel, pp. 57–78. Ames, Iowa: John Wiley & Sons.

Heumesser, C., S. Thaler, M. Schonhart, and E. Schmid. 2013. Current state and future potential of global food production and consumption. In: *Improving Water and Nutrient Use Efficiency in Food Production Systems,* ed., Z. Rengel, pp. 3–19, Ames, Iowa: John Wiley & Sons.

Hirel, B. and P. J. Lea. 2011. The molecular genetics of nitrogen use efficiency in crops. In: *The Molecular and Physiology Basis of Nutrient Use Efficiency in Crops*, eds., M. J. Hawkesford and P. Barraclough, pp. 139–164. Hoboken, New Jersey: Wiley & Sons.

Hirel, B., J. L. Gouis, and A. Gallais. 2007. The challenge of improving nitrogen use efficiency in crop plants: Towards a more central role for genetic variability and quantitative genetics within integrated approaches. *J. Exp. Bot.* 58:2369–2387.

Holt, J. A. and R. J. Mayer. 1998. Changes in soil microbial biomass and protease activities of soil associated with long-term sugar cane monoculture. *Biol. Fertil. Soils* 27:127–131.

Hong, Y., B. R. Glick, and J. J. Pastternak. 1991. Plant-microbial interaction under gnotobiotic conditions: A scanning microscope study. *Curr. Microbiol.* 23:111–114.

Howarth, C. J. 1991. Molecular response of plants to an increased incidence of heat shock. *Plant Cell Environ.* 14:831–841.

Hue, N. V. 1992. Correcting soil acidity of a highly weathered Ultisol with chicken manure and sewage sludge. *Commun. Soil Sci. Plant Anal.* 23:241–264.

Hue, N. V. and D. L. Licudine. 1999. Amelioration of subsoil acidity through surface application of organic manures. *J. Environ. Qual.* 28:623–632.

Hue, N. V., G. R. Craddock, and F. Adams. 1986. Effect of organic acids on aluminum toxicity in subsoil. *Soil Sci. Soc. Am. J.* 50:28–34.

Hussain, S., B. L. Ma, M. F. Saçeem, S. A. Anjum, A. Saeed, and J. Iqbal. 2012. Abscisic acid spray on sunflower acts differently under drought and irrigation conditions. *Agron. J.* 104:561–568.

Impa, S. M., S. Nadaradjan, P. Boominathan, G. Shashidhar, H. Bindumadhava, and M. S. Sheshshaycc. 2005. Carbon isotope discrimination accurately reflects variability in WUE measured at a whole plant level in rice *Crop Sci.* 45:2517–2522.

Isfan, D. 1993. Genotypic variability for physiological efficiency index of nitrogen in oats. *Plant Soil* 154:53–59.

Islam, M. S., M. M. Akhter, A. E. Sabagh, L. Y. Liu, N. T. Nguye, and A. Ueda. 2011. Comparative studies on growth and physiological response to saline and alkaline stresses of foxtail millet and proso millet. *Aust. J. Crop Sci.* 5:1269–1277.

Jabro, J. D., W. M. Iversen, R. G. Evans, and W. B. Stevens. 2012. Water use and water productivity of sugarbeet, malt barley, and potato as affected by irrigation frequency. *Agron. J.* 104:1510–1516.

Jaggard, K. W., A. Qi, and E. S. Ober. 2010. Possible changes to arable crop yields by 2050. *Phil. Trans. R. Soc. B.* 365:2835–2851.

Jenny, H. 1941. *Factors of Soil Formation*. New York: McGraw Hill Book Co.

Jensen, E. S. 1997. Nitrogen immobilization and mineralization during initial decomposition of ^{15}N-labelled pea and barley residues. *Biol. Fertil. Soils* 24:39–44.

Jiang, G. H., Y. Q. He, and C. G. Xu, 2004. The genetic basis of stay-green in rice analyzed in a population of doubled haploid lines derived from an indica by japonica cross. *Theo. Appl. Gen.* 108:688–698.

Johnson, M. S. and R. T. Leah. 1990. Effects of superabsorbent polyacrylamides on efficacy of water use by crop seedlings. *J. Sci. Food Agric.* 52:431–434.

Johnson, J. W. and R. E. Wilkinson. 1993. Wheat growth responses of cultivars to H^+ concentration. In: *Genetic Aspects of Plant Mineral Nutrition*, ed., P. J. Randall, pp. 69–73. Dordrecht: Kluwer Academic Publishers.

Jokela, W. E. and G. W. Randall. 1989. Corn yield and residual soil nitrate as affected by time and rate of nitrogen application. *Agron. J.* 81:720–726.

Jones, M. J. and Z. Arous. 1999. Effects of time of harvest of vetch (*Vicia sativa* L.) on yield of subsequent barley in a dry Mediterranean environment. *J. Agron. Crop Sci.* 182:291–294.

Jones, M. J. and M. Singh. 2000. Long term yield patterns in barley based cropping systems in northern Syria, 2. The role of feed legumes. *J. Agric. Sci.* 135:237–249.

Joppa, L. R., C. Du., and G. E. Hart. 1997. Mapping genes for grain protein in tetraploid wheat using a population of recombinant inbred chromosome lines. *Crop Sci.* 37:1586–1589.

Kaiser, D. E. and K. I. Kim. 2013. Soybean response to sulfur fertilizer applied as a broadcast or starter using replicated strip trials. *Agron. J.* 105:1189–1198.

Kaiser, D. E. and J. C. Rubin. 2013a. Corn nutrient uptake as affected by in-furrow starter fertilizer for three soils. *Agron. J.* 105:1199–1210.

Kaiser, D. E. and J. C. Rubin. 2013b. Maximum rates of seed placed fertilizer for corm for three soils. *Agron. J.* 105:1211–1221.

Kaiser, D. E., A. P. Mallarino, and M. Bermudez. 2005. Corn grain yield, early growth, and early nutrient uptake as affected by broadcast and in furrow starter fertilization. *Agron. J.* 97:620–623.

Kamprath, E. J., R. H. Moll, and N. Rodriguez. 1982. Effects of nitrogen fertilization and recurrent selection on performance of hybrid populations of corn. *Agron. J.* 74:955–958.

Kanampiu, F. K., W. R. Raun, and G. V. Johnson. 1997. Effect of nitrogen rate on plant nitrogen loss in winter wheat varieties. *J. Plant Nutr.* 20:389–404.

Kandeler, E., D. Tscherko, and H. Spiegel. 1999. Long-term monitoring of microbial biomass, N mineralization and enzyme activities of a chernozem under different tillage managements. *Biol. Fertil. Soils* 28:343–351.

Kannan, S. 1986. Foliar absorption and translocation of organic nutrients. *CRC Crit. Rev. Plant Sci.* 4:341–375.

Kant, S., Y. M. Bi, and S. J. Rothstein. 2010. Understanding plant response to nitrogen limitation for the improvement of crop nitrogen use efficiency. *J. Exp. Bot.* 6:1–11.

Karimi, A. and M. Naderi. 2007. Yield and water use efficiency of forage corn as influenced by superabsorbent polymer application in soils with different texture. *Agric. Res.* 3:187–198.

Karlen, D. L., N. C. Wollenhaupt, D. C. Erbach, E. C. Berry, J. B. Swan, N. S. Eash, and J. L. Jordahl. 1994. Long terms soil effects on soil quality. *Soil Tillage Res.* 32:313–327.

Kasper, T. C., H. J. Brown, and E. M. Kassmeyer. 1991. Corn root distribution as affected by tillage, wheel traffic, and fertilization placement. *Soil Sci. Soc. Am. J.* 55:1390–1394.

Kay, B. D. 1990. Rates of changes of soil structure under different cropping systems. *Adv. Soil Sci.* 12:1–51.

Kedziora, A. and Z. W. Kundzewicz. 2013. Translating water into food: How water cycles in natural and agricultural landscapes. In: *Improving Water and Nutrient Use Efficiency in Food Production Systems,* ed., Z. Rengel, pp. 33–56. Ames, Iowa: John Wiley & Sons.

Kichey, T., B. Hirel, and E. Heumez. 2007. In winter wheat, post-anthesis nitrogen uptake and remobilization to the grain correlates with agronomic traits and nitrogen physiological markers. *Field Crop Res.* 102:22–32.

Kim, K., D. E. Kaiser, and J. A. Lamb. 2013. Corn response to starter fertilizer and broadcast sulfur evaluated using strip trials. *Agron. J.* 105:401–411.

Kim, H. S., C. H. Sneller, and B. W. Diers. 1999. Evaluation of soybean cultivars for resistance to *Sclerotinia* stem rot in field environments. *Crop Sci.* 39:64–68.

Kiniry, J. R., C. A. Jones, J. C. O'Toole, R. Blancher, M. Cabelguenne, and D. A. Spanel. 1989. Radiation use efficiency in biomass accumulation prior to grain filling for five grain crop species. *Field Crops Res.* 20:51–64.

Kirchmann, H. 1988. Shoot and root growth and nitrogen uptake by six green manure legumes. *Acta Agric. Cand.* 38:25–31.

Kirchmann, H. and R. Bergqvist. 1989. Carbon and nitrogen mineralization of white clover plants (*Trifolium repens*) of different age during aerobic incubation with soil. *Zeitschrift Pflanzenernahrung und Bodenkunde* 152:283–288.

Kirchner, M. J., A. G. Wollum, and L. D. King. 1993. Soil microbial populations and activities in reduced chemical input agroecosystems. *Soil Sci. Soc. Am. J.* 57:1289–1295.

Kochain, L. V. 1995. Cellular mechanisms of aluminum toxicity and resistance in plants, *Annu. Rev. Plant Physiol. Plant Mol. Biol.* 46:237–260.

Koerselman, W., M. B. Van Kerkhoven, and J. T. A. Verhoeven. 1993. Release of inorganic N, P, and K in peat soils: Effect of temperature, water chemistry and water level. *Biogeochemistry* 20:63–81.

Kramer, H. H. 1967. *Plant Protection and World Crop Production.* Leverkusen: Bayer Pflanzenschutz.

Krupinsky, J. M., D. L. Tanaka, M. T. Lares, and S. D. Merrill. 2004. Leaf spot diseases of barley and spring wheat as influenced by preceding crops. *Agron. J.* 96:259–266.

Krupinsky, J. M., D. L. Tanaka, S. D. Merrill, M. A. Liebig, and J. D. Hanson. 2006. Crop sequence effects of 10 crops in the northern Great Plains. *Agric. Syst.* 88:227–254.

Kumar, K. and K. M. Goh. 2000. Crop residues and management practices: Effects on soil quality, soil nitrogen dynamics, crop yield, and nitrogen recovery. *Adv. Agron.* 68:197–319.

Kuo, S. and E. J. Jellum. 2002. Influence of winter cover crop and residue management on soil nitrogen availability and corn. *Agron. J.* 94:501–508.

Kuo, S. and U. M. Sainju. 1998. Nitrogen mineralization and availability of mixed leguminous and non-leguminous cover crop residue in soil. *Biol. Fertil. Soils* 26:346–355.

Laboski, C. A. M. 2008. Understanding salt index of fertilizers. In: *Proceedings of the Wisconsin Fertilizer, Aglime, and Pest Management Conference,* Madison, Wisconsin, 15–17 January 2008. Vol. 47. Madison: University of Wisconsin Coop. Ext.

Lafitte, H. R. and G. O. Edmeades. 1994. Improvement for tolerance to low soil nitrogen in tropical maize. II. Grain yield, biomass production, and N accumulation. *Field Crops Res.* 39:15–25.

Lafond, G. P., F. Walley, W. E. May, and C. B. Holzapfel. 2011. Long-term impact of no-till soil properties and crop productivity on the Canadian prairies. *Soil Tillage Res.* 117:110–123.

Lal, R. 1991. Tillage and agricultural sustainability. *Soil Tillage Res.* 20:133–146.

Lal, R. 1977. Soil management systems and erosion control. In: *Soil Conservation and Management in the Humid Tropics,* eds., D. Greenland, and R. Lal, pp. 81–86. Chichester, New York: John Wiley & Sons.

Lal, R. 2005. World crop residue production and implications of its use as a biofuel. *Environ. Int.* 31: 575–584.

Larney, F. J. and H. H. Janzen. 1996. Restoration of productivity to a desurfaced soil with livestock manure, crop residue, and fertilizer amendments. *Agron. J.* 88:921–927.

Lauchli, A. and U. Luttge. 2002. Salinity in the soil environment. In: Salinity: *Environment Plants—Molecules*, ed., K. K. Tanji, pp. 21–23. Boston: Kluwer Academic Publishers.

Lavelle, P., T. Decaens, M. Abuert, S. Barot, M. Blouin, F. Bureau, P. Margerie, P. Mora, and J. P. Rossi. 2006. Soil invertebrates and ecosystem services. *Eur. J. Soil Biol.* 42:3–15.

Leclerc, M. L., M. C. Nolin, D. Cluis, and R. R. Simard. 2001. Grouping soils of Montreal lowlands (Quebec) according to fertility and P sorption and desorption characteristics. *Can. J. Soil Sci.* 81:71–83.

Legere, A., R. R. Simard, and C. Lapierre. 1994. Response of spring barley and weed communities to lime, phosphorus and tillage. *Can. J. Plant Sci.* 74:421–428.

Lemaire, G., X. Charrier, and Y. Hebert. 1996. Nitrogen uptake and capacity of maize and sorghum crops in different nitrogen and water supply conditions. *Agronomie* 16:231–246.

Lentz, R. D. and R. F. Sojka. 1994. Field results using polyacrylamide to manage furrow erosion and infiltration. *Soil Sci.* 158:274–282.

Li, C., X. Li, W. Cong, Y. Wu, and J. Wang. 2010. Effect of monoculture soybean on soil microbial community in the Northeast China. *Plant Soil* 330:423–433.

Liang, B., B. McConkey, C. Campbell, D. Curtin, G. Lafond, S. Brandt, and A. Moulin. 2004. Total and labile soil organic nitrogen as influenced by crop rotations and tillage in Canadian prairie soils. *Biol. Fertil. Soils* 39:249–257.

Lima, J. E. F. W., R. S. A. Ferreira, and D. Christofidis. 1999. O uso da irrigação no Brasil. In: *O estado das águas no Brasil*, ed., M. A. V. Freitas, pp. 73–101. Brasília: MME, MMA/SRH, OMM.

Lin, Y. R., K. F. Schertz, and A. H. Paterson. 1995. Comparative analysis of QLTs affecting plant height and maturity across the poaceae, in reference to an interspecific sorghum population. *Genetics* 141:391–411.

Lipiec, J. and W. Stepniewski. 1995. Effects of soil compaction and tillage systems on uptake and losses of nutrient. *Soil Tillage Res.* 35:37–52.

Liu, J. and N. V. Hue. 1996. Ameliorating subsoil acidity by surface application of calcium fulvates derived from common organic materials. *Biol. Fertil. Soils* 21:264–270.

Logan, T. J., R. Lal., and W. A. Dick. 1991. Tillage systems and soil properties in North America. *Soil Tillage Res.* 20:241–270.

Lopes, A. S., M. C. Silva, and L. R. G. Guilherme. 1991. Soil acidity and liming. Technical Bulletin 1, São Paulo, Brazil: National Association for Diffusion of Fertilizers and Agricultural Amendments.

Lopez-Bellido, L., R. J. Lopez-Bellido, and F. J. Lopez-Bellido. 2006. Fertilizer nitrogen efficiency in durum wheat under rainfed Mediterranean conditions: Effect of split application. *Agron. J.* 98:55–62.

Lowery, B. and R. T. Schuler. 1991. Temporal effects of subsol compaction on soil strength and plant growth. *Soil Sci. Soc. Am. J.* 55:216–223.

Lu, S., X. Wang, and Z. Guo. 2013. Differential responses to chilling in *Stylosanthes guianesis* (Aublet) Sw. and its mutants. *Agron. J.* 105:377–382.

Ludlow, M. M. and R. C. Muchow. 1990. A critical evaluation of traits for improving crop yields in water-limited environments. *Adv. Agron.* 43:107–153.

Lupwayi, N. Z. and A. C. Kennedy. 2007. Grain legumes in northern Great Plains: Impacts on selected biological soil processes. *Agron. J.* 99:1700–1709.

Lupwayi, N. Z., G. W. Clayton, J. T. O'Donovan, K. N. Harker, T. K. Turkington, and W. A. Rice. 2004. Soil microbial properties during decomposition of crop residues under conventional and zero tillage. *Can. J. Soil Sci.* 84:411–419.

Lupwayi, N. Z., G. P. Lafond, W. E. May, C. B. Holzapfel, and R. L. Lemke. 2012. Intensification of field pea production: Impact on soil microbiology. *Agron. J.* 104:1189–1196.

Lv, B. S., X. W. Li, H. Y. Ma, Y. Sun, L. X. W., C. J. Jiang, and Z. W. Liang. 2013. Differences in growth and physiology of rice in response to different saline-alkaline stress factors. *Agron. J.* 105:1119–1128.

Lynch, J. and N. S. Rodriguez. 1994. Photosynthetic nitrogen-use efficiency in relation to leaf longevity in common bean. *Crop Sci.* 34:1284–1290.

Lynch, J. and J. W. White. 1992. Shoot nitrogen dynamics in tropical common bean. *Crop Sci.* 32:392–397.

Lynch, R. E. and T. P. Mack. 1995. Biological and biotechnical advances for insect management in peanut. In: *Advances in Peanut Science*, eds., H. E. Pattee, and H. T. Stalker, pp. 95–159. Stillwater, OK: Am. Peanut Res. and Edu. Soc.

Makinde, E. A. and A. A. Agboola. 2002. Soil nutrient changes with fertilizer type in cassava based cropping system. *J. Plant Nutr.* 25:2303–2313.

Malik, A. I. and Z. Rengel. 2013. Physiology of nitrogen use efficiency. In: *Improving Water and Nutrient Use Efficiency in Food Production Systems,* ed., Z. Rengel, pp. 105–121. Ames, Iowa: John Wiley & Sons.

Marquez, L. M., R. S. Redman, R. J. Rodrigues, and J. Roosinck. 2007. A virus in a fungus in a plant: Three way symbiosis required for thermal tolerance. *Science* 315:513–515.

Marschner, H. 1991. Mechanisms of adaptation of plants to acid soils. *Plant Soil* 134:1–20.

Marschner, H. 1995. *Mineral Nutrition of Higher Plants*, 2nd edition. New York: Academic Press.

Martins, R. E., G. P. Asner, R. J. Ansley, and A. R. Moiser. 2003. Effects of woody vegetation enrichment on soil nitrogen oxide emissions in a temperate savanna. *Ecol. Appl.* 4:897–910.

Masclaux-Daubresse, C., I. Quillere, and A. Gallais. 2001. The challenge of remobilization in plant nitrogen economy. A survey of physio-agronomic and molecular approaches. *Ann. Appl. Biol.* 138:69–81.

Masclaux-Daubresse, C., F. Daniel-Vedele, and J. Dechorgnat. 2010. Nitrogen uptake, assimilation and remobilization in plants: Challenges for sustainable and productive agriculture. *Ann. Bot.* 105:1141–1157.

Matson, P. A., C. Billow, and S. Hall. 1996. Fertilization practices and soil variations control oxide emissions from tropical sugarcane. *J. Geophy. Res.* 101:18533–18545.

Matson, P. A., W. J. Parton, and A. G. Power. 1997. Agricultural intensification and ecosystem properties. *Science* 277:504–509.

Matus, A., D. A. Derksen, F. L. Walley, H. A. Loeppky, and C. Van Kessel. 1997. The influence of tillage and crop rotation on nitrogen fixation in lentil and pea. *Can. J. Plant Sci.* 77:197–200.

McConkey, B. G., C. A. Campbell, R. P. Zentner, F. B. Dyck, and F. Selles. 1996. Long-term tillage effects on spring wheat production on three soil textures in the brown soil zone. *Can. J. Plant Sci.* 76:747–756.

McCullough, D. E., P. Girardin, M. Mihajlovic, A. Aguilera, and M. Tollenaar. 1994. Influence of N supply on development and dry matter accumulation of an old and a new maize hybrid. *Can. J. Plant Sci.* 74:471–477.

Mchunu, C. N., S. Lorentz, G. Jewitt, A. Manson, and V. Chaplot. 2011. No-till on soil organic carbon erosion under crop residue scarcity in Africa. *Soil Sci. Soc. Am. J.* 75:1503–1512.

McLellan, C. A., T. J. Turbyville, E. M. K. Wijeratne, E. V. Kerschen, C. Queitsch, L. Whitesell, and A. A. L. Gunatilaka. 2007. A rhizosphere fungus enhances *Arabidopsis* thermo tolerance through production of an HSP90 inhibitor. *Plant Physiol.* 145:174–182.

McMaster, G. S., T. R. Green, R. H. Erskine, D. A. Edmunds, and C. Ascough. 2012. Spatial interrelationships between wheat phenology, thermal time, and terrain attributes. *Agron. J.* 104:1110–1121.

Mengel, D. B., S. E. Hawkins, and P. Walker. 1988. Phosphorus and potassium placement for no-till spring plowed corn. *J. Fert. Issues* 5:31–36.

Mengel, K., E. A. Kirkby, H. Kosegarten, and T. Appel. 2001. *Principles of Plant Nutrition*, 5th edition. Dordrecht: Kluwer Academic Publishers.

Menzies, N. W. 2003. Toxic elements in acid soils: Chemistry and measurement. In: *Handbook of Soil Acidity*, ed., Z. Rengel, pp. 267–296. New York: Marcel Dekker.

Mesfin, T., G. B. Tesfahunegn, C. S. Wortmann, M. Mamo, and O. Nikus. 2010. Skip-row planting and tie-ridging for sorghum production in semiarid areas of Ethiopia. *Agron. J.* 102:745–750.

Meyers, R. J. K. 1992. Management of nitrogen in cropping systems in Indonesia: A review of studies using ^{15}N. *Soil Manage. Abstr.* 4:109–137.

Mickelson, S., D. See, and F. D. Meyer. 2003. Mapping of QTL associated with nitrogen storage and remobilization in barley leaves. *J. Exp. Bot.* 54:801–812.

Mikha, M. M. and C. W. Rice. 2004. Tillage and manure effects on soil and aggregate associated carbon and nitrogen. *Soil Sci. So. Am. J.* 68:809–816.

Mikkelsen, R. L. 1994. Using hydrophilic polymers to control nutrient release. *Fert. Res.* 38:53–59.

Miller, P. R., B. McConkey, G. W. Clayton, S. A. Brandt, J. A. Staricka, and A. M. Johnston. 2002. Pulse crop adaptation in the northern Great Plains. *Agron. J.* 94:261–271.

Milroy, S. P., M. P. Bange, and A. B. Hearn. 2004. Row configuration in rainfed cotton systems: Modification of the OZCOT simulation model. *Agric. Syst.* 82:1–16.

Miyazawa, M., M. A. Pavan, and A. Calegari. 1993. The effect of plant materials on soil acidity. *Rev. Bras. Ci. Solo.* 17:411–416.

Moll, R. H. and E. J. Kamprath. 1977. Effects of population density upon agroeconomic traits associated with genetic increases in yield of *Zea mays* L. *Agron. J.* 69:81–84.

Moll, R. H., E. J. Kamprath, and W. A. Jackson. 1982. Analysis and interpretation of factors which contribute to efficiency of nitrogen utilization. *Agron. J.* 74:562–564.

Moll, R. H., E. J. Kamprath, and W. A. Jackson. 1987. Development of nitrogen efficient prolific hybrids of maize. *Crop Sci.* 27:181–186.

Monfreda, C., N. Ramankutty, and J. A. Foley. 2008. Farming the planet: 2. Geographic distribution of crop areas, yields, physiological types, and net primary production in the year 2000. *Global Biogeochem. Cycles* 22:GB1022.

Monnig, S. 2005. Water saturated superabsorbent polymers used in high strength concrete. *Otto Graf J.* 16:193–202.

Montagu, K. D., J. P. Conroy, and B. J. Atwell. 2001. The position of localized soil compaction determines root and subsequent shoot growth responses. *J. Exp. Bot.* 52:2127–2133.

Mosier, A. and Kroeze, C. 1998. A new approach to estimate emissions of nitrous oxide from agriculture and its implications for the global N$_2$O budget. *Global Change Newsl.* 34:8–13.

Mubarak, A., A. Rosenani, A. Anuar, and S. Zauyah. 2002. Decomposition and nutrient release of maize stover and groundnut haulm under tropical field conditions of Malaysia. *Commun. Soil Sci. Plant Anal.* 33:609–622.

Muchow, R. C. and T. R. Sinclair. 1994. Nitrogen response of leaf photosynthesis and canopy radiation use efficiency in field grown maize and sorghum. *Crop Sci.* 34:721–727.

Mulder, C. P. H., A. Jumpponen, P. Hogberg, and K. Huss-Danell. 2002. How plant diversity and legumes affect nitrogen dynamics in experimental grassland communities. *Oecologia* 33:412–421.

Muruli, B. I. and G. M. Paulsen. 1981. Improvement of nitrogen use efficiency and its relationship to other traits in maize. *Maydica* 26:63–73.

Myers, R. J. K. 1992. Management of nitrogen in cropping systems in Indonesia: A review of studies using [15]N. *Soil Manage. Abstr.* 4:109–137.

Myers, R. J. K., M. A. Foale, G. A. Thomas, A. V. French, and B. Hall. How row spacing affects water use and root growth of grain sorghum. In: *Proc. Aust. Sorghum Conf.*, 1st, Gatton, QLD. 4–6 February 1986. eds., M. A. Foale and R. G. Hensell, pp. 586–591. Organising Committee of the Aust. Sorghum Conf. Gatton, QLD.

Nair, S., J. Johnson, and C. Wang. 2013. Efficiency of irrigation water use: Review from the perspectives of multiple disciplines. *Agron. J.* 105:351–363.

Nash, P. R., K. A. Nelson, and P. P. Motavalli. 2013. Corn yield response to timing of strip-tillage and nitrogen source applications. *Agron. J.* 105:623–630.

Nayyar, A., H. Hamel, G. Lafond, B. D. Gossen, K. Hansen, and J. Germida. 2009. Soil microbial quality associated with yield reduction in continuous pea. *Appl. Soil Ecol.* 43:115–121.

Nelson, K. A., K. A. Renner, and R. Hammerschmidt. 2002. Cultivar and herbicide selection affects soybean development and the incidence of selerotinia stem rot. *Agron. J.* 94:1270–1281.

Niehues, B. J., R. E. Lamond, C. B. Godsey, and C. J. Olsen. 2004. Starter fertilizer management for continuous no-till production. *Agron. J.* 96:1412–1418.

Nielsen, D. C., M. F. Vigil, and R. L. Anderson. 2002. Cropping system influence on planting water content and yield of winter wheat. *Agron. J.* 94:962–967.

Niu, X., R. A. Bressan, P. M. Hasegawa, and J. M. Pardo. 1995. Ion homeostasis in NaCl stress environments. *Plant Physiol.* 109:735–742.

Nolin, M. C., C. Wang, and M. J. Caillier. 1989. Fertility grouping of Montreal lowlands soil mapping units based on selected soil characteristics of the plow layer. *Can. J. Plant Sci.* 69:525–541.

Novoa, B. and R. S. Loomis. 1981. Nitrogen and plant production. *Plant Soil* 58:177–204.

Nyiraneza, J., M. H. Chantigny, A. N'Dayegamiye, and M. R. Laverdiere. 2010. Long-term manure application and forage reduce nitrogen fertilizer requirements of silage-corn cereal cropping systems. *Agron. J.* 102:1244–1251.

Nyiraneza, J., A. N. Cambouris, N. Ziadi, N. Tremblay, and M. C. Nolin. 2012. Spring wheat yield and quality related to soil texture and nitrogen fertilization. *Agron. J.* 104:589–599.

Oberle, S. L. and D. R. Keeney. 1990. Factors influencing corn fertilizer N requirements in the Northern U. S Corn Belt. *J. Prod. Agric.* 3:527–534.

Oerke, E. C. 2006. Crop losses to pests. *J. Agric. Sci.* 144:31–43.

Oerke, E. C. and W. Dehne. 1997. Global crop production and efficacy of crop protection: Current situation and future trends. *Eur. J. Plant Pathol.* 103:203–215.

Oosterhuis, D. M., B. Zhu, and S. D. Wullschleger. 1989. The uptake of foliar applied nitrogen in cotton. In: *Proceedings of Arkansas Cotton Research Meeting and Summaries of Cotton Research in Progress*, ed., D. M. Oosterhuis, pp. 23–25. Fayetteville: Arkansas Agriculture Experimental Station, Special Report 138.

Oosterhuis, D. M., R. E. Hampton, and S. D. Wullschleger. 1991. Water deficit effects on cotton leaf cuticle and the efficiency of defoliants. *J. Prod. Agric.* 4:260–265.

Osborne, S. L. and W. E. Riedell. 2006. Starter nitrogen fertilizer impact on soybean yield and quality in the northern Greta Plains. *Agron. J.* 98:1569–1574.

Oussible, M., R. K. Crookstoon, and W. F. Larson. 1992. Subsurface compaction reduces the root and shoot growth and grain of wheat. *Agron. J.* 84:34–38.

Pan, W. L., E. J. Kamprath, R. H. Moll, and W. A. Jackson. 1984. Prolificacy in corn: Its effects on nitrate and ammonium uptake and utilization. *Soil Sci. Soc. Am. J.* 48:1101–1106.

Pandey, R. K., J. W. Maranville, and Y. Bako. 2001. Nitrogen fertilizer response and efficiency for three cereal crops in Niger. *Commun. Soil Sci. Plant Anal.* 32:1465–1482.

Pang, J., X. Wang, Y. Mu, Z. Ouyang, and W. Liu. 2009. Nitrous oxide emissions from an apple orchard soil in the semiarid loess plateau of China. *Biol. Fertil. Soils* 46:37–44.

Pankhurst, C. E., G. R. Stirling, R. C. Magarey, B. L. Blair, J. A. Holt, M. J. Bell, and A. L. Garside. 2005. Quantification of the effects of rotational breaks on soil biological properties and their impact on yield decline in sugarcane. *Soil Biol. Biochem.* 37:1121–1130.

Parr, M., J. M. Grossman, S. Reberg-Horton, C. Brinton, and C. Crozier. 2011. Nitrogen delivery from legume cover crops in no-till organic corn production. *Agron. J.* 103:1578–1590.

Paterson, A. H., Y. R. Lin, Z. Li, K. F. Schertz, and J. F. Doebley. 1995. Convergent domestication of cereal crops by independent mutations at corresponding genetic loci. *Science* 269:1714–1718.

Paul, E. A. and F. E. Clark. 1996. *Soil Microbiology and Biochemistry*, 2nd edition. San Diego, California: Academic Press.

Pereira, M. G. and M. Lee. 1995. Identification of genomic regions affecting plant height in sorghum and maize. *Theor. Appl. Genetic.* 90:380–388.

Peters, R. D., A. V. Sturz, M. R. Carter, and J. B. Sanderson. 2003. Developing disease-suppressive soils through crop rotation and tillage management practices. *Soil Tillage Res.* 72:181–192.

Peterson, G. A., A. D. Halvorson, J. L. Havlin, O. R. Jones, D. G. Lyon, and D. I. Tanaka. 1998. Reduced tillage and increasing cropping intensity in the Great Plains conserve soil carbon. *Soil Tillage Res.* 47:207–218.

Pimentel, D., C. Harvey, P. Resosudarmo, K. Sinclair, D. Kurz, and M. McNair. 1995. Environmental and economic costs of soil erosion and conservation benefits. *Science* 267:1117–1123.

Pimentel, D., U. Stachow, D. A. Takacs, H. W. Brubaker, A. R. Dumas, J. J. Meaney, J. A. S. O'Neil, D. E. Onsi, and D. B. Corzilius. 1992. Conserving biological diversity in agricultural/forestry systems. *Bioscience* 42:354–362.

Pollmer, W. G., D. Eherhard, D. Klein, and B. S. Dhillon. 1979. Genetic control of nitrogen uptake and translocation in maize. *Crop Sci.* 19:82–86.

Powlson, D. S., P. C. Brookes, and B. T. Christensen. 1987. Measurement of soil microbial biomass provides an early indication of changes in total soil organic matter due to straw incorporation. *Soil Biol. Biochem.* 19:159–164.

Prakash , V., R. Bhattacharyya, and A. K. Srivastva. 2004. Effect of tillage management on yield and soil properties under soybean based cropping system in mid-hills of north-western Himalayas. *Indian J. Agric. Sci.* 81:78–83.

Presterl, T., G. Seitz, M. Landbeck, E. M. Thiemt, W. Schmidt, and H. H. Geiger. 2003. Improving nitrogen use efficiency in European maize: Estimation of quantitative genetic parameters. *Crop Sci.* 43:1259–1265.

Raij, B. V., N. M. Silva, O. C. Bataglia, J. A. Quaggio, R. Hiroce, H. Cantarella, Bellinazzi, A. R. Dechen, and P. E. Trani. 1985. Fertilizer and lime recommendations for the State of São Paulo, Brazil. Technical Bulletin 100, Campinas, Brazil: Agronomy Institute.

Rami, J. F., P. Dufour, G. Trouche, G. Fliedel, and C. Mestres. 1998. Quantitative traits for grain quality, productivity, morphological and agronomical traits in sorghum. *Theor. Appl. Genetic.* 97:605–616.

Randall, G. W. and J. A. Vetsch. 2005. Corn production on a subsurface drained Mollisol as affected by fall versus spring application of nitrogen and nitrapyrin. *Agron. J.* 97:472–478.

Randall, G. W., J. A. Vetsch, and J. R. Huffman. 2003. Corn production on a subsurface-drained Mollisol as affected by time of nitrogen application and nitrapyrin. *Agron. J.* 95:1213–1219.

Rannells, N. N. and M. J. Wagger. 1996. Nitrogen release from grass and legume cover crop monocultures and bicultures. *Agron. J.* 88:777–782.

Raun, W. R. and G. V. Johnson. 1999. Improving nitrogen use efficiency for cereal production. *Agron. J.* 91:357–363.

Raun, W. R., D. H. Sander, and R. A. Olson. 1986. Emergence of corn as affected by source and rate of solution fertilizers applied with the seed. *J. Fert. Issues* 3:18–24.

Raven, J. A. 1975. Transport of indoleacetic acid in plant cells in relation to pH and electrical potential gradients, and its significance for polar IAA transport. *New Phyt.* 74:163–175.

Reddy, K. N., R. M. Zablotowicz, M. A. Locke, and C. H. Koger. 2003. Cover crop, tillage, and herbicide effects on weeds, soil properties, microbial populations, and soybean yield. *Weed Sci.* 51:987–994.

Redman, R. S., K. B. Sheehan, R. G. Stout, R. J. Rodrigues, and J. M. Henson, 2002. Thermotolerance generated by plant/fungal symbiosis. *Science* 298:1581.

Reed, A. J., F. E., Below, and R. H. Hageman. 1980. Grain protein accumulation and relationship between leaf nitrate reductase and protease activities during grain development in maize: I. Variation between genotypes. *Plant Physiol.* 66: 164–170.

Rehm, G. W. 1999. Use of banded fertilizer for corn production. Ext. Publ. FO-7425-F. University of Minnesota Ext. St. Paul.

Reis, T. C. and A. A. Rodella. 2002. Dynamics of organic matter degradation and pH variation of soil under different temperatures. *Rev. Bras. Ci. Solo* 26:619–626.

Rengel, Z. and P. M. Damon. 2008. Crops and genotypes differ in efficiency of potassium uptake and use. *Physiol. Plantaram* 133:624–636.

Rengel, Z. and P. Marschner 2005. Nutrient availability and management in the rhizosphere: Exploiting genotypic differences. *New Phytol.* 168:305–312.

Robertson, G. P. 1997. Nitrogen use efficiency in row-crop agriculture: Crop nitrogen use and soil nitrogen loss. In: *Ecology in Agriculture*, ed., L. E. Jackson, pp. 347–365. New York: Academic Press.

Rooney, W. L., J. Blumenthal, B. Bean, and J. E. Mullet. 2000. Designing sorghum as a dedicated bioenergy feedbacks. *Biofuel, Bioproducts Biorefining* 1:103–110.

Rose, T. and B. Bowden. 2013. Matching soil nutrient supply and crop demand during the growing season. In: *Improving Water and Nutrient Use Efficiency in Food Production Systems,* ed., Z. Rengel, pp. 93–103. Ames, Iowa: John Wiley & Sons.

Rosielle, A. A. and J. Hamblin. 1981. Theoretical aspects of selection for yield in stress and non-stress environments. *Crop Sci.* 21:943–946.

Rostampour, M. F., M. Yarnia, F. Rahimzadeh, M. J. Seghatoleslami, and G. R. Moosavi. 2013. Physiological response of forage sorghum to polymer under water deficit conditions. *Agron. J.* 105:951–959.

Sa, T. M. and D. W. Israel. 1998. Phosphorus deficiency effects on response of symbiotic N_2 fixation and carbohydrate status in soybean to atmospheric CO_2 enrichment. *J. Plant Nutr.* 21:2207–2218.

Sainju. U. M. and B. P. Singh. 1997. Winter cover crops for sustainable agricultural systems: Influence on soil properties, water quality, and crop yields. *HortScience* 32:21–28.

Sainju. U. M. and B. P. Singh. 2001. Tillage, cover crop, and kill planting date effect on corn yield and soil nitrogen. *Agron. J.* 93:878–886.

Sainju, U. M., B. P. Singh, and S. Yaffa. 2002. Soil organic matter and tomato yield following tillage, cover cropping, and nitrogen fertilization. *Agron. J.* 94:594–602.

Sainju, U. M., A. W. Lenssen, and J. L. Barsotti. 2013. Dryland malt barley yield and quality affected by tillage, cropping sequence, and nitrogen fertilization. *Agron. J.* 105:329–340.

Sainju, U. M., T. Caesar-Ton That, A. W. Lenssen, R. G. Evans, and R. Kolberg. 2009. Tillage and cropping sequence impacts on nitrogen cycling in dryland farming in eastern Montana, USA. *Soil Tillage Res.* 103:332–341.

Sainju, U. M., A. W. Lenssen, T. Caesar-Tonthat, and J. Waddell. 2006. Tillage and crop rotation effects on dryland and soil residue carbon and nitrogen. *Soil Sci. Soc. Am. J.* 67:1533–1543.

Sainz, R. H. R., H. E. Echeverria, F. H. Andrade, and G. A. Studdert. 1997. Effect of urease inhibitors and fertilization time on nitrogen uptake and maize grain yield under no-tillage. *Rev. Fac. Agron. La Plata* 102:129–136.

Sainz, R. H. R., H. E. Echeverria, and P. A. Barbieri. 2004. Nitrogen balance as affected by application time and nitrogen fertilizer rate in irrigated no-tillage maize. *Agron. J.* 96:1622–1631.

Salisbury, A. and M. H. Stolt. 2011. Estuarine subaqueous soil temperature. *Soil Sci. Soc. Am. J.* 75:1584–1587.

Salsac, L., S. Chaillou, J. F. Morot-Gaudry, C. Lesaint, E. and Jolivoe. 1987. Nitrate and ammonium nutrition in plants. *Plant Physiol. Biochem.* 25:805–812.

Sanders, J. H., B. I. Shapiro, and S. Ramaswamy. 1996. *The Economics of Agricultural Technology in Semi-Arid Sub-Saharan Africa.* Baltimore, MD: Johns Hopkins University Press.

Savant, N. K., De Datta, S. K., and E. T. Craswell. 1982. Distribution patterns of ammonium nitrogen and ^{15}N uptake by rice after deep placement of urea supergranules in wetland soil. *Soil Sci. Soc. Am J.* 46:567–573.

Schimel, D. S., B. H. Braswell, E. A. Holland, R. McKeown, D. S. Ojima, T. H. Painter, W. J. Parton, and A. R. Townsend. 1994. Climatic, edaphic, and biotic controls over storage and turnover of carbon in soils. *Global Biogeochem. Cycles* 8:279–293.

Schjoerring, J. K., J. G. H. Bock, L. Gammelvind, C. R. Jensen, and V. O. Mogensen. 1995. Nitrogen incorporation and remobilization in different shoot components of field grown winter oilseed rape (*Brassica napus* L.) as affected by rate of nitrogen application and irrigation. *Plant Soil* 177:255–264.

Schoenau, J. J., R. Bolton, and C. Baan. 2008. Soil fertility changes under long-term direct seeding. In: *Fuelling the Farm: SSCA 2008 Annual Conf. Proc.*, pp. 11–15. Regina, SK, Canada. www.ssca.ca/conference/conference2008/Schoenau.pdf (accessed 10 February 2012).

Schon, N. L., A. D. Mackay, M. J. Hedley, and M. A. Minor. 2011. Influence of soil faunal communities on nitrogen dynamics in legume based mesocosms. *Soil Res.* 2:1012.

Schon, N. L., A. D. Mackay, and M. A. Minor. 2012. Relationship between food resource, soil physical condition, and invertebrates in pastor soils. *Soil Sci. Soc. Am. J.* 76:1644–1654.

Schultz, T. R. and R. F. Line. 1992. High temperature, adult plant resistance to stripe rust and effects on yield components. *Agron. J.* 84:170–175.

Seifert, B., Z. W. Zhou, M. Wallbraun, G. Lohaus, and C. Mollers. 2004. Expression of a bacterial asparagines synthetase gene in oilseed rape (*Brassica napus* L.) and its effect on traits related to nitrogen efficiency. *Physiol. Plantarum* 121:656–665.

Shainberg, I., D. N. Warrington, and P. Rengasamy. 1990. Water quality and PAM interactions in reducing surface sealing. *Soil Sci.* 149:301–307.

Shainberg, I., J. M. Laflen, J. M. Bradford, and L. D. Norton. 1994. Hydraulic flow and water quality characteristics in rill erosion. *Soil Sci. Soc. Am. J.* 58:1007–1012.

Sharifi, M., B. J. Zebarth, D. L. Burton, C. A. Grant, S. Bittman, C. F. Drury, B. G. McConkey, and N. Ziadi. 2008. Response of potentially mineralizable soil nitrogen and indices of nitrogen availability to tillage system. *Soil Sci. Soc. Am. J.* 72:1124–1131.

Sherrod, L. A., G. A. Peterson, D. G. Westfall, and L. R. Ahuja. 2003. Cropping intensity enhances soil organic carbon and nitrogen in a no-till agroecosystems. *Soil Sci. Soc. Am. J.* 67:1533–1543.

Sherwood, J. L., M. K. Beute, D. W. Dickson, V. J. Elliot, R. S. Nelson, C. H. Operrman, and B. B. Shew. 1995. Biological and biotechnological advances in *Arachis* diseases. In: *Advances in Peanut Science*, eds., H. E. Pattee and H. T. Stalker, pp. 160–206. Stillwater, OK: Am. Peanut Res. and Edu. Soc.

Shiklomanov, I. 1993. World freshwater resources. In: *Water Crisis: A Guide to the World's Freshwater Resources*, ed., P. H. Gleick, pp. 13–24. New York: Oxford University Press.

Shrawat, A. K., R. T. Carroll, and M. DePauw. 2008. Genetic engineering of improved nitrogen use efficiency in rice by the tissue-specific expression of alanine aminotransferase. *Plant Biotech. J.* 6:722–732.

Sileshi, G. W., L. K. Debusho, and F. K. Akinnifesi. 2012. Can integration of legume trees increase yield stability in rainfed maize cropping systems in southern Africa. *Agron. J.* 104:1392–1398.

Sileshi, G. W., F. K. Akinnifesi, O. C. Ajayi, and B. Muys. 2011. Integration of legumes trees in maize-based cropping systems improves rainfall use efficiency and crop yield stability. *Agric. Water Manage.* 98:1364–1372.

Sims, J. L. and L. R. Frederick. 1970. Nitrogen immobilization and decomposition of corn residue in soil and sand as affected by residue particle size. *Soil Sci.* 109:355–361.

Sinclair, T. R. and T. Horie. 1989. Leaf nitrogen, photosynthesis, and crop radiation use efficiency: A review. *Crop Sci.* 29:90–98.

Singh, B. P. 2013. Physiology and genetics of biofuel crop yield. In: *Biofuel Crops: Production, Physiology and Genetics*, ed., B. P. Singh, pp. 102–134. London: CAB International.

Singh, M. and T. A. Singh. 1986. Leaching losses of nitrogen from urea as affected by application of neem-cake. *J. Indian Soc. Soil Sci.* 34:766–773.

Singh, U., J. K. Ladha, E. G. Castillo, G. Punzalan, A. Irol-Padre, and M. Duqueza. 1998. Genotypic variation in nitrogen use efficiency in medium and long duration rice. *Field Crops Res.* 58:35–53.

Six, J., H. Bossuyt, S. Degryze, and K. Denef. 2004. A history of research on the link between (micro) aggregates, soil biota, and soil organic matter dynamics. *Soil Tillage Res.* 79:7–31.

Smakhtin, V. C. Revenga, and P. Doll. 2004. A pilot global assessment of environmental water requirements and scarcity. *Water Int.* 29:307–317.

Smith, S. N. 1934. Response of inbred lines and crosses in maize to variation of nitrogen and phosphorus supplied as nutrients. *J. Am. Soc. Agron.* 26:785–804.

Smith, P., D. Martino, Z. Cai, D. Gwary, H. Janzen, P. Kumar, B. McCarl et al. 2008. Greenhouse gas mitigation in agriculture. *Phil. Trans R. Soc. B.* 363:789–813.

Smith, S. E. and D. J. Read. 2008. *Mycorrhizal Symbiosis*, 3rd edition. London: Academic Press.

Snapp, S. S., S. M. Swinton, R. Labarta, D. Mutch, J. R. Black, and R. Leep. 2005. Evaluating cover crops for benefits, costs and performance within cropping system niches. *Agron. J.* 97:322–332.

Socolow, R. H. 1999. Nitrogen management and the future of food: Lessons from the management of energy and carbon. *Proc. Natl. Acad. Sci. USA.* 96:6001–6008.

Soil Science Society of America. 2008. *Glossary of Soil Science Terms*. Madison, Wisconsin: Soil Science Society of America.

Soil Survey Staff. 1975. *Soil Taxonomy: A Basic System of Soil Classification for Making and Interpreting Soil Surveys*. Washington, DC: USDA-SCS. U.S. Govt. Print. Office.

Soil Survey Staff. 1999. *Soil Taxonomy: A Basic System of Soil Classification for Making and Interpreting Soil Surveys*, 2nd edition. Washington, DC: USDA-SCS. U.S. Govt. Print. Office.

Somda, Z. C., P. B. Ford, and W. L. Hargrove. 1991. Decomposition and nitrogen recycling of cover crops and crop residues. In: *Cover Crops for Clean Water. Proc. Int. Conf.*, Jackson, TN. 9–11 April 1991, ed., W. L. Hargrove, pp. 103–105. Ankeny, Iowa: Soil and Water Conserv. Soc.

Somerville, C. and J. Briscoe. 2001. Genetic engineering and water. *Science* 292:2217.

Soon, Y. K., G. W. Clayton, and W. A. Rice. 2001. Tillage and previous crop effects on dynamics of nitrogen in a wheat-soil system. *Agron. J.* 93: 842–849.

Sousa, D. M. G., L. N. Miranda, and E. Lobato. 1996. Evaluation methods of lime requirements in cerrado soils. Cerrado Center of EMBRAPA Technical Circular 33, Planaltina, Brazil: Cerrado Center of EMBRAPA.

Spratt, E. D. and J. K. R. Gasser. 1970. Effect of ammonium and nitrate form of nitrogen and restricted water supply on growth and nitrogen uptake of wheat. *Can. J. Soil Sci.* 50:263–273.

Springfield, G. and R. Salter. 1934. Differential response of corn varieties to fertility levels and seasons. *J. Agric. Res.* 49:991–1000.

Stanford, G. 1973. Rationale for optimum nitrogen fertilization in corn production. *J. Environ. Qual.* 2:159–166.

Stevenson, F. C. and C. Van Kessel. 1996. The nitrogen and non-nitrogen benefits of pea to succeeding crops. *Can. J. Plant Sci.* 76:735–745.

Stute, J. K. and J. L. Posner. 1995. Synchrony between legume nitrogen release and corn demand in the upper Midwest. *Agron. J.* 87:1063–1069.

Sumner, M. E. and A. D. Noble. 2003. Soil acidification: The world story. In: *Handbook of Soil Acidity*, ed., Z. Rengel, pp. 1–28. New York: Marcel Dekker.

Sung, F. J. M. and W. S. Lo. 1990. Growth response of rice in ammonium based nutrient solution with variable calcium supply. *Plant Soil* 125:239–244.

Svecnjak, Z. and Z. Rengel. 2005. Canola cultivars differ in nitrogen utilization efficiency at vegetative stage. *Field Crops Res.* 97:221–226.

Sylvia, D. M., L. C. Hammond, J. M. Bennett, J. H. Haas, and S. B. Linda. 1993. Field response of maize to a VAM fungus and water management. *Agron. J.* 85:193–198.

Tan, K. H., J. H. Edwards, and O. L. Bennett. 1985. Effect of sewage sludge on mobilization of surface applied calcium in a Greenville soil. *Soil Sci.* 139:262–268.

Tanaka, D. L., R. L. Anderson, and S. C. Rao. 2005. Crop sequencing to improve use of precipitation and synergize crop growth. *Agron. J.* 97:385–390.

Tanner, C. B. and T. R. Sinclair. 1983. Efficient water use in crop production: Research or research. In: *Limitations to Efficient Water Use in Crop Production*, ed., H. M. Taylor, pp. 1–27. Madison, Wisconsin: ASA, CSSA, and SSSA.

Tao, Y. Z., R. G. Henzell, D. R. Jordan, D. G. Butler, and A. M. Kelley. 2000. Identification of genomic regions associated with stay green in sorghum by testing RILs in multiple environment. *Theor. Appl. Genetic.* 100:1225–1232.

Tardieu, F. 1994. Growth and functioning of roots and root systems subjected to compaction: Towards a system with multiple signaling. *Soil Tillage Res.* 30:217–243.

Teasdale, J. R. and C. L. Mohler. 1993. Light transmittance, soil temperature, and soil moisture under residue of hairy vetch and rye. *Agron. J.* 85:673–680.

Thind, H. S., Bijay-Singh, and R. P. S. Pannu. 2010. Relative performance of need (*Azadirachta indica*) coated urea vis-à-vis ordinary urea applied to rice on the basis of soil test or following need based nitrogen management using leaf color chart. *Nutr. Cycl. Agroecosyst.* 87:1–8.

Thomason, W. E., W. R. Raun, G. V. Johnson, K. W. Freeman, K. J. Wynn, and R. W. Mullen. 2002. Production system techniques to increase nitrogen use efficiency in winter wheat. *J. Plant Nutr.* 25:2261–2283.

Thompson, R. K., E. B. Jackson, and J. R. Gebert. 1975. Irrigated wheat production response to water and nitrogen fertilizer. University of Arizona Agriculture Experimental Station, Technical Bulletin 229.

Tilman, D., K. Cassman, and P. Matson. 2002. Agricultural sustainability and intensive production practices. *Nature* 418:671–677.

Timmusk, S. and E. G. H. Wagner. 1999. The plant growth promoting rhizobacterium *Paenibacillus polymyxa* induces changes in *Arabidopsis thaliana* gene expression: A possible connection between biotic and abiotic stress responses. *Mol. Plant-Microb. Interact.* 12:951–959.

Tisdale, S. L., W. L. Nelson, J. D. Beaton, and J. L. Havlin. 1993. *Soil Fertility and Fertilizers*, 5th edition. New York: Macmillan Publ. Co.

Topp, G. C., B. Dow, M. Edwards, E. G. Gregorich, W. E. Curnoe, and F. J. Cook. 2000. Oxygen measurements in the root zone facilitated by TDR. *Can. J. Soil Sci.* 80:33–41.

Traore, A. and J. W. Maranville. 1999. Nitrate reductase activity of diverse grain sorghum genotypes and its relationship to nitrogen use efficiency. *Agron. J.* 91:863–869.

Traxler, G. 2006. The GMO experience in North and South America. *Int. J. Technol Globalization* 2:46–64.

Tsai, C. Y., I. Dweikat, D. M. Huber, and H. L. Warren. 1992. Interrelationship of nitrogen nutrition with maize (*Zea mays* L.) grain yield, nitrogen use efficiency and grain quality. *J. Sci. Food Agric.* 58:1–8.

Van Kessel, C. and C. Hartley. 2000. Agricultural management of grain legumes: Has it led to an increase in nitrogen fixation? *Field Crops Res.* 65:165–181.

Vargas Gil, S. V., J. Meriles, C. Conforto, M. Basanta, V. Radl, A. Hagn, M. Schloter, and G. J. March. 2011. Response of soil microbial communities to different management practices in surface soils of a soybean agroecosystem in Argentina. *Eur. J. Soil Biol.* 47:55–60.

Varvel, G. E. and W. W. Wilhelm. 2008. Soil carbon levels in irrigated western Corn Belt rotations. *Agron. J.* 100:1180–1184.

Vasilas, B. L., J. O. Legg, and D. C. Wolf. 1980. Foliar fertilization of soybeans: Absorption and translocation of ^{15}N labeled urea. *Agron. J.* 72:271–275.

Vaughan, K. L., M. C. Rbenhorst, and B. A. Needelman. 2009. Saturation and temperature effects on the development of reducing condition in soils. *Soil Sci. Soc. Am. J.* 73:663–667.

Vaughn, T., T. Cavato, G. Brar, T. Coombe, T. DeGooyer, and S. Ford. 2005. A method of controlling corn rootworm feeding using a *Bacillus thuringiensis* protein expressed in transgenic maize. *Crop Sci.* 45:931–938.

Verma, V., M. J. Foulkes, and A. J. World. 2004. Mapping quantitative trait loci for flag leaf senescence as a yield determinant in winter wheat under optimal drought-stress environments. *Euphytica* 135:255–263.

Vetsch, J. A. and G. W. Randall. 2004. Corn production as affected by nitrogen application timing and tillage. *Agron. J.* 96:502–509.

Vigil, M. F., R. A. Anderson, and W. F. Beard. 1997. Base temperature growing degree-hour requirements for emergence of canola. *Crop Sci.* 37:844–849.

Voorhees, W. B., R. R. Allmares, and C. E. Johnson. 1981. Alleviating temperature stress. In: *Modifying the Root Environment to Reduce Crop Stress*, eds., G. F. Arkin and H. M. Taylor, pp. 217–266. St. Joseph, Michigan: Am. Soc. Agric. Eng.

Wang, J., W. Z. Liu, T. H. Dang, and U. M. Sainju. 2013. Nitrogen fertilization effect on soil water and wheat yield in the Chinese loess plateau. *Agron. J.* 105:143–149.

Wang, H., Z. Wu, Y. Chen, C. Yang, and D. Shi. 2011. Effects of salt and alkali stresses on growth and ion balance in rice. *Plant Soil Environ.* 57:286–294.

Wang, Y. C., E. L. Wang, D. L. Wang, S. M. Huang, Y. B. Ma, C. J. Smith, and L. G. Wang. 2010. Crop productivity and nutrient use efficiency as affected by long-term fertilization in North China Plain. *Nutr. Cycling Agroecosyst.* 86:105–119.

Weil, R. R., K. A. Lowell, and H. M. Shade. 1993. Effects of intensity of agronomic practices on a soil ecosystem. *Am. J. Alternate Agric.* 8:5–14.

West, T. O. and W. M. Post. 2002. Soil organic sequestration rates by tillage and crop rotation: A global data analysis. *Soil Sci. Soc. Am. J.* 66:1930–1946.

Wicks, G. A., D. A. Crutchfield, and O. C. Burnside. 1994. Influence of wheat (*Triticum aestivum* L.) straw mulch and metolachlor on corn (*Zea mays* L.) growth and yield. *Weed Sci.* 42:141–147.

Wiesler, F. 1998. Comparative assessment of efficiency of various fertilizers. In: *Nutrient Use in Crop Production*, ed., Z. Rengel, pp. 81–114. New York: The Haworth Press.

Wiesler, F., T. Behrens, and W. J. Horst. 2001a. The role of nitrogen efficient cultivars in sustainable agriculture. *Scientific World J.* 1:61–69.

Wiesler, F., T. Behrens, and W. J. Horst. 2001b. Nitrogen efficiency of contrasting rape ideotypes. In: *Plant Nutrition: Food Security and Sustainability of Agro-Ecosystems through Basic and Applied Research.*, eds., W. J. Horst, M. K. Schenk, A. Burkert, N. Claassen, H. Flessa, W. B. Frommer, H. Goldbach et al. pp. 60–61. Dordrecht: Kluwer Academic Publishers.

Wilhelm, W., J. Johnson, J. Hatfield, W. Voorhees, and D. Linden. 2004. Crop and soil productivity response to corn residue removal: A literature review. *Agron. J.* 96:1–17.

Wilkinson, R. E. and R. R. Duncan. 1989. Sorghum seedling root growth as influenced by H, Ca, and Mn. *J. Plant Nutr.* 12:1379–1394.

Willert, F. J. V. and R. C. Stehouwer. 2003. Compost, limestone, and gypsum effects on calcium and aluminum transport in acidic minespoil. *Soil Sci. Soc. Am. J.* 67:778–786.

Williams, M. M., D. A. Mortensen, and J. W. Doran. 1998. Assessment of weed and crop fitness in cover crop residues for integrated weed management. *Weed Sci.* 46:595–603.

Wittwer, S. H. 1975. Food production: Technology and the resource base. *Science* 188:578–584.

Wong, M. T. F. and R. S. Swift. 2003. Role of organic matter in alleviating soil acidity. In: *Handbook of Soil Acidity*, ed., Z. Rengel, pp. 337–358. New York: Marcel Dekker.

Wong, M. T. F., S. Asseng, and H. Zhang. 2005. Precision agriculture improves efficiency of nitrogen use and minimizes its leaching at within-field to farm scales. In: *5th European Conference on Precision Agriculture*, ed., J. V. Stafford, pp. 969–976. Wageningen, The Netherlands: Wageningen Academic Publishers.

Wong, M. T. F., S. Nortcliff, and R. S. Swift. 1998. Method for determining the acid ameliorating capacity of plant residue compost, urban waste compost, farmyard manure and peat applied to tropical soils. *Commun. Soil Sci. Plant Anal.* 29:2927–2937.

Wortman, S. E., C. A. Francis, M. L. Bernards, R. A. Drijber, and J. L. Lindquist. 2012. Optimizing cover crop benefits with diverse mixtures and an alternative termination method. *Agron. J.* 104:1425–1435.

Wuest, S. B., and W. F. Schillinger. 2011. Evaporation from high residue no-till versus tilled fallow in a dry summer climate. *Soil Sci. Soc. Am. J.* 75:1513–1519.

Xin, Z. R. M. Aiken, and J. J. Burke. 2009. Genetic diversity of transpiration efficiency in sorghum. *Field Crops Res.* 111:74–80.

Xue, Y. F., S. C. Yue, Y. Q. Zhang, Z. L. Cui, X. P. Chen, F. C. Yang, I. Cakmak, S. P. McGrath, F. S. Zhang, and C. Q. Zou. 2012. Grain shoot zinc accumulation in winter wheat affected by nitrogen management. *Plant Soil* 361:153–163.

Yamada, Y. 1962. Studies on foliar absorption of nutrients using radioisotopes. PhD dissertation. Kyoto: University of Kyoto, Japan.

Yang, C. W., D. C. Shi, and D. L. Wang. 2008. Comparative effects of salt and alkali stresses on growth, osmotic adjustment and ionic balance of an alkali-resistant halophyte *Suaeda glauca*. *Plant Growth Regul.* 56:179–190.

Yang, C. W., J. N. Chong, C. Y. Li, C. M. Kim, D. C. Shi, and D. L. Wang. 2007. Osmotic adjustment and ion balance traits of an alkali resistant halophyte kochia sieversiana during adaption to salt and alkali conditions. *Plant Soil* 294:263–276.

Yang, X. B., P. Lundee, and M. D. Uphoff. 1999. Soybean varietal response and yield loss caused by *Scleriotinia sclerotiorum*. *Plant Dis.* 83:456–461.

Yang, Z. M., M. Sivaguru, W. J. Horst, H. Matsumoto, and Z. M. Yang. 2000. Aluminum tolerance is achieved by exudation of citric acid from roots of soybean (*Glycine max*). *Physiol. Plantarum* 110:72–77.

Yau, S. K., M. Bounejmate, J. Ryan, A. Nassar, R. Baalbaki, and R. Maacaroun. 2003. Barley-legumes rotations for semi-arid areas of Lebanon. *Eur. J. Agron.* 19:599–610.

Yau, S. K. and J. Ryan. 2012. Does growing safflower before barley reduce barley yields under Mediterranean conditions? *Agron. J.* 104:1493–1500.

Zakeri, H., G. P. Lafond, J. J. Schoenau, M. H. Pahlavani, A. Vandenberg, W. E. May, C. B. Holzapfel, and R. A. Bueckert. 2012. Lentil performance in response to weather, no-till duration, and nitrogen in Saskatchewan. *Agron. J.* 104:1501–1509.

Zhang, F., A. F. Mackenzie, and D. L. Smith. 1993. Corn yield and shifts among corn quality constituents following application of different nitrogen fertilizer sources at several times during corn development. *J. Plant Nutr.* 16:1317–1337.

Zhang, Z. H., H. X. Song, Q. Liu, X. M. Rong, C. Y. Guan, J. W. Peng, G. X. Xie, and Y. P. Zhang. 2010. Studies on differences of nitrogen efficiency and root characteristics of oilseed rape (*Brassica napus* L.) cultivars in relation to nitrogen fertilization. *J. Plant Nutr.* 33:1448–1459.

Index

A

AAPFCO, *see* American Association of Plant Food Control Officials (AAPFCO)
Acetylene reduction assay, 289–290
Acidity-neutralizing reactions, 334
Adenosine triphosphate (ATP), 31, 144, 188
Adequate zone, 131
Adsorbed ionic form, *see* Exchangeable ionic form
AE, *see* Agronomic efficiency (AE)
Aeration, 80
    poor, 295
    of soil, 71
Aerobic rice, *see* Upland rice
AGF, *see* Agroforestry (AGF)
Agricultural technologies, modern, 327
Agriculture, 363, 364, *see also* Modern agriculture
    animal, 72
    CO_2 and N_2O emissions, 75
    conservation tillage in, 362
    dryland, 82
    fertilizers and, 4, 5
    global food security and, 217
    importance of legumes, 283
    monoculture agriculture systems, 191
    NO_3 leaching, 90–91
    research in, 157
    salt-affected soils effect, 279
    soil fertility and plant nutrition, 149–150
    texture use in, 179
Agroforestry (AGF), 91
Agronomic efficiency (AE), 220, 221; *see also* Physiological efficiency (PE); Utilization efficiency (UE)
    of lowland rice, 221, 222
    of upland rice, 223
    of wheat, 330
Agrophysiological efficiency (APE), 220
    as economic production, 223
    of lowland rice, 224
Allelochemicals, 186
Allelopathy, 185–186
Aluminum toxicity, 184–185
American Association of Plant Food Control Officials (AAPFCO), 272
AMF, *see* Arbuscular mycorrhizal fungi (AMF)
4-amino-1-, 3-, 4-triacole (ATC), 72
Ammonia (NH_4^+), 68, 235
    combined with organic acids, 318
    formation, 287
    gas losses, 74
    reducing floodwater concentration, 226
Ammonia volatilization, 72
    consumption of H ion, 74
    flooded rice soils, 73
    from foliage, 77
    N loss from soil–plant systems, 72

    surface-applied urea, 74
Ammonification, 70, 74, 194
Ammonium absorption, 92
Animal agriculture, 72
APE, *see* Agrophysiological efficiency (APE)
Apparent recovery efficiency (ARE), 220, 225; *see also* Physiological efficiency (PE); Utilization efficiency (UE)
    agronomic and recovery efficiencies, 227
    of lowland rice, 225
    N fertilizer management, 228
    N recovery efficiency, 226
    quantity of N, 225
Arabidopsis, 236
Arbuscular mycorrhizal fungi (AMF), 365
ARE, *see* Apparent recovery efficiency (ARE)
Assimilation efficiency, 234
ATC, *see* 4-amino-1-, 3-, 4-triacole (ATC)
ATP, *see* Adenosine triphosphate (ATP)
Atrazine, 200
AuTophaGy genes (ATG genes), 236

B

Balanced nutrient supply, 344
Biochar, 86
Biological nitrogen fixation
    acetylene reduction assay, 289–290
    conservation tillage, 311
    in cropping system, 284
    crop rotation, 311–312
    enhancing strategies, 307
    legumes, 283
    by legumes, 287–288
    methods of assessment, 288
    microorganisms, 284
    ^{15}N abundance, 293
    N balance determination, 290
    N difference method, 290
    ^{15}N enrichment method, 292
    N equivalence, 290–291
    ^{15}N isotope dilution technique, 292–293
    N_2 reduction, 283
    NFE, 291
    N isotopic technique, 291–292
    nitrogenase enzyme, 287
    nodule evaluation, 290
    nodule formation, 284–286
    pest and disease control, 312
    plant breeding and selection, 312–317
    planting high N2-fixing crop species, 312
    plant nutrition, 312
    quantity of N fixed, 289
    soil amelioration, 312
    strains for inoculation selection, 308–311
Blue-baby syndrome, *see* Methemoglobinemia

Printed and bound by CPI Group (UK) Ltd, Croydon, CR0 4YY

18/10/2024

01776204-0017